About Island Press

Island Press is the only nonprofit organization in the United States whose principal purpose is the publication of books on environmental issues and natural resource management. We provide solutions-oriented information to professionals, public officials, business and community leaders, and concerned citizens who are shaping responses to environmental problems.

In 2004, Island Press celebrates its twentieth anniversary as the leading provider of timely and practical books that take a multidisciplinary approach to critical environmental concerns. Our growing list of titles reflects our commitment to bringing the best of an expanding body of literature to the environmental community throughout North America and the world.

Support for Island Press is provided by the Agua Fund, Brainerd Foundation, Geraldine R. Dodge Foundation, Doris Duke Charitable Foundation, Educational Foundation of America, The Ford Foundation, The George Gund Foundation, The William and Flora Hewlett Foundation, Henry Luce Foundation, The John D. and Catherine T. MacArthur Foundation, The Andrew W. Mellon Foundation, The Curtis and Edith Munson Foundation, National Environmental Trust, National Fish and Wildlife Foundation, The New-Land Foundation, Oak Foundation, The Overbrook Foundation, The David and Lucile Packard Foundation, The Pew Charitable Trusts, The Rockefeller Foundation, The Winslow Foundation, and other generous donors.

The opinions expressed in this book are those of the author(s) and do not necessarily reflect the views of these foundations.

Agroforestry and
Biodiversity Conservation
in Tropical Landscapes

Agroforestry and Biodiversity Conservation in Tropical Landscapes

EDITED BY

Götz Schroth, Gustavo A. B. da Fonseca,
Celia A. Harvey, Claude Gascon,
Heraldo L. Vasconcelos, and
Anne-Marie N. Izac

ISLAND PRESS

Washington • Covelo • London

Library of Congress Cataloging-in-Publication Data

Agroforestry and biodiversity conservation in tropical landscapes / edited by Gotz Schroth ... [et al.].
          p.   cm.
    ISBN 1–55963–356–5 (cloth : alk. paper) — ISBN 1–55963–357–3 (pbk. : alk. paper)
      1. Agroforestry—Tropics.   2. Agrobiodiversity—Tropics.
    3. Agroforestry—Environmental aspects—Tropics.   4. Agrobiodiversity conservation—Tropics.   I. Schroth, G. (Goetz)

S494.5.A45A3535 2004
634.9'9'0913—dc22   2003023124

British Cataloguing-in-Publication Data available.

Book design by Brighid Willson

Printed on recycled, acid-free paper ♲

Manufactured in the United States of America
09 08 07 06 05 04 03 02   8 7 6 5 4 3 2 1

Contents

Preface

Agroforestry is increasingly recognized as a useful and promising approach to natural resource management that combines goals of sustainable agricultural development for resource-poor tropical farmers with greater environmental benefits than less diversified agricultural systems, pastures, or monoculture plantations. Among these expected benefits is the conservation of a greater part of the native biodiversity in human-dominated landscapes that retain substantial and diversified tree cover. Although the protection of natural habitat remains the backbone of biodiversity conservation strategies, promoting agroforestry on agricultural and other deforested land could play an important supporting role, especially in mosaic landscapes where natural habitat has been highly fragmented and forms extensive boundaries with agricultural areas.

A substantial amount of information on the effects of different agroforestry practices on biodiversity conservation has accumulated in recent years. However, land managers, researchers, and proponents of tropical land use and natural resource management lack a readily usable and comprehensive source of information to guide their efforts toward the creation of more biodiversity-friendly tropical landscapes. This book attempts to fill this gap by exploring the roles of agroforestry practices in conserving biodiversity in human-dominated tropical landscapes and synthesizing the current state of knowledge. It has been edited by a team of conservation biologists and tropical land use specialists and includes contributions from a variety of disciplines (e.g., resource economics, rural sociology, agroforestry, wildlife biology, and conservation genetics), reflecting the interdisciplinary nature of its subject. Contributions are based on many decades of field experience in the tropics of Central and South America, Africa, Asia, and Australia of 46 authors from 13 countries.

This book was made possible through the technical input and support from the Center for Applied Biodiversity Science at Conservation International, Washington, DC, and the Brazilian National Council for Scientific and Technological Development (CNPq) through the Biological Dynamics of Forest Fragments Project at the National Institute for Research in the Amazon, Manaus, Brazil. Numerous people have contributed to this book at all stages of its development. We would particularly like to thank a number of colleagues for their thoughtful and constructive reviews, which have greatly improved the quality of the individual chapters: Andrew Bennett, Elizabeth Bennett, Emilio Bruna, Chris Dick, Gareth Edwards-Jones, Paulo Ferraro, Bryan Finegan, Hubert de Foresta, Karen Garrett, Luadir Gasparotto, Andy Gillison, Jim Gockowski, Colin Hughes, Norman Johns, David Lamb, Nadia Lepsch-Cunha, Gary Luck, Jeff McNeely, Jean-Paul Metzger, Lisa Naughton, Alex Pfaff, Robert Rice, Jim Sanderson, Nigel Tucker, Louis Verchot, Jeff Waage, Bruce Williamson, and Sven Wunder. Barbara Dean and her team at Island Press accompanied the book through its development and greatly improved its style and consistency.

Introduction: The Role of Agroforestry in Biodiversity Conservation in Tropical Landscapes

Götz Schroth, Gustavo A. B. da Fonseca, Celia A. Harvey, Heraldo L. Vasconcelos, Claude Gascon, and Anne-Marie N. Izac

In the tropics, as in the temperate zone, agricultural land use almost always takes place at the expense of natural ecosystems and their biodiversity. For several millennia, humans have attempted to domesticate tropical ecosystems and landscapes in order to channel a larger share of primary production toward their own consumption. Initially they often did this in a subtle way by enriching forests close to campsites with useful plant species or clearing small patches of forest or savanna with primitive tools and fire. But as human populations and their technological capabilities increased and markets for tropical agricultural products developed, the impact of agriculture on tropical ecosystems and landscapes became more dramatic. The devastation of the Brazilian coastal rainforest by European immigrants for growing sugarcane, coffee, cocoa, and other commodities is but one example of wasteful agricultural use of a biodiversity-rich ecosystem in the tropics (Dean 1995). With the rapid increase of tropical populations and global markets in the twentieth century, human impacts on tropical and global ecosystems have reached new dimensions (McNeill 2000).

However, the degree to which tropical ecosystems and landscapes have been transformed through human land use differs dramatically between regions. Depending on their natural resource base, population density, land use history, proximity to urban markets, and many other factors, human-dominated tropical landscapes may be areas completely devoid of tree cover, largely forested mosaics of extractively used primary and secondary forests with small clearings for annual crops, homegardens, and habitations, or anything in between. The concept of agroforestry embraces many intermediate-intensity land use forms, where trees still cover a significant proportion of the landscape and influence microclimate, matter and energy cycles, and biotic processes.

1

In the last three decades, agroforestry has been widely promoted in the tropics as a natural resource management strategy that attempts to balance the goals of agricultural development with the conservation of soils, water, local and regional climate, and, more recently, biodiversity (Izac and Sanchez 2001). Agroforestry practices such as homegardens, crop-fallow rotations, and the use of timber trees in tree crop plantations are being studied at national and international research centers, and courses in agroforestry are being taught at colleges and universities all over the world. As a consequence, a large body of scientific information and practical experiences is available on the effects of trees on soil fertility and carbon stocks, the matching of crop and tree species for different site conditions, tree management and related agronomic-technical issues (e.g., Young 1997; Schroth and Sinclair 2003). Information on complex biotic interactions such as the importance of diversified tree cover in pest and disease dynamics on plot and landscape scales is less available (Schroth et al. 2000; Swift et al. in press). However, a comprehensive review of information on the biodiversity associated with different agroforestry practices and the landscapes of which they are part has not been conducted. This lack of information is felt both in practical conservation and development projects in the field and in university and college courses teaching tropical agroforestry, conservation biology, and related topics.

This book attempts to fill this gap by reviewing the present knowledge of the potential role of agroforestry in conserving tropical biodiversity and by identifying knowledge gaps that warrant further research. More specifically, its objectives are to explore the potential of agroforestry for landscape-scale biodiversity conservation in the tropics; discuss benefits related to the biodiversity of agroforestry systems and the landscapes of which they are part, which could increase private and public support for the use of agroforestry in conservation strategies; identify some of the ecological, socioeconomic, and political constraints on biodiversity-friendly land use systems; and present some practical examples of the use of agroforestry in biodiversity conservation projects in the tropics.

Agroforestry in Tropical Landscapes

Agroforestry is a summary term for practices that involve the integration of trees and other large woody perennials into farming systems through the conservation of existing trees, their active planting and tending, or the tolerance of spontaneous tree regrowth. Following a recent definition by the World Agroforestry Center (ICRAF 2000), agroforestry is defined here as a dynamic, ecologically based natural resource management practice that, through the integration of trees and other tall woody plants on farms and in the agricultural landscape, diversifies production for increased social, economic, and environmental benefits.

A landscape is defined in this book as a mosaic of ecosystems or habitats, present over a kilometer-wide area. Landscapes are composed of individual elements (e.g., forests, agricultural or agroforestry plots, wooded corridors, or pasture areas) that in turn make up the patches, corridors, and matrix elements of the landscape (Forman 1995). Landscapes are also characterized by their relief, including hills, plateaus, and valleys, which influence the flow and distribution of energy and matter and biotic processes (Sanderson and Harris 2000). In many tropical landscapes, the presence of agroforestry systems (e.g., shaded tree crops, fallow areas, or crop and pasture areas with trees) influences ecological processes and characteristics such as the presence and dispersal of fauna and flora, water and nutrient flows, microclimate, and disease and pest dynamics within the landscape. Such landscapes are appropriately called agroforestry landscapes, reflecting the common view in landscape ecology, conservation biology, and agroforestry that certain important effects of agroforestry on biodiversity conservation, water and nutrient cycling, and soil conservation cannot be fully appreciated by merely looking at the individual plot or system because their most significant impacts may occur at the landscape scale. Furthermore, a given agroforestry system does not exist in isolation in that farmers may manage forest gardens or shaded tree crop plantations together with shifting cultivation plots, irrigated rice fields, or pastures, which therefore occur together in the same landscape and jointly determine its properties.

What agroforestry means and how agroforestry practices influence the structure and composition of tropical landscapes are best illustrated with some examples (note that an agroforestry practice or system is not synonymous with an agroforest, which includes the most complex, forest-like types of agroforestry systems). Tropical smallholder farmers often grow staple food crops such as upland rice, maize, and cassava in slash-and-burn systems in rotation with natural tree fallows, which may vary in length from a few years to several decades. This shifting cultivation (or swidden agriculture), which results in a mosaic of crop fields and plots with secondary forest or savanna regrowth in the landscape, is one of the oldest and most extensive forms of agroforestry, although it has often been excluded from the concept of agroforestry on the faulty assumption that all shifting cultivation is unsustainable or inefficient as a land management strategy. Several specific agroforestry practices have evolved in different tropical regions from their common origin in shifting cultivation. In the West African savanna, for example, it is common for farmers to retain useful trees (which may also be difficult to fell and resistant to fire) when preparing a plot for cropping, thereby creating parklike landscapes of scattered trees between crop fields and rangelands that are typical of this region (Figure I.1; Boffa 1999). In the lowlands of Sumatra and Kalimantan (Indonesia), smallholder farmers have modified the traditional crop-fallow rotation by introducing rubber trees into their cropping systems together with annual and short-lived perennial crops. Through a prolonged fallow cycle of

Figure I.1. Parklike landscape with scattered trees in pastures and crop fields in the northern Côte d'Ivoire, West Africa.

several decades and tolerance of spontaneous forest regrowth, these systems gradually evolve into a type of managed secondary forest enriched with rubber trees, the so-called jungle rubber (Gouyon et al. 1993; de Jong 2001). Similar systems have been described from the central Amazon (Figure I.2; Schroth et al. 2003).

Highly complex systems also arise from practices found in southeast Asia and some parts of the Amazon, where farmers plant a food crop (e.g., upland rice) and intercrop it with one or two timber or fruit tree species that have a tall canopy. After harvesting the crop, they plant other timber and fruit tree species with intermediate-level canopies, to be followed by other tree species with lower canopies, creating systems that have an appearance almost similar to that of a natural forest. These systems, which include the damar (*Shorea robusta*) and durian (*Durio zibethinus*) gardens of Sumatra, have appropriately been called agroforests (Figure I.3; Michon and de Foresta 1999). In parts of Latin America and West Africa, coffee and cocoa (both shade-tolerant crops) traditionally are established under an open canopy of remnant trees that were retained when a forest plot was cleared (Johns 1999; de Rouw 1987), resulting in another type of complex agroforest. Similar tea-based systems have been described from northern Thailand and Myanmar (Preechapanya et al. in press).

Throughout the tropics, smallholders commonly plant trees in small homegardens for shade and various products such as fruits and medicinal products (Figure I.4; Torquebiau 1992; Coomes and Burt 1997). They may

Figure I.2. Rubber (*Hevea brasiliensis*) agroforest in the lower Tapajós region, central Amazon, Brazil.

Figure I.3. Complex agroforest with durian (*Durio zibethinus*) and cinnamon (*Cinnamomum burmanii*) trees in Sumatra; in the foreground is a rice field.

Figure I.4. Homegarden in Sumatra.

also retain, plant, or allow the spontaneous regeneration of trees in their pastures for shade, fodder, and timber production and as living fenceposts, as is common in Costa Rica (Harvey and Haber 1999). Furthermore, trees may occur on farms as hedges along boundaries, riparian strips along rivers, palm groves in swampy areas, shelterbelts on wind-exposed sites, and woodlots on slopes, low-fertility sites, and places of cultural and spiritual value.

What Can Agroforestry Contribute to Biodiversity Conservation?

Agroforestry systems and the heterogeneous mosaic landscapes of which they are part have recently attracted the interest of conservation biologists and other investigators working on the interface between integrated natural resource management and biodiversity conservation (e.g., Gajaseni et al. 1996; Perfecto et al. 1996; Rice and Greenberg 2000). On both theoretical and empirical grounds, increased biodiversity has been suggested as making plant communities more resilient (McCann 2000) and thus as having a direct link with productivity gains in the long run. More importantly, as natural ecosystems shrink and remaining patches of natural vegetation are increasingly reduced to isolated habitat islands (protected or not in parks) in a matrix of agricultural land, it becomes crucial to understand what land use systems replace the natural ecosystems and the nature of the matrix surrounding the remaining fragments. In these fragmented landscapes, agroforestry could play a role in helping to maintain a higher level of biodiversity, both within and outside protected areas, when compared with the severe negative effects resulting from more drastic land transformations. Where landscapes have been denuded through inadequate land use or degraded agricultural areas have been abandoned, revegetation with agroforestry practices can promote biodiversity conservation.

It can be rightly argued that all agroforestry systems, however forest-like they may appear, ultimately displace natural ecosystems, either through outright clearing and replanting with crop and tree species or through variable degrees of "domestication" of the original landscape and ecosystem. However, when compared with other nonforest land use options, such as modern, intensively managed monocultures of coffee, rubber, or oil palm with little genetic and structural diversity, or even vast stretches of pasture or annual crops with little tree cover or none at all, agroforestry systems may offer a greater potential as auxiliary tools for biodiversity conservation strategies while attaining production goals.

What forms the basis for the expectation that agroforestry practices can help conserve biodiversity in human-dominated landscapes? Can this expectation be empirically justified? Answering these questions is a central goal of this book. Here, we present three hypotheses of how agroforestry could contribute to biodiversity conservation in human-dominated tropical landscapes. These hypotheses are explored in detail in the chapters and evaluated in the Conclusion at the end of this book.

The Agroforestry-Deforestation Hypothesis

Agroforestry can help reduce pressure to deforest additional land for agriculture if adopted as an alternative to more extensive and less sustainable land use practices, or it can help the local population cope with limited availability of forest land and resources, for example near effectively protected parks.

This hypothesis is based largely on the assumption that certain agroforestry practices, if profitable and sustainable, may occupy the available labor force and satisfy the needs of a given population on a smaller land area than extensive land use practices such as cattle pasture, thereby reducing the need to deforest additional land. Extensive land use practices are common in agricultural frontier regions because of the often low land prices and poor market access. More intensive agricultural practices, where economically viable, may be able to bring area needs per household or unit of produce lower than agroforestry practices can but may expose farmers to unacceptable economic and ecological risks (Johns 1999). Furthermore, agroforestry practices may be more sustainable and therefore allow the use of deforested plots over a longer time period than alternative land use methods, such as pure annual cropping (which may rapidly degrade the soil, especially on erosion-prone and low-fertility sites) and tree crop monocultures (which may be more susceptible to pest and disease outbreaks than agroforestry plantings; Schroth et al. 2000). Consequently, the adoption of agroforestry may reduce the need to deforest new areas. However, it should be stressed that sustainability is not an intrinsic characteristic of agroforestry practices. Sustainability has both biological and socioeconomic dimensions, and even if it is technically possible to manage a certain land use system sustainably, it may be more advantageous for a farmer not to do so if land for new fields and plantations is readily available or if there is an advantage to occupying a large land area (e.g., acquiring property or land use rights). The agroforestry-deforestation hypothesis is analyzed from a socioeconomic and historical viewpoint in Part II of this volume.

The Agroforestry-Habitat Hypothesis

Agroforestry systems can provide habitat and resources for partially forest-dependent native plant and animal species that would not be able to survive in a purely agricultural landscape.

The biodiversity of agroforestry systems, and of agroecosystems in general, consists of planned and unplanned components. By their very nature, agroforestry systems contain more planned diversity (i.e., more planted and selected plant species) than the corresponding monoculture crops, although not necessarily more than some traditional mixed cropping systems (Thurston et al. 1999). Certain agroforestry systems such as tropical homegardens, which may contain several dozen species and varieties of trees and crops, are seen as important reservoirs of tropical tree and crop germplasm (Torquebiau 1992). However, not all agroforestry systems have much planned diversity; for example, certain shaded coffee plantations essentially consist of one crop and a single, sometimes exotic shade tree species, and live fences typically consist of only a handful of tree species.

Of similar or greater importance for the conservation value of agroforestry systems than their planned diversity is their unplanned diversity, that is, the plants and animals that colonize or use the structure and habitat formed by the planted species. Structurally heterogeneous perennial vegetation can provide more niches for native flora and fauna than structurally simpler monocultures and pastures (Thiollay 1995). A humus-rich soil that is not regularly disturbed by tillage and the permanent litter layer that usually develops under agroforestry may also provide appropriate habitat for a diverse soil fauna and microflora that may not be present in simpler and regularly disturbed agricultural systems, although little is known about such belowground biodiversity benefits of complex land use systems (Lavelle et al. 2003).

The role of agroforestry systems as refugia for forest-dependent species is most relevant in landscapes that are largely devoid of natural vegetation. In such deforested and often densely populated landscapes, agroforestry systems may maintain more species of plants, animals, and microorganisms from the original ecosystems than corresponding agricultural monocultures and pastures and therefore could be a better compromise between production goals and biodiversity conservation (Thiollay 1995). It should be stressed that one cannot evaluate this role for an agroforestry system by simply counting the species present because these will invariably include species that are adapted to disturbed conditions and may not need special protection. Instead, it is necessary to determine whether forest-dependent and threatened species use the agroforestry areas, the degree to which they depend on these areas for habitat or food, and whether their populations are viable over the long term. Parts III, IV, and V of this volume explore this hypothesis in greater detail.

The Agroforestry-Matrix Hypothesis

In landscapes that are mosaics of agricultural areas and natural vegetation, the conservation value of the natural vegetation remnants (which may or may not be protected) is greater if they are embedded in a landscape dominated by agroforestry elements than if the surrounding matrix consists of crop fields and pastures largely devoid of tree cover.

This hypothesis refers to the larger-scale properties that agroforestry elements may confer to landscapes with respect to their suitability as habitat for native fauna and flora, that is, effects that reach beyond the limits of an individual agroforestry system and extend to the entire landscape. In tropical land use mosaics, ecological processes and characteristics such as microclimate, water and nutrient fluxes, pest and disease dynamics, and the presence and dispersal of fauna and flora may be significantly influenced by agroforestry elements. For example, strategically placed agroforestry systems may serve as biological corridors between patches of natural vegetation or act as stepping stones that facilitate animal movement. Where two forest fragments are separated by a tree crop plantation with a diversified shade canopy of rainforest remnant trees, it should be easier for arboreal forest fauna to disperse from one fragment to the other than if they had to cross an open pasture, and this may help to reduce problems of small populations in the individual fragments by maintaining biotic connectivity. Insects, birds, and bats, crossing from one forest patch to another via a riparian strip or using remnant trees in a pasture as stepping stones, may pollinate trees that occur at low densities in the individual patches. Birds may carry seeds from one fragment to the next, moving along live fences, hedges, and windbreaks or flying from one isolated tree to another, thereby enhancing seed dispersal in fragmented landscapes. Where agroforestry systems adjoin forest areas, they may also buffer them against the stronger winds and harsher microclimate of open agricultural fields and pastures, thereby increasing the size of the core area available to certain sensitive forest interior species. Such agroforestry buffer zones may also protect forests from fire, which is a frequently used management tool for growers of annual crops and pastoralists but anathema to owners of valuable tree crops and timber trees. The potential role of agroforestry in increasing the conservation value of forest fragments and parks through such landscape-scale processes has been little explored but could be of tremendous importance for landscape conservation strategies in heavily but not totally deforested regions (Center for Applied Biodiversity Science 2000). The available evidence in support of this hypothesis is also reviewed in Parts III, IV, and V of this volume.

Audience and Structure of the Book

This book has been written for students and practitioners of tropical agriculture, forestry and agroforestry, conservation biology, landscape ecology, natural resource management, ecological economics, and related disciplines. In accord with the interdisciplinary nature of the subject and the heterogeneity of the targeted audience, an effort has been made to keep the language as simple and universally understandable as possible.

The book is divided into five parts. Part I provides a background in conservation biology and landscape ecology that will help nonspecialists understand later chapters. It also gives an update of recent concepts and research results in these fields. Part II focuses on socioeconomic issues related to biodiversity-friendly land use practices. After reviewing approaches to quantifying the economic value of the environmental services of agroforestry, it discusses whether and to what extent agroforestry can help reduce pressures on natural ecosystems, using both historical and present-day perspectives. Conservation concessions are introduced as a complementary approach to agroforestry in conservation strategies.

Part III reviews the potential of selected agroforestry practices to promote biodiversity conservation by serving as habitats, biological corridors, and buffer zones for protected areas and by increasing connectivity and genetic exchange within landscapes. The floristic, structural, and management aspects that increase the value of agroforestry systems for biodiversity conservation on the plot and landscape scales are a particular focus of this section.

The objective of Part IV is to analyze the trade-offs between conservation and production goals in diversified tropical land use mosaics. Such assessment is crucial for avoiding conflict and forging alliances between farmers and conservationists. Biodiversity benefits for farmers include timber and nontimber products, hunting opportunities, and protection from pest and disease outbreaks through biological control mechanisms; costs may include wildlife damage to crops, livestock, and humans and pest and disease transfer between native vegetation and crops. Risks associated with the use of exotic and potentially invasive tree species in agroforestry for the biodiversity of natural habitat are reviewed. The question of how wildlife can be managed sustainably in tropical land use mosaics is also addressed.

Part V reviews practical examples of the use of agroforestry and farm forestry in conservation strategies, including both traditional and more recent approaches, and provides advice on selecting tree species for agroforestry programs. The section also addresses the potential of agroforestry to buffer natural ecosystems against changing climate. The book's Conclusion synthesizes the information presented in the volume, provides recommendations, and identifies research needs.

References

Boffa, J. M. 1999. *Agroforestry parklands in sub-Saharan Africa.* Rome: Food and Agriculture Organization of the United Nations.

Center for Applied Biodiversity Science. 2000. *Designing sustainable landscapes: the Brazilian Atlantic Forest.* Washington, DC: Center for Applied Biodiversity Science at Conservation International and Institute for Social and Environmental Studies of Southern Bahia.

Coomes, O. T., and G. J. Burt. 1997. Indigenous market-oriented agroforestry: dissecting local diversity in western Amazonia. *Agroforestry Systems* 37:27–44.

Dean, W. 1995. *With broadax and firebrand: the destruction of the Brazilian Atlantic Forest.* Berkeley: University of California Press.

de Jong, W. 2001. The impact of rubber on the forest landscape in Borneo. Pages 367–381 in A. Angelsen and D. Kaimowitz (eds.), *Agricultural technologies and tropical deforestation.* Wallingford, UK: CAB International.

de Rouw, A. 1987. Tree management as part of two farming systems in the wet forest zone (Ivory Coast). *Acta Oecologica Oecologia Applicata* 8:39–51.

Forman, R. T. T. 1995. *Land mosaics: the ecology of landscapes and regions.* Cambridge, MA: Cambridge University Press.

Gajaseni, J., R. Matta-Machado, and C. F. Jordan. 1996. Diversified agroforestry systems: buffers for biodiversity reserves, and landbridges for fragmented habitats in the tropics. Pages 506–513 in R. C. Szaro and D. W. Johnston (eds.), *Biodiversity in managed landscapes: theory and practice.* Oxford, UK: Oxford University Press.

Gouyon, A., H. de Foresta, and P. Levang. 1993. Does "jungle rubber" deserve its name? An analysis of rubber agroforestry systems in southeastern Sumatra. *Agroforestry Systems* 22:181–206.

Harvey, C. A., and W. A. Haber. 1999. Remnant trees and the conservation of biodiversity in Costa Rican pastures. *Agroforestry Systems* 44:37–68.

ICRAF. 2000. *Paths to prosperity through agroforestry. ICRAF's corporate strategy, 2001–2010.* Nairobi: International Centre for Research in Agroforestry.

Izac, A.-M. N., and P. A. Sanchez. 2001. Towards a natural resource management paradigm for international agriculture: the example of agroforestry research. *Agricultural Systems* 69:5–25.

Johns, N. D. 1999. Conservation in Brazil's chocolate forest: the unlikely persistence of the traditional cocoa agroecosystem. *Environmental Management* 23:31–47.

Lavelle, P., B. K. Senapati, and E. Barros. 2003. Soil macrofauna. Pages 303–323 in G. Schroth and F. L. Sinclair (eds.), *Trees, crops and soil fertility: concepts and research methods*. Wallingford, UK: CAB International.

McCann, K. S. 2000. The diversity-stability debate. *Nature* 405:228–233.

McNeill, J. 2000. *Something new under the sun: an environmental history of the twentieth century*. London: Penguin.

Michon, G., and H. de Foresta. 1999. Agro-forests: incorporating a forest vision in agroforestry. Pages 381–406 in L. E. Buck, J. P. Lassoie, and E. C. M. Fernandes (eds.), *Agroforestry in sustainable agricultural systems*. Boca Raton, FL: Lewis.

Perfecto, I., R. A. Rice, R. Greenberg, and M. E. van der Voort. 1996. Shade coffee: a disappearing refuge for biodiversity. *BioScience* 46:598–608.

Preechapanya, P., J. R. Healy, M. Jones, and F. L. Sinclair. In press. Retention of forest biodiversity in multistrata tea gardens in northern Thailand. *Agroforestry Systems*.

Rice, R. A., and R. Greenberg. 2000. Cacao cultivation and the conservation of biological diversity. *Ambio* 29:167–173.

Sanderson, J., and L. D. Harris. 2000. Landforms and landscapes. Pages 45–55 in J. Sanderson and L. D. Harris (eds.), *Landscape ecology: a top-down approach*. Boca Raton, FL: Lewis.

Schroth, G., P. Coutinho, V. H. F. Moraes, and A. K. M. Albernaz. 2003. Rubber agroforests at the Tapajós river, Brazilian Amazon: environmentally benign land use systems in an old forest frontier region. *Agriculture, Ecosystems and Environment* 97:151–165.

Schroth, G., U. Krauss, L. Gasparotto, J. A. Duarte Aguilar, and K. Vohland. 2000. Pests and diseases in agroforestry systems of the humid tropics. *Agroforestry Systems* 50:199–241.

Schroth, G., and F. L. Sinclair. 2003. *Trees, crops and soil fertility: concepts and research methods*. Wallingford, UK: CAB International.

Swift, M. J., A.-M. N. Izac, and M. van Noordwijk. In press. Biodiversity and ecosystem services in agricultural landscapes: are we asking the right questions? *Agriculture, Ecosystems and Environment*.

Thiollay, J.-M. 1995. The role of traditional agroforests in the conservation of rain forest bird diversity in Sumatra. *Conservation Biology* 9:335–353.

Thurston, H. D., J. Salick, M. E. Smith, P. Trutmann, J.-L. Pham, and R. McDowell. 1999. Traditional management of agrobiodiversity. Pages 211–243 in D. Wood and J. M. Lenné (eds.), *Agrobiodiversity: characterization, utilization and management*. Wallingford, UK: CAB International.

Torquebiau, E. 1992. Are tropical agroforestry home gardens sustainable? *Agriculture, Ecosystems and Environment* 41:189–207.

Young, A. 1997. *Agroforestry for soil management*. Wallingford, UK: CAB International.

PART I

Conservation Biology and Landscape Ecology in the Tropics: A Framework for Agroforestry Applications

This part of the book introduces some major concepts of conservation biology and landscape ecology for application in tropical landscapes. Its intention is to provide the necessary background knowledge in conservation science with a focus on landscape-scale issues so that nonspecialist readers can easily follow the discussions of the biodiversity effects of different types of agroforestry in later chapters. For readers who are familiar with the concepts, it provides an update of recent progress in these fields.

Chapter 1 outlines the current threats to biodiversity in the tropics, including habitat loss, fragmentation, overexploitation of ecosystems, and invasions by exotic plant and animal species. It discusses different conservation strategies and stresses the need for strategies comprising landscapes, regions, and larger scales. It points to the role in local, regional, and global conservation strategies that agroforestry can and cannot play: although protected areas and conservation set-asides are the irreplaceable backbone of any sensible conservation strategy, agroforestry can play an important supporting role by linking and buffering reserves and by maintaining or reintroducing a modest level of biodiversity in biologically degraded areas from which natural vegetation has been lost through human land use.

Chapters 2 and 3 focus on landscape processes that could be influenced by agroforestry practices. Chapter 2 discusses the demographic and genetic consequences of fragmentation of natural ecosystems through human land use for plant and animal populations and the key landscape features (area, edge, matrix, and distance effects) that affect fragmented populations. It also addresses the possibility of agroforestry land uses partially mitigating some of the negative effects of habitat fragmentation by reducing edge effects, increasing fragment connectivity, providing food or shelter for fragmented wildlife populations, and reducing the use of fire.

Chapter 3 discusses the potential role that agroforestry elements in the agricultural matrix could play in increasing landscape connectivity by serving as biological corridors for fauna and flora between remnant forest fragments. As experiences from corridors of natural vegetation show, the effectiveness of corridors for different plant and animal groups depends greatly on their size, structure, and floristic composition and on the biology of the target plant or animal species, and such background information must be taken into account in evaluating and designing agroforestry corridors.

Chapter 1

Biodiversity Conservation in Deforested and Fragmented Tropical Landscapes: An Overview

Claude Gascon, Gustavo A. B. da Fonseca, Wes Sechrest, Kaycie A. Billmark, and James Sanderson

Our planet is in the midst of a sixth mass extinction. The earth is losing its biological resources at an ever-increasing rate, a trend that began with the emergence of humans. The majority of the earth's land surface has been colonized over the last few tens of thousands of years and was increasingly affected by the agricultural revolution around 10,000 years BP and the industrial revolution in more recent times. If this trajectory is maintained, many of the planet's biological resources will disappear. There is a need for a more thorough scientific understanding of natural systems and their functioning as a base for crucial global, regional, and local conservation decisions. The earth's tropical regions, in particular, are highly vulnerable to human impact. The wealth and distinctiveness of their biodiversity, combined with the multifaceted threats that they face make these regions an urgent priority for biodiversity conservation. Current scientific research efforts in tropical areas have yielded insight into many important biological questions. Conservation actions, including the implementation of protected areas and corridors, and attention to the surrounding matrix of agricultural and degraded land must be integrated into cohesive regional plans. The application of more conservation-friendly land uses, such as agroforestry, for improving biodiversity conservation in tropical landscapes can contribute to such landscape-scale conservation strategies. The implementation of these efforts is an important step in translating science into effective conservation action.

The goal of this chapter is to provide an overview of important global biodiversity conservation issues, with special attention to terrestrial tropical ecosystems. Additionally, this chapter provides a framework for the discussions in later chapters with regard to biodiversity threats and conservation strategies and applications, including agroforestry.

Tropical Ecosystems

Tropical ecosystems cover a large part of the earth's surface and contain more than half of all terrestrial species (Myers and Myers 1992). These ecosystems have played a unique role in the evolution of the planet's biodiversity. Tropical environments, especially humid forests, were once much more widespread than at present. Today, approximately half of all tropical regions are forests, with the remainder savannas and deserts. Worldwide, there are about 3.87 billion ha of forest, 5 percent of which are forest plantations (FAO 2001). World forests may be categorized as tropical, subtropical, temperate, or boreal (Figure 1.1a). Tropical forests consist of tropical rain, tropical moist deciduous, tropical dry, and tropical mountain forests (Figure 1.1b).

All forests are affected on some level by direct and indirect human activity, although there are no accurate global assessments of forest conditions.

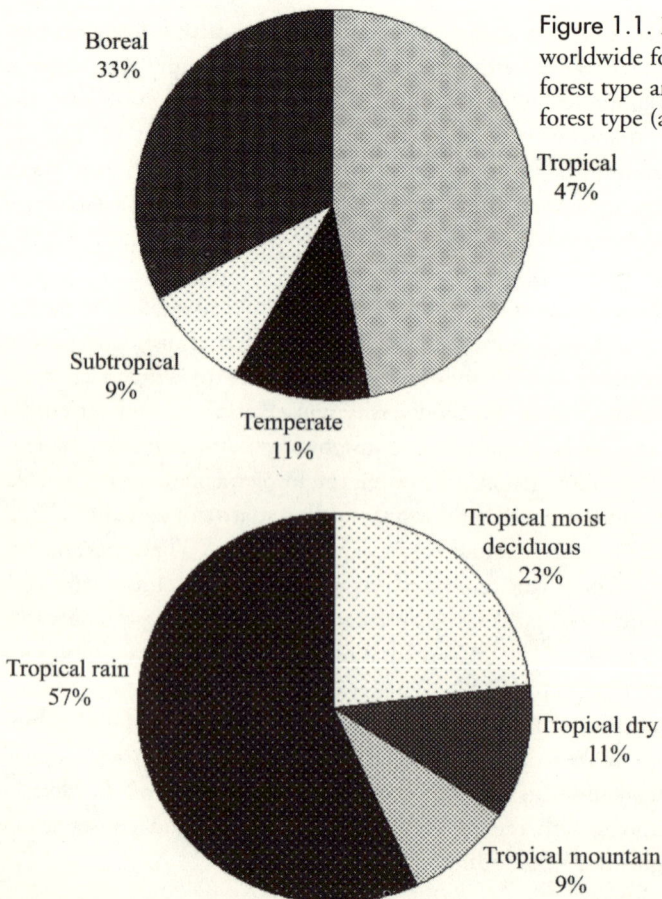

Figure 1.1. Distribution of worldwide forests by (a) general forest type and (b) tropical forest type (after FAO 2001).

Between 1990 and 2000, 14.2 million ha per year of tropical forest were deforested, with an additional 1 million ha per year converted to forest plantations. Natural forest expansion over this time was 1 million ha per year, with an additional 0.9 million ha per year afforested by humans as forest plantations. This deforestation occurred differently on regional and local scales. For instance, during this 10-year time period, the country of Burundi in Central Africa lost 9 percent of its remaining forest per year. This significant percentage loss is of great importance to national policymakers in Burundi, but actual deforestation rates of 15,000 ha per year were much lower than in other parts of the world and therefore are less important from a global perspective. The largest actual loss in Africa occurred in the Sudan, with 959,000 ha deforested each year. Indonesia deforested a staggering 1,312,000 ha per year over this time period (FAO 2001). If left unchecked, the clearing, burning, logging, and fragmentation of forest will destroy most of the world's tropical forests in our lifetime. The planet's forested areas have already decreased by almost 2 billion ha since the beginning of the agricultural revolution (Noble and Dirzo 1997). The impacts of this destruction on any geographic scale are not yet fully understood. In addition to the release of CO_2 via biomass combustion and microbial activity, soil erosion, and hydrological cycle disturbance, this destruction also results in the extinction of numerous known populations and species and the loss of undiscovered species, each with a unique history and habits never to be known.

One important tool for mitigating tropical deforestation is the establishment of tropical agroforested areas or protected parks. Parks are effective in preventing deforestation and thereby protect biodiversity despite the fact that many are underfunded and experience substantial land use pressure (Bruner et al. 2001). Within the matrix surrounding tropical parks, other methods, such as agroforestry, can be used to protect biodiversity and help alleviate the negative effects of deforestation and associated edge effects. By simulating to some extent natural forest cover through the cultivation of tree species with agricultural crops, agroforestry areas may serve as biodiversity corridors between protected areas and nonprotected remnants of natural vegetation while providing sustainable crop and wood harvests.

The Tropical Biodiversity Crisis

Biodiversity is not simply a measure of the world's species; rather, it also encompasses genetic variability within and between populations, species' evolutionary histories, and other measures of the diversity of life. Biodiversity patterns vary between regions. This variability results both from the present ecology and past evolutionary history of species and from habitat type, habitat availability, and physical qualities such as climatic conditions and geological and hydrological patterns, all varying over space and time. The future preservation of biodiversity requires intricate knowledge of the patterns and processes that affect ecosystem function. The tropics, particularly tropical forests, are expansive biodiversity reservoirs (Stevens 1989). Many species in the tropics are limited in distribution, and the spatial turnover of species is high among many taxonomic groups (Condit et al. 2002). Species distribution patterns are not uniform across the globe; most groups of organisms show a strong increase in species richness, or number of species per unit area, nearer to the equator. Additionally, the number of species in most terrestrial and freshwater groups is greater at lower than at higher elevations and greater in forests than in deserts (Gaston 2000). These general patterns suggest that tropical environments are favorable to the evolution of new species and the persistence of existing species. High diversity in the tropics is generally attributed to high productivity, low environmental variance (e.g., seasonality), persistent predation and competition, lower historical climatic change impacts, and differential speciation and extinction rates. Recognizing that these attributes tend to support high diversity in the tropics, it is important to note that there are significant intratropical diversity patterns and that lower-diversity regions can also be found in the tropics.

Conservation efforts have focused much attention on tropical forests because they are the richest strongholds of terrestrial biodiversity. Therefore, exploitation of natural resources in the tropics results in the destruction of large genetic reservoirs. Incalculable benefits are gained from maintaining species numbers and the current diversity of organisms. Much of the research on ecological and evolutionary benefits is new, and more research must be conducted to determine broad patterns and processes. Research has shown that on local scales, the lower the species diversity within a system, the more vulnerable it is to species and population extinctions as a result of nonnative species invasions (Levine 2000). One can conclude that the maintenance of high diversity could reduce the number of invading species, thereby greatly reducing the negative impacts of these species (Kennedy et al. 2002). Other biodiversity effects on ecosystem processes have also been demonstrated (Cardinale et al. 2002). For example, plant diversity of European grasslands positively influences plant primary production (Loreau and Hector 2001). Additionally, diverse areas tend not only to have more functional components (more species with diverse ecologies) but also to maintain more predictable ecological processes (McGrady-Steed et al. 1997).

Unfortunately, short-term economic gains driven by increasing human populations usually influence the decision-making process that leads to resource overuse. High population growth rates in tropical countries create socioeconomic difficulties. Environmental constraints, such as climate, often compound prevalent problems such as malnutrition and famine. This situation, combined with the need of tropical countries to rely on more advanced countries for technical assistance and for the development of their own resources, often leads to exploitive rather than sustainable use. Poverty, war, and social inequality generate environmental degradation, which further drives socioeconomic crises in a continuous feedback loop. These underlying drivers of environmental degradation and biodiversity loss must be addressed for successful conservation of tropical ecosystems.

Threats to Tropical Forest Ecosystems

Environmental degradation is driven by several major threats, including habitat loss and fragmentation, exploitation, pollution, introductions of nonnative species, and human-induced global change. For tropical ecosystems, land use is ranked as the major driver affecting these regions for the next 100 years (Sala et al. 2000). In this section we briefly review these threats and point to the potential role of agroforestry that will be discussed in more detail in subsequent chapters.

Habitat Fragmentation

Although human presence affects landscape biodiversity in many ways, one of the most visible and widespread effects is habitat fragmentation (Gascon et al. 2003). Because of the dynamic nature of landscapes, fragmentation alters the behavior of natural interactions within the landscape and the functioning of the entire landscape. For example, the species composition and diversity of a tropical landscape differ near a treefall as compared with a dense canopy. However, the temporal recovery of treefalls over an entire tropical landscape results in areas at all stages of natural forest growth. These areas provide a varying but consistent species composition and diversity for the entire landscape. Conversely, in fragmented landscapes, the number of areas at different stages of forest growth is lower, and the average functioning of the landscape becomes less predictable. If a substantial portion of a tropical landscape undergoes deforestation, the ecological function of the fragmented landscape can be permanently altered from its natural state. These changes in the biodiversity and integrity of fragmented landscapes argue in favor of the construction of conservation corridors, where biodiversity-friendly land uses such as agroforestry can be integrated with fragments of natural habitat in interconnected networks that help restore functional aspects of the landscape.

Fragmentation alters not only the functioning of the landscape but also the behavior and dynamics of populations in the fragmented system (Bierregaard et al. 2001; Chapter 2, this volume). The response of populations to landscape changes often is very negative. If no patches exist that are habitable for a particular population, then that population is likely to be lost. Forest fragmentation can result in species population survival or extinction, depending on many factors such as how easily the species can disperse between forest patches and whether the species can use the modified landscape and find resources. For instance, nocturnal species may be better able to survive fragmentation than their diurnal counterparts because of the greater similarity of ambient conditions between forest fragments and the surrounding matrix at night (Daily and Ehrlich 1996). Fragmentation has also been shown to decrease aboveground biomass, especially on the fragment edges (Laurance et al. 1997). A study in Brazil showed that large canopy trees in tropical rainforests experience a higher mortality rate when they are in a heavily fragmented system (Laurance et al. 2000). Fragmentation also affects the reproduction of species that remain in the forest patches. For example, species of dipterocarp trees that inhabit lowland forests of Borneo exhibit seed dispersal events that coincide with El Niño–Southern Oscillation events. Because these dipterocarp species are dominant canopy species, their dispersal and reproduction are strongly affected by local and regional logging, which can disrupt their timed reproduction (Curran et al. 1999).

Finally, tropical forest fragmentation can differentially affect species dispersal mechanisms on a landscape scale (Aldrich and Hamrick 1999; see also Chapter 3, this volume). Metapopulation dynamics between habitat patches result in local population extinctions, causing diversity losses in patches that are often unrecoverable in large expanses of degraded areas. Genetic isolation between widely isolated or dispersal-limited populations leads to loss of overall genetic diversity between populations and increasing vulnerability to deleterious genetic effects, such as susceptibility to pathogens. Landscape-scale strategies must use research on a broad base of ecosystems, species, and populations. For example, Madagascar, which holds a high amount of unique biodiversity, has lost more than 90 percent of its primary forest. Threats on the island have not abated, and forest losses continue in the few remaining fragments. The medium-term existence of many tropical forest species is threatened by widespread forest loss and fragmentation.

Introduced Species

A biodiversity concern related to fragmentation is that of introduced species. Tropical regions have a large number of endemic species that are unique to a particular area or region, usually because of genetic isolation created by physical barriers (e.g., water in the case of island species). Often in the case of disturbed areas, such as in fragmented systems, local endemic species are replaced by wide-ranging species, including those tolerant of disturbed habitats (Tocher et al. 2001). Successful nonnative species often are ones that range over wide areas and tolerate disturbance well. Globally, almost all areas are affected by these introduced species, with island biota being especially vulnerable. Changes in complex ecological systems, such as introduction of prey species, can have cascading effects on fauna (Roemer et al. 2002). Invasive species are homogenizing the global flora and fauna, which has led to extinctions and population reductions of native species (Lovel 1997).

This negative impact on native species is sometimes masked by an increase in species richness. With the influx of competing species, species numbers in a fragmented system can increase, which creates a situation in which further biodiversity degradation can occur through species displacements and more local extinctions. To mitigate these problems, direct preventive measures are needed in addition to increases in connectivity, area-to-perimeter ratios, buffer zones, and improvements to the matrix around existing reserves (Gascon et al. 2000). The use of agroforestry outside protected areas may play a role in such strategies by increasing connectivity and serving as buffers but may also pose additional threats if invasive alien tree species are used (see Chapter 15, this volume).

Exploitation

Exploitation of the natural environment has always been a part of human culture. Increases in the human population have likewise increased demands on natural resources. These demands have reached levels that cannot be maintained without permanently damaging natural ecosystems and processes. For instance, subsistence hunting in Amazonian Brazil is estimated to affect more than 19 million individual animals per year. This hunting, coupled with wildlife trade and demand for wildlife products such as pelts, ivory, and organs, places serious pressure on native fauna (Harcourt and Sayer 1996). New roads, which provide access to previously inaccessible areas for colonization, have increased human-induced threats. In fact, even in Brazilian Amazonia, every nature reserve was found to be 40 to 100 percent accessible by roads or navigable rivers (Peres and Terborgh 1995). Landscape planners must use knowledge of the cascading and synergistic effects of road building and settlement on biodiversity and must place greater value on wildlife and natural habitats to reduce exploitation. Agroforestry land uses including fallows and secondary forests may help to avoid overexploitation of the timber and non-timber resources of natural habitats and thus contribute to integrated strategies of natural resource management and forest conservation (see Chapter 14, this volume).

Global Change

Anthropogenic physical changes also threaten tropical systems. One of the most important is the alteration of biogeochemical cycles. Carbon, nitrogen, phosphorus, and other nutrients are cycled through natural systems. Through industrial emissions, anthropogenic biomass burning, mining, and agriculture runoff, among others, humans artificially increase nutrient and pollutant loads in air, land, and water (Garstang et al. 1996; Tilman 1999; Tilman et al. 2001). This has caused direct and indirect effects on global climate and biogeochemical cycling that, without abatement, will lead to a drastically altered environment.

The threat of human-induced climate change to the planet is well documented: global average surface temperatures have increased by approximately 0.6°C since the end of the nineteenth century (Houghton et al. 2001). Greenhouse gas emissions are still accelerating, and we need to keep the remaining forests intact to mitigate CO_2 release. Indeed, tropical deforestation releases about 2 Giga-tons (Gt) of carbon per year; in the 1980s this was estimated to make up 25 percent of carbon emissions from human activity (FAO 2001). Shukla et al. (1990) simulated the hydrological cycle over Amazonia and found that rapid deforestation could result in a longer dry season. This disruption in precipitation patterns would have widespread ecological implications, such as increases in fire frequencies and disruption of the life cycles of pollination vectors. So severe are the potential changes that if large areas of Amazon tropical forests are destroyed, they may not return (Shukla et al. 1990).

The protection of tropical ecosystems is a cornerstone of global climate change solutions. The effect of human-induced climate change on biodiversity will be profound. Species ranges will track climatic patterns, including temperature and precipitation patterns. The heterogeneous nature of climate change over time and space makes it difficult to predict the effects on local or even regional scales. In general, in the warming climate species ranges will independently shift toward the poles and upwards in altitude, although there is no general linear pattern (Peters 1991). Protected areas must not only serve the flora and fauna within their borders but also permit natural migrations and climate-induced range shifts. The surrounding matrix will be a key to mitigating biodiversity losses from global climate change as landscapes undergo rapid temporal changes. Protecting biodiversity cannot be achieved on static spatial scales, and matrix areas must be used to conserve biodiversity. Agroforestry practices may help to create a permeable matrix that allows such migrations (Chapter 20, this volume) and may also make a certain contribution in reducing carbon emissions after forest conversion. Practices such as riparian strips and contour plantings may also help to reduce nutrient and sediment losses from agricultural lands and thereby limit the effects of agriculture on biogeochemical cycling.

Conservation Strategies

Recent scientific knowledge about how the tropical rainforests are affected by fragmentation, logging, road building, and encroaching agricultural frontiers suggests that much of the resulting ecological degradation (postfragmentation) can be accounted for by just a few factors. These factors include the size and shape of forest fragments, the presence and extent of abrupt forest edges, and the activities in the surrounding matrix. All else being equal, smaller forest patches contain fewer species per unit area than larger ones (Brown and Hutchings 1997; Didham 1997; Tocher et al. 1997; Warburton 1997). Smaller patches also contain more edge relative to area than larger patches. Abrupt forest edges also affect most ecological variables and indicators of forest dynamics, such as species distributions, tree mortality and recruitment, biomass loss, and community composition of trees. According to some recent estimates of the extent of edge-affected processes, only the largest forest fragments (>50,000 ha) are immune from detectable ecological effects of isolation (Curran et al. 1999).

The activities and intensity of use of the matrix habitat surrounding isolated forest patches can have profound and irreversible effects on the sustainability of the patches (Gascon et al. 1998, 2000). For example, species that are able to use the modified matrix habitat are those that will be preferentially maintained in the habitat patches. Therefore, the management of landscapes should take these considerations into account through their translation into public policy at all levels. This may include the promotion of agroforestry in areas that are critical for the connectivity of habitat fragments (for examples see Chapters 17 and 18, this volume).

Global Conservation Strategies

Two main global strategies are commonly used in conservation efforts, one that incorporates threats and one that uses ecological representation. The first type of global conservation strategy focuses attention on the areas and biota that are most threatened and most distinctive. The hotspot approach of Conservation International is an example of this type of global conservation strategy (Mittermeier et al. 2000; Myers et al. 2000). Hotspots are land areas with more than 0.5 percent of all vascular plant species endemic to them and with at least a 70 percent loss of their natural primary habitats. Plant diversity is used as a surrogate for the diversity of ecosystems and other taxonomic groups. There are 25 identified hotspots (Figure 1.2), which cover 11.8 percent of the earth's land surface, but because of habitat destruction, natural primary habitat in these areas covers only 1.4 percent of the earth's land surface. These areas provide the only remaining habitat for an estimated 44 percent of all species of vascular plants and 35 percent of all species of mammals, birds, reptiles, and amphibians. Many species in the hotspots are extremely vulnerable, with diminished populations, highly fragmented habitat, and pressures from numerous human sources. Since 1800, close to 80 percent of all bird species that have gone extinct were lost from the biodiversity hotspots (Myers et al. 2000). Additionally, Conservation International has designated three main

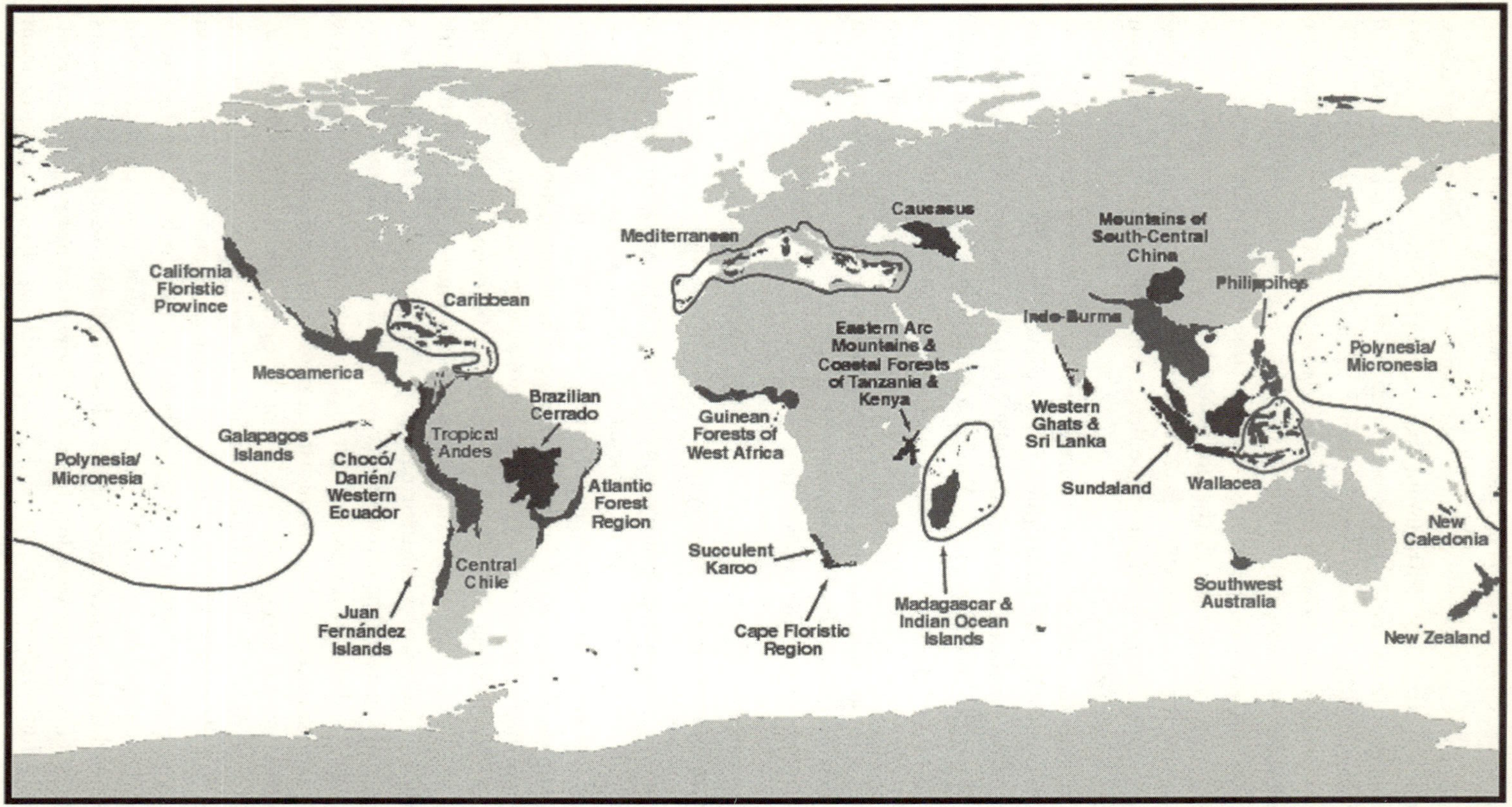

Figure 1.2. Global biodiversity hotspots (adapted from Myers et al. 2000). Major tropical wilderness areas are in the Amazon and Congo basins and New Guinea.

Major Tropical Wilderness Areas, which have much of their primary habitat still intact and contain high amounts of biodiversity.

By defining conservation priority areas based on threatened and distinctive biota, the hotspot approach evaluates specific threats in manageable land areas. Although threats vary, ubiquitous to all hotspots are disproportionately high human population pressures. An estimated 1 billion or more people, or close to 20 percent of the world population, live in hotspot areas, which cover less than 12 percent of the earth's land surface. The human population growth rate is 1.8 percent per year in hotspots and 1.3 percent outside hotspots. Human demand for resources in and around hotspots may be significantly higher than in other areas. Even in the three major tropical wilderness areas (New Guinea and Melanesian islands, upper Amazonia, and the Congo River basin), which support low population densities of about eight people per km^2 (including several urban areas), population growth rates are well above the average current global growth rate of 1.3 percent per year (Cincotta et al. 2000).

A second major global conservation strategy uses a representative approach. A descriptive example of this conservation strategy, used by World Wildlife Fund, is the ecoregion approach. This approach seeks to focus efforts on conserving representative areas in major ecosystem and habitat types (Olson and Dinerstein 1998). Some areas, which have maintained isolation for long time periods, such as oceanic islands, mountain ranges, karst, and caves, often are reservoirs of incredible amounts of biodiversity. The evolution of flora and fauna in these regions has created unique and rare organisms, often found nowhere else. These areas therefore are top priorities for conservation.

Landscape and Local Conservation Strategies

Smaller-scale conservation efforts often use a landscape approach to conserving biodiversity (landscape scale, which includes conservation corridors, is defined here as tens of thousands of square kilometers). This approach is most easily incorporated into predictive computer models and therefore is used to predict changes or shifts of ecosystems caused by environmental and anthropogenic factors such as human population increase and climate change. Landscapes are made up of spatially heterogeneous areas where biodiversity exists and interacts dynamically between areas. Biodiversity on the landscape scale consists of the composition of these areas and the dynamic interactions between areas and landscape elements. Interactions can occur through the flow of nutrients, water, energy, organisms, and other resources. Detailed location-specific data collection and knowledge of the pattern of spatial interactions, such as biodiversity effects, are needed to capture the dynamic nature of landscapes. This approach can be applied anywhere without the constraint of focusing on specific biodiversity-rich regions. Furthermore, this approach is critical in maintaining reserve areas with established corridors and evaluating complex topography and regions surrounding reserves, all for the purpose of mitigating threats to biodiversity. A better understanding of the patterns and processes of ecosystems across different landscapes will allow the more accurate prediction of impacts of human activity on landscape structure and the possibility of mitigation through land use practices such as agroforestry.

Regardless of which conservation strategy is used to determine priority areas and the scale at which that strategy is applied, use of comprehensive data is paramount. Collecting and integrating data on species distribution, habitat associations, and abundances should be a focal point of conservation networks because both the amount of data and the technology for integrating and compiling data have improved (van Jaarsveld et al. 1998).

Understanding biodiversity patterns is essential in establishing science-based conservation strategies. Quantifying patterns of endemism, rarity, and endangerment can be accomplished using a coordinated global framework. One important effort has been undertaken by the International Union for the Conservation of Nature and Natural Resources (IUCN), an organization with 900 Institutional Members (governments, government agencies, and non-government organizations), supported by a network of approximately 10,000 scientists and other conservation specialists. The IUCN is developing a freely accessible database of biodiversity information, coordinated by the Species Survival Commission (SSC). Information gathered by the IUCN SSC includes species identity, distribution, and conservation status. The success of this and other initiatives will allow conservation managers to make more scientifically informed decisions. A systematic evaluation of the conservation status of species, through the IUCN Red List, has been accomplished for the majority of terrestrial vertebrates, and there are ongoing efforts to include plants, invertebrates, and marine organisms that have not yet been evaluated (Hilton-Taylor 2000). This systematic designation of the conservation status of individual species allows conservation efforts to focus on species of immediate concern, such as the critically endangered muriqui (*Brachyteles arachnoides*) of the Brazilian Atlantic Forest and the Ethiopian wolf (*Canis simensis*), limited to several grassland areas in Ethiopia, and to prioritize conservation efforts.

Conservation Implementation

In the past decade, conservation research has produced an important body of knowledge that has reshaped practical conservation efforts from a narrow focus of isolated protected areas to a set of integrated actions at the landscape scale (Gascon et al. 2000). Although we are now scientifically literate in the effects of many types of land use on biodiversity, such as logging and agriculture, we have yet to translate much of this information into concrete actions to counter and mitigate these negative impacts. Large networks of protected areas connected through reforestation and agroforestry projects (see examples in Part V, this volume) and the promotion of less destructive uses of land surrounding protected areas (such as elimination of pesticides and controlled use of fire) are but some of the guidelines that should be part of a comprehensive practical conservation plan. Unfortunately, many of these landscape conservation guidelines have not been translated into integrated public policy in countries where biodiversity is rich and at risk. The absence of legislation that links the most recent scientific advances to land use and economic development policies and regulations puts any sustainable conservation strategy at risk.

Once geographic priority areas are established, the challenge becomes implementing effective conservation in these regions. Although the baseline of site conservation has been establishing protected areas, many other components are needed for long-term ecosystem and biodiversity protection. These include the use of sustainable development projects and other innovative proposals such as conservation concessions (Rice et al. 2001; Chapter 7, this volume), landscape corridors for conservation, and improved use of the landscape matrix surrounding less degraded areas. Sustainable agroforestry practices can play an important role in such strategies.

Traditionally, conservationists have focused only on patch-scale landscape dynamics. Unfortunately, geographic limitations of scale, which place conservation and development goals in competition with each other, have impeded many past efforts to combine conservation and development objectives. Therefore, in addition to these conservation approaches, efforts are needed to broaden conservation applications to a landscape scale and expand the focus of conservation planning to promote conservation and development goals together and address both ecological and economic needs (Fonseca et al. in preparation).

Protected areas provide a foundation for long-term conservation by directly securing biodiversity. Criticisms of reserves include many cases of ineffective protection from human activities such as logging and hunting. In the current environmental crisis, landscape conservation must be viewed in light of the major global changes including global climate change, pollution, invasive species, and other human-related problems. The scale of human disturbance is such that almost no area is unaffected, which means that the conservation value of most areas can be improved. This includes severely degraded landscapes, such as fallow agricultural fields, which can be integrated into the overall landscape conservation of a region. Sustainable development projects can prove useful for conservation efforts, although many limitations exist, such as with forestry programs.

It is evident from research on protected areas in tropical countries that although governments are effective in their conservation efforts despite their low level of funding and significant human pressures (Bruner et al. 2001), more parks and more effective parks are necessary. Management in and around parks is critical to their success. In one study, approximately 70 percent of 93 protected areas in 22 tropical countries contained humans residing within park boundaries (Bruner et al. 2001). Protected areas are not the final objective for conservation; the land surrounding strict reserves plays vital roles in maintaining diversity and ecosystem function. Tailoring the landscape surrounding reserves to increase conservation utility will also improve the medium- and long-term benefits from human land use. By taking advantage of natural processes, areas devoted to agriculture and agroforestry can improve productivity while providing conservation gains (see Parts IV and V, this volume).

The purpose of landscape corridors is to reconcile conservation actions, such as enhancing dispersal of plant and animal populations, with inevitable economic development. Landscape corridors provide a way to subdivide large areas into biologically and ecologically relevant subregional spaces that allow conservation planning and implementation. Planners can appropriate the subdivisions within landscape corridors so that biodiversity and economic goals are met. For instance, planners may place critical biodiversity areas under strict protection, allocate important areas to economic development, and allow other areas with mixed goals to be used accordingly. Therefore, a landscape corridor comprises an integrated and physically connected network of parks, reserves, and other areas of less intensive use whose management is integrated into the landscape matrix. In this way, landscape corridors maximize survival of existing biodiversity without conflicting with urgent economic development needs (Fonseca et al. in preparation).

Landscape-scale conservation allows the optimal allocation of resources to conserve biodiversity at the least economic cost to society. This cannot be accomplished through planning at the scale of individual parks and buffer zones. Long-term trends and changes in ecological and economic dynamics are more adequately addressed at the landscape scale. Finally, landscape-scale conservation allows the designation of patch-scale mosaics that occur in the landscape. These mosaic patches can be defined such that they are mutually beneficial to both conservation and development goals, such as protected areas to conserve watersheds and tourism resources and compatible development to promote species movement between protected areas or to provide important buffers (Fonseca et al. in preparation).

Landscape Management

Human-dominated landscapes can be managed in a manner that benefits conservation. Scientific knowledge accumulated in the last several decades must be incorporated into management of agricultural areas, including areas devoted to agroforestry.

The underlying concept of agroforestry, or the practice of cultivating tree species and agricultural crops together or in sequence, has been in practice throughout human agricultural history and was used to maintain soil fertility while supporting crop growth. For example, before modern agricultural developments, a common practice was to clear and burn forests before cultivating crops, and this is still the method of choice in many tropical regions today (see Chapter 8, this volume). Trees are then often planted with the agricultural crops (see Chapter 10, this volume). Today, in central and equatorial South America, combinations of plants with different growth habits, such as coconut, bananas, coffee, and maize, enhance agricultural landscapes. In Zambia, cassava is grown in small cleared plots within the larger matrix of the Miombo woodland. Other agroforestry practices include contour hedgerows for soil and water conservation, trees in croplands, improved fallows, and shaded perennial crops. Hedgerow, trees in cropland (see Chapter 11, this volume), and shaded perennial crop systems (see Chapter 9, this volume) combine trees and crops simultaneously in the same field, whereas fallow systems (see Chapter 8, this volume) involve crop and tree rotation over time.

Besides offering some secondary habitat, agroforestry can be used as an indirect conservation tool to protect natural areas from exploitation. Reforestation of corridors between protected areas is necessary to improve connectivity between patches (see Chapter 3, this volume). There are many areas where corridors would be useful and where protected forests are subjected to intense human activity such as fuelwood collection. When unprotected forests are depleted of fuelwood, protected areas often are targeted. Agroforestry systems can be integrated into such corridors, where they would play a conservation role by producing timber and nontimber forest products and thereby minimizing the exploitation of protected areas. Similarly, managed forest plantations and forest mixed with agriculture can be planned and managed for increased conservation value.

The indirect value of agroforestry systems can also extend to other environmental benefits, such as carbon sequestration, watershed maintenance, and buffering against climate change biome shifts (see Chapter 20, this volume). Furthermore, nutrient cycling in natural forest systems often is highly conservative as nutrients are quickly and efficiently recycled within the system, whereas agricultural systems often exist at the other extreme with high nutrient losses. Agroforestry may help to maintain a sustainable agriculture-forest coupled system in which nutrients are conserved within the system.

The use of science to guide the search for innovative agroforestry systems that integrate production objectives with environmental services can complement a solid biodiversity conservation strategy anchored around protected areas and therefore help to mitigate biodiversity losses.

Conclusions

Many of the present threats to tropical biodiversity have been played out in temperate regions in the past few centuries. Current technologies have enabled habitat destruction to spread from primarily temperate regions to tropical areas at an unprecedented rate and magnitude. Global impacts of tropical area degradation should be argument enough for all countries to work toward global conservation goals, with local, national, and regional peoples and governments working in concert. There is a critical need for wealthy countries, organizations, and individuals to contribute to tropical conservation efforts because of the mutual benefit to all countries (Barrett et al. 2001). However, current state policies are instead focused on practices that bring short-term gain; for example, an estimated $1.5 trillion per year is spent on subsidies that are both economically and environmentally destructive (Myers 1999). The full use of environmental sciences for conservation of tropical ecosystems can provide the basis for strong social, economic, and political decisions to best protect tropical biodiversity.

References

Aldrich, P. R., and J. L. Hamrick. 1999. Reproductive dominance of pasture trees in a fragmented tropical forest mosaic. *Science* 281:103–105.

Barrett, C. B., K. Brandon, and C. Gibson. 2001. Conserving tropical biodiversity amid weak institutions. *BioScience* 51:497–502.

Bierregaard, R. O., Jr., C. Gascon, T. E. Lovejoy, and R. Mesquita. 2001. *The ecology and conservation of a fragmented forest: lessons from Amazonia.* New Haven, CT: Yale University Press.

Brown, K. S., and R. W. Hutchings. 1997. Disturbance, fragmentation, and the dynamics of diversity in Amazonian forest butterflies. Pages 91–110 in W. F. Laurance and R. O. Bierregaard (eds.), *Tropical forest remnants: ecology, management, and conservation of fragmented communities.* Chicago: University of Chicago Press.

Bruner, A. G., R. E. Gullison, and R. E. Rice. 2001. Effectiveness of parks in protecting tropical biodiversity. *Science* 291:125–128.

Cardinale, B. J., M. A. Palmer, and S. L. Collins. 2002. Species diversity enhances ecosystem functioning through interspecific facilitation. *Nature* 415:426–429.

Cincotta, R. P., J. Wisnewski, and R. Engelman. 2000. Human population in the biodiversity hotspots. *Nature* 404:990–992.

Condit, R., N. Pitman, E. G. Leigh Jr., J. Chave, J. Terborgh, R. B. Foster, P. Núñez, S. Aguilar, R. Valencia, G. Villa, H. C. Muller-Landau, E. Losos, and S. P. Hubbell. 2002. Beta-diversity in tropical forest trees. *Science* 295:666–669.

Curran, L. M., I. Caniago, G. D. Paoli, D. Astianti, M. Kusneti, M. Leighton, C. E. Nirarita, and H. Haeruman. 1999. Impact of El Niño and logging on canopy tree recruitment in Borneo. *Science* 286:2184–2188.

Daily, G. C., and P. R. Ehrlich. 1996. Nocturnality and species survival. *Proceedings of the National Academy of Sciences* 93:11709–11712.

Didham, R. 1997. The influence of edge effects and forest fragmentation on leaf litter invertebrates in central Amazonia. Pages 55–70 in W. F. Laurance and R. O. Bierregaard (eds.), *Tropical forest remnants: ecology, management, and conservation of fragmented communities*. Chicago: University of Chicago Press.

FAO. 2001. *State of the world's forests*. Rome: Food and Agriculture Organization of the United Nations.

Fonseca, G. A. B., A. Bruner, R. A. Mittermeier, K. Alger, C. Gascon, and R. E. Rice. In preparation. *On defying nature's end.*

Garstang, M., P. D. Tyson, R. J. Swap, M. Edwards, P. Kallberg, and J. A. Lindesay. 1996. Horizontal and vertical transport of air over southern Africa. *Journal of Geophysical Research* 101:23721–23736.

Gascon, C., W. F. Laurance, and T. E. Lovejoy. 2003. Forest fragmentation and biodiversity in central Amazonia. In G. A. Bradshaw and P. A. Marquet (eds.), *How landscapes change. Ecological Studies* 162. Berlin: Springer.

Gascon, C., T. E. Lovejoy, R. O. Bierregaard, J. R. Malcolm, P. C. Stouffer, H. Vasconcelos, W. F. Laurance, B. Zimmerman, M. Tocher, and S. Borges. 1998. Matrix habitat and species persistence in tropical forest remnants. *Biological Conservation* 91:223–229.

Gascon, C., G. B. Williamson, and G. A. B. da Fonseca. 2000. Receding forest edges and vanishing reserves. *Science* 288:1356–1358.

Gaston, K. J. 2000. Global patterns in biodiversity. *Nature* 405:220–227.

Harcourt, C., and J. Sayer. 1996. *The conservation atlas of tropical forest: the Americas*. New York: Simon & Schuster.

Hilton-Taylor, C. (ed.). 2000. *2000 IUCN Red List of threatened species*. Gland, Switzerland: International Union for the Conservation of Nature and Natural Resources.

Houghton, J. T., T. Ding, D. J. Griggs, M. Noguer, P. J. van der Linden, and D. Xiaosu (eds.). 2001. *Climate change 2001: the scientific basis*. Cambridge, MA: Cambridge University Press.

Kennedy, T. A., S. Naeem, K. M. Howe, J. M. H. Knops, D. Tilman, and P. Reich. 2002. Biodiversity as a barrier to ecological invasion. *Nature* 417:636–638.

Laurance, W. F., P. Delamônica, S. G. Laurance, H. L. Vasconcelos, and T. E. Lovejoy. 2000. Conservation: rainforest fragmentation kills big trees. *Nature* 404:836.

Laurance, W. F., S. G. Laurance, L. V. Ferreira, J. M. Rankin-de Merona, C. Gascon, and T. Lovejoy. 1997. Biomass collapse in Amazonian forest fragments. *Science* 278:1117–1118.

Levine, J. M. 2000. Species diversity and biological invasions: relating local process to community pattern. *Science* 288:852–854.

Loreau, M., and A. Hector. 2001. Partitioning selection and complementarity in biodiversity experiments. *Nature* 412:72–76.

Lovel, G. L. 1997. Global change through invasion. *Nature* 388:627–628.

McGrady-Steed, J., P. M. Harris, and P. J. Morin. 1997. Biodiversity regulates ecosystem predictability. *Nature* 390:162–165.

Mittermeier, R. A., N. Myers, P. R. Gil, and C. G. Mittermeier. 2000. *Hotspots*. Mexico City: Cemex.

Myers, N. 1999. Lifting the veil on perverse subsidies. *Nature* 392:327–328.

Myers, N., R. A. Mittermeier, C. G. Mittermeier, G. A. B. da Fonseca, and J. Kent. 2000. Biodiversity hotspots for conservation priorities. *Nature* 403:853–858.

Myers, N. 1984. *The primary source: tropical forests and our future*. New York: W. W. Norton.

Noble, I. R., and R. Dirzo. 1997. Forests as human-dominated ecosystems. *Science* 277:522–525.

Olson, D., and E. Dinerstein. 1998. The Global 200: a representation approach to conserving the earth's most biologically valuable ecoregions. *Conservation Biology* 12:502–515.

Peres, C. A., and J. W. Terborgh. 1995. Amazonian nature reserves: an analysis of the defensibility status of existing conservation units and design criteria for the future. *Conservation Biology* 9:34–46.

Peters, R. L. 1991. Consequences of global warming for biological diversity. Pages 99–118 in R. L. Wyman (ed.), *Global climate change and life on Earth*. New York: Routledge Chapman & Hall.

Rice, R. E., C. Sugal, S. M. Ratay, and G. A. B. da Fonseca. 2001. *Sustainable forest management: a review of conventional wisdom*. Advances in Biodiversity Science 3. Washington, DC: Conservation International.

Roemer, G. W., C. J. Donlan, and F. Courchamp. 2002. Golden eagles, feral pigs, and insular carnivores: how exotic species turn native predators into prey. *Proceedings of the National Academy of Science* 99:791–796.

Sala, O. E., F. S. Chapin III, J. J. Armesto, E. Berlow, J. Bloomfield, R. Dirzo, E. Huber-Sanwald, L. F. Huenneke, R. B. Jackson, A. Kinzig, R. Leemans, D. M. Lodge, H. A. Mooney, M. Oesterheld, N. L. Poff, M. T. Sykes, B. H. Walker, M. Walker, and D. H. Wall. 2000. Global biodiversity scenarios for the year 2100. *Science* 287:1770–1774.

Shukla, J., C. Nobre, and P. Sellers. 1990. Amazon deforestation and climate change. *Science* 247:1322–1325.

Stevens, G. C. 1989. The latitudinal gradient in geographical range: how so many species coexist in the tropics. *American Naturalist* 133:240–256.

Tilman, D. 1999. Global environmental impacts of agricultural expansion: the need for sustainable and efficient practices. *Proceedings of the National Academy of Sciences* 96:5995–6000.

Tilman, D., J. Fargione, B. Wolff, C. D'Antonio, A. Dobson, R. Howarth, D. Schindler, W. H. Schlesinger, D. Simberloff, and D. Swackhamer. 2001. Forecasting agriculturally driven global environmental change. *Science* 292:281–284.

Tocher, M., C. Gascon, and J. Meyer. 2001. Community composition and breeding success of frogs in a modified landscape. Pages 235–247 in R. O. Bierregaard, C. Gascon, T. E. Lovejoy, and R. Mesquita (eds.), *Lessons from Amazonia: the ecology and conservation of a fragmented forest*. New Haven, CT: Yale University Press.

Tocher, M., C. Gascon, and B. L. Zimmerman. 1997. Fragmentation effects on a central Amazonian frog community: a ten-year study. Pages 124–137 in W. F. Laurance and R. O. Bierregaard (eds.), *Tropical forest remnants: ecology, management, and conservation of fragmented communities*. Chicago: University of Chicago Press.

van Jaarsveld, A. S., S. Freitag, S. L. Chown, C. Muller, S. Koch, H. Hull, C. Bellamy, M. Kruger, S. Endrody-Younga, M. W. Mansell, and C. H. Scholtz. 1998. Biodiversity assessment and conservation strategies. *Science* 279:2106–2108.

Warburton, N. H. 1997. Structure and conservation of forest avifauna in isolated rainforest remnants in tropical Australia. Pages 190–208 in W. F. Laurance and R. O. Bierregaard (eds.), *Tropical forest remnants: ecology, management, and conservation of fragmented communities*. Chicago: University of Chicago Press.

Chapter 2

Ecological Effects of
Habitat Fragmentation in the Tropics

William F. Laurance and Heraldo L. Vasconcelos

This chapter provides an overview of the ecological consequences of habitat fragmentation on tropical biota. We begin by describing the demographic and genetic effects of fragmentation on individual populations, then discuss key landscape factors that affect fragmented populations, particularly area, edge, matrix, and distance effects. We then consider briefly the interactions of habitat fragmentation with other simultaneous environmental changes that often occur in human-dominated tropical landscapes, such as fire, logging, and overhunting. We conclude by proposing some ways in which the deleterious effects of habitat fragmentation could be partially ameliorated by agroforestry and reforestation.

Why Isolated Populations Are Vulnerable

Forest fragmentation proceeds as intact forest blocks are subdivided and reduced in size. This also reduces and subdivides natural populations, often greatly increasing the rate of local species extinction. Such losses occur for several reasons. First, small populations are highly vulnerable to random demographic events (Shafer 1981). Consider, for example, the fate of a population of 20 short-lived animals that, simply by chance, had two consecutive breeding seasons in which few females were born into the population. The reproductive capacity of the population would be drastically reduced, and it could easily disappear. In large populations such chance events are of little importance, but simple random fluctuations in births and deaths can have dire impacts on small populations.

Such events probably are important in nature. Many species appear to exist in metapopulations, that is, a series of small subpopulations, each of which is partially isolated from other such subpopulations (Hanski and Gilpin 1996).

33

These subpopulations may disappear frequently because of random demographic events but are generally reestablished by immigrants from nearby subpopulations. Although no individual subpopulation is stable over the long term, the overall metapopulation is likely to persist almost indefinitely (Smith et al. 1978; Harrison 1989; Hecnar and M'Closkey 1996; Wahlberg et al. 1996). In fragmented habitats, however, immigration for most species is halted or drastically reduced. Small populations may then falter and disappear, never to be replenished.

Second, small, isolated populations are also vulnerable to inbreeding and genetic drift. Inbreeding occurs because individuals are forced to breed with close relatives, lowering genetic heterozygosity and often reducing fecundity and offspring viability (Ralls et al. 1986). (Genetic heterozygosity is the degree to which an organism has more than a single form of each gene; these variants are called alleles; outbred individuals have greater heterozygosity than those that are inbred and generally suffer fewer genetic problems.) As a result, inbred populations may grow more slowly and be increasingly prone to random demographic events (Mills and Smouse 1994). Genetic drift (the random loss of alleles) is also amplified in small populations, and the resulting loss of genetic variability may reduce a population's resistance to new diseases or environmental challenges (Nei et al. 1975; Allendorf and Leary 1986).

Finally, natural environmental variations and local catastrophes often compound the effects of random demographic events and genetic problems (Leigh 1981). Environmental changes such as adverse weather conditions, increasing densities of predators or competitors, or pathogen outbreaks may drive a small population down to a critically low level. Once the population falls below a certain threshold, the interacting and potentially reinforcing effects of random demographic events, genetic problems, and environmental variations can become a powerful driving force of extinction (Gilpin and Soulé 1986).

Fragmentation Effects on Tropical Biota

Forest fragmentation affects tropical species and ecosystems in many ways. Here we describe its most important consequences and the interactions of fragmentation with other simultaneous environmental changes (such as logging, fires, and hunting) that commonly occur in human-dominated landscapes.

Area Effects

Large habitat fragments usually contain more species overall (greater species richness) and a higher density of species per unit area than do smaller fragments (Figure 2.1). This occurs for at least four reasons.

First, large fragments are less strongly influenced by sample effects. Simply

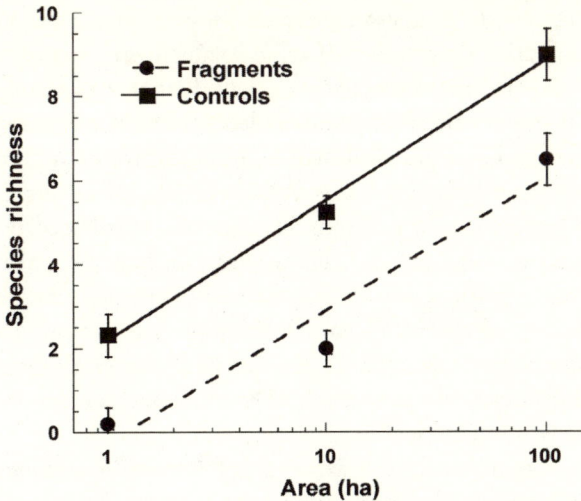

Figure 2.1. Species-area relationships for nine species of terrestrial insectivorous birds (mean ± *SE*) in central Amazonia, illustrating that large forest fragments typically sustain greater species richness than do smaller fragments and that forest fragments have fewer species than do equal-sized tracts of intact forest (controls) (after Stratford and Stouffer 1999).

by chance, small patches of forest inevitably sample fewer species than do larger forest patches. In the tropics this phenomenon tends to be amplified by the fact that many forest organisms have patchy distributions and complex patterns of endemism (Gentry 1986; Zimmerman and Bierregaard 1986; Vasconcelos 1988; Rankin-de Merona et al. 1992; Laurance et al. 1998c). Another key attribute of tropical rainforests is that many species are locally rare throughout all or much of their geographic range (Hubbell and Foster 1986; Pittman et al. 1999). For example, predators and large-bodied animals generally are rarer than herbivores and small-bodied species; a single jaguar (*Panthera onca*), for instance, can have a home range spanning hundreds of square kilometers (Rabinowitz 2000). In fragmented landscapes, rarity can have a strong influence on whether species ultimately persist. Even if a rare species is present when a fragment is isolated, its population size may be so tiny that it has little chance of surviving in the long term (Laurance et al. 1998c).

Second, large fragments usually support a wider range of habitats than smaller fragments, and this generally means more species will be present. Habitat diversity is important in the tropics, where many species need specialized food resources or microhabitats (Zimmerman and Bierregaard 1986; Brown and Hutchings 1997). For example, many herbivorous insects feed on only one or a few closely related plant species, and numerous birds have unique foraging specializations, such as following swarms of army ants to capture fleeing insects, feeding exclusively on flower nectar, or foraging only in clusters of suspended dead leaves. If critical habitats are missing or poorly represented, the dependent species probably will disappear too.

Third, big fragments are proportionally less affected by edge effects, the physical and biological changes associated with the abrupt boundary between forests and adjoining modified habitats. Area and edge effects are difficult to distinguish, and few studies have effectively done this, usually because samples in small fragments are near edges, whereas those in large fragments are far from edges, creating a strong correlation between edge and area predictors. However, it is becoming clear that many population and community changes in habitat fragments that were commonly attributed to area effects are in fact the result of edge effects (Lovejoy et al. 1986; Laurance et al. 2002). Edge effects are discussed in detail later in this chapter.

Finally, large fragments have lower rates of population extinction than small fragments, especially for species that need large territories, are sensitive to edge effects, are unable to cross even small clearings, or cannot tolerate conditions in the surrounding modified habitats. An intriguing example is the specialized ant-following birds of the neotropics, which accompany marauding swarms of army ants in order to capture fleeing insects. Each ant colony raids extensive areas of up to 30 ha, and the birds' home ranges must encompass two or three colonies because every colony spends several weeks per month in an inactive phase (Harper 1989). Because their ants need large areas and because they must have access to several colonies, the specialized ant followers are especially prone to extinction in small fragments (Harper 1989; Stouffer and Bierregaard 1995b).

Despite these factors—and contrary to predictions of the island biogeography theory (MacArthur and Wilson 1967)—larger habitat fragments do not always support more species than smaller fragments. In certain taxonomic groups, species richness can actually increase in fragments when there is an influx of species from the surrounding modified habitats (Brown and Hutchings 1997; Didham 1997b; Tocher et al. 1997) or when conditions near edges become more favorable for a particular species or guild of species (Stouffer and Bierregaard 1995a). Species that proliferate near edges or in the matrix can include both nonforest species and those that were formerly limited to naturally disturbed forest patches (Stouffer and Bierregaard 1995a; Brown and Hutchings 1997).

Distance Effects

Interfragment distance can affect the movement of animals and plant propagules between fragments, and even remarkably small clearings can become impassable barriers for many rainforest organisms. In the Amazon, many terrestrial insectivorous birds have disappeared from forest fragments and failed to recolonize even those isolated by only 80 m from nearby forest tracts (Stratford and Stouffer 1999). Clearings of just 15–100 m are insurmountable barriers for certain dung and carrion beetles (Klein 1989), euglossine bees (Powell and Powell 1987), and arboreal mammals (Malcolm 1991; Gilbert and Setz 2001). Peccaries (*Tayassu* spp.; Offerman et al. 1995) and many insect-gleaning bats (Kalko 1998) are also highly reluctant to enter clearings. Even an unpaved road only 30–40 m wide dramatically alters the community structure of understory birds and inhibits the movements of many species (S. G. Laurance 2000).

Some species cross small clearings but are inhibited by larger expanses of degraded land. For example, woodcreepers (Dendrocolaptidae) were induced by translocations to move between Amazonian forest fragments and nearby (80–150 m) forest tracts (Harper 1989) but have disappeared from slightly more isolated areas such as Barro Colorado Island in Panama (Robinson 1999). Large predators such as jaguars and pumas (*Puma concolor*) are capable of traversing pastures and regrowth but tend to avoid these areas if hunters are present or human density is not low (Rabinowitz 2000).

Rainforest animals avoid clearings for many reasons. Most understory species have had little reason to traverse clearings in their evolutionary history, so the avoidance of such areas probably is an innate response (Greenberg 1989). Other species are constrained by morphology or physiology; for instance, strictly arboreal species will find even a small pasture an impenetrable barrier. Specialized habitat needs probably limit others (Stratford and Stouffer 1999). A final factor that limits interfragment movements, at least in rainforest birds, is that few species are migratory. In temperate forests, even truly isolated fragments can be colonized in the breeding season by migratory species (Blake and Karr 1987), but rainforest birds appear far less likely to do so.

Edge Effects

Habitat fragmentation inevitably leads to the creation of edges where previously there were none. However, these edges are different from natural transition zones (ecotones) because they are abrupt and artificial. Both physical and biological changes occur along fragment edges. The importance and magnitude of these changes depend to some extent on the contrast between the fragmented habitat and the adjoining modified habitat; in general, the greater the contrast, the stronger the edge effect (Mesquita et al. 1999).

Recent evidence indicates that tropical forest fragments are particularly prone to edge effects (Lovejoy et al. 1986; W. F. Laurance 1991b, 2000). Increased insolation and wind penetration along newly formed forest edges affect forest microclimate, which becomes warmer and drier (Kapos 1989; Williams-Linera 1990). After a few years, however, these microclimatic alterations may decline in importance as edges are partly sealed by a profusion of second growth (Williams-Linera 1990; Kapos et al. 1997), making them less permeable to lateral light penetration and the penetration of hot, dry winds from adjoining agricultural lands.

Forest structure is also markedly altered, especially by an increase in tree mortality rates. When a new edge is created, some trees simply drop their leaves and die standing, possibly because of sudden shifts in relative humidity, temperature, or soil moisture that exceed their physiological tolerances (Lovejoy et al. 1986). Other trees are snapped or felled by winds, which accelerate over cleared land and then strike forest edges (Ferreira and Laurance 1997; Laurance 1997). Finally, lianas (woody vines)—important structural parasites that reduce tree growth, survival, and reproduction—increase markedly near edges and may further elevate tree mortality (Laurance et al. 2001).

This abrupt rise in tree mortality fundamentally alters canopy gap dynamics (Williams-Linera 1990; Ferreira and Laurance 1997; Laurance et al. 1998a), in turn affecting forest composition and diversity (Brokaw 1985; Hubbell and Foster 1986; Denslow 1987; Viana and Tabanez 1996). Smaller fragments often become hyperdisturbed, leading to progressive changes in floristic composition. In the Amazon, new trees regenerating within 100 m of forest edges are significantly biased toward disturbance-loving pioneer and secondary species and against old-growth, forest interior species (Laurance et al. 1998b). Pioneer species such as *Cecropia sciadophylla* can increase in density by several thousand percent in fragmented landscapes (Laurance et al. 2001). Because pioneer trees have a higher leaf turnover (Coley 1983), rates of litter fall tend to increase near forest edges (H. L. Vasconcelos, unpublished data, 2002). Litter depth typically is greater within 100 m of the forest edge than in the forest interior (Carvalho and Vasconcelos 1999; Didham and Lawton 1999). Changes in litter cover along forest edges not only have important effects on plant (Bruna 1999) and animal communities (Carvalho and Vasconcelos 1999) but also make forests vulnerable to devastating surface fires during droughts (Cochrane et al. 1999).

Despite an initial increase in tree seedling recruitment especially of pioneer species (Sizer 1992), just after forest edges are created, seedling density tends to decrease near edges (Benitez-Malvido 1998). The reasons for this are not completely clear but probably involve reductions in seed rain and dispersal and greater seed and seedling mortality near edges. In particular, the development of a dense layer of secondary vegetation along forest edges may increase seedling mortality by diminishing light availability and increasing damage from heavy litterfall (Benitez-Malvido 1998). Microclimatic changes, especially reduced soil moisture, may also be involved. Studies of the understory shrub *Heliconia acuminata* show that even when seedlings are protected from litter damage, survival is still lower near fragment edges than in interiors (Bruna 1999).

Changes in forest structure along edges have diverse consequences for forest fauna. Some species are insensitive to these changes and readily use edge habitats (Gascon 1993; Kotze and Samways 2001). Others, including many insect species, respond rapidly to edge-related factors. In the Amazon, insect abundance and diversity usually increase in the understory near edges, probably because of increased understory plant density and productivity (Fowler et al. 1993; Malcolm 1997b). However, insect abundance is lower in the upper forest strata, presumably because the density of overstory vegetation is reduced by recurring canopy disturbances (Malcolm 1997b). Similarly, species adapted for humid, dark forest interiors, including certain beetles (Didham et al. 1998), ants (Carvalho and Vasconcelos 1999), and butterflies (Brown and Hutchings 1997), decline in abundance near edges.

Birds that forage in treefall gaps, such as some arboreal insectivores, hummingbirds, and habitat generalists, often become abundant near edges (Stouffer and Bierregaard 1995a, 1995b; Dale et al. 2000). However, a number of insectivorous understory birds avoid edges (Quintela 1985; Dale et al. 2000), especially solitary species, obligatory ant followers, and those that forage in mixed-species flocks (S. G. Laurance 2000). Animals that nest or forage on fallen dead trees, including wood-decomposing insects (Souza and Brown 1994) and certain marsupials (Malcolm 1991) and rodents (Ready et al. 1983), are also favored and increase in abundance near edges. The abundance and species richness of small mammals increase in Amazonian fragments, presumably in response to greater availability of insect prey along the edge (Malcolm 1997a). However, an opposite response was detected in Brazilian Atlantic forests (Fonseca 1988), tropical Australia (Laurance 1994), and Thailand (Lynam 1997). Similarly, although ants generally increase in abundance near forest edges in central Amazonian rainforests (Didham 1997a; Carvalho and Vasconcelos 1999), they decline in dry tropical forests of Madagascar (Olson and Andriamiadana 1996), suggesting that local climatic factors can affect species responses to habitat fragmentation. In at least some cases, the nature of the edge response depends on edge age. In Colombian montane forests, for example, new and old forest edges had different fruit abundance and different communities of fruit-eating birds (Restrepo et al. 1999).

Given the great diversity of edge effects, it is not surprising that different edge phenomena penetrate to varying distances inside fragments. In central Amazonia, different kinds of edge effects have been shown to penetrate anywhere from 10 m to at least 400 m into fragment interiors (Figure 2.2). The penetration distance (d) of an edge effect is a key parameter because if determined empirically it can be used with a mathematical core-area model (Laurance and Yensen 1991) to predict the vulnerability of any fragment to that particular edge effect. In the central Amazon, the furthest-penetrating edge effect documented to date is wind damage to forests (Figure 2.2), detectable up to 400 m from edges (Laurance et al. 1998a; Lewis 1998). However, recent evidence reveals that certain other edge effects, such as destructive fires and invasions of feral animals, can penetrate at least several thousand meters into trop-

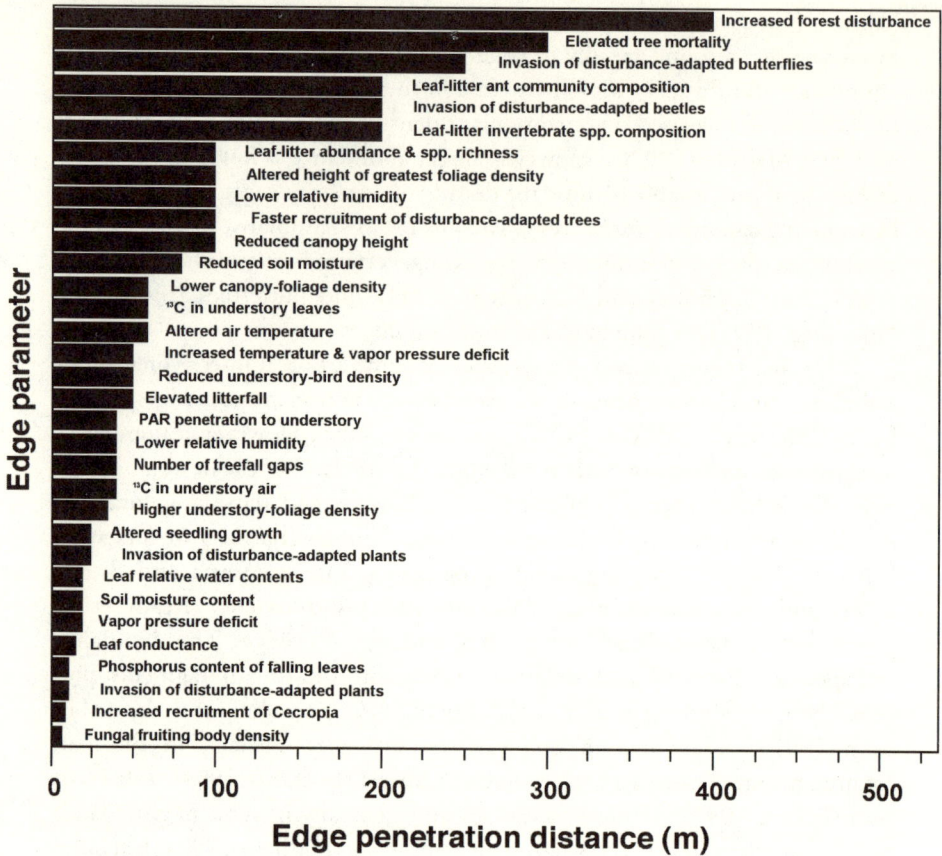

Figure 2.2. Penetration distances of different edge effects into Amazonian forest remnants (after Laurance et al. 2002). PAR = Photosynthetically active radiation.

ical forests in Indonesian Borneo (Curran et al. 1999), peninsular Malaysia (Peters 2001), and eastern Amazonia (Cochrane and Laurance 2002).

Matrix Effects

The matrix is the mosaic of modified habitats, such as pastures, crops, plantations, and secondary forest, that surrounds habitat fragments (Forman 1995). Different matrix habitats can have a major influence on the ecology of fragmented forests. In the Amazon, forest fragments surrounded by 5- to 10-m-tall regrowth forest experienced less intensive changes in microclimate (Didham and Lawton 1999) and had lower edge-related tree mortality (Mesquita et al. 1999) than did similar fragments adjoined by cattle pastures. Edge avoidance by mixed-species bird flocks was also lower when fragments were surrounded by regrowth forest rather than cattle pastures (Stouffer and Bierregaard 1995b).

Table 2.1. Total number of species recorded and percentage of species shared with mature forest (in parentheses) in three types of matrix habitat surrounding Amazonian forest fragments.

	Taxon		
Matrix Habitat	Ants[a]	Birds[b]	Frogs[c]
Pasture	36 (19.4–33.3%)	No data	25 (76.0%)
Vismia-dominated regrowth	62 (43.5–51.6%)	123 (86.2%)	40 (90.0%)
Cecropia-dominated regrowth	83 (48.2–51.8%)	141 (97.2%)	43 (81.4%)

Sources: [a]Vasconcelos (1999); [b]Borges and Stouffer (1999); [c]Tocher (1998).

Of even more significance is that the matrix strongly influences fragment connectivity. Several species of Amazonian primates (Gilbert and Setz 2001), antbirds, obligate flocking birds (Stouffer and Bierregaard 1995b), and euglossine bees (Becker et al. 1991) that disappeared soon after fragment isolation recolonized fragments when regrowth regenerated in the surrounding landscape. Among small mammals, bats, birds, and frogs in tropical Australia and Amazonia, matrix-avoiding species are much more likely to decline or disappear in fragments than are those that use the matrix (Laurance 1991a; Offerman et al. 1995; Stouffer and Bierregaard 1995b; Kalko 1998; Borges and Stouffer 1999; Gascon et al. 1999; Stratford and Stouffer 1999).

Some matrix habitats are more suitable for rainforest fauna than others. In the Amazon, regrowth dominated by Cecropia trees, which tends to be tall and floristically diverse with a closed canopy (Williamson et al. 1998), is used by more rainforest bird, frog, and ant species than more open Vismia-dominated regrowth, and almost any kind of regrowth is better than cattle pastures (Table 2.1). Forest-dependent dung and carrion beetles are also far more likely to cross a matrix of regrowth than one that has been completely clearcut (Klein 1989). In general, the more closely the matrix approximates the structure and microclimate of primary forest, the more likely fragmentation-sensitive species are to be able to use it.

Synergistic Effects

In tropical anthropogenic landscapes, forests are rarely just fragmented; they are also subjected to logging, overhunting, incursions of fire, and other human disturbances. Such simultaneous environmental changes can interact additively or synergistically, leading to even greater impacts on fragmented populations (Laurance and Cochrane 2001).

For example, recent studies demonstrate that forest fragmentation dramatically increases the vulnerability of Amazonian forests to fire (Cochrane et al. 1999; Nepstad et al. 1999). Farmers burn their cattle pastures at 1- to 2-year intervals to control weeds and promote a flush of green grass, and these fires commonly burn into the understory of adjoining forests. Although of low intensity, such surface fires can cause dramatic mortality of rainforest trees and vines, which have thin bark and thus are poorly protected from flames (Uhl and Kauffman 1990; Kauffman 1991). Once-burned forests become highly vulnerable to recurring fires of even greater intensity because the mortality of many plants reduces canopy cover and increases the amount of dry litter on the forest floor (Cochrane et al. 1999). In two fragmented landscapes of eastern Amazonia, surface fires were significantly more common within 2,400 m of forest edges (Figure 2.3; Cochrane and Laurance 2002). As a result of repeated fire incursions, the edges of rainforest remnants may recede over time, leading to fragment "implosion" (Gascon et al. 2000). Unless fires are controlled, especially in seasonal areas of the tropics, rainforest vegetation could be largely replaced in many areas by fire-adapted savannas or scrubby regrowth (Cochrane and Laurance 2002).

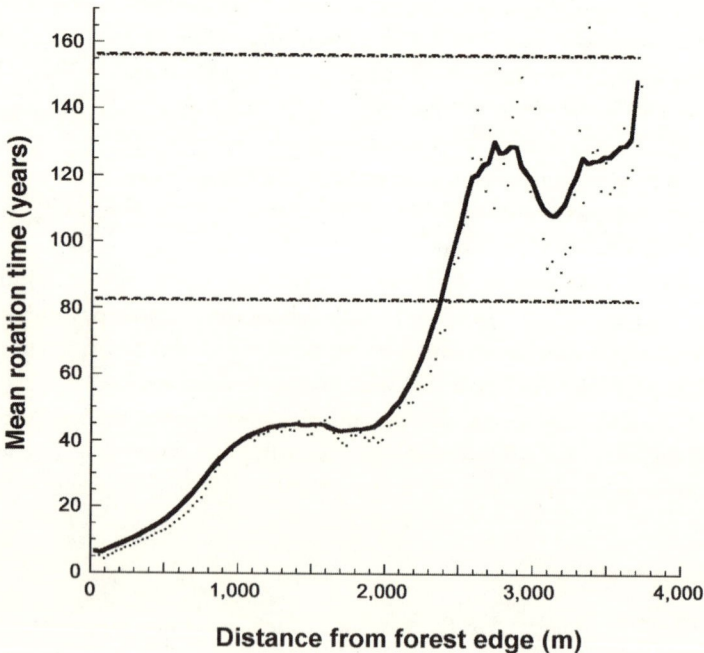

Figure 2.3. Estimated fire rotation times (number of years between successive fires) as a function of distance from forest edge for a fragmented landscape in the eastern Brazilian Amazon. Data were estimated using 14 years of remote-sensing data. The curve was fitted with a smoothing function. Dotted lines show the 95% range of variation (mean ± 1.96 standard deviations) for forest interiors (2,500 m from edge) (after Cochrane and Laurance 2002).

Conclusions

If it does not replace natural vegetation but rather is established in already-deforested lands, agroforestry has the potential to benefit fragmented populations and ecosystems in several ways. First, it is apparent that creating dense, tall-statured vegetation near the margins of forest fragments can reduce many of the deleterious edge effects discussed in this chapter (Gehlhausen et al. 2000). Forest fragments surrounded by tree plantations or agroforestry are likely to suffer far less severe alterations in microclimate and less wind turbulence than do fragments encircled by cattle pastures or herbaceous crops. Given the obvious importance of edge effects in fragmented tropical landscapes, reducing the intensity of such effects could be a major benefit of enlightened agroforestry practices.

Second, agroforestry methods have the potential to increase fragment connectivity and species survival, especially if used in concert with landscape design principles such as corridor systems (see Chapter 3, this volume). Forest-dependent animals are much more likely to traverse modified lands that contain substantial tree cover than areas that have been denuded of trees (Laurance and Bierregaard 1997). Nevertheless, the most vulnerable species in fragmented landscapes often avoid modified habitats altogether (Laurance 1991a; Gascon et al. 1999; Laurance and Laurance 1999), and for these species the potential benefits of agroforestry may be limited.

Third, under some circumstances agroforestry can be used to provide food or shelter for fragmented wildlife populations. In tropical Queensland, Australia, the use of "framework" tree species that attract a wide variety of frugivorous birds and bats has been advocated as a strategy for landscape restoration (Lamb et al. 1997). These mobile frugivores deposit the seeds of many other plant species beneath the planted trees, accelerating the process of forest regeneration. Likewise, almost all tropical forests experience strong seasonal or interannual declines in fruit availability (Terborgh 1986; Wright et al. 1999). These declines can cause severe famines in fragmented populations (Van Shaik et al. 1993; Wright et al. 1999), which are unable to migrate to areas with greater fruit abundance. The provision of plants near fragments that have continuous fruit crops, such as figs (*Ficus* spp.) and many palms, or that reproduce during periods of annual fruit scarcity could have important benefits for fragmented wildlife populations.

A final and obvious benefit of agroforestry is that it reduces the use of fire. Fires are anathema to farmers who rely on perennial plants such as fruit trees and timber plantations. Given the striking vulnerability of fragmented forests to fire, the reduction of burning may be one of the most important environmental benefits of agroforestry systems in the tropics.

Acknowledgments

We thank Jim Sanderson, Götz Schroth, Susan Laurance, and an anonymous reviewer for commenting on a draft of this chapter. The NASA-LBA program, A. W. Mellon Foundation, and Smithsonian Tropical Research Institute provided support. This is publication number 376 in the BDFFP technical series.

References

Allendorf, F. W., and R. F. Leary. 1986. Heterozygosity and fitness in natural populations of animals. Pages 57–76 in M. E. Soulé (ed.), *Conservation biology: the science of scarcity and diversity.* Sunderland, MA: Sinauer.

Becker, P., J. B. Moure, and F. J. A. Peralta. 1991. More about euglossine bees in Amazonian forest fragments. *Biotropica* 23:586–591.

Benitez-Malvido, J. 1998. Impact of forest fragmentation on seedling abundance in a tropical rain forest. *Conservation Biology* 12:380–389.

Blake, J. G., and J. R. Karr. 1987. Breeding birds of isolated woodlots: area and habitat relationships. *Ecology* 68:1724–1734.

Borges, S., and P. C. Stouffer. 1999. Bird communities in two types of anthropogenic successional vegetation in central Amazonia. *Condor* 101:529–536.

Brokaw, N. V. L. 1985. Gap-phase regeneration in a tropical forest. *Ecology* 66:682–687.

Brown, K. S., Jr., and R. W. Hutchings. 1997. Disturbance, fragmentation, and the dynamics of diversity in Amazonian forest butterflies. Pages 91–110 in W. F. Laurance and R. O. Bierregaard (eds.), *Tropical forest remnants: ecology, management, and conservation of fragmented communities.* Chicago: University of Chicago Press.

Bruna, E. M. 1999. Seed germination in rainforest fragments. *Nature* 402:139.

Carvalho, K. S., and H. L. Vasconcelos. 1999. Forest fragmentation in central Amazonia and its effects on litter-dwelling ants. *Biological Conservation* 91:151–158.

Cochrane, M. A., A. Alencar, M. D. Schulze, C. M. Souza, D. C. Nepstad, P. Lefebvre, and E. Davidson. 1999. Positive feedbacks in the fire dynamics of closed canopy tropical forests. *Science* 284:1832–1835.

Cochrane, M. A., and W. F. Laurance. 2002. Fire as a large-scale edge effect in Amazonian forests. *Journal of Tropical Ecology* 18:311–325.

Coley, P. D. 1983. Herbivory and defensive characteristics of tree species in a lowland tropical forest. *Ecological Monographs* 53:209–233.

Curran, L. M., I. Caniago, G. D. Paoli, D. Astianti, M. Kusneti, M. Leighton, C. E. Nirarita, and H. Haeruman. 1999. Impact of El Niño and logging on canopy tree recruitment in Borneo. *Science* 286:2184–2188.

Dale, S., K. Mork, R. Solvang, and A. J. Plumptre. 2000. Edge effects on the understory bird community in a logged forest in Uganda. *Conservation Biology* 14:265–276.

Denslow, J. S. 1987. Tropical rainforest gaps and tree species diversity. *Annual Review of Ecology and Systematics* 18:431–451.

Didham, R. K. 1997a. The influence of edge effects and forest fragmentation on leaf-litter invertebrates in central Amazonia. Pages 55–70 in W. F. Laurance and R. O. Bierregaard (eds.), *Tropical forest remnants: ecology, management, and conservation of fragmented communities.* Chicago: University of Chicago Press.

Didham, R. K. 1997b. An overview of invertebrate responses to habitat fragmentation. Pages 303–320 in A. D. Watt, N. E. Stork, and M. D. Hunter (eds.), *Forests and insects.* London: Chapman and Hall.

Didham, R. K., P. M. Hammond, J. H. Lawton, P. Eggleton, and N. E. Stork. 1998. Beetle species responses to tropical forest fragmentation. *Ecological Monographs* 68:295–303.

Didham, R. K., and J. H. Lawton. 1999. Edge structure determines the magnitude of changes in microclimate and vegetation structure in tropical forest fragments. *Biotropica* 31:17–30.

Ferreira, L. V., and W. F. Laurance. 1997. Effects of forest fragmentation on mortality and damage of selected trees in central Amazonia. *Conservation Biology* 11:797–801.

Fonseca, G. A. B. 1988. *Patterns of small species diversity in the Brazilian Atlantic Forest.* Ph.D. thesis, University of Florida, Gainesville.

Forman, R. T. T. 1995. *Land mosaics.* Cambridge: Cambridge University Press.

Fowler, H. G., C. A. Silva, and E. Ventincinque. 1993. Size, taxonomic and biomass distributions of flying insects in central Amazonia: forest edge vs. understory. *Revista de Biologica Tropical* 41:755–760.

Gascon, C. 1993. Breeding habitat use by Amazonian primary-forest frog species at the forest edge. *Biodiversity and Conservation* 2:438–444.

Gascon, C., T. E. Lovejoy, R. O. Bierregaard, J. R. Malcolm, P. C. Stouffer, H. Vasconcelos, W. F. Laurance, B. Zimmerman, M. Tocher, and S. Borges. 1999. Matrix habitat and species persistence in tropical forest remnants. *Biological Conservation* 91:223–229.

Gascon, C., G. B. Williamson, and G. A. B. Fonseca. 2000. Receding edges and vanishing reserves. *Science* 288:1356–1358.

Gehlhausen, S. M., M. W. Schwartz, and C. K. Augspurger. 2000. Vegetation and microclimatic edge effects in two mixed-mesophytic forest fragments. *Plant Ecology* 147:21–35.

Gentry, A. H. 1986. Endemism in tropical versus temperate plant communities. Pages 153–181 in M. E. Soulé (ed.), *Conservation biology: the science of scarcity and diversity.* Sunderland, MA: Sinauer.

Gilbert, K. A., and E. Z. F. Setz. 2001. Primates in a fragmented landscape: six species in central Amazonia. Pages 262–270 in R. O. Bierregaard, C. Gascon, T. E. Lovejoy, and R. Mesquita (eds.), *Lessons from Amazonia: ecology and conservation of a fragmented forest.* New Haven, CT: Yale University Press.

Gilpin, M. E., and M. E. Soulé. 1986. Minimum viable populations: processes of species extinction. Pages 19–34 in M. E. Soulé (ed.), *Conservation biology: the science of scarcity and diversity.* Sunderland, MA: Sinauer.

Greenberg, R. 1989. Neophobia, aversion to open space, and ecological plasticity in song and swamp sparrows. *Canadian Journal of Zoology* 67:1194–1199.

Hanski, I., and M. E. Gilpin (eds.). 1996. *Metapopulation biology: ecology, genetics, and evolution.* London: Academic Press.

Harper, L. H. 1989. The persistence of ant-following birds in small Amazonian forest fragments. *Acta Amazonica* 19:249–263.

Harrison, S. 1989. Long-distance dispersal and colonization in the bay checkerspot butterfly. *Ecology* 70:1236–1243.

Hecnar, S. J., and R. T. M'Closkey. 1996. Regional dynamics and the status of amphibians. *Ecology* 77:2091–2097.

Hubbell, S. P., and R. B. Foster. 1986. Commonness and rarity in a neotropical forest: implications for tropical tree conservation. Pages 205–232 in M. E. Soulé (ed.), *Conservation biology: the science of scarcity and diversity.* Sunderland, MA: Sinauer.

Kalko, E. K. V. 1998. Organization and diversity of tropical bat communities through space and time. *Zoology: Analysis of Complex Systems* 101:281–297.

Kapos, V. 1989. Effects of isolation on the water status of forest patches in the Brazilian Amazon. *Journal of Tropical Ecology* 5:173–185.

Kapos, V., E. Wandelli, J. L. Camargo, and G. Ganade. 1997. Edge-related changes in environment and plant responses due to forest fragmentation in central Amazonia. Pages 33–44 in W. F. Laurance and R. O. Bierregaard (eds.), *Tropical forest remnants: ecology, management, and conservation of fragmented communities.* Chicago: University of Chicago Press.

Kauffman, J. B. 1991. Survival by sprouting following fire in the tropical forests of the eastern Amazon. *Biotropica* 23:219–224.

Klein, B. C. 1989. Effects of forest fragmentation on dung and carrion beetle communities in central Amazonia. *Ecology* 70:1715–1725.

Kotze, D. J., and M. J. Samways. 2001. No general edge effects for invertebrates at Afromontane forest/grassland ecotones. *Biodiversity and Conservation* 10:443–446.

Lamb, D., J. Parrota, R. Keenan, and N. Tucker. 1997. Rejoining habitat remnants: restoring degraded rainforest lands. Pages 366–387 in W. F. Laurance and R. O. Bierregaard (eds.), *Tropical forest remnants: ecology, management, and conservation of fragmented communities.* Chicago: University of Chicago Press.

Laurance, S. G. 2000. Effects of linear clearings on movements and community composition of Amazonian understory birds. Abstract. Brisbane, Australia.

Laurance, S. G. W., and W. F. Laurance. 1999. Tropical wildlife corridors: use of linear rainforest remnants by arboreal mammals. *Biological Conservation* 91:231–239.

Laurance, W. F. 1991a. Ecological correlates of extinction proneness in Australian tropical rainforest mammals. *Conservation Biology* 5:79–89.

Laurance, W. F. 1991b. Edge effects in tropical forest fragments: application of a model for the design of nature reserves. *Biological Conservation* 57:205–219.

Laurance, W. F. 1994. Rainforest fragmentation and the structure of small mammal communities in tropical Queensland. *Biological Conservation* 69:23–32.

Laurance, W. F. 1997. Hyper-disturbed parks: edge effects and the ecology of isolated rainforest reserves in tropical Australia. Pages 71–83 in W. F. Laurance and R. O. Bierregaard (eds.), *Tropical forest remnants: ecology, management, and conservation of fragmented communities.* Chicago: University of Chicago Press.

Laurance, W. F. 2000. Do edge effects occur over large spatial scales? *Trends in Ecology and Evolution* 15:134–135.

Laurance, W. F., and R. O. Bierregaard (eds.). 1997. *Tropical forest remnants: ecology, management, and conservation of fragmented communities.* Chicago: University of Chicago Press.

Laurance, W. F., and M. A. Cochrane. 2001. Synergistic effects in fragmented landscapes. *Conservation Biology* 15:1488–1489.

Laurance, W. F., L. V. Ferreira, J. M. Rankin-de-Merona, and S. G. Laurance. 1998a. Rain forest fragmentation and the dynamics of Amazonian tree communities. *Ecology* 79:2032–2040.

Laurance, W. F., L. V. Ferreira, J. M. Rankin-de-Merona, S. G. Laurance, R. Hutchings, and T. E. Lovejoy. 1998b. Effects of forest fragmentation on recruitment patterns in Amazonian tree communities. *Conservation Biology* 12:460–464.

Laurance, W. F., C. Gascon, and J. M. Rankin-de-Merona. 1998c. Predicting effects of habitat destruction on plant communities: a test of a model using Amazonian trees. *Ecological Applications* 9:548–554.

Laurance, W. F., T. E. Lovejoy, H. L. Vasconcelos, E. M. Bruna, R. K. Didham, P. C. Stouffer, C. Gascon, R. O. Bierregaard, S. G. Laurance, and E. Sampiao. 2002. Ecosystem decay of Amazonian forest fragments: a 22-year investigation. *Conservation Biology* 16:605–618.

Laurance, W. F., D. Perez-Salicrup, P. Delamônica, P. M. Fearnside, S. D'Angelo, A. Jerozolinski, L. Pohl, and T. E. Lovejoy. 2001. Rain forest fragmentation and the structure of Amazonian liana communities. *Ecology* 82:105–116.

Laurance, W. F., and E. Yensen. 1991. Predicting the impacts of edge effects in fragmented habitats. *Biological Conservation* 55:77–92.

Leigh, E. G., Jr. 1981. The average lifetime of a population in a varying environment. *Journal of Theoretical Biology* 90:213–239.

Lewis, S. 1998. *Treefall gaps and regeneration: a comparison of continuous forest and fragmented forest in central Amazonia.* Ph.D. thesis, University of Cambridge, England.

Lovejoy, T. E., R. O. Bierregaard, A. B. Rylands, J. R. Malcolm, C. E. Quintela, L. Harper, K. Brown, A. H. Powell, G. V. N. Powell, H. O. Schubart, and M. B. Hays. 1986. Edge and other effects of isolation on Amazon forest fragments. Pages 257–285 in M. E. Soulé (ed.), *Conservation biology: the science of scarcity and diversity.* Sunderland, MA: Sinauer.

Lynam, A. J. 1997. Rapid decline of small mammal diversity in monsoon evergreen forest fragments in Thailand. Pages 222–240 in W. F. Laurance and R. O. Bierregaard Jr. (eds.), *Tropical forest remnants: ecology, management, and conservation of fragmented communities.* Chicago: University of Chicago Press.

MacArthur, R. O., and E. O. Wilson. 1967. *The theory of island biogeography.* Princeton, NJ: Princeton University Press.

Malcolm, J. R. 1991. *The small mammals of Amazonian forest fragments: pattern and process.* Ph.D. thesis, University of Florida, Gainesville.

Malcolm, J. R. 1997a. Biomass and diversity of small mammals in Amazonian forest fragments. Pages 207–221 in W. F. Laurance and R. O. Bierregaard (eds.), *Tropical forest remnants: ecology, management, and conservation of fragmented communities.* Chicago: University of Chicago Press.

Malcolm, J. R. 1997b. Insect biomass in Amazonian forest fragments. Pages 510–533 in N. E. Stork, J. Adis, and R. K. Didham (eds.), *Canopy arthropods.* London: Chapman & Hall.

Mesquita, R., P. Delamônica, and W. F. Laurance. 1999. Effects of surrounding vegetation on edge-related tree mortality in Amazonian forest fragments. *Biological Conservation* 91:129–134.

Mills, L. S., and P. E. Smouse. 1994. Demographic consequences of inbreeding in remnant populations. *American Naturalist* 144:412–431.

Nei, M., T. Maruyama, and R. Chakraborty. 1975. The bottleneck effect and genetic variability in populations. *Evolution* 29:1–10.

Nepstad, D. C., A. Verissimo, A. Alencar, C. Nobre, E. Lima, P. Lefebre, P. Schlesinger, C. Potter, P. Moutinho, E. Mendoza, M. Cochrane, and V. Brooks. 1999. Large-scale impoverishment of Amazonian forests by logging and fire. *Nature* 398:505–508.

Offerman, H. L., V. H. Dale, S. M. Pearson, R. O. Bierregaard, Jr., and R. V. O'Neill. 1995. Effects of forest fragmentation on neotropical fauna: current research and data availability. *Environmental Reviews* 3:191–211.

Olson, D. M., and A. Andriamiadana. 1996. The effects of selective logging on the leaf litter invertebrate community of a tropical dry forest in western Madagascar. Pages 175–187 in J. U. Ganzhorn and J. P. Sorg (eds.), *Ecology and economy of a tropical dry forest in Madagascar.* Primate Report 46-1. Göttingen, Germany: German Primate Center (DPZ).

Peters, H. A. 2001. *Clidemia hirta* invasion at the Pasoh Forest Reserve: an unexpected plant invasion in an undisturbed tropical forest. *Biotropica* 33:60–68.

Pittman, N. C. A., J. Terborgh, M. R. Silman, and P. Nuñez. 1999. Tree species distributions in an upper Amazonian forest. *Ecology* 80:2651–2661.

Powell, A. H., and G. V. N. Powell. 1987. Population dynamics of male euglossine bees in Amazonian forest fragments. *Biotropica* 19:176–179.

Quintela, C. E. 1985. *Forest fragmentation and differential use of natural and man-made edges by understory birds in central Amazonia.* M.S. thesis, University of Chicago.

Rabinowitz, A. 2000. *Jaguar.* Washington, DC: Island Press.

Ralls, K., P. H. Harvey, and A. M. Lyles. 1986. Inbreeding in natural populations of birds and mammals. Pages 35–56 in M. E. Soulé (ed.), *Conservation biology: the science of scarcity and diversity.* Sunderland, MA: Sinauer.

Rankin-de Merona, J. M., G. Prance, R. W. Hutchings, M. F. Silva, W. A. Rodrigues, and M. E. Uehling. 1992. Preliminary results of a large-scale inventory of upland rain forest in the central Amazon. *Acta Amazonica* 22:493–534.

Ready, P. D., R. Laison, and J. J. Shaw. 1983. Leishmaniasis in Brazil. XX. Prevalence of enzootic rodent leishmaniasis (*Leishmania mexicana amazonensis*), and apparent absence of "pian bois" (*L. braziliensis guyanensis*), in plantations of introduced tree species and in other non-climax forests in eastern Amazônia. *Transactions of the Royal Society for Tropical Medicine and Hygiene* 77:775–785.

Restrepo, C., N. Gómez, and S. Heredia. 1999. Anthropogenic edges, treefall gaps, and fruit-frugivore interactions in a neotropical montane forest. *Ecology* 80:668–685.

Robinson, W. D. 1999. Long-term changes in the avifauna of Barro Colorado Island, Panama, a tropical forest isolate. *Conservation Biology* 13:85–97.

Shafer, M. L. 1981. Minimum population sizes for species conservation. *BioScience* 31:131–134.

Sizer, N. 1992. *The impact of edge formation on regeneration and litterfall in a tropical rain forest fragment in Amazonia.* Ph.D. thesis, University of Cambridge, UK.

Smith, M. H., M. N. Manlove, and J. Joule. 1978. Spatial and temporal dynamics of the genetic organization of small mammal populations. *Populations of Small Mammals under Natural Conditions, Pymatuning Laboratory of Ecology Special Publication* 5:99–113.

Souza, O. F. F., and V. K. Brown. 1994. Effects of habitat fragmentation on Amazonian termite communities. *Journal of Tropical Ecology* 10:197–206.

Stouffer, P. C., and R. O. Bierregaard Jr. 1995a. Effects of forest fragmentation on understory hummingbirds in Amazonian Brazil. *Conservation Biology* 9:1085–1094.

Stouffer, P. C., and R. O. Bierregaard Jr. 1995b. Use of Amazonian forest fragments by understory insectivorous birds. *Ecology* 76:2429–2445.

Stratford, J. A., and P. C. Stouffer. 1999. Local extinctions of terrestrial insectivorous birds in a fragmented landscape near Manaus, Brazil. *Conservation Biology* 13:1416–1423.

Terborgh, J. 1986. Keystone species in the tropical rainforest. Pages 330–334 in M. Soulé (ed.), *Conservation biology: the science of scarcity and diversity.* Sunderland, MA: Sinauer.

Tocher, M. 1998. A comunidade de anfíbios da Amazônia central: diferenças na composição específica entre a mata primária e pastagens. Pages 219–232 in C. Gascon and P. Moutinho (eds.), *Floresta Amazônica: dinâmica, regeneração e manejo.* Manaus, Brazil: Instituto Nacional de Pesquisas da Amazônia.

Tocher, M., C. Gascon, and B. L. Zimmerman. 1997. Fragmentation effects on a central

Amazonian frog community: a ten-year study. Pages 124–137 in W. F. Laurance and R. O. Bierregaard (eds.), *Tropical forest remnants: ecology, management, and conservation of fragmented communities*. Chicago: University of Chicago Press.

Uhl, C., and J. B. Kauffman. 1990. Deforestation effects on fire susceptibility and the potential response of the tree species to fire in the rainforest of the eastern Amazon. *Ecology* 71:437–449.

Van Shaik, C. P., J. Terborgh, and S. J. Wright. 1993. The phenology of tropical forests: adaptive significance and consequences for primary consumers. *Annual Review of Ecology and Systematics* 24:353–377.

Vasconcelos, H. L. 1988. Distribution of *Atta* (Hymenoptera, Formicidae) in "terra-firme" rain forest of central Amazonia: density, species composition and preliminary results on effects of forest fragmentation. *Acta Amazonica* 18:309–315.

Vasconcelos, H. L. 1999. Effects of forest disturbance on the structure of ground-foraging ant communities in central Amazonia. *Biodiversity and Conservation* 8:409–420.

Viana, V. M., and A. A. J. Tabanez. 1996. Biology and conservation of forest fragments in the Brazilian Atlantic Moist Forest. Pages 151–167 in J. Schelhas and R. Greenberg (eds.), *Forest patches in tropical landscapes*. Washington, DC: Island Press.

Wahlberg, N., A. Moilanen, and I. Hanski. 1996. Predicting the occurrence of endangered species in fragmented landscapes. *Science* 273:1536–1538.

Williams-Linera, G. 1990. Vegetation structure and environmental conditions of forest edges in Panama. *Journal of Ecology* 78:356–373.

Williamson, G. B., R. Mesquita, K. Ickes, and G. Ganade. 1998. Estratégias de árvores pioneiras nos Neotrópicos. Pages 131–144 in C. Gascon and P. Moutinho (eds.), *Floresta Amazônica: dinâmica, regeneração e manejo*. Manaus, Brazil: Instituto Nacional de Pesquisas da Amazônia.

Wright, S. J., C. Carrasco, C. Calderón, and S. Paton. 1999. The El Niño Southern Oscillation, variable fruit production, and famine in a tropical forest. *Ecology* 80:1632–1642.

Zimmerman, B. L., and R. O. Bierregaard Jr. 1986. Relevance of the equilibrium theory of island biogeography and species-area relations to conservation with a case from Amazonia. *Journal of Biogeography* 13:133–143.

Chapter 3

Landscape Connectivity and Biological Corridors

Susan G. W. Laurance

Natural habitats in the tropics are being converted to agricultural land faster than in any other biome (Whitmore 1997). The results of such rapid clearing will be apparent in the next few decades, when most of the remaining tropical forest will occur as isolated remnants (Myers 1984). The type of habitat that surrounds these remnants may play a crucial role in their conservation. Adjoining habitats that are more similar to the remnants in terms of structure and floristic composition (e.g., agroforestry lands rather than pasture or open crop fields) will be the most beneficial to the long-term preservation of biodiversity.

In addition to supporting native species of plants and animals, agroforestry areas may contribute to the conservation of biodiversity by increasing the connectivity of populations, communities, and ecological processes in fragmented landscapes. Habitats that can maintain this connectivity across the landscape are commonly called biological corridors or simply corridors. Corridors can consist of various types of habitat, but by definition they differ from the surrounding vegetation and link habitat remnants that were once originally connected (Saunders and Hobbs 1991).

When using agroforestry to increase landscape connectivity, it is important to understand which characteristics of a corridor will make it effective for a given organism. Because few studies have been carried out on the movement of wildlife through agroforestry corridors, this chapter reviews the relevant literature on tropical forest corridors. It describes how rainforest animals select and use linear habitat remnants and which features appear to be most important for corridor effectiveness. The chapter also considers some of the relevant research on wildlife use of agroforestry systems to discuss their usefulness in connecting landscapes.

Landscape connectivity is a function of both the environmental features of a corridor and the behavior of wildlife species that may attempt to use the corridor (Merriam 1984). The general assumption underlying the value of landscape connectivity is that a fragmented landscape that is interconnected is more likely to support viable faunal and floral populations and intact ecological processes than a landscape that is made up of only isolated fragments (Harris 1984; Bennett 1998).

This assumption is based on two theoretical concepts: island biogeography theory (MacArthur and Wilson 1967) and metapopulation models (Levins 1970). Island biogeography theory proposed that the number of species contained on an isolated community (such as an oceanic island or forest fragment) is the result of a dynamic equilibrium between opposing forces of colonization and extinction. The theory predicts that with increasing isolation between forest fragments, there will be a decreasing rate of immigration by species unable to traverse the modified habitat. Numerous studies have documented that land bridge islands (i.e., islands artificially created by flooding a piece of land) and forest fragments lose species after isolation, a phenomenon called species relaxation (Diamond 1972; Terborgh 1974; see also Chapter 2, this volume).

Rather than examining entire species assemblages, metapopulation models consider the population of a single species, which occurs in spatially separate subpopulations that are connected by dispersal (Levins 1970; Forman 1997). Although there are various types of metapopulation models, there are two general forms. The first identifies a major source population that disperses outward to smaller sink populations. This refers to a situation in which small habitat fragments are partly separated from a larger area of intact habitat. The second is a population that is patchily distributed throughout the landscape and connected by dispersal, as would occur when only small forest fragments remain in a formerly forested landscape. Small populations in the models are prone to local extinction, and the movement of individuals between patches can both bolster dwindling populations via their genetic and demographic contributions (called the rescue effect; Brown and Kodric-Brown 1977) and result in the recolonization of local patches where the species has gone extinct.

In both island biogeography theory and metapopulation models, the extinction rate of a species depends on the size and quality of the remnant habitats, whereas recolonization depends on the level of landscape connectivity. Various landscape features may affect connectivity, such as the distance between remnant patches, the type of surrounding matrix (modified habitats), and the presence of corridors or small habitat patches that can function as stepping stones.

Corridors were first proposed for conservation planning in 1975, based on fragmentation and island studies (Diamond 1975; Wilson and Willis 1975). Since that time many studies have demonstrated the major benefits of corridors, which include facilitating wildlife movement, providing habitat, and

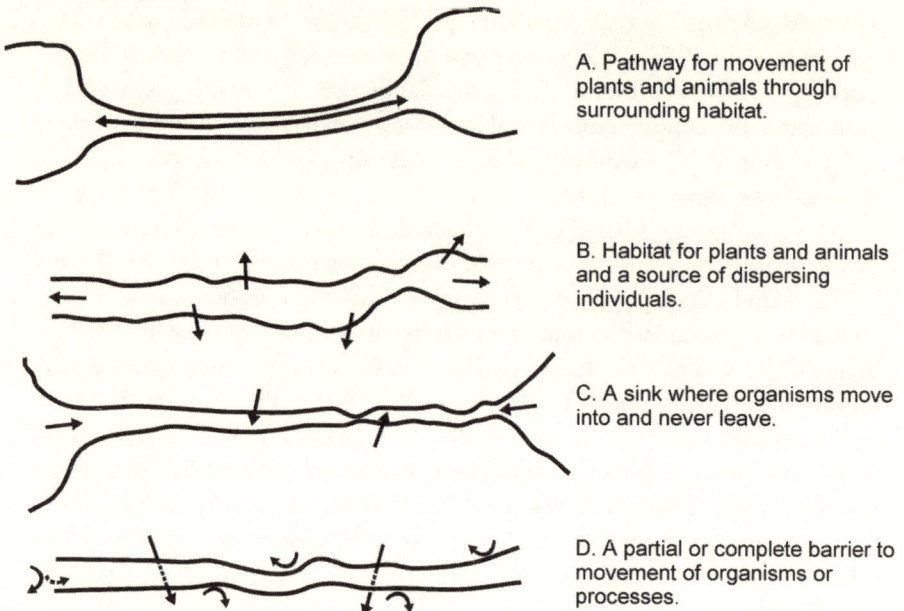

A. Pathway for movement of plants and animals through surrounding habitat.

B. Habitat for plants and animals and a source of dispersing individuals.

C. A sink where organisms move into and never leave.

D. A partial or complete barrier to movement of organisms or processes.

Figure 3.1. An illustrative summary of four major landscape functions of corridors: pathway, habitat, population sink, and barrier.

benefiting ecosystem processes (Bennett 1998). Alternatively, some studies have warned of the potential costs of corridors, such as the risk of spread of biotic and abiotic disturbances to remnant populations and habitats, the potential for increased wildlife mortality in corridors, and insufficient information on whether the financial costs of corridors could be better invested in other conservation initiatives (e.g., purchasing land). The benefits and costs of corridors are considered in the following sections (Figure 3.1).

Benefits of Corridors

Major landscape functions of corridors include facilitating movements of wildlife through the landscape, providing habitat, and aiding ecosystem processes.

Facilitating Different Types of Movement

An array of studies has demonstrated that habitat corridors can facilitate the movement of wildlife. Three types of movement have been described: local, migratory, and dispersal (Bennett 1990). Although few studies have been conducted in tropical rainforest, there is evidence that some species of mammals, birds, butterflies, and beetles will undertake local movements through corridors within their home range or between foraging and nesting sites (Lovejoy et al. 1986; Isaacs 1995; and see Bennett 1998 for review).

Migratory movements of wildlife through corridors have been observed in mammals, birds, and amphibians (Newmark 1993; Powell and Bjork 1995; Forman and Deblinger 2000). Wildlife species in these studies have been observed preferentially moving through forested corridors rather than the surrounding agricultural matrix. One study in Costa Rica, for example, showed that large tropical frugivores (e.g., resplendent quetzal, *Pharomachrus mocinno*) needed forested corridors from montane to lower elevational forests so that birds could follow seasonal changes in their food supply (Powell and Bjork 1995).

Dispersal movements normally are described as the one-way movements of young animals seeking unoccupied territories in which to breed. Corridor studies have demonstrated that the dispersal movements of mammals, birds, butterflies, and plants can be assisted by habitat linkages (Bennett 1998). Dispersal movements are important for population dynamics because they allow individuals to immigrate to new populations or to recolonize locally extinct populations. One of the key challenges concerning the effectiveness of corridors is to demonstrate that dispersing individuals not only move through corridors but also become established in fragment populations. Only in this way can immigrants reduce the negative consequence of insularization on fragment populations. Genetic evidence is beginning to accumulate that confirms that gene flow is occurring between fragmented faunal populations linked by corridors, thereby demonstrating the successful establishment of immigrants in the population (Mech and Hallett 2001).

Providing Habitat for Resident Species

Depending on the shape, habitat structure, and floristic composition of corridors, a range of wildlife species may reside in them. Edge and generalist species probably are the most common occupants of corridors, predominating in narrow habitat strips that occur along roadsides, riparian areas, and windbreaks (Crome et al. 1994; Hill 1995; Forman 1997; Laurance and Laurance 1999; de Lima and Gascon 1999), although some forest species may also be present (Harvey 2000). These types of corridors often are just slender strips of edge habitat with little or no interior. Rare and endangered species usually avoid such areas and are more likely to reside in wider corridors with a higher-quality (or interior) habitat (Laurance and Laurance 1999).

Residency in corridors is the most effective way of maintaining population connectivity, particularly for less mobile species or those that will move long distances (Bennett 1990). For such species, corridors must provide adequate resources such as food and shelter. If habitat is suitable for residency, then population continuity (and gene flow) will be maintained by both the movements

of individuals in and out of the corridor and by their movements and reproduction within the corridor (Figure 3.1; Bennett 1990). In addition to
dispersing animals, reproduction in corridors, as detected in the rainforest corridors of northern Australia and the central Amazon (Laurance 1996; de Lima
1998), also provides an additional source of individuals for the corridor population. For some Nearctic migrant birds, shade coffee and cocoa plantations
in Central America provide useful and even crucial habitats (Robbins et al.
1992; Perfecto et al. 1996; Greenberg et al. 1997).

Aiding Ecosystem Processes

Habitat corridors can also provide additional landscape services by aiding
ecosystem processes by protecting watersheds (Karr and Schlosser 1978) and
providing windbreaks, for example. Riparian vegetation along streams can
reduce soil erosion and maintain water quality and stream flows by shading
streams and thereby reducing the excessive growth of aquatic plants, including exotics (Parendes and Jones 2000; Chapter 18, this volume). Furthermore,
adequate streamside vegetation can reduce the inflow of agrochemicals and
nutrients, thereby helping to maintain water quality and inhibit the growth of
aquatic algae. Windbreaks and fencerows can reduce windspeeds and consequently help protect pastures, crops, livestock, and natural habitats from exposure.

Costs of Corridors

Wildlife corridors have their critics. A number of potential detrimental effects
have been suggested (Simberloff and Cox 1987; Simberloff et al. 1992; Hess
1994) that should be considered when recommending corridors as a component of a regional conservation strategy (see also Figure 3.1).

Spread of Biotic and Abiotic Disturbances

First, as a result of increased immigration, wildlife corridors could facilitate the
spread of diseases, exotic species, weeds, and undesirable species (Simberloff
and Cox 1987; Hess 1994). Increased immigration might also disrupt local
adaptations of species and even decrease the level of genetic variation by causing outbreeding depression, which occurs with the mating of highly dissimi-
lar individuals (Simberloff and Cox 1987). However, such events are unlikely
to occur when corridors are being used to reconnect habitats that have been
isolated by human land uses rather than connecting naturally unconnected
habitats (Noss 1987).

Second, corridors might facilitate the spread of abiotic disturbances such
as fire (Simberloff and Cox 1987). In tropical landscapes, ignition sources of

fire most commonly occur in managed pastures or slash-and-burn fields. Agroforestry areas normally are less susceptible to fire than pastures and fields because the closed or semiclosed tree canopy creates a dark, humid environment with low fuel loads and, especially, because agroforesters tend to protect their investment in tree planting (e.g., through firebreaks). Therefore, agroforestry areas could act as potential fire buffers to remnant habitats and are more likely to inhibit rather than promote the spread of fire.

Population Sinks

Corridors could act as population sinks by attracting individuals to areas where they experience reduced survival or reproduction. By aiming to promote wildlife movement across the landscape, corridors may funnel wildlife through private lands, increasing their exposure to hunters, poachers, predators, and domestic animals (see Chapter 13, this volume). Moreover, corridors may maneuver organisms into an environment with limited resources and potentially superior competitors such as generalist and edge species, which could reduce their reproductive success (e.g., nest predation; Angelstam 1986) and compromise their survival (Simberloff and Cox 1987). To date most research into increased mortality in corridors has been extrapolated from edge effect studies in fragmented habitats (Gates and Gysel 1978), and little evidence is available on mortality rates in corridors compared with other habitats.

Financial Costs and Benefits of Corridors

There has been some suggestion that the financial cost of corridors may outweigh their benefits and that scarce conservation dollars could be better spent on other initiatives (Simberloff et al. 1992). For example, in areas such as eastern Madagascar, where less than 5 percent of original rainforest remains (Smith 1997), conservation options might include purchasing land for protecting remnant natural habitat, creating corridors between remnant reserves (which may not function for all species), and revegetating lands that adjoin reserves to provide additional habitat area. Given that resources for conservation are highly limited, funding corridors at the expense of other initiatives may not invariably be the best option.

Corridor Features That Facilitate Faunal Movements and Plant Dispersal

As mentioned earlier, the efficacy of movement corridors often is assessed not only as their ability to facilitate individual movement but also as the successful establishment of dispersing individuals as breeding members in a new population. Dispersal may result from density-dependent factors such as limited resources (food, shelter, mates) or density-independent factors such as inbreeding avoidance (not breeding with closely related individuals; Howard 1960). Individuals can increase their chances of being accepted in a new population by becoming receptive to breeding upon their arrival (Lidicker 1975).

How capable a species is of moving through a corridor may depend on its food and area needs, vagility, denning needs, social behavior, and other factors. Bennett (1990) identified three types of species-specific movement along corridors. First, movement can be a single motion along the entire length of the corridor, which could be seasonal in nature. This type of movement usually is made by large, highly mobile species. For example, Grimshaw and Foley (1991 in Newmark 1993) found that forest elephants on Mount Kilimanjaro in East Africa made seasonal use of the remaining forest link between the mountain and Amboseli National Park.

Second, corridor movement may be punctuated by pauses for food or shelter that may last from hours to days or even weeks. This situation may be typical of smaller species that have high energy needs and limited mobility. Finally, individuals may reside in a corridor, resulting in immigration and gene flow through the resident population. Recent studies have demonstrated that many species reside in corridors, including sensitive wildlife such as rainforest birds and arboreal marsupials (Isaacs 1995; Laurance and Laurance 1999).

Many of the corridor features that facilitate animal movement across the landscape also are beneficial to plant pollination and dispersal (Lamont and Southall 1982; Chapter 12, this volume). The movement of plant pollen, seeds, spores, and other propagules can occur via vectors such as wind, water, and flying and terrestrial animals. Although plant pollination and dispersal in habitat corridors have received little attention (Forman 1997; Lamont and Southall 1982; Loney and Hobbs 1991), two movement patterns have been detected. First, short-distance movements of some plant species have been observed via vegetative spread or adjacent seed dispersal. Second, the most common movements have been wind or animal dispersal of seeds some distance along the corridor (Hascova 1992). Plants that need wind pollination and dispersal probably will not be as sensitive to corridor features such as habitat quality and structure as plant species that clearly depend on an animal's ability to move through the landscape. The successful establishment of plants, irrespective of their dispersal mechanism, into an appropriate environment and their subsequent reproduction allows species to spread further across the landscape (Forman 1997).

In addition to the types of species-specific movements there are four major physical corridor features that will influence species use and movement in a corridor. These are habitat quality, corridor width, length, and continuity.

Habitat Quality

Habitat quality is rarely uniform in landscapes because of natural variation in topography, soils, and vegetation (Foster 1980). Habitat quality is a critical feature of corridor effectiveness because, irrespective of the mobility of species, survival rates are predicted to be much greater in high-quality habitats than in marginal or poor habitats (Henein and Merriam 1990). Although the diversity of species and their needs make it impossible to focus on individual species, there do appear to be some general patterns. For tropical rainforests, high-quality habitat is that which most closely resembles primary forest. Structural features such as canopy height, canopy connectivity, canopy and understory structure, and floristic composition may all be important, depending on the dispersing species.

In a study of 36 rainforest corridors or linear remnants in northern Australia, for example, Laurance and Laurance (1999) found that only corridors of primary rainforest supported the entire assemblage of arboreal mammals. Mature (tall and floristically diverse) regenerating forests supported more arboreal mammal species than young or less diverse regrowth, but they still did not support the sensitive forest species that were at greatest risk of extinction in forest fragments (Laurance and Laurance 1999).

Yet even a little forest cover is significantly better for wildlife movement than none. At the Biological Dynamics of Forest Fragments Project near Manaus, Brazil, the deleterious effects of fragmentation on some faunal communities were significantly alleviated once the pasture areas that surround the fragments were abandoned to forest regrowth (Gascon et al. 1999). For example, the abundances of many understory bird species recovered significantly in small fragments when forest regrowth provided continuous cover to undisturbed rainforest (Stouffer and Bierregaard 1995).

In Indonesia and Central America, the traditional agroforestry systems often are refuges for high biological diversity (Thiollay 1995; Perfecto et al. 1996). A number of comparative studies have demonstrated that traditional rubber and coffee plantations harbor more species than the more simplified commercial plantations (Thiollay 1995; Perfecto et al. 1996; see also Chapters 9 and 10, this volume). The higher structural complexity of these plantations supports many forest species (Thiollay 1995), including rare forest trees (Purata and Meave 1993 in Perfecto et al. 1996), orchids (Nir 1988), invertebrates (Perfecto et al. 1996), and migrant birds (Greenberg et al. 1997).

Similarly, the floristic diversity of both natural habitat corridors (Laurance 1996) and agroforestry plantations (Perfecto et al. 1996) has been found to be an important contributing factor to high biodiversity. A diverse array of flowering and fleshy-fruited plant species in the canopy and understory can support many resident and dispersing wildlife species. Furthermore, the presence of plant species (e.g., fig trees and palms) that offer fruit for long periods or year-round (i.e., keystone species) is especially important in corridors because their small area significantly limits the availability of resources.

Corridor Width

Corridor width is also a vital feature because it helps determine the area of available habitat, the diversity of resources for wildlife, and the vulnerability of the corridor to potentially adverse edge effects (Janzen 1986; Lovejoy et al. 1986). Edge effects encompass a diverse array of ecological and microclimatic changes that can occur on and near the forest-agricultural boundary (Lovejoy et al. 1986; Chapter 2, this volume). In tropical rainforests the penetration distance of various edge effects can be up to 200 m (Laurance et al. 1997). Furthermore, because corridors generally are long and linear in design, they tend to have a high ratio of edge to interior habitat.

An effective corridor will promote the survival of the most extinction-prone species in a reserve or landscape (Newmark 1993). Forest interior species often are targeted for corridor studies because many have declined dramatically in fragmented habitats (Laurance 1990; Stouffer and Bierregaard 1995). One key correlate of some extinction-prone species is avoidance of edge-affected habitat (Laurance 2001). Therefore, wide corridors usually are far better at supporting forest interior species than narrower corridors. Only corridors more than 250 m wide were found to support the most sensitive of the arboreal mammals in tropical Australia (Laurance and Laurance 1999). Similarly, many understory rainforest birds in central Amazonia had reduced abundances within at least 70 m of forest edges, and some probably exhibited even stronger edge avoidance (Laurance 2001). For this bird community, a 250-m-wide corridor might provide only about 100 m of forest interior habitat and could still fail to support very sensitive species.

Corridor Length

Corridor length can also be a key factor. For slow-moving species with a short lifespan, a long corridor may be a population sink. Successful dispersal through a corridor probably will be negatively related to corridor length. A long corridor must contain all the habitat needs of a species. If this is not possible, then the presence of larger habitat patches serving as stepping stones along the corridor route may help meet critical habitat needs for dispersing individuals (Newmark 1993).

Canopy and Corridor Continuity

Closed-canopy ecosystems such as tropical forests may be more strongly affected by habitat discontinuities than other habitats. Canopy connectivity can be an important factor for the movement of canopy and understory species. A discontinuous canopy may act as a barrier for species that need cover from predators or for arboreal species that move mainly through upper forest strata.

Some corridors may be bisected by roads or rivers, which can be a barrier to animal movements. In tropical forests, even narrow road and powerline clearings (less than 80 m wide) can significantly impede movements by sensitive understory birds (Laurance 2001), small mammals (Goosem and Marsh 1997; Goosem 2000), and arboreal marsupials (Wilson 2002). Therefore, such habitat discontinuities should be avoided where possible by facilitating wildlife movement via replanting gaps and structural corridors such as highway underpasses.

Agroforestry Corridors and the Wildlife at Risk

Agroforestry systems may not be effective corridors for all wildlife. There are two groups that will need special attention. The first is rainforest specialists that avoid disturbed habitat. In two comparative studies of complex agroforests and primary forests, for example, the bird species that were not detected in agroforests were the large, understory, or terrestrial insectivores (Thiollay 1995; Greenberg et al. 1997). Therefore, these species must become one of the target groups for future research and monitoring because they are sensitive to extinction in fragmented landscapes, even in the presence of agroforestry corridors.

The second wildlife group at risk is game and large predator species. Agroforestry corridors could become critical population sinks for game and predator species that are actively hunted for food, for income, or to protect domestic animals (see Chapters 13 and 14, this volume). On central Amazon farms, for example, large predators such as jaguars and harpy eagles often are attracted to domestic animals and are therefore killed by farmers (S. G. Laurance, pers. obs.).

Conclusions

It has been 25 years since wildlife corridors were first suggested as a means to mitigate the effects of habitat fragmentation. Despite a rapidly increasing body of research on corridors, for some scientists there is still too little information available to justify the inclusion of corridors in regional conservation strategies. Yet natural forested areas are diminishing in size, and as a result there are fewer opportunities to acquire or protect remnant habitats that can maintain habitat connectivity. Under these circumstances a wait-and-see approach could be disastrous. Acting now to protect and establish wildlife corridors might entail some risks, but it will be far easier to remove a corridor in the future than to create one where the original habitat has been destroyed.

Agroforestry systems have the potential to increase the movements of plants and animals across the landscape and thereby contribute to biodiversity conservation. There is a great variation in the habitat quality, structure, and natural dynamics of different agroforestry systems that will affect the wild species that use them (see Part III of this volume). For example, a windbreak in a pasture may provide habitat for edge and generalist species such as insects and rodents, whereas riparian corridors may contain remnant vegetation and be more beneficial for forest interior species.

One feature that many agroforestry systems share is that they tend to be situated on high-quality sites (Perfecto et al. 1996), whereas forest remnants are most likely to be preserved on infertile soils or on steep topography. These productive areas may once have supported populations of native plants and animals in high densities and been important sources of dispersing individuals (Thiollay 1995). For this reason, agroforestry lands could play a key role in wildlife conservation if plantations can attempt to follow some simple recommendations that are aimed to facilitate the movement and persistence of animal and plant species.

First, remnant primary and riparian forests should be protected, and any patches of primary rainforest should be incorporated into the plantation. Small remnants may harbor locally endemic species and act as stepping stones for faunal and floral dispersal. Second, traditional shade plantations that maintain original canopy species or plant mixed canopies should be encouraged and some natural recruitment allowed. Third, agroforestry plantings should be diverse and include native fruiting and flowering plants that produce large fruit crops for long periods (e.g., fig trees or large palms). Fourth, linear plantings such as windbreaks should be as wide as possible and have a complex structure rather than a single tree row. Fifth, tree plantings should fill in canopy gaps and maintain canopy connectivity. And finally, domestic animals should be confined where possible and hunting controlled, especially for extinction-prone species. Agroforestry methods that attempt to achieve these goals will have the greatest benefits for species that are highly vulnerable to habitat loss, disturbance, and fragmentation. Such methods should be an integral part of regional conservation strategies in the tropics.

References

Angelstam, P. 1986. Predation on ground-nesting birds' nests in relation to predator densities and habitat edge. *Oikos* 47:365–373.

Bennett, A. F. 1990. *Habitat corridors: their role in wildlife management and conservation.* Victoria, Australia: Department of Conservation and Environment.

Bennett, A. F. 1998. *Linkages in the landscape: the role of corridors and connectivity in wildlife conservation.* Gland, Switzerland: International Union for the Conservation of Nature and Natural Resources.

Brown, J. H., and A. Kodric-Brown. 1977. Turnover rates in insular biogeography: effect of immigration on extinction. *Ecology* 58:445–449.

Crome, F., J. Isaacs, and L. Moore. 1994. The utility to birds and mammals of remnant riparian vegetation and associated windbreaks in the tropical Queensland uplands. *Pacific Conservation Biology* 1:328–343.

de Lima, M. G. 1998. *Composição da communidade de pequenos mamíferos e sapos de liteira em remanescentes lineares no Projeto Dinâmica Biológica de Fragmentos Florestais (PDBFF)*. M.S. thesis, National Institute for Research in the Amazon (INPA), Manaus, Brazil.

de Lima, M. G., and C. Gascon. 1999. The conservation value of linear forest remnants in central Amazonia. *Biological Conservation* 91:241–247.

Diamond, J. M. 1972. Biogeographic kinetics: estimation of relaxation times for avifaunas of southwest Pacific islands. *Proceedings of the National Academy of Science (USA)* 69:3199–3203.

Diamond, J. M. 1975. The island dilemma: lessons of modern biogeography studies for the design of natural reserves. *Biological Conservation* 7:129–146.

Forman, R. T. T. 1997. *Land mosaics: the ecology of landscapes and regions.* Cambridge, MA: Cambridge University Press.

Forman, R. T. T., and R. D. Deblinger. 2000. The ecological road-effect zone of a Massachusetts suburban highway. *Conservation Biology* 14:36–46.

Foster, R. B. 1980. Heterogeneity and disturbance in tropical vegetation. Pages 75–92 in M. E. Soulé and B. A. Wilcox (eds.), *Conservation biology: an evolutionary-ecological perspective.* Sunderland, MA: Sinauer.

Gascon, C., T. E. Lovejoy, R. O. Bierregaard Jr., J. R. Malcolm, P. C. Stouffer, H. L. Vasconcelos, W. F. Laurance, B. Zimmerman, M. Tocher, and S. Borges. 1999. Matrix habitat and species richness in tropical forest remnants. *Biological Conservation* 91:223–230.

Gates, J. E., and I. W. Gysel. 1978. Avian nest dispersion and fledging success in field-forest ecotones. *Ecology* 59:871–83.

Goosem, M. 2000. Effects of tropical rainforest roads on small mammals: edge changes in community composition. *Wildlife Research* 27:151–163.

Goosem, M., and H. Marsh. 1997. Fragmentation of a small mammal community by a powerline corridor through tropical rainforest. *Wildlife Research* 24:613–629.

Greenberg, R., P. Bichier, and J. Sterling. 1997. Bird populations in rustic and planted shade coffee plantations of eastern Chiapas, Mexico. *Biotropica* 29:501–514.

Harris, L. D. 1984. *The fragmented forest.* Chicago: University of Chicago Press.

Harvey, C. A. 2000. The colonization of agricultural windbreaks by forest trees: effects of connectivity and remnant trees. *Ecological Applications* 10:1762–1773.

Hascova, J. 1992. The role of corridors for plant dispersal in the landscape. Pages 88–99 in *Ecological stability of landscape: ecological infrastructure and management.* Kostelec, Czechoslovakia: Institute of Applied Ecology.

Henein, K., and G. Merriam. 1990. The elements of connectivity where corridor quality is variable. *Landscape Ecology* 4:157–170.

Hess, G. R. 1994. Conservation corridors and contagious disease: a cautionary note. *Conservation Biology* 8:256–262.

Hill, C. J. 1995. Linear strips of rainforest vegetation as potential dispersal corridors for rainforest insects. *Conservation Biology* 9:1559–1566.

Howard, W. E. 1960. Innate and environmental dispersal of individual vertebrates. *The American Midland Naturalist* 63:152–161.

Isaacs, J. L. 1995. *Species composition and movement of birds in riparian vegetation in a tropical agricultural landscape.* M.S. thesis, James Cook University, Townsville, Australia.

Janzen, D. H. 1986. The eternal external threat. Pages 286–303 in M. E. Soulé (ed.), *Conservation biology: the science of scarcity and diversity.* Sunderland, MA: Sinauer.

Karr, J. E., and I. J. Schlosser. 1978. Water resources and the land-water interface. *Science* 201:229–234.

Lamont, B. B., and K. J. Southall. 1982. Biology of mistletoe *Anyema preissii* on road verges and undisturbed vegetation. *Search* 13:87–88.

Laurance, S. G. W. 1996. *The utilisation of linear rainforest remnants by arboreal marsupials in north Queensland.* M.S. thesis, University of New England, Armidale, Australia.

Laurance, S. G. W. 2001. *The effects of roads and their edges on the movement patterns and community composition of understorey rainforest birds in central Amazonia.* Ph.D. thesis, University of New England, Armidale, Australia.

Laurance, S. G. W., and W. F. Laurance. 1999. Tropical wildlife corridors: use of linear rainforest remnants by arboreal marsupials. *Biological Conservation* 91:231–239.

Laurance, W. F. 1990. Comparative responses of five arboreal marsupials to tropical forest fragmentation. *Journal of Mammalogy* 71:641–653.

Laurance, W. F., R. O. Bierregaard Jr., C. Gascon, R. K. Didham, A. P. Smith, A. J. Lynam, V. M. Viana, T. E. Lovejoy, K. E. Seiving, J. W. Sites, M. Anderson, M. Tocher, E. A. Kramer, C. Restrepo, and C. Moritz. 1997. Tropical forest fragmentation: synthesis of a diverse and dynamic discipline. Pages 502–514 in W. F. Laurance and R. O. Bierregaard, Jr. (eds.), *Tropical forest remnants: ecology, management, and conservation of fragmented communities.* Chicago: University of Chicago Press.

Levins, R. 1970. Extinction. Pages 75–107 in *Some mathematical questions in biology,* Vol. 2. Providence, RI: American Mathematical Society.

Lidicker, W. Z., Jr. 1975. The role of dispersal in demography of small mammals. Pages 103–128 in K. Petrysewicz and L. Ryszkowski (eds.), *Small mammals their productivity and population dynamics.* London: Cambridge University Press.

Loney, B., and R. J. Hobbs. 1991. Management of vegetation corridors: maintenance, rehabilitation and establishment. Pages 299–311 in D. A. Saunders and R. J. Hobbs (eds.), *Nature conservation 2: the role of corridors.* Sydney: Surrey Beatty and Sons.

Lovejoy, T., R. O. Bierregaard Jr., A. B. Rylands, J. R. Malcolm, C. Quintela, L. H. Harper, K. S. Brown, A. H. Powell, G. V. N. Powell, H. O. Schubert, and M. Hays. 1986. Edge and other effects of isolation on Amazon forest fragments. Pages 257–285 in M. Soulé (ed.), *Conservation biology: the science of scarcity and diversity.* Sunderland, MA: Sinauer.

MacArthur, R. H., and E. O. Wilson. 1967. *The theory of island biogeography.* Princeton, NJ: Princeton University Press.

Mech, S. G., and J. G. Hallett. 2001. Evaluating the effectiveness of corridors: a genetic approach. *Conservation Biology* 15:467–474.

Merriam, G. 1984. Connectivity: a fundamental ecological characteristic of landscape pattern. Pages 5–15 in M. Ruzicka, T. Hrnciarova, and L. Miklos (eds.), *Proceedings of the First International Seminar on Methodology in Landscape Ecological Research and Planning.* Roskilde, Denmark: International Association for Landscape Ecology.

Myers, N. 1984. *The primary source: tropical forests and our future.* New York: W. W. Norton.

Newmark, W. D. 1993. The role and design of wildlife corridors with examples from Tanzania. *Ambio* 22:500–504.

Nir, M. A. 1988. The survivors: orchids on a Puerto Rican coffee finca. *American Orchid Society Bulletin* 57:989–995.

Noss, R. F. 1987. Corridors in real landscapes: a reply to Simberloff and Cox. *Conservation Biology* 1:159–164.

Parendes, L. A., and J. A Jones. 2000. Role of light availability and dispersal in exotic plant invasion along roads and streams in the H. J. Andrews experimental forest, Oregon. *Conservation Biology* 14:64–75.

Perfecto, I., R. A. Rice, R. Greenberg, and M. E. Van de Voort. 1996. Shade coffee: a disappearing refuge for biodiversity. *BioScience* 46:598–608.

Powell, G. V. N., and R. Bjork. 1995. Implications of intratropical migration in reserve design: a case study using *Pharomachrus moccino. Conservation Biology* 9:354–362.

Robbins, C. S., B. A. Dowell, D. K. Dawson, J. S. Colon, R. Estrade, A. Sutton, R. Sutton, and D. Weyer. 1992. Comparison of neotropical migrant landbirds populations wintering in tropical forest, isolated forest fragments, and agricultural habitats. Pages 207–220 in J. M. Hagan III and D. W. Johnson (eds.), *Ecology and conservation of neotropical migrant landbirds.* Washington, DC: Smithsonian Institution Press.

Saunders, D. A., and R. J. Hobbs. 1991. The role of corridors in conservation: what do we know and where do we go? Pages 421–427 in D. A. Saunders and R. J. Hobbs (eds.), *Nature conservation 2: the role of corridors.* Sydney: Surrey Beatty and Sons.

Simberloff, D. S., and J. Cox. 1987. Consequences and costs of conservation corridors. *Conservation Biology* 1:63–71.

Simberloff, D. S., J. A. Farr, J. Cox, and D. W. Mehlam. 1992. Movement corridors: conservation bargains or poor investments? *Conservation Biology* 6:493–504.

Smith, A. P. 1997. Deforestation, fragmentation, and reserve design in western Madagascar. Pages 415–441 in W. F. Laurance and R. O. Bierregaard Jr. (eds.), *Tropical forest remnants: ecology, management, and conservation of fragmented communities.* Chicago: University of Chicago Press.

Stouffer, P. C., and R. O. Bierregaard Jr. 1995. Use of Amazonian forest fragments by understory insectivorous birds. *Ecology* 76:2429–2443.

Terborgh, J. 1974. Preservation of natural diversity: the problem of extinction prone species. *BioScience* 24:715–722.

Thiollay, J. M. 1995. The role of traditional agroforests in the conservation of rain forest bird diversity in Sumatra. *Conservation Biology* 9:335–353.

Whitmore, T. C. 1997. Tropical forest disturbance, disappearance, and species loss. Pages 3–12 in W. F. Laurance and R. O. Bierregaard Jr. (eds.), *Tropical forest remnants: ecology, management, and conservation of fragmented communities.* Chicago: University of Chicago Press.

Wilson, E. O., and E. O. Willis. 1975. Applied biogeography. Pages 522–534 in M. L. Cody and J. M. Diamond (eds.), *Ecology and evolution of communities.* Cambridge, MA: Belknap Press.

Wilson, R. 2002. The effects of powerline clearings on movements of arboreal marsupials. *Proceedings of the Second Canopy Conference,* June 2002, Cairns, Australia.

PART II

The Ecological Economics of Agroforestry: Environmental Benefits and Effects on Deforestation

Tropical farmers influence biodiversity through their land use decisions, such as whether to use a piece of land for a diversified agroforestry system or for a slash-and-burn plot or monoculture plantation or whether to invest in the intensification of existing agricultural fields or the clearing of forest land for new fields. One must understand and take such decisions into account when attempting to promote biodiversity-friendly land use practices. The objective of this section is to analyze the economic bases of such decisions, focusing on the valuation of the environmental services of diversified, tree-dominated land uses and the factors that influence deforestation or conservation of natural ecosystems by tropical farmers.

The section opens with a discussion of the economics of land use choices, focusing on the environmental benefits of biodiversity-friendly land use practices such as agroforestry for land users and society (Chapter 4). Using case studies from The Sudan, Nigeria, and Peru, the chapter shows that economic studies have not fully embraced the multiplicity of benefits of agroforestry systems to farmers and society and that new incentive schemes are needed to encourage agroforestry and compensate for nonmarket benefits such as biodiversity conservation.

Chapter 5 addresses the question of whether and under which conditions agroforestry is likely to reduce deforestation. Earlier claims that adopting 1 ha of agroforestry leads to a reduction of 5 (or 10) ha of deforestation are rejected as unfounded. Instead, a thorough economic analysis is offered of the factors that may lead to a decrease or increase in deforestation after agroforestry adoption, considering different scenarios with respect to land and forest availability,

65

labor and capital constraints, characteristics of the agroforestry technology including labor and capital intensity, riskiness and sustainability, markets for agricultural produce and labor, and land tenure. These ideas are illustrated in Chapter 6 with a historical analysis of the contradictory role that one tropical tree crop, cocoa, has played in tropical deforestation and forest conservation over the centuries. On one hand, cocoa is grown in several tropical regions in complex agroforests that are among the most forest-like agricultural ecosystems and that nourish hopes for a profitable tropical agriculture that conserves many of the environmental services of the forest. On the other hand, cocoa has often been grown in an unsustainable way and has acted as a major driver of tropical deforestation. This chapter looks at the factors that determine how cocoa was and is being grown, with a focus on the Côte d'Ivoire, Brazil, and Cameroon, and tries to isolate the lessons that can help agronomists and resource managers promote sustainability in tropical tree crop agriculture.

The concepts and historical experiences laid out in Chapters 5 and 6 show clearly that even land use practices such as agroforestry that could be used in a sustainable, forest-conserving way will not necessarily be so used as long as forest is freely available for further agricultural expansion. A solution to this dilemma could be legal protection of forest areas and its strict enforcement, but this often meets difficulties in tropical countries with weak institutions. Conservation concessions are an emerging concept in conservation economics and may be an ideal complement to agroforestry practices in conservation strategies (Chapter 7). Conservation concessions may involve direct payments to land users or investments in health or educational infrastructure in exchange for conservation set-asides or the adoption of sustainable land use practices such as agroforestry. Using case studies from Colombia and Bahia, Brazil, the chapter shows how this approach can be used to integrate conservation set-asides and biodiversity-friendly agroforestry practices in regional conservation plans.

Chapter 4

The Economic Valuation of Agroforestry's Environmental Services

David Pearce and Susana Mourato

Agroforestry practices vary, but their essential feature is that they are explicitly designed to be multifunctional, that is, to produce multiple products and services. Included in these outputs may be conservation outcomes, such as biodiversity conservation. Each of these products and services has an economic dimension. It is no exaggeration to say that all ecological services have an economic counterpart, that is, there is an economic value attached to them. Eliciting what these values are and how they compare with alternative land use systems, especially potentially destructive systems such as some forms of slash-and-burn agriculture, pasture, or monoculture plantations, is crucial to providing incentives for the expansion of agroforestry.

This chapter outlines the links between economics and biodiversity. It sets out the basic economics of land use choice, of which agroforestry is one option, investigates the economic benefits of agroforestry, and illustrates them in three case studies for northern Nigeria, the Sudan, and Peru. Most attention is paid to the Peru study because it reveals important insights into how agroforestry practices should be appraised.

Economics and Biodiversity

To date economic studies of agroforestry systems have largely neglected the biodiversity gains from agroforestry. Part III of this volume shows clearly that agroforestry improves the biodiversity profile compared with less diversified and land-degrading alternatives. However, there are several reasons why the economic studies lag behind the ecological studies. First, economists have made substantial efforts to place economic values on many of the outputs and services supplied by natural and sustainably managed systems. However, biodiversity typically has been construed to mean wildlife services (e.g., the provision of meat) and broader ecosystem benefits, such as the economic value of genetic material for pharmaceuticals (Pearce and Moran 2002). These have been studied in the context of tropical forests but not for agroforestry. Second, it is useful to distinguish biological resources from biological diversity. The ecological benefits of diversity should include the role it plays in ensuring the resilience of an ecosystem against stress or shocks such as climate change (see Chapter 20, this volume). Valuing biological resources in economic terms is fairly well established in the literature, but valuing broader ecosystem functions such as resilience has not been attempted. Third, until recently there has been only a limited incentive to place economic values on biological diversity. The ultimate purpose in finding economic values is to develop markets that capture the economic values. An example would be payments for watershed protection services provided by upstream forest owners or payments for carbon stored in the biomass. Such markets are rapidly emerging, but markets in biodiversity have tended to lag behind. The earliest markets in biodiversity conservation were debt-for-nature swaps. Such swaps involved the purchase by conservationists of secondary debt owed by an indebted country. The conservationist would then offer to retire the debt or, more usually, have it converted from foreign exchange to domestic currency, in return for an agreement to conserve a biodiverse area. Though popular in the 1980s, such swaps went out of fashion in the 1990s but show some signs of returning again. More recent examples of markets include conservation concessions or payments. These are being championed by bodies such as Conservation International and simply involve payments to landowners or land users to forgo environmentally destructive land uses (Rice 2002; see Chapter 7, this volume).

Clearly, it will be important to improve this deficiency in economic research into agroforestry. It may well be that many forms of agroforestry can pay their way, relative to other land uses, with more readily identified values such as carbon storage. But biodiversity benefits must be valued in economic terms to ensure that agroforestry schemes showing economic returns below those of some less diversified and potentially unsustainable land uses are not sacrificed on the basis of incomplete economic analysis.

The Economics of Land Use Choice

Agroforestry is a managed use of land. If land is not used in this way, it will be managed in other ways: for pasture, slash-and-burn (which can actually be viewed as a form of agroforestry; see Chapter 8, this volume), logging under various management regimes, or outright preservation in which no consumptive uses of the land are permitted. If the net economic returns to agroforestry are less than the net economic returns from these alternative uses of the land, then there will be little incentive to engage in agroforestry. In fact, agroforestry will not be favored if any single alternative use of the land has a higher net economic return. Showing that agroforestry has multiple benefits, then, is a vital first step in arguing the case for agroforestry.

To provide some substance to this apparently simple argument we need to investigate further precisely what is meant by *net economic benefits*. The picture quickly becomes more complicated. First, we need to distinguish between economic returns to the farmer and economic returns to society as a whole. The farmer tends to focus on what the economist would call private economic returns. These are essentially the financial revenues obtained from the given use of the land minus any financial costs. Revenues and costs are not timeless; they occur over specified periods of time. The horizon for these costs and benefits tends to be set by the farmer, but also relevant are any biological factors affecting the horizon. For example, if land use involves the taking of natural tree species or growing of plantation species, the time horizon could be influenced by the period of the rotation or the time until optimal production. The farmer's time horizon may or may not coincide with such biological horizons. Although there are exceptions, in poor societies time horizons tend to be short; people do not look very far ahead (Poulos and Whittington 2000; Cuesta et al. 1997). In the economist's language, poor people are said to discount the future highly. Their focus is on what they can secure this year, next year, and a few years hence. Looking far into the future is a luxury that the poor cannot often afford.

Other factors also determine the degree of discounting. For example, in traditional slash-and-burn systems, the time horizon for a specific piece of land may appear very short and is determined by the period over which crop yields can be sustained from the initial (often poor) capital endowment of fertile soil, plus the nutrient base derived from the initial burn. There may be no concern to look beyond this period, perhaps 5–10 years, if it is known that there is further frontier land that can be colonized. There is evidence that if the frontier is closed (i.e., substitute land is not available), farmers will take more care of existing land and will seek sustainable use of it (Tiffen et al. 1994). Similarly, land for which there are secure property rights is far more likely to be farmed sustainably than land with insecure property rights. In the latter case farmers can easily be dispossessed by more powerful agents, including governments. The rate at which farmers discount the future in this case tends to be very high because the discount rate incorporates the risk of dispossession. We return to this discounting argument shortly. For the moment, we note that the less secure property rights are and the more frontier land is available, the higher the rate of discount is likely to be. High discount rates are consistent with mining the nutrients of the land (i.e., treating land as an exhaustible resource rather than as a sustainable, renewable resource).

The revenues and costs to the farmer must be distinguished from the flow of economic benefits and costs to society as a whole. Because society includes the farmers, their private costs and benefits are subsets of the wider social costs and benefits. Society here need not be constrained to be the sum total of citizens within a national border. To see why, consider the role agroforestry may play in conserving watershed functions. Compared with tree removal, agroforestry tends to prevent forest soils from being washed into rivers. Because sediment both settles and travels in water, it can quickly affect downstream fisheries and other economic activities. Agroforestry therefore may generate what are known as external benefits that accrue to others in society. The benefit is external because the farmer does not receive any payment for this protective function. (As it happens, this institutional feature of protective land use systems is beginning to change as downstream users better appreciate the benefits they are receiving from upstream soil conservation practices.) This is a simple example of how social benefits can be greater than private benefits. The converse is that the social costs of destructive or damaging land use systems are greater than the private costs. But social benefits and costs may not be confined to the nation. If a land use system yields benefits in the form of biodiversity conservation, then those benefits may accrue not just to local people or to the nation but to the world as a whole. That the world values biodiversity is evidenced in the existence of treaties such as the Convention on Biological Diversity. Therefore, social benefits may comprise national social benefits and global social benefits.

In the economic analysis so far, nothing has been said about the ways in which these costs and benefits can be valued nor whether they are associated with actual cash flows. This brings us to the second complication of the simple rate of return argument outlined earlier. To the economist, cash flows do not define a benefit or cost, contrary to widespread misunderstanding of the point. An economic benefit is defined as any gain in human well-being. An economic cost is any loss in well-being. In turn, well-being is defined with reference to human preferences. To say that my well-being is higher in context A than in context B is to say that I prefer A to B. Economic appraisal therefore is preference-based, reflecting the basic democratic value judgment that human preferences should count. It is important to understand that preferences may have many different motivations.

A further major misunderstanding of economics is that preferences are alleged to be based on self-interest, that is, on what the individual prefers for himself or herself. This is incorrect. Preferences may reflect pure self-interest, a concern for what an immediate family group wants, a concern for children, grandchildren, and future generations generally, a concern for the environment in some intrinsic sense, and so on. Although there is a scientific debate within economics about the precise ways in which these values can be aggregated without double counting, the essential point is that an individual's preference can have varying motivations. This turns out to be important, as we shall see.

Thus, social costs and benefits may or may not be associated with a cash flow. The downstream benefits of upstream soil conservation may have a cash flow associated with them, such as the value of a commercial fishery that is protected. But the cash flow does not accrue to the farmer protecting the soil. It therefore remains an externality. The biodiversity benefits accruing to the world as a whole are unlikely to be associated with cash flows. They might accrue via ecological tourism expenditures, for example. But if people in the United States or Europe value biodiversity in Asia, then that is an economic benefit even though no cash flow is involved. The important distinction, then, is between marketed and nonmarketed benefits (and costs).

The formal model for deciding which land use is best from an economic standpoint can be set out as follows:

Social benefits = Private benefits + External benefits
Social costs = Private costs + External costs

External benefits and costs may be national or global (rest of world). All costs and benefits accrue over some period of time so that private benefits in a given period, t, are $B_{P,t}$. Similarly, private costs in period t are $C_{P,t}$. Using the subscripts E for external and P for private, we can write

$$[B_{P,t} + B_{E,t} - C_{P,t} - C_{E,t}] > 0$$

as the condition for any land use to be worthwhile and

$$[B_{AP,t} + B_{AE,t} - C_{AP,t} - C_{AE,t}] > [B_{iP,t} + B_{iE,t} - C_{iP,t} - C_{iE,t}]$$

as the condition for agroforestry (subscript A) to be chosen over any alternative land use system (subscript i). In turn, B_E and C_E can be divided into national and global. The distinction is not pursued here but it is relevant to the issue of how to design incentives so that farmers take account of external benefits and costs in their decisions.

The obvious problem with this formal equation is that it does not incorporate time. With time included, the requirement that the net social returns from agroforestry (the left-hand side of the preceding inequality) be greater than the net social returns from any alternative land use becomes

$$\sum_t \frac{B_{A,t} - C_{A,t}}{(1 + r)^t} > \sum_t \frac{B_{i,t} - C_{i,t}}{(1 + r)^t}.$$

Here t refers to time, B conflates both private and external benefits, and C conflates both private and external costs. The new element is the discount factor:

$$\frac{1}{(1 + r)^t},$$

where r is the discount rate. Although farmers do not use the terminology of discounting, the term accurately describes the mental process of attaching a lower weight to future gains and losses compared with gains and losses today. The discount factor is the weight that the farmer or society assigns to each period, reflecting the greater importance attached to having net benefits now rather than later. A discount rate of 10 percent, say, would be consistent with valuing next year's benefits at 91 percent of this year's benefits, benefits in 10 years at 38 percent of today's benefits, and so on. However, if the discount rate is high—reflecting a strong concern to have benefits now, as would be consistent with behavior in very poor societies—then future net benefits will be given a very low weight. For example, a discount rate of 25 percent would assign a weight of only 11 percent to net benefits in year 10. This is why many poor farmers are not interested in soil conservation measures, which generate long-term rather than short-term benefits (Cuesta et al. 1997; Poulos and Whittington 2000). This is again relevant to agroforestry if the relevant practices hold out promise of long-term rather than short-term benefits.

Finding Total Economic Value

Efficient decisions should be based on what is good for society as a whole. Therefore, it is the social rather than the private benefits and costs that matter, and it is these that should be used to determine the optimal configuration of land uses. Nonetheless, it is easy to see why, even if social benefit analysis produces the result that agroforestry is best, it may not become the chosen land use. Private costs and benefits determine actual land uses. Social appraisals provide a measure of how the land should be used, but if the farmer is in a position of power over that specific land use, then the private costs and benefits dictate the actual use. By *power* here we mean that institutions designed to reflect the wider social concerns may not function in such a way as to alter farmers' decisions. This is especially true where monitoring and policing of land use is weak, as is usually the case where the frontier is large (e.g., Indonesia, Brazil) and where public resources are very limited (almost all low-income developing counties). In such circumstances land use tends to follow open access solutions (i.e., land is not owned by anyone in the sense of property rights being enforced). The same point can be made differently. Farmers ignore the external costs and benefits of their land use because they do not receive any cash or resource flows corresponding to the nonmarket flows of benefits and costs. For example, if farmers were paid to conserve biodiversity or to store carbon in trees rather than release it as carbon dioxide, then they would change their revenue and cost flows to reflect the nonmarket benefits.

This last result is critical. It is one thing to appraise land use options and to declare that one use is socially better than another. It is quite another thing to devise systems of incentives to capture the nonmarket costs and benefits in such a way that they influence private land use decisions. This process of designing incentives is a major focus of concern in environmental economics. There are many examples. Hydroelectric companies may pay upstream forest owners to conserve their forests rather than harvest them for timber, as is done in Costa Rica. Logging would open up the canopies and allow rain to wash away some forest soils, producing sediment in the rivers, which then produces silt in the reservoir, reducing hydroelectric output. The value of the lost electricity is the upper limit of the hydro company's willingness to pay to avoid the damage. In this way, the externality has been valued in monetary terms, and an incentive system (direct payment) has been found to incorporate the external costs and benefits into forest owners' decisions. Other examples of payments include the Watershed Conservation Fund in Ecuador and tax concessions on money invested in biodiversity-friendly projects (as in the Netherlands). For an overview, see Bayon et al. (2000).

Carbon storage has been the subject of several hundred bilateral agreements since 1989. In these agreements, an agent seeks to offset the damage done by its own carbon dioxide emissions by sequestering carbon, or storing carbon that would otherwise be released, in another location. It therefore pays the incremental cost of storing carbon and collects a paper credit to the effect that what it has stored has offset its own emissions. These carbon offset deals are formally sanctioned in the Kyoto Protocol under the arrangements for the Clean Development Mechanism, which enables rich countries to pay poor countries for carbon emission reduction. Again, the externality has been internalized, and it has also been valued because the storage is worth whatever the rich country is willing to pay for it.

Such ecosystem service payment arrangements are important because the context for agroforestry is highly likely to be one in which the net financial returns to farmers from agroforestry are less than the net returns from some unsustainable land use, and the social rate of return from agroforestry is higher than the alternatives. Essentially, there is a mismatch between social and private returns. This can be overcome only if farmers are compensated for forgoing the net financial gains from unsustainable land use, with such compensation coming from resource flows associated with the nonmarket benefits of agroforestry.

At the moment, agreements to pay farmers in agroforestry schemes for generating national and global benefits are rare. But the Global Environment Facility is funding several projects in Nicaragua, Colombia, and Costa Rica in which payments are made for farm practices that adopt silvopastoral systems that benefit biodiversity and carbon sequestration (see also Chapter 19, this volume). How are the external benefits of sustainable land use systems to be estimated? As discussed earlier, we are interested in estimating the total economic value of agroforestry to society, which includes both the private values to farmers and the social external values to other members of society, nationally or internationally. Figure 4.1 illustrates these value components.

As suggested in Figure 4.1, the total economic value of agroforestry systems is a combination of use values (i.e., the benefits to its users) and what are

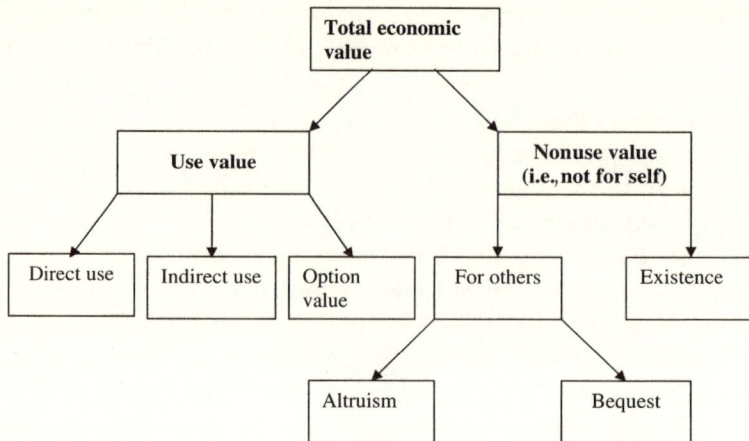

Figure 4.1. Components of the total economic value of agroforestry.

usually known as nonuse values, which are benefits that are unrelated to any form of use. To illustrate the meaning of these concepts for agroforestry, direct use benefits include revenues to farmers from the sale of agricultural and forest products and things such as provision of shade and protection from the wind. Indirect use benefits encompass most of the ecological services provided by forests preserved under agroforestry, such as watershed protection that may indirectly benefit others in society. Option value is the value of preserving biodiversity and forests through agroforestry for possible future benefit such as the potential pharmaceutical use of a species or substance or the development of new crop varieties.

In addition, various groups in society (including farmers) may also derive a benefit from the simple knowledge that ecosystems, forests, and biodiversity are being protected by agroforestry systems when compared with more damaging land uses. Economists call this an existence value. Society may also value the fact that these ecosystems, forests, and biodiversity are being protected for the use of others (altruistic value) or for future generations (bequest value). As noted earlier, these nonuse values are not selfishly motivated and reflect concern for others, for future generations, and for the environment in general.

The question is, How can these values of an agroforestry system be measured in economic terms? It may seem like an impossible task given the multiple dimensions involved. But a number of techniques can be used to place a monetary value on the various components of Figure 4.1, with some techniques being more appropriate to estimate use values and others being particularly suited to measure nonuse benefits. Although estimating the total value is necessarily a complex and time-consuming process, possibly involving a range of techniques, the estimation of subcomponents of that total value is more straightforward.

Typically, use values are easier to estimate than nonuse values. For example, the economic returns from crops, timber, and other forest products (direct use values) or the indirect gains of fishers and hydroelectric companies benefiting from forest conservation (indirect use values) might be simply estimated using market prices or, in a less straightforward but superior way, using shadow prices, which are prices corrected for internal distortions that could cause them to deviate from true economic costs (see the Sudan case study later in this chapter for more details on shadow pricing). If the area where the agroforestry system exists has potential for ecotourism, then the tourist benefits can be assessed indirectly by looking at how much visitors are spending to travel to the site and how much they spend during their visit. This is called the travel cost method. Needless to say, what these alternatives are also depends on the scale of the areas involved. If the analysis is related to a small area, alternative land uses such as ecotourism may be severely limited. This suggests that, to maximize options, fairly large areas must be combined to exploit fully conservation economic values.

Only a few years ago, nonuse values were commonly called intangibles precisely because they are naturally harder to measure. But economists have developed a number of sophisticated techniques to estimate the full economic value of environmental resources, including nonuse values. These techniques are mostly survey based and are generally known as stated preference methods. Of these, contingent valuation (CV) is the most frequently used method (Carson 2004 lists more than 5,000 CV studies covering a wide range of subject areas and more than 100 countries). CV works by asking people directly how much they value a particular environmental change that is described via a specially designed questionnaire. In particular, they might be asked for their willingness to pay (WTP) to secure the environmental improvements arising from agroforestry; alternatively, the question might be how much individuals would be willing to accept (WTA) to incur any losses arising from the adoption of agroforestry (vis-à-vis alternative land uses). These WTP and WTA measures are monetary estimates of the total value people place on the land use change of interest. The reasons why people are prepared to pay for an environmental improvement (or require compensation for a deterioration) are diverse and might include considerations of personal gains in parallel with a preoccupation with the benefits that might befall other members of their family, third parties, or future generations or simply concerns with the environment itself; therefore, WTP and WTA measures reflect both use and nonuse values. To draw a parallel, the contingent valuation method works in a similar way as survey-based market research studies that assess people's preferences and willingness to pay for new market products and services. There is a fundamental twist, however: whereas market research typically deals with private goods most of whose benefits accrue to users (e.g., cars, toothpaste, hamburgers), CV focuses on goods and services that have a value not only to users but to society in general, and part of the value is not related to any form of present or future use.

The next three sections present illustrations of how the methods described here have been used successfully to measure the economic value of agroforestry systems in a variety of contexts and regions. More information about these and other methods can be found in Bateman et al. (2002).

Agroforestry Systems in Practice I: *Acacia senegal* in the Sudan

Acacia senegal grows in the Sahelian-Sudanian zone of Africa. When cut, the tree releases a sticky gum in defense of the wound, and the gum, known as gum arabic, has multiple uses ranging from confectionery, lithography, and beverages to pharmaceuticals and pesticides. It is superior to the gum released by *Acacia seyal,* which produces a gum known as talha. Collecting gum arabic is an important feature of farming systems in the Sudan. It is especially important for small farmers in the "gum belt" of central and western Sudan, where gum gardens are part of a bush fallow rotation with other crops. At any one time, farmers devote some land to crops, some to *A. senegal,* and some to fallow.

A. senegal is remarkable for the multiplicity of benefits it produces. Apart from gum arabic, these include leaves and seed pods for livestock fodder, fuelwood, conservation of soil because of deep tap roots and lateral root systems, nitrogen fixation with consequent effects on grass growth near the trees, microclimatic protection from shelterbelts, and dune fixing. The roots of the tree have been used to line water wells.

Gum arabic systems have been analyzed in a number of studies. Pearce (1988) estimated an internal rate of return[1] of 36 percent for combined gum, fodder, and fuelwood production, which is extremely high for investments in developing countries. Barbier (1992) notes that any analysis cannot be truly representative of gum agroforestry practices because of the sheer diversity of the various combinations that are practiced in the gum belt. In addition, *A. senegal* varies in its production of gum by age and type of soil. Barbier therefore analyzes returns to gum and crop production for several regions (Table 4.1).

The results are shown in two forms: in financial terms (i.e., by looking at revenues and costs in Sudanese pounds) and in shadow price terms (often called economic returns). The idea of shadow pricing is that domestic prices may not represent true economic costs. For example, a worker paid a wage, if not employed in this occupation, may be otherwise unemployed. The opportunity cost of the worker then approaches zero because he or she would not

Table 4.1. Economic returns to mixed *Acacia senegal* and crop production in the Sudan.[a]

Area	A. senegal[b]	Sorghum	Millet	Groundnut	Sesame	Total
Blue Nile	334	589	233	—	—	1,156
White Nile	259	−1,507	−1,088	−231	20	−2,547
North Kordofan	56	—	−117	—	307	246
South Kordofan	81	−586	—	601	1,375	1,471
North Darfur	87	—	−408	−910	—	−1,231
South Darfur	95	−1,180	−404	−726	—	−2,215
All systems average	152	−671	−357	−316	567	−520
Average at shadow prices[c]	2,065	2,412	2,330	9,461	9,468	16,656

Source: Barbier (1992).

[a]Net present value in Sudanese pounds per feddan = 0.4 ha.
[b]Gum fodder and fuelwood.
[c]See text for explanation of shadow prices.

otherwise be producing anything. The shadow wage would be the wage that the worker could get in the next best occupation. If that alternative occupation does not exist, then the shadow wage approaches zero. The idea here is that employing someone has a positive social benefit and that projects with this effect have larger social profits than first appears to be the case. A second illustration relates to foreign trade. If crops are not grown indigenously, they would have to be imported. Similarly, any crop grown and consumed domestically forgoes export revenues of the crop that could have been exported. The relevant shadow price therefore becomes the price that the crop would have gotten if it were internationally traded, its so-called border price.

Barbier's analysis is conducted both for conventional financial costs and revenues and also in terms of shadow prices. Looking at the financial analysis first, we see that, on average, all crops other than sesame actually make financial losses, whereas *A. senegal* makes a profit. Second, mixed farming systems as a whole make profits in only three regions: Blue Nile, North Kordofan, and South Kordofan. In contrast, *A. senegal* is profitable in all regions. Sesame is also seen to be very profitable in the regions where there is evidence of farm income. This raises the issue of why multiple crops, including gum trees, are grown in regions where sesame is profitable. There appear to be two possible answers. First, mixed outputs are a risk aversion strategy: events adversely affecting one crop may not affect others. Second, the crops serve both a market strategy and subsistence strategy. Sorghum and millet are grown primarily for subsistence. Third, there are interdependencies between the crops and gum arabic. These interdependencies reflect the environmental benefits of gum arabic, that is, the returns to cropping would not be what they are but for the external benefits of gum arabic. But they also reflect the fact that gum arabic from *A. senegal* can be harvested at different times of the year than crops, so that there is no competition for labor time. Fourth, the data in Table 4.1 are best thought of as snapshots of actual situations. In practice, because some crops are more profitable than others, there will indeed be a process of substitution away from less profitable crops.

Once the shadow pricing exercise is conducted, all systems become socially profitable. This results almost entirely from the effect of pricing outputs at border prices because these are well above domestic prices. Whereas the financial returns to gum arabic exceed those of all crops other than sesame, the economic (shadow priced) returns to gum arabic are the lowest. The reason for this is that the financial returns already reflect border prices because almost all gum arabic is exported (of course, this does not apply to the fuelwood and fodder components of *A. senegal* benefits), whereas the other crops generally are not exported.

Barbier was unable to isolate the wider environmental benefits of *A. senegal*, so that the figures shown in Table 4.1 may well understate the true social returns to *A. senegal* agroforestry. A subtle issue is how far these additional benefits extend. For example, if they result in increased crop productivity, then they are already reflected in the crop output benefits; they have been internalized. To count them again would be double counting. Therefore, it could be that many of the environmental benefits have already been accounted for. Three central roles of gum arabic are thus suggested: it acts as a reasonably secure and stable form of income, it generates positive net financial and economic benefits, and it interacts as a support system for other crops. Barbier's major caution is that gum arabic makes financial sense only if the international price of gum arabic is sustained. This is a complex issue because there are available substitutes from synthetic starches, and the political instability in the Sudan threatens supply stability.

Agroforestry Systems in Practice II: Mixed Crops and Trees in Kano, Nigeria

A second example of a detailed cost-benefit assessment of agroforestry is provided by Anderson (1987, 1989) for rural afforestation schemes in the Kano region of northern Nigeria. Anderson notes that trees on African farmlands often have not been given any priority, with the result that tree loss has exacerbated soil erosion. The background to this may lie in land tenure problems: tree growing assumes some form of longer-term tenure, low priority is generally given to the agricultural sector in macroeconomic policies in Africa, the wider benefits of tree growing are not perceived, and there is the classic externality issue introduced before. Anderson simulated two forms of tree growing: trees as shelterbelts on the edge of farmland and farm forestry (i.e., the mix of crops and trees), both of which qualify as agroforestry. Simply looking at timber and crop yields fails to capture the wide-ranging benefits of agroforestry.

Table 4.2. Net benefits of agroforestry in northern Nigeria.

Type of System	Shelterbelt Benefit-to-Cost Ratio	Farm Forestry Benefit-to-Cost Ratio
Base case	2.2	4.5
Low yield, high cost[a]	1.7	2.3
High yield	2.6	—
No erosion	1.8	2.9
More rapid erosion	1.8	2.5
Soil restored to initial condition plus yield jump[b]	2.9	6.1
Wood benefits only	0.3	0.6

Source: Anderson (1987).
[a]Sorghum, millet, cowpeas, and groundnuts.
[b]Assumes favorable ecological conditions.

The benefits Anderson lists are reductions in crop loss through avoided reductions in soil fertility, increases in crop yield caused by better moisture retention and nutrient cycling, increases in livestock productivity caused by the availability of dry season fodder from trees, and the value of the tree products themselves as fuelwood, poles, and fruits.

A large part of the analysis involves the physical estimation of response functions, that is, how crop yields and fodder supply respond to tree planting. In terms of economic valuation, the procedures are simple because in all cases market prices are available to value the changes in productivity and output. Table 4.2 shows the resulting benefit-cost ratios (i.e., the ratio of the present value of benefits to the present value of costs). This is somewhat more meaningful than citing absolute figures for net present values.

The base case allows the tree-growing project to generate benefits against the backdrop of an assumed 1 percent per annum decrease in soil fertility. The "no erosion" and "wood benefits only" cases assume no erosion, whereas the low- and high-yield cases also assume 1 percent erosion, as does the "soil restored" case. In contrast, the rapid erosion case assumes a 2 percent per annum decline in soil fertility.

For a project to be prima facie worthwhile, the benefit-cost ratio should exceed 1. If projects were looked at in the traditional way (i.e., in terms of the timber benefits only), then the project would fail a benefit-cost test (ratios are 0.3 and 0.6, so costs exceed benefits). But once the additional benefits are included, the ratios quickly rise above 1. Moreover, for farm forestry the ratios are high, systematically above 2 and with the potential for 4–6. These are very high rates of return. The only apparent anomaly is the fact that if erosion rates are higher, the rate of return to the planting program is lower. Anderson argues that this is because the benefits of improving seriously degraded soil are less than the benefits of improving less degraded soil. The implication for policy is a familiar but often neglected one: the temptation is to invest in areas where soil is most seriously degraded, but it is often better first to protect soil that is marginally degraded.

No follow-up assessment to Anderson's cost-benefit appraisal is yet available, so it is uncertain whether such projects have succeeded. Anderson is clear that success is contingent on high participation rates among farmers, raising another important issue. Unless farmers cooperate in agroforestry schemes, they are likely to fail. In turn this means that such schemes have a game theory context: each farmer must be assured that the other farmers will cooperate. For any one farmer there is an incentive not to participate, thus avoiding the costs of tree planting while securing the benefit of tree planting if all others undertake the project. This produces a classic free-riding potential to such schemes and underlines the importance of community agreement and participation in such schemes, with well-designed incentives to stay in such schemes and punishments for defecting from them.

It is also worth noting that Anderson's study excludes credit for biodiversity and carbon storage. As noted earlier, economic studies involving the estimation of biodiversity benefits are not available. Zelek and Shively (2002) secured estimates of the price of carbon for agroforestry systems in Bukidnon, Northern Mindanao, Philippines. The carbon price is the price that would have to be paid to farmers to compensate them for switching out of traditional crops (maize and vegetables) to agroforestry schemes. If the land would otherwise be fallow, an agroforestry scheme takes $6–$10 per hectare compensation for the carbon stored in the agroforestry system. However, if the alternative is maize growing, then the necessary compensation is $58–$61 per hectare, and for forgoing vegetables it is $211–$283 per hectare. Although the compensation for sacrificing vegetable growing appears high, the estimates are in present value terms, that is, discounted annual returns that are then summed. The equivalent annual payments would be modest, suggesting that paying farmers for carbon sequestration would be inexpensive and sufficient to switch them into agroforestry.

Estimating the Benefits to a Hypothetical Land Use Change to Agroforestry in Peru

Slash-and-burn agriculture by small-scale farmers is estimated to account for about one-third of the deforestation in tropical America (Houghton et al. 1991; see also Chapter 8, this volume). In a slash-and-burn system, farmers typically clear the land for agriculture and plant crops during 1 or 2 years, after which the land is left fallow for varying periods while another part of the farm is cleared for agriculture. As a result, within a few decades after slash-and-burn colonists move into an area, often only small areas of primary forest are left.

Mourato and Smith (2002) estimated the environmental benefits and financial costs of adopting agroforestry systems for slash-and-burn farmers in the Peruvian Amazon.

Satellite images from 1993 show that 70 percent of the study area in the Peruvian Amazon (the District of Campo Verde) was deforested (Fujisaka 1996). Multistrata agroforestry combines crops with useful tree species that mimic natural succession (Nair 1993). Thus, by reducing deforestation and encouraging tree planting, agroforestry systems could generate significant benefits locally and even globally: timber and nontimber forest products; increases in soil fertility, watershed protection, and local climate regulation; provision of carbon sequestration services and biodiversity protection; and existence and bequest values.

Data from the Peruvian Amazon show that under controlled experimental conditions, slash-and-burn agriculture gives a net present value of –$2,176 per hectare over a time horizon of 10 years (Labarta 1996). By contrast, for the same time horizon, multistrata agroforestry gives a net present value of $1,137 per hectare at a discount rate of 25 percent, considered to be appropriate for smallholders (Cuesta et al. 1997). However, only about a fifth of the local farmers practice agroforestry in a small area (0.5 ha on average) of their farms (around 30 ha on average). Farmers pointed out that this was because only returns in the first few years were relevant for their decision making. Adjusting the data to a time horizon of 2 years and lowering yields for annual and semiperennial crops in agroforestry to reflect agronomic conditions in farmers' fields show that agroforestry gives lower returns than slash-and-burn (Table 4.3). In addition to these financial considerations, other studies have shown that seasonal labor constraints, market risk, lack of technical knowledge about trees, and abundance of natural forest products also impede the adoption of agroforestry (Current et al. 1985).

These results show that farmers are unlikely to change environmentally damaging agricultural practices without external incentives. In this context, Mourato and Smith (2002) used the contingent valuation method to elicit the compensation required by Peruvian farmers to switch from slash-and-burn agriculture to agroforestry. A questionnaire was administered to 214 farmers, presenting them with a possible future project in which utility companies in developed countries, driven by the possibility of emission reduction legislation, were willing to compensate farmers who adopted multistrata agroforestry systems. A fixed annual payment would be made for each hectare of agroforestry (payments would cease if the agroforestry area was cleared for slash-and-burn). Farmers were then asked for their minimum annual WTA compensation to convert 1 ha of land from slash-and-burn agriculture to multistrata agroforestry, taking into consideration the potential financial impacts of the proposed land use change in terms of investment, labor, yields, and available products. Results show that the average compensation requested

Table 4.3. Returns to slash-and-burn agriculture and multistrata agroforestry during the first 2 years (in US$ per hectare).

Production System	Year 1	Year 2
SLASH-AND-BURN:		
Value of output:[a]		
Rice	366	0
Cassava	0	300
Plantain	0	862
Labor[b]	663	459
Gross margin	−297	703
MULTISTRATA AGROFORESTRY:		
Value of output:[a]		
Rice	366	0
Cassava	0	300
Plantain	0	862
Labor[c]	683	527
Planting material[d]	200	0
Gross margin	−517	635
Difference in gross margins		
(agroforestry − slash-and-burn)	−220	−68

Source: Adapted from Labarta (1996).

[a]Yields: rice = 1.7 ton ha^{-1}, cassava = 8.7 ton ha^{-1}, plantain = 10 ton ha^{-1}.

[b]Deforestation: 60 labor days/ha^{-1}; other: 135 labor days/ha^{-1}.

[c]Additional planting and weeding: first year = 11 labor days/ha^{-1}; second year = 20 labor days/ha^{-1}.

[d]Fruit trees = 50%; timber species = 50%.

to convert agricultural land to agroforestry is $138 per hectare per year. Reassuringly, this result is very close to the average difference in returns between slash-and-burn and agroforestry in the first 2 years ($144), depicted in Table 4.3, indicating that the compensation amounts estimated with the CV method reflect expected financial losses.

The compensation amount estimated here took into account only the financial losses experienced by farmers from the adoption of agroforestry. However, the survey showed that Peruvian farmers also benefit from the environmental services provided by forests preserved or planted under agroforestry. Almost all respondents (96 percent) claimed it was important to preserve forests. Forests were seen primarily as a source of construction materials and nontimber forest products (mainly game animals, medicinal plants, and firewood), but despite the dominance of these consumptive uses, nonconsumptive forest services such as air purification, shade, wind shelter, and water quality were frequently mentioned as well. And, remarkably, farmers also advocated bequest values and option values (in the form of biodiversity preservation for potential future commercial value) as reasons to promote forest conservation. To measure the value of these nonmarket environmental forest services, Peruvian farmers were asked to state the yearly compensation amount that they would require to adopt agroforestry practices, this time specifically taking into account the perceived environmental benefits associated with the proposed land use change. The average compensation was found to be only $97 per year and per hectare. In other words, the implicit value of forest environmental services was estimated to be about $41, expressed in terms of a forgone compensation (i.e., the difference between the two compensation amounts, or $138 − $97). This striking result shows not only that Peruvian farmers derive positive benefits from environmental services associated with forest conservation via agroforestry practices but also that the value attributed to these environmental externalities is surprisingly high, at about 30 percent of the total compensation required to change land use.

Mourato and Smith (2002) also investigated whether forest carbon markets may enable slash-and-burn farmers to capture some of the positive global externalities of agroforestry while encouraging land use change. The cost of carbon for the study area, based on farmers' required compensation, was calculated to range from $8 to $31 per ton of carbon, depending on the discount rate used. These estimates are toward the higher end of cost estimates from comparable forestry-based carbon sequestration projects in developing countries: in an analysis of eight developing country carbon sequestration forestry projects, the average cost per ton of carbon was $12, ranging from $3 to $35 (Ridley 1998). However, previous cost estimates made no adjustments for the typically shorter duration of projects relative to the residency time of carbon in the atmosphere. Although some very low-cost projects clearly exist in the energy sector, the carbon costs estimated for agroforestry conversion in the Peruvian Amazon are much lower than the average cost of reducing carbon emissions through fuel-switching projects in nine countries, estimated at an average cost of $165 per ton of carbon (Ridley 1998). This implies that there might be scope for gains from trade, even given the limitations and uncertainties associated with these estimates (such as lack of consideration of transaction and implementation costs, difficulty in proving additionality, and quality of carbon data used in the calculations).

Conclusions

The case studies discussed in this chapter suggest important features of agroforestry land uses when viewed from an economic standpoint. First, agroforestry systems provide a multiplicity of benefits to farmers and local and national populations. These include not only yields from crops, timber production, and collection of other forest products but also ecological services such as watershed protection, soil improvement, and protection from the wind. Some benefits may also extend across borders, for example, in the case of biodiversity protection and carbon sequestration. Many of these benefits are not selfishly motivated and are unrelated to any particular use of the resources; they reflect concerns for the environment and a desire to preserve environmental and forest resources for others and for future generations. To date, economic studies have not embraced attempts to value the biodiversity gains from agroforestry. This reflects partly the difficulties of using economic valuation techniques for valuing biodiversity, but also, we argue, it reflects the lack of an incentive to value biodiversity because the potential for capturing its value through market creation has, until recently, been small.

Second, economists have developed methods capable of estimating in monetary terms the wide range of benefits associated with agroforestry land uses, including nonuse values. These range from more straightforward procedures such as using market prices to calculate financial flows from agroforestry systems to more sophisticated survey techniques (such as contingent valuation) that are able to measure nonuse values. Application of these techniques tends to confirm the intuition that once nonmarket benefits are accounted for, the economics of agroforestry can be transformed (see also Tomich et al. 1998). The additional stage that is needed for actual policy aimed at encouraging agroforestry involves the design of incentives and institutions for the capture of these nonmarket benefits.

Third, farmers are unlikely to preserve primary forest on their farms, even when it is part of a larger land use strategy that includes a shift to multistrata agroforestry as a way of providing agricultural and forest products. The initial costs of establishing agroforestry systems are perceived to be excessively high, and farmers typically have short time horizons, so that the long-term benefits of agroforestry are heavily discounted. This suggests that the development of less intensive alternatives to multistrata agroforestry and improved systems, which build on farmers' current practices, may have a higher likelihood of adoption. For example, Smith et al. (1999) show that farmers in the Peruvian Amazon tend to diversify their agricultural systems with small areas of perennial crops and to regenerate significant areas of secondary forests to recuperate degraded areas as part of their slash-and-burn practices. Enrichment of these secondary forests may provide many of the economic and environmental benefits of agroforestry systems while involving much lower investment costs.

Fourth, the importance farmers attach to the environmental externalities from agroforestry systems and nonuse values indicates that the common perception that smallholders are interested only in short-term survival may be misplaced and that intergenerational issues and environment-related factors should be given greater emphasis in the design of improved land use systems. The implication is that land uses such as agroforestry that encompass forest protection and regeneration may benefit not only the world community but also slash-and-burn farmers.

Finally, the potential for farmers to be compensated through carbon trading for the losses they incur by switching to agroforestry land uses should be explored further. Carbon trading could increase the likelihood of adoption of more sustainable land use systems and is one of several environmental improvement strategies that could be pursued without subsidies or regulations while benefiting resource-poor forest dwellers.

Endnote

1. The internal rate of return is found by setting the present value of costs equal to the present value of benefits and solving for the discount rate. The resulting discount rate is known as the internal rate of return and must be compared with some cutoff rate. Typical cutoff rates in developing countries are 8–15 percent.

References

Anderson, D. 1987. *The economics of afforestation: a case study in Africa.* Baltimore, MD: Johns Hopkins University Press.

Anderson, D. 1989. Economic aspects of afforestation and soil conservation. Pages 172–184 in G. Schramm and J. Warford (eds.), *Environmental management and economic development.* Baltimore, MD: Johns Hopkins University Press.

Barbier, E. 1992. Rehabilitating gum arabic systems in Sudan: economic and environmental implications. *Environmental and Resource Economics* 2:341–352.

Bateman, I., R. T. Carson, B. Day, M. Hanemann, N. Hanley, T. Hett, M. Jones-Lee, G. Loomes, S. Mourato, E. Ozdemiroglu, D. W. Pearce, R. Sugden, and J. Swanson. 2002. *Economic valuation with stated preference techniques: a manual.* Cheltenham, UK: Edward Elgar.

Bayon, R., J. S. Lovink, and W. Veening. 2000. *Financing biodiversity conservation.* Washington, DC: Inter-American Development Bank.

Carson, R. 2004. *Contingent valuation: a comprehensive bibliography and history.* Cheltenham, UK: Edward Elgar.

Cuesta, M., G. Carlson, and E. Lutz. 1997. *An empirical assessment of farmers' discount rates in Costa Rica and its implications for soil conservation.* Washington, DC: World Bank, Environment Department (mimeo).

Current, D., E. Lutz, and S. Scherr (eds.). 1985. *Costs, benefits and farmer adoption of agroforestry. Project experience in Central America and the Caribbean.* Washington, DC: World Bank Environment Paper No. 14.

Fujisaka, S. 1996. *La dinamica del uso y manejo de tierras Amazonicas: Brazil y Peru.* Paper presented at the Mesa Redonda sobre las Modalidades de Aplicacion de Concepto de Desarrollo Sostenible en la Region Amazonica, Instituto Italo-Latino Americano, Rome.

Houghton, R., D. Lefkowitz, and D. Skole. 1991. Changes in the landscape of Latin America between 1850 and 1985: I. progressive loss of forests. *Forest Ecology and Management* 38:143–172.

Labarta, R. 1996. *Analisis economico-financiero de 5 sistemas experimentales de uso de la tierra en Yurimaguas.* Yurimaguas, Peru: International Center for Research in Agroforestry.

Mourato, S., and J. Smith. 2002. Can carbon trading reduce deforestation by slash-and-burn farmers? Evidence from the Peruvian Amazon. Pages 358–376 in D. W. Pearce, C. Pearce, and C. Palmer (eds.), *Valuing the environment in developing countries: case studies.* Cheltenham, UK: Edward Elgar.

Nair, P. K. R. 1993. *An introduction to agroforestry.* Dordrecht, the Netherlands: Kluwer.

Pearce, D. W. 1988. Natural resource management and anti-desertification policy in the Sahel-Sudan zone: a case study of gum arabic. *Geojournal* 17:349–355.

Pearce, D. W., and D. Moran. 2002. *The economic value of biological diversity.* Paris: Organization for Economic Cooperation and Development.

Poulos, C., and D. Whittington. 2000. Time preferences for life-saving programs: evidence from six less developed countries. *Environmental Science and Technology* 34:1445–1455.

Rice, R. 2002. *Conservation concessions: our experience to date.* Paper presented to the Annual Meeting of the Society for Conservation Biology, Canterbury, UK, July 2002.

Ridley, M. A. 1998. *Lowering the cost of emission reduction: joint implementation in the framework Convention on Climate Change.* Dordrecht, the Netherlands: Kluwer.

Smith, J., P. van de Kop, K. Reategui, I. Lombardi, C. Sabogal, and A. Diaz. 1999. Dynamics of secondary forests in slash-and-burn farming: interactions among land use types in the Peruvian Amazon. *Agriculture, Ecosystems and Environment* 76:85–98.

Tiffen, M., M. Mortimore, and F. Gichuki. 1994. *More people, less erosion. Environmental recovery in Kenya.* New York: Wiley.

Tomich, T., M. van Noordwijk, S. Vosti, and J. Witcover. 1998. Agricultural development with rainforest conservation: methods for seeking best bet alternatives to slash and burn, with applications to Brazil and Indonesia. *Agricultural Economics* 19:159–174.

Zelek, C., and G. Shively. 2002. *The economics of carbon sequestration in low-income agriculture.* Lafayette, IN: Department of Agricultural Economics, Purdue University (mimeo).

Chapter 5

Is Agroforestry Likely to Reduce Deforestation?

Arild Angelsen and David Kaimowitz

Is agroforestry likely to reduce deforestation? Most agroforesters for the past 15 years have said, "Yes," with some adding, "By about 5 ha of reduced deforestation per hectare of agroforestry adopted." We argue that the answer is "It depends," and in many cases it is likely to be "No." Our aim is to discuss key factors that condition the agroforestry-deforestation link. That is, what makes a "Yes" more likely than a "No"? Our reformulated question is, "Which types of agroforestry under what conditions are likely to reduce the conversion of natural forest?"

The key argument behind the promotion of agroforestry as a strategy to reduce deforestation is that it is a more sustainable land use solution than the alternatives (e.g., unsustainable shifting cultivation practices or extensive cattle pasture). "People deforest in the search for new lands to replace nutrient-depleted land. . . . Profitable agroforestry systems, through increasing returns to land in existing agricultural areas, may deflect deforestation on the remaining patches of primary forest" (Sanchez et al. 2001, 342–343). One might label this argument the land degradation–deforestation hypothesis: land-degrading agricultural practices force farmers to clear new forestland to sustain a living.

However, introducing agroforestry practices has contradictory effects on farmers' incentives and opportunities to convert more natural forest to agriculture or agroforestry. Farmers could make forest conversion more profitable by using agroforestry, which in turn could give them an incentive for further forest encroachment. Better profitability can also attract new migrants, further multiplying the effects. Higher output increases farm surplus and relaxes capital constraints, which may enable farmers to put additional resources into forest clearing.

This chapter provides a systematic review of arguments for and against agroforestry as a means to reduce deforestation. We start off by reviewing the agroforestry-deforestation debate. Based on a study from Peru, originally published in Sanchez and Benites (1987), a common claim is that 1 ha of new agroforestry systems saves 5 ha of forest. We demonstrate how this single case study has been misused and extrapolated to global predictions about how agroforestry will reduce tropical deforestation. The reasoning ignores the existence of market repercussions and the fact that farmers are rational and will take advantage of new opportunities. We then outline three typical situations of agroforestry adoption and describe how the deforestation impact is likely to vary between them. Then we discuss the subsistence logic underlying the land degradation–deforestation hypothesis and question why farmers should *not* expand their agricultural land into forests if a new and profitable agroforestry technology becomes available. In the following sections we consider in more detail three broad sets of conditioning factors that shape the agroforestry-deforestation link: the characteristics of the farmer, technology, and market and land tenure conditions. We use several examples to illustrate our arguments.

This chapter draws on earlier work on how new agricultural technologies affect the rate of tropical deforestation (Angelsen and Kaimowitz 2001). Viewing agroforestry as one type of technological change, we can apply the developed framework and examine some of the case studies to answer the question raised in the title.

The reader should note that we focus on the impact of agroforestry on the conversion of natural forests to agroforestry and thus on off-site biodiversity. This chapter does not discuss the question of on-site biodiversity, which is covered by several other chapters in this book (see Part III). However, a key issue is possible trade-offs between on-site and off-site biodiversity, which is only superficially dealt with in this chapter. Furthermore, we focus on deforestation and not on the use of forest products and the impact of agroforestry on forest product dependence (see Murniati et al. 2001 for a case from Sumatra).

Agroforestry Research and the Deforestation Debate

In a review of agroforestry research and debate over the past couple of decades, a few observations relevant to the topic of this chapter stand out. First, as Mercer and Miller (1998) note, "biophysical studies continue to dominate agroforestry research while other important areas have not received the attention they deserve." (177) They found that only 22 percent of the articles published in *Agroforestry Systems* between 1982 and 1996 dealt mainly with socioeconomic issues. Among these articles, quantitative economic studies constitute more than half, with cost-benefit analysis being the most popular method. Nair (1998) reported that the share of socioeconomic research articles published in four leading agroforestry journals (1991–1996) was much lower: only 10 percent. Second, the focus has been almost exclusively on small-scale forestry (i.e., plot, field, or farm level). Nair (1998) found that less than 5 percent of the articles in his sample were based on global, regional, or watershed scales. Issues that have to be assessed on larger scales have been neglected. Given these findings, it is not surprising that very little research has been conducted to directly assess the impact of agroforestry on deforestation.

A third and interesting observation is how the claim that 1 ha of agroforestry saves 5 to 10 ha of forest has become almost an established truth in the agroforestry community. It is therefore fascinating to track the origin of this 1:5 ratio and how it has been absorbed in the agroforestry literature. This figure appears to be based on a single study from Yurimaguas, Peru, originally reported in the prestigious journal *Science* by Sanchez and Benites (1987). The article does not discuss the impact of deforestation directly but states that "to produce the grain yields reported in Table 1, a shifting cultivator would need to clear about 14 ha in 3 years, in comparison to clearing 1 ha once, by means of the low input system." (1526) The low-input system described in the article was not agroforestry but rather an improved cropping system with selected varieties of rice, cowpea, and chemical weed control.

In subsequent articles, Sanchez and colleagues applied their results more directly to the deforestation and global warming debates. Sanchez (1990) and Sanchez et al. (1990) claim that "for every hectare put into these sustainable soil management technologies by farmers, five to ten hectares per year of tropical rainforests will be saved from the shifting cultivator's axe, because of their higher productivity." (378; 218) Brady et al. (1993), presenting the Alternatives to Slash and Burn (ASB) Programme, make a similar extrapolation: "Research has suggested that for every hectare converted to sustainable soil management technologies, 5 to 10 ha/year of tropical rain forest will be saved from unsustainable slash and burn." (5)

The link from higher yields to reduced deforestation is explicit in these articles, but none of the articles states explicitly that rainforest will be saved by agroforestry but rather by "sustainable soil management technologies." Both of the Sanchez articles present tables that list the hectares saved from deforestation for various management options. But no figure is presented for the number of hectares saved from deforestation by agroforestry systems; it is "not determined." Although no actual yield or revenue figure for agroforestry was available from the Peruvian study, agroforestry is presented as one of the principal sustainable management options and alternatives to slash-and-burn, along with paddy rice production on alluvial soils, low-input cropping, continuous cultivation, and legume-based pastures (Sanchez et al. 1990).

This distinction between the actual production systems studied in the Peruvian case and agroforestry disappears in articles by other authors on agroforestry and global warming in the 1990s. Schroeder et al. (1993) discuss the potential of agroforestry to reduce atmospheric CO_2, including reducing the level of forest loss. They note that "direct evidence of this potential is limited, but one research study indicated that a low input agroforestry system, involving the rotation of acid-tolerant crops, produced agricultural products on a single hectare equivalent to the volume normally produced on 5 to 10 ha under slash-and-burn agriculture," (53) citing Sanchez and Benites (1987). Later in the article they use an average, 7.5 ha (now citing Sanchez 1990), to conclude that by establishing 1 million ha per year of agroforestry, 7.5 million ha of forest would be saved.

Dixon (1995) assesses agroforestry in relation to greenhouse gases and uses Sanchez and Benites (1987) to calculate reduced carbon emissions from agroforestry: "If it is assumed that in low-latitude nations the establishment of one hectare agroforestry can provide products that would otherwise require 5 ha of deforestation . . . and that 2×10^6 ha of new agroforestry systems are established annually . . . then the amount of low latitude forest potentially conserved ranges up to 10×10^6 ha (108)." Dixon et al. (1993) make a similar unqualified claim: "Sanchez and Benites estimated that 1 ha of agroforestry could offset 5–10 ha of deforestation." (164) Finally, in a classic text on agroforestry, Young (1997) uses Dixon (1995) and the 1:5 ratio, although he is more modest about the potential of agroforestry: "An optimistic but plausible establishment of 1 Mha of new agroforestry systems annually could therefore potentially reduce the need to clear 5 Mha of forest a year, compared with actual current clearance of 15 Mha." (257)

In short, this is the story about how a study on yield differences (not impact on deforestation), comparing shifting cultivation and different low-input production systems (not agroforestry) in one particular location in Peru, has been used to make global predictions about the impact of agroforestry on deforestation.

It raises a number of questions. How representative is the Peruvian case study for the tropics? Are farmers really giving up forest-clearing activities when adopting agroforestry or other agricultural systems, such that their income is kept constant? How will widespread adoption of agroforestry affect markets, and what effects will this have on deforestation? Will successful agroforestry adoption attract more migrants, and will this increase the pressure on forests? These questions are central to this chapter.

Typical Cases of Agroforestry Adoption

Technological progress simply means getting more physical output for the same amount of inputs (or the same output with less inputs). Taking a dynamic perspective, diversification of a cropping system so that total output from different products remains more stable over time (reduced downside risk) and measures that increase the longevity (sustainability) of a system therefore also qualify as technological progress. Based on this definition, introducing agroforestry practices in most cases would qualify as technological progress. Technological progress often is linked to agricultural intensification, defined as more inputs (or output) per hectare, but the concepts are different. Technological progress may or may not imply intensification, and intensification can take place (and often does) without any change in the underlying technology.

Agroforestry may be introduced at various stages in agricultural development. In Table 5.1 we have singled out three typical cases of agroforestry adoption. Because the cases differ, they will help structure the discussion about the likely impact of agroforestry on deforestation.

The first case is when tree crops are introduced into shifting cultivation systems, typically fruit or multipurpose trees. They tend to increase the labor inputs per hectare and are more labor intensive than shifting cultivation or pastures but not more than continuous cropping of annuals. The main driver behind the development is population growth (higher land scarcity and the need to provide more food from a limited land area), and agroforestry is a low-cost intensification in response to the need to supplement subsistence food production (e.g., fruits or protein banks as fodder supplement for cattle) and products traditionally collected from the forests.

The context for the second case is in many respects similar to the first one in that land is abundant and the forest frontier is still open. The principal difference is that the driver behind the adoption of agroforestry is the desire to produce commercial tree crops for an outside market. The trigger can either be new market outlets (e.g., a new road) or a new technology or production system being introduced by government extension agencies, commercial companies, or entrepreneurial individuals. Commercial tree crops can be introduced and modify existing systems, such as rubber agroforestry in Indonesia, which developed from the introduction of rubber trees into the traditional shifting cultivation system (see Chapter 10, this volume). Eventually, commercial tree crops might become so dominant that the system cannot be classified as agroforestry anymore. Sunderlin et al. (2001) found evidence of such dominance in the case of rubber, cocoa, and coffee in Indonesia. An extensive survey of more than 1,000 households in the outer islands of Indonesia suggested that among those clearing forests, almost one-third did so for sedentary agriculture of mainly tree crops (more than half chose rubber), another third for 1–2 years of annual food crops only, and the remaining combined the two.

The third case presents a different situation in which scarcities of land and forest products are major driving factors for implementing agroforestry on farmland to provide forest products. The demand for these forest products typically is from local or regional markets, not international ones (unlike the second case, in which markets can be national or international). In this

Table 5.1. Three typical cases of agroforestry adoption.

Case	Land Availability	Driving Factor	Typical Examples	Possible Impact on Deforestation
Introducing agroforestry in shifting cultivation systems	Land abundance	Population, knowledge of new technologies	Fruit trees for household consumption, cattle fodder	Probably small effects (limited market opportunities) or a reduction of deforestation
Introducing agroforestry with commercial tree crops	Land abundance	Markets (national and international), technology	Cocoa, rubber	Potentially significant increase in deforestation, particularly if market outlets and migration exist
Using agroforestry on farmland to substitute for forest-derived products	Land scarcity	Markets (local and regional), forest scarcity	Fuelwood, building materials, woodlots	No direct effects because natural forests are scarce; possible positive indirect effects (output, labor markets)

scenario, because most of the forest has disappeared, the effect of agroforestry adoption on deforestation is indirect through output and labor market effects, as discussed later in this chapter.

The Economic Logic

The key arguments in what we have called the land degradation–deforestation hypothesis (Angelsen and Kaimowitz 2001) are that land-degrading and pro-ductivity-reducing agricultural practices force farmers to clear new forestland to make a living, that agroforestry is a sustainable practice and allows farmers to generate more food and income over time from the same amount of land than previously, and that agroforestry therefore reduces the need to convert natural forests to agriculture.

The counter-hypothesis is that if a new farming practice or technology is more profitable than previous land uses, farmers may expand their farmland to make more money. Consider a farmer who wants to maximize the surplus (net income) from the farm. He has access to as much labor as he wants in the local labor market, can freely borrow money from the local bank to finance any investments, and can sell as much of the produce as he wants at a fixed price. He can increase his agricultural area by clearing forest, which is either open access or part of his farm. A new technology becomes available or known to him, and he adopts it because it will increase his income. Will he also clear more forest? Certainly!

Here we made at least four key assumptions, and we will examine them briefly. Each of them can modify or even reverse the conclusion.

First, expansion might not be an option. For example, the remaining land could be protected by land use regulations (e.g., a park or wildlife reserve), inaccessible or unsuitable for crop production, or already occupied by agricul-ture or other land uses. This is the typical situation of a closed frontier, as in the third case in Table 5.1. Obviously, from a research viewpoint the interest-ing case to study is when expansion is a real option. But from a practical and policy viewpoint, the distinction between open and closed frontiers is crucial, with important policy implications.

Second, farmers might have "full belly" preferences, that is, they aim for a specific subsistence target and, having reached that, they prefer leisure or social activities. The assumption that farmers lose all interest in increasing their income and consumption once a subsistence target has been reached seems quite unrealistic, although evidence of such cases can be found. However, we argue that "full belly" preferences are not as common as often assumed in many deforestation analyses and in development and conservation interven-tions. Indeed, basing policies and projects on this assumption is risky (see Angelsen and Luckert in press).

Third, labor and capital constraints prevent farmers from enlarging their

Table 5.2. Conditioning factors in the link from agroforestry (AF) to deforestation.

Variable	Impact on Deforestation	Relevance to AF
FARMER CHARACTERISTICS		
Labor and capital constraints	Reduces the opportunity for expansion or the chances for reduced deforestation if technology is labor or capital intensive.	Important because AF often is labor intensive but might also constrain adoption in the first place.
Poverty (income)	Can limit capacity but increases incentives for expansion. Might favor labor-intensive rather than capital-intensive technologies.	AF often more labor than capital intensive and therefore suitable for poor farmers.
Farmer managing several subsystems on the same farm	Offer higher flexibility and fewer labor constraints for particular systems.	Makes labor constraints in AF less serious.
TYPE OF TECHNOLOGY		
Labor intensity	Labor-intensive technologies are likely to result in less deforestation as farmers hit labor constraints or local wages rise.	AF typically is labor intensive, thus more likely to reduce deforestation than other types of technological change.
Capital intensity	Similar to above.	AF is generally less capital intensive and has a less decisive role than above.
Risk	Risk-reducing practices reduce the need to cultivate excessive land as insurance.	AF generally reduces both yield and market risk, with favorable effects on deforestation.
Sustainability	High degree of sustainability reduces the need to abandon land and clear new forests.	AF more sustainable than most alternative systems.
MARKET AND TENURE CONDITIONS		
Output market	Inelastic demand (small local markets) makes supply increases lower prices, which constrain expansion.	Tree crops often sold in international markets, making this price-dampening effect small and deforestation more likely.
Labor market	Small, isolated local labor markets make labor shortage or higher wages constrain expansion.	Highly relevant because AF tends to be labor intensive, limiting area cultivated per farm.
Secure land tenure	Contradictory effects; important to distinguish between how rights are obtained and tenure security.	Insecure tenure can prevent investments in trees, but trees also increase tenure security.

agricultural land. Particularly when the new technology is labor or capital intensive, as often is the case, such constraints at the household or village level might significantly limit the scope for converting forest. We elaborate this point later.

Fourth, technological changes are unlikely to involve only one household. If a large number of farmers adopt a new technology, this will have economic repercussions in the output and labor markets. Large changes in output might affect the price of the farmers' produce. Changes in the demand for labor are likely to affect local wages and migration flows.

Farmer characteristics, agroforestry technology, and market and tenure conditions are possible checks on agricultural expansion that we shall discuss in the following sections. Various aspects of each of these are listed in Table 5.2.

Farmer Characteristics

Farmers allocate their scarce resources (land, labor, and capital) to meet their objectives, such as ensuring family survival, maximizing income, or minimizing risk. Available technology, assets, market conditions, land tenure, and other factors constrain the choices farmers have available. Technological change may modify these constraints and provide incentives to encourage farmers to allocate their resources in a different manner. To understand farmers' response to technological change, one must understand farmers' constraints and incentives.

Farmers in developing countries are generally constrained, particularly in regard to labor and cash supplies, and the markets in which they engage are far from perfect. Consider a situation in which the farm household cannot sell its labor in a nonfarm labor market, nor can it hire labor to work on the farm. Assume that with the old technology the available family labor allows the household to cultivate 3 ha of land. If they adopt an agroforestry technology for all agricultural land and that technology is 50 percent more labor intensive (labor days per hectare), they can now cultivate only 2 ha; the remaining 1 ha reverts to secondary forest. Thus, introduction of labor-intensive technologies in the presence of labor constraints can reduce deforestation rates.

Farmers range from poor, isolated, and subsistence-oriented peasants to rich, commercially oriented landowners. Each type of farmer tends to specialize in different crops and production systems, making certain innovations relevant only for particular groups of farmers. Farmers respond differently to new technological innovations in terms of both technology adoption and forest impact. Smallholders tend to be more cash constrained, and this might prevent them from using certain technological innovations. For example, an agroforestry practice might require purchase of expensive tree crop seedlings, and fruit tree agriculture might rely on expensive transportation of the harvest to an urban market.

Capital-intensive technologies can have negative impacts on already poor farmers in several ways: they may not be able to afford the new technologies, they might suffer from lower wages and output prices, and deforestation may reduce forest-based incomes and environmental services. The main asset of the poor normally is their own labor. One might therefore argue that poor farmers have a comparative advantage in labor-intensive technologies such as many agroforestry practices. But rich farmers might be in a better position to mobilize (hired) labor on a large scale. For the very poor, the quick cash argument may make them rely more on short-term wage labor and harvesting of forest products, making it difficult to undertake medium- to long-term investments that agroforestry normally entails (see also Chapter 6, this volume).

Tropical farmers normally engage in several types of production systems, which makes analysis of the impact on deforestation more complex. These systems interact; they compete for family labor and produce food and cash income to cover the family's needs. Consider a simple case in which a farming family operates one intensive system with lowland rice cultivation and one extensive slash-and-burn system on the upland. A fixed amount of labor is to be allocated between the two systems, and a new labor-intensive rice technology is introduced (e.g., a new rice variety). This will clearly pull labor out of the extensive system and reduce deforestation.

Consider next a labor-intensive agroforestry technology for the uplands with the introduction of nitrogen-fixing legume trees or interplanting of the annual food crops with tree crops. If we consider this system in isolation, the result will be less deforestation because of the household's labor constraints. But with two systems, farmers can switch resources between them, and the result might be an expansion of the labor-intensive agroforestry technology. The lesson is that labor constraints for particular systems and deforestation activities often are flexible because farmers can shift resources to the more profitable activities.

It is also critical to identify where the change occurs. Generally, technologies suitable for more intensive agriculture—normally located far from the forest frontier—have better potential to reduce deforestation because of their effects in both the labor market (absorb labor) and the output market (compete with frontier crops). Agroforestry practices occur in both forest-abundant and forest-scarce situations (see Table 5.1), but a significant share of agroforestry adoption falls under the first two cases in the table. Therefore, one should be careful in stating that agroforestry is analogous to the green revolution as a deforestation-reducing strategy, as Sanchez et al. (1990) claim. Whereas green revolution technologies are targeted at intensive agricultural systems and therefore tend to pull resources away from forest-rich areas, agroforestry technologies often do not (Cattaneo 2001).

Agroforestry Technologies

The characteristics of new agroforestry technologies are important in assessing their impact on deforestation, but such analysis is complex. The impact is determined by the technology characteristics in combination with the farmer and market characteristics. Generally, as we move away from the economic textbooks' "perfect markets"—the world where farmers have perfect information and can sell or buy as much as they want at a fixed price—the technology characteristics become more important.

The labor and capital needs (factor intensities) are key features of new agroforestry practices. We assume here that these are fixed for a given technology, ignoring that farmers can to some extent vary the labor and capital inputs in each technology or production system (e.g., how often and how thoroughly weeding is done). But in practice the technology chosen clearly constrains the relevant range of inputs. Because most farmers are capital or labor constrained, how new technologies affect total capital and labor demand determines how much land farmers can cultivate. In particular, when markets are imperfect, the households' endowments of labor and cash critically influence the deforestation outcome. A labor- or capital-intensive agroforestry technique is more likely to promote forest conservation when markets are imperfect and farmers are constrained. When farmers are not capital or labor constrained or these markets are functioning well, labor and capital intensities are less important for the deforestation outcome.

Most agroforestry technologies appear to be labor intensive, although some practices such as the use of tree shade to reduce weed pressure (and replace hand weeding) in cropping systems aim to reduce the labor needs (but compared with herbicides, tree shade tends to be more labor intensive). In particular, compared with traditional shifting cultivation, pasture, and slash-and-burn annual cropping systems, permanent agroforestry practices entail more labor per unit area. In fact, labor shortage often is a reason why agroforestry practices are not adopted. Interestingly, a technology characteristic (labor intensiveness) that makes farmers reluctant to adopt agroforestry practices is the same characteristic that makes the practice, once adopted, less likely to lead to primary forest encroachment.

But there are exceptions. Kudzu (*Pueraria phaseoloides*) is a leguminous vine that fixes nitrogen and makes more nutrients available to the soil, speeding up soil recuperation. It also suppresses weeds, reducing the demand for labor for clearing and weeding. Kudzu therefore permits shorter fallow periods. This should reduce the stock of fallow land, allowing a larger forest area. This is a low-cost, labor-saving technology that increases yields and could potentially save forests. What more could you wish for? It is therefore not surprising that kudzu is explicitly mentioned in the Sanchez and Benites (1987) article as one of the promising species for managed fallows. But no one can guarantee the forest outcome. Yanggen and Reardon (2001) reported from a study of 220 farm households in Pucallpa, Peru, that farmers who use kudzu fallows can clear substantially less forest to cultivate the same land area (traditional secondary forest fallow uses 40–116 percent more land). But higher productivity and labor savings pull in the opposite direction. The authors' econometric analysis shows that kudzu reduces primary forest clearing but boosts secondary forest clearing, with the net effect being a modest rise in total forest clearing. This study illustrates a major point of this chapter: higher yield can in principle reduce deforestation, but higher benefits (increased yields or lower costs) of a technique can more than offset this effect by providing an incentive to cultivate a larger total area. And because kudzu saves labor, the farmers will have the capacity to do so.

One should be very cautious in drawing general conclusions about improved fallows based on a single study. First, kudzu is not a tree, although introducing it in some cases can be seen as a technological change in an agro-forestry context. Second, improved tree fallows might be more labor intensive than kudzu fallows; clearing certain planted tree fallows takes more work than clearing a spontaneous fallow (see Franzel 1999 and other articles in the same issue). The general lesson is that the impact on natural forests depends on the characteristics of the agroforestry practice in question, in combination with farmer and market characteristics of the area, as well as government policies.

Agroforestry could reduce the pressure on forest if it reduces ecological and economic risks for the farmer. Agriculture, particularly rain-fed tropical agri-culture, is a risky business, and risk considerations are central in farmers' decision-making processes. Farmers may overexploit natural resources as (short-term) insurance against yield and price risk, thereby ensuring that their income, even in a bad year, is above their subsistence needs. Coxhead et al. (2001) found evidence of this in their study from northern Mindanao, in the Philippines. Risk-reducing technologies therefore should enable farmers to convert less forest to agriculture.

Another aim of agroforestry is to make the system more resilient to ecolog-ical shocks such as dry years or pest outbreaks, which will not affect all species in a mixed system or not to the same extent (Schroth et al. 2000; see also Chapter 16, this volume). The multi-output nature of the system is also insur-ance against fluctuating market prices; all eggs are not put in the same basket. On this account, mixed tree crop systems (multistrata agroforestry), compet-ing with monoculture systems, should be favorably placed to reduce the pres-sure on natural forests.

Not all systems reduce risk, however. Introducing cash tree crops (e.g., fruit trees) at the expense of crops for domestic use (e.g., cassava) or cattle exposes the farmer to higher market risks. Moreover, when a number of farmers simul-taneously adopt new tree crops, the increased supply puts downward pressure on prices and could easily make them unprofitable. Furthermore, we need to consider the sustainability of the technology and how suitable it is for cultiva-tion on recently cleared forests. Ruf and Schroth (Chapter 6, this volume) point out that cocoa is particularly suitable for recently cleared forestland and that cocoa farmers enjoy a "forest rent" (i.e., a higher profit compared with planting on previously cultivated land) in recently converted forests. Suitability also refers to the necessary infrastructure. Perishable products, such as fruits, necessitate proximity to markets and good infrastructure (something normally not found at the forest frontier), whereas nonperishable products such as rub-ber and cattle are less dependent on regular transportation.

A final aspect of agroforestry technology that is relevant for the deforestation impact is the extent to which a new practice or technology consists of fixed investments. The land degradation–deforestation hypothesis is more applicable to annual than perennial crops. If farmers have invested a lot in their land, by planting perennials or making land improvements (e.g., terracing), they would be more reluctant to move on and clear new forestland. In Nicaragua, for example, coffee planting has helped stabilize the forest frontiers (D. Kaimowitz, pers. obs., 1991).

Overall, our discussion is based largely on the question about what happens when farmers adopt new agroforestry practices, ignoring the issue of technology adoption. Generally, farmers prefer to adopt technologies that increase their opportunities rather than limit them. This was suggested long ago by the work of Boserup (1965) on the demographic determinants of agricultural intensification and of Hayami and Ruttan (1985) on induced innovation based on the relative factor scarcity. Thus, if farmers face serious labor constraints, they will be reluctant to adopt labor-intensive practices or technologies. Similarly, if land is abundant, they have limited incentives to adopt land-saving practices or technologies. As we saw before, labor- and capital-intensive practices that bind the farmers' resources on small, intensively managed plots are those with the greatest potential to reduce pressures on the forest, yet labor (and capital) shortage and land abundance are typical characteristics of the situation at the forest frontier. Thus, the paradox is that the practices or technologies that have the highest forest-saving potential are more likely to be adopted once the forest is gone.

Sometimes farmers adopt these labor-intensive and land-saving practices or technologies in forest-abundant situations if they are very profitable or have other desirable characteristics such as reducing risk or fitting in well with the farmers' seasonal labor needs. Coffee adoption among smallholder settlers in Ecuador illustrates this point (Pichon et al. 2001). More generally, adoption of agroforestry practices is also determined by a number of factors, classified by Franzel (1999) as feasibility, profitability, and acceptability. Nevertheless, the general point is still valid: encouraging farmers to adopt practices or technologies that save resources that are not perceived as scarce, such as forestland in an open frontier situation, is difficult.

Market and Tenure Conditions

The prevailing market conditions are important for the deforestation outcome of a particular technological change. A large number of farmers adopting agroforestry practices will change the demand and supply in various markets and alter the prices of the commodities in these markets. Such effects—general equilibrium effects, in economic jargon—can in some cases be crucial for the final forest outcomes of agroforestry.

Output Markets

The idea that technological progress increases supplies, which lowers output prices and sometimes even reduces farmer incomes, is often called the tread-mill effect. Because the demand for food is generally inelastic, small increases in supply can lead to significant price declines, so net consumers win but net producers lose.

The magnitude of this price effect is an empirical question and is the product of two factors: overall market demand elasticity and the relative increase in supply. If a yield-increasing agroforestry practice introduced in frontier agriculture is very locale specific and adopted by only a small fraction of producers, the price effect will be small. Similarly, if the crops in question are exported and each country has only a small share of the global market, price declines will not dampen the expansion. Commodity booms involving export crops therefore can lead to a rapid increase in cropped area and corresponding deforestation. One example is the rapid increase in cocoa production by small-holders in West Africa in the twentieth century, although this was more by monoculture cocoa plantation than cocoa agroforestry (see Chapter 6, this volume).

However, frontier agriculture often is characterized by high transaction costs, poor infrastructure, and limited market access. Some of the cash crops reach only local markets, which easily become saturated if supply increases. If the main outputs from agroforestry are for the local markets, any expansion will quickly choke off because of depressed prices, and little or no additional deforestation will occur. The issue of price responses to supply increases presents us with a puzzle or trade-off. From a rural development (farm income) view one ought to go for crops sold in markets that can absorb an increase in supply, that is, large domestic (urban) or export markets. But these are exactly the type of markets that can lay the foundation for new technologies, including agroforestry practices, to remove large tracts of forests.

Labor Markets and Migration

In isolated forest-rich economies, one can expect labor-intensive agroforestry practices to have a positive or minimal impact on forests. Labor shortages and high wages quickly constrain any expansion. On the other hand, if regional or national labor markets function reasonably well and there is high labor mobility (migration), labor shortages are less likely to limit expansion. The extent of interregional flows of labor and capital therefore play a crucial role in determining how much the agricultural sector expands, particularly over the long term.

When labor-intensive technological change occurs outside the frontier areas, active labor markets can help curb deforestation. Employment opportunities outside the frontier will attract labor away from forest-clearing activities in the uplands, as illustrated by a Philippine irrigation study by Shively and Martinez (2001). Labor-saving technologies, on the other hand, foster greater migration to the frontier. Ruf (2001) reports that green revolution technologies (e.g., mechanization) in Sulawesi were labor saving and spurred the conversion of forests to cocoa holdings in the uplands.

A comparison of rubber agroforestry in selected sites in Borneo and Sumatra in Indonesia illustrates the critical role of the labor market in determining the forest outcome. Many analysts have blamed the introduction of rubber into shifting cultivation systems in Southeast Asia for provoking large-scale forest conversion. The ASB Programme has conducted extensive research in Sumatra on rubber agroforestry. Although rubber agroforestry has many attractive features, it has not stopped conversion of primary forest to smallholder rubber holdings for exactly the reasons explained earlier: high profitability in combination with large in-migration of labor (Angelsen 1995; Tomich et al. 2001). The situation is a typical example of the second case in Table 5.1. This result is contrasted with a study by de Jong (2001), who found that rubber agroforestry contributed little to encroachment into primary forest in areas of West Kalimantan (Indonesia) and Sarawak (Malaysia). De Jong's study found that incorporating rubber gardens (or rubber-enriched fallows) and additional tree cover in lands previously used for slash-and-burn agriculture produced both economic and ecological benefits. Several factors explain this difference, such as better forest management in the Borneo cases. An important difference was the remoteness of and low in-migration to the Borneo sites. In Sumatra the adoption of rubber was accompanied by substantial in-migration from Java and other parts of Sumatra, so labor shortages did not dampen the conversion of forests to rubber agroforestry. The situation was very similar to that of cocoa agroforestry in West Africa, where expansion was possible because of in-migration from the savanna zone into the rainforest (see Chapter 6, this volume). Moreover, the expansion had limited impact on the world market price of rubber.

Land Tenure

So far we have not dealt much with issues of property rights and land tenure. Insecure land rights and open access situations often are noted as key underlying causes of deforestation and act as a disincentive for investments in land, including agroforestry (Wachter 1992). However, the empirical evidence is more complex than the simple theory suggests. The forest impact of technological change generally, and agroforestry specifically, depends critically on the existing property regime. Generally, open access situations might encourage investments by clearing more forests, whereas contexts with well-defined property rights should lead to investments in the resource base by better management of existing land (Otsuka and Place 2001). Although this might hold true as a general proposition, there are several qualifications. Private forest owners with reasonably secure rights to their land might decide to convert some of their forest to crops or pasture. However, reasonably secure tenure, which is normally equated with individual land rights, provides no incentives for taking the externalities of land use into account.

It is commonly argued that poorly defined or insecure property rights reduce the incentives to invest in agroforestry. Again, there are a number of qualifications to this generalization. The meaning of "secure tenure" often is wrongly perceived. Customary land tenure typically gives individual use and income rights, whereas the transfer rights rest with the lineage, chief, or community (Ensminger 1996). This system often provides sufficient tenurial incentives for agroforestry investments. Detailed analysis often points to investments in land conservation and agroforestry as being constrained by other factors, such as labor and capital constraints and high discount rates. In a study from Benin, Neef and Heidhues (1994) conclude that "the key issues holding up agroforestry investments in Benin would not be addressed by land titling programmes (158)." However, they point out that land tenure can become a key factor in the success of agroforestry programs, particularly in densely populated areas.

A farmer's right to the land is not fixed by the institutional environment but is influenced by personal decisions. In extensive forms of tropical agriculture such as shifting cultivation, planting tree crops tends to increase tenure security. Tree planting becomes a strategy to claim land rights. Therefore, the conventional argument that tenure insecurity causes deforestation is turned upside down: insecurity becomes a reason for planting trees and investing in land because this will boost the farmer's claim to the land. This effect has been documented among rubber smallholders in Sumatra (Suyanto et al. 2001). In fact, when there is de facto open access to natural forests and land rights are established or strengthened by planting tree crops, there are incentives for both deforestation (chopping down natural forest) and reforestation (planting trees rather than annuals) on the cleared land.

To summarize, the property regime is important, but its impact on the forest cover is not straightforward. More attention must be given to how land rights are established and strengthened. Researchers and policymakers should not only take into account changes in outputs and inputs when comparing different systems but also assess their impact on tenure security. Better tenure security should stimulate investments in existing agricultural land, reducing the need for land expansion. But the higher expected profitability provided by agroforestry compared with shifting agriculture also makes the investment in forest conversion more profitable. The latter effect could further stimulate land races (Angelsen 1999).

Conclusions

Agroforestry researchers have not paid much attention to the impact of agro-forestry on deforestation. To the extent that they have, a Peruvian case study on yield differences has been taken as a global measure of deforestation impacts. This has created a widespread belief that agroforestry reduces defor-estation by a ratio of about 1:5. Extensive research, taking the farmers and market responses into account, demonstrates that technological change often leads to more, rather than less, deforestation, and this may also apply to agro-forestry if it increases the profitability of land use unless there are factors such as labor, capital, or market constraints that limit agricultural expansion into forest areas.

The study by Sanchez and Benites (1987) formed an important pillar of the ASB Programme, with earlier articles using their results to demonstrate its deforestation-reducing potential (ASB 1994). It is therefore interesting to see how the ASB-Indonesia studies conclude, based on their findings in Sumatra: "It is naïve to expect that productivity increases necessarily slow forest conver-sion or improve the environment. Indeed quite the opposite is possible. . . . ASB research in Indonesia has shown that land use change normally involves tradeoffs between global environmental concerns and the objectives of poverty alleviation and national development" (Tomich et al. 2001, 242).

This is not to deny that agroforestry in some cases can curb deforestation. Our message is that a general claim that agroforestry reduces deforestation (including the 1:5 ratio) is wrong, basing deforestation policies and programs on unqualified assumptions will not help reduce deforestation and can lead to misallocation of development and research efforts, and the impact of intro-ducing agroforestry practices is conditioned by the type of practice, farmer characteristics, and market and tenure conditions. In short, the win-win situ-ations in which agroforestry can meet both local development and forest conservation objectives are characterized by technologies that are suited specif-ically for forest-poor areas; labor-intensive technologies when labor is scarce and in-migration limited; promotion of intensive systems when farmers are involved in extensive, low-yielding practices; and technologies that raise the aggregate supply significantly when demand is inelastic (causing price decline).

We have also pointed out that trade-offs might be more common than often assumed. When agroforestry practices are successful and adopted on a large scale and forest areas still are accessible, one is often presented with a trade-off: land under agroforestry practices has desirable ecological character-istics compared with alternative land uses, but the area of primary forest is being reduced. And the types of agroforestry techniques that are most likely to be adopted and therefore to be most beneficial to the farmers are those that save labor and produce crops for large national or international markets (elastic demand), but they are also the ones with the greatest potential for increased deforestation.

So is agroforestry likely to reduce deforestation? After criticizing those who gave an unconditional "Yes" to this question, we will avoid going to the other extreme by saying "No." Rather, we argue that it depends on the particular case, and our aim has been to point to a number of factors that determine net effects on deforestation.

References

Angelsen, A. 1995. Shifting cultivation and "deforestation": a study from Indonesia. *World Development* 23:1713–1729.

Angelsen, A. 1999. Agricultural expansion and deforestation: modelling the impact of population, market forces and property rights. *Journal of Development Economics* 58:185–218.

Angelsen, A., and D. Kaimowitz (eds.). 2001. *Agricultural technologies and tropical deforestation.* Wallingford, UK: CAB International.

Angelsen, A., and M. Luckert. In press. Limited or unlimited wants in the presence of limited means? Inquiries into the role of satiation in affecting deforestation. In D. J. Zarin, J. Alavalapati, F. E. Putz, and M. C. Schmink (eds.), *Working forests in the American tropics: conservation through sustainable management?* New York: Colombia University Press.

ASB. 1994. *Alternatives to slash-and-burn. A global strategy.* Nairobi: International Center for Research in Agroforestry.

Boserup, E. 1965. *The conditions for agricultural growth. The economics of agrarian change under population pressure.* London: Allen and Unwin; Chicago: Aldine.

Brady, D. E., D. P. Garrity, and P. A. Sanchez. 1993. The worldwide problem of slash-and-burn agriculture. *Agroforestry Today* 5:2–6.

Cattaneo, A. 2001. A general equilibrium analysis of technology, migration and deforestation in the Brazilian Amazon. Pages 69–90 in A. Angelsen and D. Kaimowitz (eds.), *Agricultural technologies and tropical deforestation.* Wallingford, UK: CAB International.

Coxhead, I., G. Shively, and X. Shuai. 2001. Agricultural development policies and land expansion in a southern Philippine watershed. Pages 347–366 in A. Angelsen and D. Kaimowitz (eds.), *Agricultural technologies and tropical deforestation.* Wallingford, UK: CAB International.

de Jong, W. 2001. The impact of rubber on the forest landscape in Borneo. Pages 367–383 in A. Angelsen and D. Kaimowitz (eds.), *Agricultural technologies and tropical deforestation.* Wallingford, UK: CAB International.

Dixon, R. K. 1995. Agroforestry systems: sources or sinks of greenhouse gases. *Agroforestry Systems* 31:99–116.

Dixon, R. K., J. K. Winjum, and P. E. Schroeder. 1993. Conservation and sequestration of carbon: the potential of forest and agroforest management practices. *Global Environmental Change* 3:159–173.

Ensminger, J. 1996. Culture and property rights. Pages 179–203 in S. Hanna, C. Folke, and K.-G. Mäler (eds.), *Rights to nature: ecological, economic, cultural, and political principles of institutions for the environment.* Washington, DC: Island Press.

Franzel, S. 1999. Socioeconomic factors affecting the adoption potential of improved tree fallows in Africa. *Agroforestry Systems* 47:305–321.

Hayami, Y., and V. W. Ruttan. 1985. *Agricultural development. An international perspective.* Baltimore, MD: Johns Hopkins University Press.

Mercer, D. E., and R. P. Miller. 1998. Socioeconomic research in agroforestry: progress, prospects and priorities. *Agroforestry Today* 38:177–193.

Murniati, D. P. Garrity, and N. Gintings. 2001. The contribution of agroforestry systems to reducing farmers' dependence on resources from adjacent national parks: a case study from Sumatra, Indonesia. *Agroforestry Systems* 52:171–184.

Nair, P. K. R. 1998. Directions in agroforestry research: past, present and future. *Agroforestry Systems* 38:223–245.

Neef, A., and F. Heidhues. 1994. The role of land tenure in agroforestry. Lessons from Benin. *Agroforestry Systems* 27:145–161.

Otsuka, K., and F. Place. 2001. *Land tenure and natural resource management. A comparative study of agrarian communities in Asia and Africa.* Baltimore, MD: Johns Hopkins University Press (for IFPRI).

Pichon, F., C. Marquette, L. Murphy, and R. Bilsborrow. 2001. Land use, agricultural technology, and deforestation among settlers in the Ecuadorian Amazon. Pages 153–166 in A. Angelsen and D. Kaimowitz (eds.), *Agricultural technologies and tropical deforestation.* Wallingford, UK: CAB International.

Ruf, F. 2001. Tree crops as deforestation and reforestation agents: the case of cocoa in Côte d'Ivoire and Sulawesi. Pages 291–316 in A. Angelsen and D. Kaimowitz (eds.), *Agricultural technologies and tropical deforestation.* Wallingford, UK: CAB International.

Sanchez, P., C. A Palm, and T. J. Smyth. 1990. Approaches to mitigate tropical deforestation by sustainable soil management practices. Pages 211–220 in H. W. Scharpenseel, M. Schomaker, and A. Ayoub (eds.), *Soils on a warmer Earth.* Developments in Soil Science 20. Amsterdam: Elsevier.

Sanchez, P. A. 1990. Deforestation reduction initiative: an imperative for world substantiality in the twenty-first century. Pages 375–382 in A. F. Bouwman (ed.), *Soils and the greenhouse effect.* New York: Wiley.

Sanchez, P. A., and J. R. Benites. 1987. Low-input cropping for acid soils of the humid tropics. *Science* 238:1521–1527.

Sanchez, P. A., B. Jama, A. I. Niang, and C. A. Palm. 2001. Soil fertility, small-farm intensification and the environment in Africa. Pages 342–344 in D. Lee and C. Barrett (eds.), *Tradeoffs or synergies? Agricultural intensification, economic development, and the environment.* Wallingford, UK: CAB International.

Schroeder, P. E., R. K. Dixon, and J. K. Winjum. 1993. Forest management and agroforestry to sequester and conserve atmospheric carbon dioxide. *Unasylva* 44:52–60.

Schroth, G., U. Krauss, L. Gasparotto, J. A. Duarte Aguilar, and K. Vohland. 2000. Pests and diseases in agroforestry systems in the humid tropics. *Agroforestry Systems* 50:199–241.

Shively, G., and E. Martinez. 2001. Deforestation, irrigation, employment and cautious optimism in southern Palawan, the Philippines. Pages 335–346 in A. Angelsen and D. Kaimowitz (eds.), *Agricultural technologies and tropical deforestation.* Wallingford, UK: CAB International.

Sunderlin, W., A. Angelsen, D. P. Resosudarmo, A. Dermawan, and E. Rianto. 2001. Economic crisis, small farmer well-being, and forest cover change in Indonesia. *World Development* 29:767–782.

Suyanto, S., T. P. Tomich, and K. Otsuka. 2001. Land tenure and farm management efficiency: the case of smallholder rubber production in customary land areas of Sumatra. *Agroforestry Systems* 52:145–160.

Tomich, T., M. van Noordwijk, S. Budidarsono, A. Gillison, T. Kusumanto, D. Murdiyarso, F. Stolle, and A. M. Fagi. 2001. Agricultural intensification, deforestation, and the environment: assessing tradeoffs in Sumatra, Indonesia. Pages 221–244 in D. Lee and C. Barrett (eds.), *Tradeoffs or synergies? Agricultural intensification, economic development, and the environment.* Wallingford, UK: CAB International.

Wachter, D. 1992. *Land titling for land conservation in developing countries.* Divisional Working Paper, Environment Department. Washington, DC: World Bank.

Yanggen, D., and T. Reardon. 2001. Kudzu improved fallows in the Peruvian Amazon. Pages 213–230 in A. Angelsen and D. Kaimowitz (eds.), *Agricultural technologies and tropical deforestation.* Wallingford, UK: CAB International.

Young, A. 1997. *Agroforestry for soil management.* 2nd edition. Guildford & King's Lynn, Norwich, UK: Biddles Ltd.

Chapter 6

Chocolate Forests and Monocultures: A Historical Review of Cocoa Growing and Its Conflicting Role in Tropical Deforestation and Forest Conservation

François Ruf and Götz Schroth

An American scientist who visited southern Bahia on the southeastern coast of Brazil in the 1950s captured the impression that the cocoa cropping systems of that region, locally known as cabruca cocoa, made on him in the following words: "Only slowly does the initiate become aware that this 'forest,' and the 'forest' that had appeared as formidable to him in the latter stages of his trip into the cacao region is that same huge orchard which he had sought from the air and from the truck window. He learns to recognise the tall trees as jungle trees left during the clearing of the land as shade for the low cocoa trees" (Leeds 1957, 41). These chocolate forests (Johns 1999), created by under-planting selectively thinned natural forest with cocoa trees (*Theobroma cacao*), not only protect the tree crops from climatic hazards and pests and increase their longevity but also conserve some of the characteristics of the original forest, including part of its biodiversity. As agricultural land use, including cocoa cultivation, has transformed the formerly vast and highly diversified Atlantic rainforest into isolated fragments in an agriculturally dominated landscape, the potential role of the cabruca agroforests for the conservation of biodiversity has increasingly attracted the attention of conservationists and natural resource managers: "In Southern Bahia, the merits of the cabruca cacao is that the system allows economical development while maintaining a portion of the original forest diversity and thus preserving wildlife" (Alves 1990, 136).

In 1996, local authorities and the scientific community used the International Cocoa Research Conference at Salvador de Bahia to develop an image of tradition, culture, and environmental protection around the cabruca cocoa farms after the slump in cocoa prices in the late 1980s and the arrival of the witches' broom disease (*Crinipellis perniciosa*) in 1989 motivated the conversion of some of these traditional systems into pasture (Alger and Caldas 1992; Trevizan 1996). The official recognition of their potential for biodiversity conservation and ecotourism marked a fundamental change in political priorities compared with campaigns of previous decades, in Brazil and in other cocoa-growing regions, to thin these dense canopies of forest remnants in order to increase cocoa yields (Johns 1999). It reflects particularly well the dual nature of the cocoa agroforests as an agent of the conversion of natural forests into agricultural ecosystems in this part of the Brazilian Atlantic forest and as one of the most biodiversity-friendly land use options available to local farmers.

Of course, basically all upland agriculture in the humid tropics has to take place on forestland and therefore ultimately at the expense of the forest. What made the cocoa tree an important agent of the conversion of primary tropical forests over the last four centuries, and especially in the twentieth century, is a history of boom-and-bust cycles, combined with the tendency of the principal cocoa-growing regions to move from one place to another. Where these cycles started, they led to the opening up of new forests, sometimes at a tremendous speed. Where they ended, they left behind, in the best cases, disease-infested groves of low productivity in a secondary forest environment but often only poor fallows and pastures. These cycles were fueled by the access to cheap forestland and often the labor force of immigrants.

In regions such as Bahia, southern Cameroon, southwest Nigeria, eastern Ghana, and initially the Côte d'Ivoire, cocoa was grown in complex agroforests that are among the most diversified and forest-like of all agricultural systems (see Chapter 10, this volume); in other cases, such as most of the Côte d'Ivoire, western Ghana, Malaysia, and Sulawesi in Indonesia, cocoa was grown in plantations with little or no shade, often almost monocultures. It is obviously important for biodiversity, both on the plot and on the landscape scale, whether forest is replaced by a tree crop monoculture or a complex agroforest with an understory of cocoa trees under the shade of old forest trees. Even more important, however, for regional biodiversity is how these land use types affect primary forest cover in the long term. Both the longevity of a tree crop such as cocoa and the ease of replanting it on the same site are system characteristics that are influenced by the degree of shading and may influence, in turn, the long-term forest consumption by cocoa farms, as we shall see.

As attempts increase around the world to change the historical role of the cocoa tree from a consumer of tropical forests into an instrument to improve the livelihoods of tropical farmers and to conserve as much as possible of tropical forests and their biodiversity, it may be instructive to review the factors that have determined whether this crop was grown in complex agroforests or monocultures, whether these systems were sustainable, and how they responded to social, economic, and technological change. Although this chapter focuses on cocoa, some of the conclusions are also valid for other tropical tree crops that are both consumers of tropical forest and potential allies in the search for sustainability in tropical forest regions.

Continental Drifts: How the Cocoa Tree Conquered the Tropics

The center of origin of the cocoa tree probably is on the eastern equatorial slope of the Andes and undoubtedly is in the Amazon basin. The oldest real center of cultivation seems to have been Central America, where the crop has been under cultivation for more than 2,000 years (Cope 1976). Once the Spanish had learned from the Amerindians how to transform it into a palatable drink, cocoa became an economically important commodity. Cocoa trees of the *criollo* variety from Central America were planted in Venezuela and Trinidad in 1525; subsequently Jamaica, Haiti, and the Windward Islands became important producers (Cope 1976).

From this point, world cocoa production increased as new countries adopted the crop while previous production centers collapsed. The continuous increase in world production over the centuries hides a succession of national and regional boom-and-bust cycles. In the sixteenth century, Central America was the first region to develop a cocoa economy before it relinquished the lead position to the Caribbean, especially Jamaica and Venezuela. Venezuela became the world's leader in cocoa production in the eighteenth century before it declined at the beginning of the nineteenth century, when Ecuador took over and its port Guayaquil became the world's capital of cocoa export from the end of the nineteenth century until the 1920s. As cocoa production in Ecuador collapsed, its place was taken by production in Brazil and Ghana. Subsequently, Ilhéus and Salvador de Bahia in Brazil, Accra in Ghana, Lagos in Nigeria, and Abidjan in the Côte d'Ivoire became the leading cocoa export ports of the twentieth century, shipping hundreds of thousands of tons of cocoa to Europe and North America. From 1980 to the early 1990s, Malaysia started to monopolize the New York stock market's fax machines, but its cocoa cycle was one of the shortest in history; Indonesia, especially the island of Sulawesi, took over almost immediately.

These production shifts from one country to the next were reproduced by similar cycles on the subnational scale. The history of cocoa growing in the Côte d'Ivoire, discussed in detail in this chapter, and the more recent one of Sulawesi show cut-and-run cycles in regions of early adoption of the crop that were then abandoned for new pioneer fronts. Descriptions of these shifts of cocoa-growing regions from different continents and separated in time by four centuries sound surprisingly similar, underscoring a feature that characterizes much of cocoa history:

> The Sonocusco province (in Mexico) was famous for its wealth and prosperity, densely populated with Indians and much visited by Spaniard merchants for its abundant cocoa production and its important trade that followed from it. There are now very few Indians. It is said that there are less than two thousand and that cocoa trade is disappearing, moving to another province, farther on the track to Guatemala. (Alonso Ponce, 1586, quoted by Touzard 1993, 53)

> In Côte d'Ivoire, cocoa cultivation is rare today between Abidjan and Abengourou, the region where the cocoa industry was born. In the Abengourou region itself, production has declined for the last 12 years to 6,000 tonnes from approximately 22,000. One can see abandoned farms everywhere. Production is shifting to the interior toward Dimbokro and Gagnoa, where new virgin forest lands are cleared. (FAO 1957, 16–17)

One may add that Dimbokro, the heart of the Ivorian cocoa belt in the 1960s, already ended its cocoa cycle in the early 1980s, when the crop moved further to the west, mostly to Soubré. Cocoa has thus been moving around the world for the last four centuries, in most cases at the expense of tropical forests. What are the factors that drove these cycles?

Cheap Labor and Forestland: Ingredients of Cocoa Booms

Throughout the world, most tree crop booms have been made possible through a combination of migrations and deforestation. Migrations result from the presence of large and mobile populations not too far from sparsely populated forest. Such a mobile work force was available in the savanna zone of West Africa to supply the cocoa booms in the Côte d'Ivoire in the 1960s to 1980s, for example, and on the densely populated southern part of Sulawesi and Bali to supply the cocoa boom in Sulawesi in the 1990s. Cheap land in sparsely populated forest zones provides a strong pull factor to poverty-stricken farmers in the source areas of such migrations; for example, in Indonesia in the 1980s, by selling a quarter of a hectare of paddy terraces in his village in Bali, a migrant could buy at least 10 ha of land suitable for cocoa planting in the forested plains of central Sulawesi.

Access for migrants to virgin forest areas (and subsequent transport of agricultural produce to markets) is facilitated when logging companies construct roads and open tracks into the forest, especially if they are subsequently maintained by public investments (lack of these may have saved logged forests in parts of Cameroon from immigration; J. Gockowski, pers. comm., 2003). Government policies also strongly influenced the pace of migration. Before independence, both the Côte d'Ivoire and Cameroon produced approximately 100,000 metric tons of cocoa per year. After independence, totally opposite migration policies in the two countries were the decisive factor behind the astonishing 1.2 million metric tons reached by the Côte d'Ivoire by the mid-1990s and the apparent stagnation around 110,000 metric tons in Cameroon (Ruf 1985; Losch 1995). In Indonesia, the cocoa boom in Sulawesi of the 1980s and 1990s, which had been launched by spontaneous Bugis migrants from the southern part of the South Province of the island, was involuntarily enhanced by the government's transmigration program: although the program intended to resettle populations from the densely populated islands of Bali and Java in irrigated rice production schemes on Sulawesi, it took a new direction when the migrants copied the successful experiences of the Bugis and became cocoa planters.

Planting a crop after clearing primary forest can have strong economic advantages over planting it on previously used crop or fallow land, a factor that can be interpreted as a "forest rent" and that has contributed significantly to the conversion of tropical forests. This factor is not specific to cocoa but may be more important for it than for other tree crops because of the difficulties of replanting cocoa in areas where the forest has disappeared. It helps to explain why cocoa has shown such a strong tendency to follow the vanishing forest, with new plantations being established on cleared forestland rather than old and disease-infested plantations being replanted on the same site.

The differential forest rent applied to cocoa is defined as the difference in investment and production costs for a metric ton of cocoa between a plantation that was established after primary forest clearing and one established on fallow land or by replanting an older cocoa plantation (Ruf 1987). It turns out that planting on forestland nearly always trumps replanting. The reasons for this are related to the different efforts needed for forest clearing and plantation maintenance, especially weeding, differences in soil fertility and microclimatic conditions between forest and replanted sites, and biological factors such as pest and disease pressures, which in concert determine production costs, yields, and risks of tree mortality when a new plantation is established.

Planted in virgin forest soil, cocoa benefits from low weed pressure, high soil fertility, and a microclimate that is conducive to the development of these drought-sensitive understory trees. Replanting fallow land or old plantations entails more weeding, the growth of the young trees is slower, and mortality may be high, especially in the first dry seasons. In addition, as the forest disappears, timber and game resources become scarce so that housing and living costs increase.

In the Côte d'Ivoire, attempts to estimate the forest rent show an approximate doubling of the investment costs for replanting after fallow (now usually dominated by the aggressive invader *Chromolaena odorata*) or after an old, weed-infested cocoa plantation compared with planting after cleared forest.

Table 6.1. Estimate of cocoa production costs in the hills of Sulawesi.

	After Forest (cents kg⁻¹)	After Grassland (cents kg⁻¹)	Forest Rent (cents kg⁻¹)
Net input costs	8	16	7
Labor costs	28	31	3
Total	36	46	10

Source: F. Ruf, unpublished data from a survey in 1997.

For the first year, the total effort for clearing, planting, and weed control was 168 working days per hectare for replanting and 86 days per hectare for planting after forest (Ruf and Siswoputranto 1995). Another estimate of all labor investments until the cocoa started to produce put the replanting effort at 260 days per hectare, compared with 74 days per hectare for planting after forest (Oswald 1997).

In the hills of Sulawesi, cocoa planting after fallow instead of forest results in higher labor costs, and most smallholders also believe that cocoa needs more fertilizer when planted on grassland than when planted on forestland, with the total difference in production costs (consisting of net inputs such as fertilizers and labor costs) amounting to approximately US$0.10 per kilogram of cocoa (Table 6.1). This should be considered a conservative estimate because the net input costs are reduced by yields of food crops that are initially associated with the cocoa, which tend to be higher after forest than after grassland. In addition, the depreciation of the labor costs during the juvenile phase of the cocoa trees puts planting after grassland at a further disadvantage because it may delay the first cocoa yields. As a consequence, forest is still sought in the hills and uplands, but farmers in the rich alluvial plains fear the loss of the forest rent less (Ruf 2001).

For centuries, this forest rent and the availability of forestland has discouraged sustainable cocoa growing. For example, MacLeod (1973) described the wasteful use of forest land in Sonocusco, Mexico, in the sixteenth century:

> The heavy cutting and burning of forests and tall grasses caused erosion, leaching of the top soil, and flash flooding. Land was plentiful compared to labor and capital on the cocoa coast, and the Spaniards saw no reason for maintaining its quality and fertility. The restoration of eroded, leached soils for cacao plantations is an extraordinarily long and difficult task even today. The Central Americans of the sixteenth and seventeenth centuries did not have the technology and the patience to attempt it. Cattle or brush often filled the poor pasture lands left behind by the exhausted cacao growers (p. 95).

In the same tone, Delawarde (1935) wrote about Martinique Island in the seventeenth century: "In the mountains, cacao cultures grow to produce much profit. However, the factor of success, new soil, is a transitory one. The colonists did not fertilise the soil, they used it up and then planted elsewhere (p. 103)."

As mentioned before, the existence of a forest rent is not specific to cocoa. Lower labor inputs for planting on virgin forestland compared with replanting have been reported for rubber trees in Sumatra (Gouyon 1995), and early colonists of the Atlantic rainforest region of Brazil believed that coffee and even sugarcane find optimum growth conditions only on recently cleared forestland (Dean 1995). The difference is that whereas rubber, coffee, and sugarcane are routinely replanted throughout the tropics today, in many regions replanting of old cocoa remains a difficult task even for modern agronomists. This is especially so where during boom times soils unsuitable for the demanding cocoa trees have been used for planting, as has occurred in many places in the western Côte d'Ivoire, in Nigeria (Ekanade 1985), and in Sulawesi (Ruf and Yoddang 2001).

These technical difficulties of replanting old cocoa are compounded by social and economic factors. For the first generation of cocoa farmers who have arrived in a region, replanting often turns out not to be economically feasible at a time of declining returns and increasing costs caused by the aging of the plantation. Furthermore, the tree life cycle interacts with the life cycle of its owner, his or her family, and the community. Migrants involved in cocoa planting often are young, and often all planters in a particular zone have arrived together during a brief period of time, and so they all age along with their farms. When replanting time comes, the farmers lack the necessary labor force, especially if they have sent their children to school. As the yields from the aging plantations decline, family size and consumption increase, which further limits the ability to invest in replanting. Conflicts between potential inheritors often aggravate the degradation of the farms by postponing investment decisions.

These factors can be compounded by ecological change such as the arrival of new diseases and fluctuating climatic conditions. Eastern Ghana was the main cocoa belt of the country in the 1930s and still an important one in the 1950s. However, as the region was struck by the swollen shoot virus, soil exhaustion, and declining annual rainfalls, its cocoa economy collapsed, and the main center of cocoa production shifted into the virgin forests of western Ghana, whereas the former cocoa belt turned into an oil palm and citrus belt (Amanor and Diderutuah 2001).

With this background, we will now discuss in some detail the cocoa history of the world's leading cocoa producer, the Côte d'Ivoire, before considering more briefly two regions where particularly complex cocoa agroforests have developed, Bahia and Cameroon.

A Forest for a Bottle of Gin: Migrants and the Cocoa Boom of the Côte d'Ivoire

The history of cocoa in the Côte d'Ivoire began in the 1890s with a short-lived cocoa boom in the extreme southwest of the country. Although economically without much consequence, this local event is instructive because it proves that early adoption sprang up as an indigenous process, not from colonial policies (Chauveau and Léonard 1996). However, the base of the current Ivorian cocoa economy was built in the eastern region after 1900. From 1910 to the 1950s, cocoa spread in this region and some parts of the center-west, mostly through micro–pioneer fronts. As the local farmers needed workers, these first decades put the structure of migration in place. After 3–10 years of good services as workers, many migrants could obtain land and become cocoa smallholders. In turn, they also called for relatives and workers. Because of the poverty in the neighboring savanna of the Côte d'Ivoire and Burkina Faso, a large and cheap labor force was ready to be exploited.

In 1960 came the independence, with an Ivorian president who fully understood the potential of combining this foreign labor force with the vast Ivorian forest. Migrations and the opening of pioneer fronts were accelerated by policies of declassifying forest reserves, distributing information, and opening the borders for foreign labor. Logging companies and their tracks facilitated the move into the forests. The result was a sweep of the tropical forest from east to west between the mid-1960s and the early 1990s. The migrants' rush to the forest was strongly related to the low prices of forestland. In the southwest of the country, where the last large cocoa pioneer front opened in the late 1970s, some migrants could still obtain 25 or even 50 ha of primary forest for a bottle of gin, 12 bottles of beer, and one piece of cloth. According to a survey in 1998, with 1,000 cocoa farmers in the whole country, one-third were indigenous farmers, one-third Ivorian migrants (coming from the savanna in the center and north of the Côte d'Ivoire), and one-third foreign migrants, mostly from Burkina Faso, with an average cocoa area of about 5.5 ha per household for all three groups (Legrand 1999). This points to the dominating role of migration in the Ivorian cocoa boom and to the resulting potential for conflict that finally erupted in civil war in September 2002.

From Agroforests to Monocultures: Agronomists, Migrants, and Technological Change

The strong migrant component determined not only the speed with which cocoa spread through the forest zone of the Côte d'Ivoire but also the way in which cocoa was grown. Until the 1960s, most cocoa planters did not cut down the biggest forest trees, at least not all of them. The undergrowth was cut and burned, but some of the giant trees were maintained and formed the upper canopy of cocoa agroforests. Reference to these agroforests in the Côte d'Ivoire, which resembled those of the neighboring Ghana (formerly Gold Coast), can be found in reports from colonial times, along with first hints at their intensification under the influence of the colonists: "Farms in the western cocoa-growing areas are ordinarily well provided with primeval bush shade, as in the Gold Coast; but in the central and eastern districts, where the influence of the European planter is strongest, the shade for cocoa is often provided by bananas and plantains, as is done on plantations" (Schwarz 1931, 6).

Because no chainsaws or even good axes were available, an important motivation of this agroforest strategy of growing cocoa was to save labor by sparing especially trees with very hard wood or large buttresses, as described decades later for indigenous cocoa farmers in the western Côte d'Ivoire (de Rouw 1987). Of course, these forest people also knew about the different uses of their trees and retained certain useful species (again for the western Côte d'Ivoire, de Rouw mentions the edible seed–producing *Irvingia gabonensis, Ricinodendron heudelotii,* and *Coula edulis,* among others). Such heavily shaded cocoa agroforests can still be found in the eastern Côte d'Ivoire, and pockets of this agroforest tradition have also survived in the center-west, in the region of Gagnoa, where a rebellion of local residents against the government's policy of encouraging immigrants in the 1970s deterred further immigration (Figure 6.1).

Figure 6.1. Forest? No: one of the few surviving cocoa agroforests in the center-west of the Côte d'Ivoire.

An inconvenience of the traditional agroforest method is delayed returns, as heavy shade slows down the growth of the tree crops. On the other hand, shading prolongs the useful life of the cocoa farm. Also, that shading protects cocoa trees from insect pests has been known in West Africa at least since the early twentieth century, when on the islands of Fernando Pó and São Tomé attempts to increase cocoa yields by drastically thinning the shade canopy resulted in complete crop failure (Gordon 1976, cited in Johns 1999). Furthermore, the almost intact root systems of the forest trees allowed the eventual regrowth of the forest. Thirty years after the establishment of the cocoa trees, this system favored a strategy of abandoning the farm and leaving shade trees and forest regrowth to develop freely. The old cocoa farm then became a secondary forest where successful replanting was almost guaranteed. As already understood by Blankenbourg in the 1960s (cited in Ruthenberg 1980), this was nothing else than the shifting cultivation principle applied to a perennial crop. Initially it consumed forest, but once it was established for a given population, it could theoretically rotate on its own tree crop-fallow land without affecting surrounding forest. Had this type of rotational agroforest practice caught on throughout the forest belt of the Côte d'Ivoire, its landscape would now be different. Why did this not happen?

In the subsequent transformation of the cocoa-growing method, the research and extension services that favored zero-shading from the early 1970s until the late 1990s in the Côte d'Ivoire, as in most other countries (but see Cameroon later in this chapter), played a significant role. A contributing factor in this philosophy was the replacement since the early to mid-1970s of the old amelonado cocoa varieties, locally known as cacao français, by a new planting material, the upper amazons and hybrids of upper amazons, locally called cacao Ghana. The vigor of the new varieties seemed to be better expressed with little or no shade.

However, the most important driver of the changing cocoa-growing practices was the demographic and social change. Up to the mid-1960s, most cocoa farmers were indigenous forest people who applied the type of forest clearing that they knew from shifting cultivation to their tree crops. There were very few migrants in the forest zone before that time. In the 1970s and 1980s, however, rural populations in the parts of the forest zone where actual booms were taking place kept growing at rates of 10–20 percent per year through immigration (Direction des Grands Travaux 1992). This social landslide was followed by a technical one, a simultaneous adoption of a labor-saving technology to remove forest trees.

The Baoulé migrants, the most dynamic of the savanna people streaming into the forest, introduced a technique of killing big trees by gathering undergrowth around them and keeping it ablaze over a few days; the trees then fell apart over the next few years. This was much more labor efficient and far less dangerous than cutting them with axes, often from a platform

Figure 6.2. Young cocoa farm established after killing most of the forest trees in the southwest of the Côte d'Ivoire.

that allowed them to attack the stem above the buttresses, as had been done previously. Along with the cocoa trees the migrants planted their staple food crop, yams, which necessitated intensive soil tillage and further reduced the chance of the forest to regenerate, instead of upland rice, which was grown by the locals and entailed little disturbance of the forest soil (de Rouw 1987).

The new technique of forest conversion served a strategy of rapidly planting cocoa trees to mark land ownership instead of spending a lot of time cutting the forest trees and clearing the plot (Figure 6.2). The intensive burning of biomass, soil tillage, and opening of the canopy also accelerated the initial growth of the cocoa trees and provided rapid financial returns: whereas the indigenous method took 5 years until the first cocoa yield and produced 500 kg of cocoa per hectare after 10 years, with the no-shade system the tree crops started to produce within 3 years and yielded close to 1 metric ton of cocoa per hectare at 6–7 years. The migrants thought in terms of quick planting and quick returns. Also, the social and demographic pressure brought by the immigrations rapidly erased any chance of implementing the traditional, extensive tree crop–fallow rotation. Where cocoa was booming, there was no space for cocoa fallows, and abandoning a farm for 5–10 years would have provoked claims on the land by indigenous people. These factors explain why complex agroforests were not an option for the migrants in the Côte d'Ivoire when they started the cocoa cycle and also helps to explain the low adoption rates of agroforest practices in most other regions where migrants dominate the cocoa sector.

The End of the Cycle

The problem of the "Baoulé method" of forest conversion was that at the end of the first cocoa cycle the forest rent had been consumed. When a cocoa farm came to the end of its life cycle, which occurred more rapidly under unshaded conditions, it was difficult to implement the shifting cultivation principle to allow the establishment of a forest vegetation where, after some time, cocoa could be conveniently replanted. Fewer forest trees could regenerate and grow again. With or without the influence of droughts and accidental fires, these old cocoa farms often turned into fallows dominated by the invasive shrub *Chromolaena odorata,* where replanting of cocoa was difficult and mortality high.

Case studies of Baoulé villages in the center-west of the country in the 1990s illustrate the end of a cocoa cycle. In this region around the town of Gagnoa, the cocoa boom started in the mid- to late 1960s. In the 1970s and early 1980s, the Baoulé migrants coming from the savanna areas in the center-north were rightly considered the winners of the race for land and forest. In a village of Baoulé migrants, Petit Toumoudi, interviews with 10 farmers in 1995 show a picture that is representative of the region. Most of the farmers had arrived just before 1970. On the average, they received 9.7 ha of forest and planted more than 90 percent of it (8.9 ha) with cocoa associated with plantains. Severe mortality of the cocoa trees began in the drought year 1983 and continued in the following years. They tried to replant an average of 1.5 ha, half of which did not survive. In 1995, 25 years after their arrival, they ended up with 5.3 ha of low-yielding cocoa. Their comments turned around the exhaustion of soils, indicated by the mortality of plantains, which announced the upcoming mortality of the cocoa trees. They also mentioned reduced and irregular rainfall, reflecting an increased duration of the dry season rather than a decrease in total annual rainfall but certainly also the drier microclimate in the increasingly deforested region (Schroth et al. 1992; Ruf 1995; Léonard and Oswald 1996). They also complained about invasion of weeds and epiphytes, and termites were destroying the cocoa trees. Although they did not mention their age, all were older than 55 years and lacked the labor force for successful replanting.

They also lacked the technique. Techniques that were extremely efficient at forest times had become obsolete in the postforest era. Instead of efficient ways to clear forest, a technique to get rid of the weed shrub *Chromolaena odorata,* which had increasingly taken over the former space of the forest, was needed. It was also increasingly difficult to control the weed pressure in young and old plantations. Furthermore, fire had increasingly become a threat to the

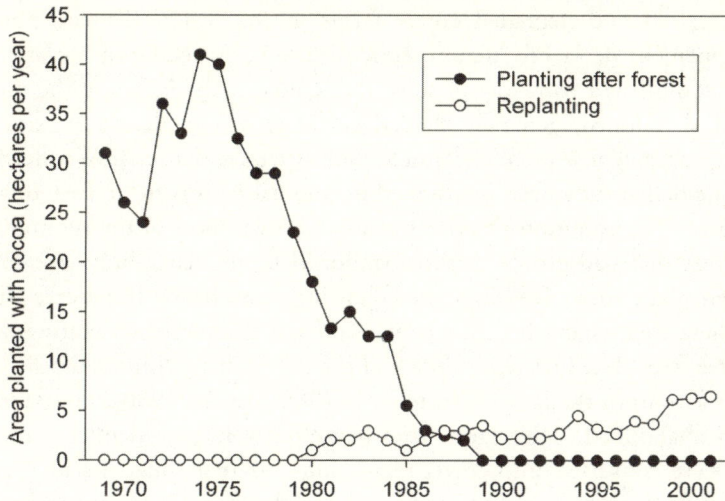

Figure 6.3. Planting and replanting of cocoa by Baoulé migrants in Konankouassikro, center-west of the Côte d'Ivoire (F. Ruf, unpublished survey results, 2001).

plantations in some regions of the country, such as the former cocoa regions of Tanda and M'Bahiakro, which had almost turned into savanna. Data from the Baoulé village of Konankouassikro, also in the Gagnoa area, show that rates of successful replanting in the 1990s remained extremely low compared with planting rates after forest in the 1970s (Figure 6.3).

Technological Innovations in the Postforest Era

The described system of cocoa growing was highly consumptive of forest. As old cocoa plantations needed to be replanted and more migrants arrived, technological innovations developed in response to the postforest conditions. The following three examples are particularly instructive in the present context because, theoretically, they offered opportunities to adopt more sustainable practices, including agroforestry, and perhaps avoid some of the difficulties described earlier. However, because extension services lacked the training and financial means to engage in the necessary dialog with the farmers and promote more sustainable practices, the opportunities were missed and farmers had to rely mostly on their own innovation and channels of information.

In the 1980s, primary forests for cocoa planting became increasingly scarce in the east and center-west regions of the Côte d'Ivoire. Both established and recently arrived migrant farmers therefore looked for alternative sites where cocoa could be planted conveniently. In the 1960s most farmers in the forest zone of the Côte d'Ivoire were still oriented toward robusta coffee, and a several-volume report from that time treated the difficulties of rehabilitating and replanting old and degraded coffee farms in the region (Bureau pour le Développement de la Production Agricole 1963). A decade or so later, the farmers found their own way to solve the "coffee crisis" by using the mostly old and abandoned, generally shaded coffee groves as alternative sites for planting cocoa, whose price was much more attractive than that of coffee (this technique had already been mentioned in colonial reports from the Congo in the 1950s). The common practice was to cut down most of the spontaneous forest trees that had grown in the abandoned farms, rehabilitate the coffee, and then plant cocoa seedlings the following years below the coffee shade. Once these were established, the conversion was completed by cutting down the coffee trees. The first clear reports of the use of this technique in the Côte d'Ivoire date from the late 1970s to early 1980s. In the 1980s, thousands of old and abandoned coffee farms that had effectively turned into secondary forests were converted into cocoa plantations. At that time, it had become clear that the forest would not last forever and that cocoa was very difficult to replant on deforested sites. This could have inspired the farmers to develop a more permanent cocoa culture on these old plantation sites by keeping some of the shade trees and spontaneous regrowth, which could later be turned into secondary forest and then replanted, thereby avoiding the replanting difficulties in the monoculture system. Instead, with the cocoa sector dominated by recently arrived migrants, increasing population pressure, and an active land market, the conversion technique, once proven successful, was adopted on a large scale by migrants, who bought abandoned, forested coffee farms from the indigenous farmers and transformed them into mostly unshaded cocoa plantations (Ruf 1981).

The second innovation occurred in the 1990s, when the Baoulé almost stopped their migrations to the forest zone once there was little forest left for planting, and replanting proved so difficult. However, young Burkinabé kept coming in numbers. Hardly surviving in their own country, they accepted to work at any cost. At that time it became increasingly clear that the future of cocoa in the Côte d'Ivoire would depend on the smallholders' ability to control the invasive shrub *Chromolaena odorata*, which invaded the plantations and dominated the fallows generated by forest clearing and cocoa aging (Ruf 1992). In this situation, many Burkinabé bought 1–2 ha of shrub fallows from indigenous people and replanted them working three times as many hours per hectare as during the previous forest time. The most successful in replanting fallows with cocoa were the recently arrived young Burkinabé migrants, who concentrated their energy on a small area rather than spreading it over a larger farm, as the indigenous farmers and the migrants who had come earlier did (Figure 6.4). Almost for the first time in the history of cocoa growing in the Côte d'Ivoire, thousands of hectares of cocoa were no longer planted after primary and secondary forests but after shrub fallows (Table 6.2). The farmers used simple associations of cocoa trees and plantains. Another possibility,

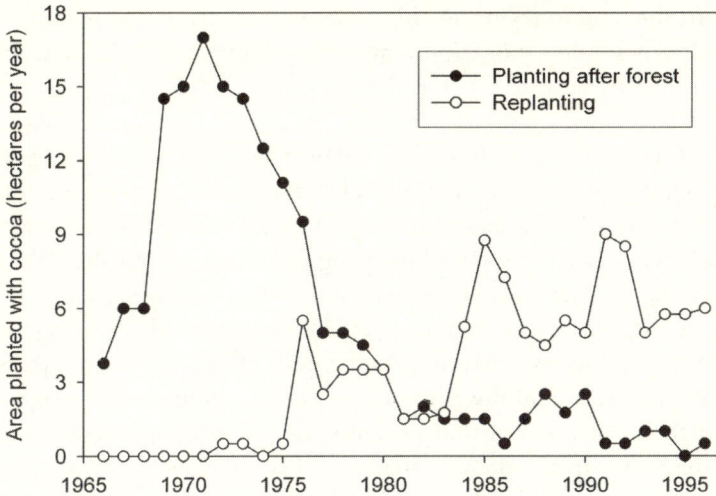

Figure 6.4. Planting and replanting of cocoa by Burkinabé migrants in the center-west of the Côte d'Ivoire (F. Ruf, unpublished survey results, 1997).

Table 6.2. Contribution of cocoa plots planted after forest or replanted to the total cocoa production of farms in the center-west of the Côte d'Ivoire.

	Planted after Forest (%)	Replanted after (Mostly Shaded) Coffee and Cocoa (%)	Replanted after Shrub Fallow (%)[a]	Total Production per Farm (kg yr[-1])
Indigenous Bété (64 farmers)	75	0	25	1,114
Baoulé and other Ivorian migrants (44 farmers)	78	3	19	3,871
Burkinabé and other foreign migrants (46 farmers)	53	11	36	2,311
TOTAL (154 farmers)	71	3	26	2,272

Source: CIRAD (F. Ruf, unpublished data from a 2001 survey).

[a]Often where a previous plantation was destroyed by fire.

more difficult but potentially leading to more permanent, shaded cocoa systems, would have been the use of tree fallows to suppress the *C. odorata* thickets, later to be underplanted with cocoa trees (Schroth et al. 1992). Whether the resulting delay in cocoa planting would have been acceptable to recently arrived immigrants is an open question; efforts to promote the principle to replant fallow lands have been initiated in the Côte d'Ivoire (N'Goran 1998).

The third technological innovation that changed the traditional ways of cocoa growing was the spontaneous adoption of mineral fertilizers by immigrant farmers in the Soubré region in the center-west of the Côte d'Ivoire in the 1990s. Already in the 1960s, agronomists had classified this region as unsuitable for cocoa because of its stony soils. Migrants did not know that and would not have cared. In the 1970s and 1980s, they rushed into the region by the tens of thousands, and in the late 1980s it became the new cocoa belt of the country, taking over from the Dimbokro-Bonguanou region further to the east. In the early 1990s, a huge number of farmers rediscovered what agronomists had said earlier: they started suffering yield declines and sharp increases in tree mortality. Poor soils in combination with strict monoculture accelerated the local cocoa cycle, and 15-year-old cocoa trees looked like trees of twice that age in the eastern region. In the mid-1990s, many cocoa plots had already disappeared. The migrants had only two alternatives: move to new forests further to the west or find a solution on site to slow down the death of their cocoa farms. It is well established that shading reduces nutritional stress in cocoa trees, so one could have again imagined the adoption of shade trees to improve the nutritional status of the cocoa trees and a gradual transition to agroforestry practices. Instead, at a time when extension services had largely disappeared, the farmers found mineral fertilizers. Although the use of fertilizers did not sustain individual farms over decades, it increased yields (possibly doubled them) and thereby gave the farmers an incentive to stay on their farms rather than move into new forests. This temporary solution to the local cocoa crisis may seem less satisfactory in environmental terms than the adoption of agroforestry practices. However, it played a decisive role in delaying a further shift of the cocoa production areas and thereby helped to increase the sustainability of cocoa growing on a regional scale.

Present Trends: Are Ivorian Cocoa Farmers Prepared to Adopt Agroforest Practices?

A trip through the former forest zone of the Côte d'Ivoire with some attention paid to landscapes shows a trend of decreasing shade density in cocoa from east to west. Preliminary survey results at the farm plot level in three villages indicate that large forest trees are present in cocoa plots at a density of about five trees per hectare in the east, two trees per hectare in the center-west, and less than one tree per hectare in the southwest (Delerue 2003). This decrease in the use of shade from east to west reflects the increasing dominance of the cocoa sector by migrants, who tend to use less shade than indigenous farmers (Table 6.3). In the east, indigenous farmers kept immigration under control, whereas in the western region migrants locally represent 80–99 percent of the farmer population.

For both indigenous farmers and migrants, most of the noncocoa trees in the farm plots are planted or spontaneous fruit trees rather than forest trees (Table 6.3), mainly oil palms, cola, orange, and avocado trees, around 20 trees

Table 6.3. Density of noncocoa trees and annual revenues in 15 cocoa farm plots in the eastern, center-west, and western regions of the Côte d'Ivoire in 2002.

	Number of Noncocoa Trees per Hectare	Number of Fruit Trees per Hectare	Number of Forest Trees per Hectare	Estimated Noncocoa Revenues per Farm[a]	Estimated Cocoa Gross Revenues per Farm[a]
Indigenous farmers	37	23	3.7	70,000	1,065,000
Migrants	21	15	1.8	125,000	1,400,000

Source: CIRAD-TERA survey (Delerue 2003).

[a]In African Financial Community francs.

per hectare (compared with about 1,500 cocoa trees per hectare). For both indigenous farmers and migrants in the survey, the revenues obtained from these fruit trees were only 5–10 percent of those obtained from cocoa. The slightly higher noncocoa revenues of the migrants could be explained by better access to the cola trade networks in their home countries, Burkina Faso and Mali, and more commercially oriented selection of noncocoa trees.

In ecological terms these fruit trees are no substitutes for the giant forest remnant trees that constituted the overstory of the traditional cocoa agroforests. However, today most Ivorian cocoa farmers, and especially the young generation, seem to have a negative perception of permanent shade from forest trees. In a 2002 survey of 65 farms focusing on the center-east and center-west regions, 70 percent of the respondents found shade from forest trees useless to cocoa, and 20 percent found it useful only as temporary shade for plantation establishment, for which (at least on suitable soils) most farmers preferred plantains. About 90 percent believed that shade trees increased not only diseases but also pest problems of the cocoa, because they allowed insects to hide in the canopies and escape insecticides, and reduce their cocoa revenues (F. Ruf, unpublished data). Although the farmers acknowledge that shade prolongs the life of their cocoa farms, this is not sufficient to prevent zero-shade plantations from spreading.

This trend tends to be reinforced by a generational change of the farm owners: when a recently arrived migrant buys an old shaded cocoa farm from an indigenous farmer, the first decision often is to eliminate the shade trees. In the indigenous population, an intergeneration transfer often provokes the same behavior. Many young cocoa farmers want immediate revenues, irrespective of the long-term impact on the cocoa farm. Because chainsaw teams are readily available, it is easy to turn the shade trees into cash. This type of behavior is even more likely if a cocoa farm is inherited by several family members together and final ownership is uncertain. Even in the most remote migrant villages close to the border of the largest forest reserve in the country, the Taï National Park, it has become very common to cut down the giant trees that are said to provide habitat to insect pests (mirids) and are considered globally harmful to cocoa (Ruf 1996).

Although more complete studies of the attitudes toward shade among Ivorian cocoa farmers and the factors that influence them are needed, evidence suggests that from an incipient tradition of cocoa growing in potentially sustainable but low-yielding rotational agroforests, with a 35-year longevity of the cocoa trees and a fair chance of successful replanting after a forest-fallow period, little has survived after four decades of dominance of the sector by migrants who had neither the tradition nor the incentives and technical assistance to adopt this type of extensive agriculture. Agricultural science and extension have also played a major part in this development by favoring low-shade systems over much of the past three decades, in the Côte d'Ivoire as elsewhere.

This history has brought the Côte d'Ivoire to the top of the list of cocoa-exporting countries in the world but has cost it not only most of its forest but also its former tradition of more conservative use of forest resources. As international efforts increase to compensate tropical countries and their inhabitants for their environmental services, such as the conservation of biodiversity and the carbon stocks of their standing forests (see Chapter 4, this volume), the farmers of the Côte d'Ivoire may find themselves at a disadvantage to their counterparts in other tropical regions, where cocoa-growing practices have been more conservative. Any efforts to move the Ivorian cocoa economy toward more sustainable practices must take into account the experiences of other countries. This brings us back to the chocolate forests of Bahia with which this chapter began and to those of southern Cameroon.

Cocoa Agroforest Traditions: Remnants of the Past or Examples for the Future?

Readers of historical descriptions of regions that are now reputed for their shaded cocoa systems, such as Grenada (Knapp 1923; Preuss 1901), Bahia in Brazil (van Hall 1914), and the eastern region of Ghana (Knapp 1923; Revue Générale de Botanique 1924), may be surprised to learn that in the early twentieth century most cocoa farms in these regions were unshaded. In countries such as São Tomé, attempts to increase cocoa yields by removing shade date back to the 1920s (Navel 1921). These cases deserve the attention of historians and agronomists because the fact that farmers in East Ghana and Bahia apparently realized benefits of shaded systems may bear lessons for the present. However, globally speaking, these cases of increased shade adoption in the twentieth century seem to be exceptions. Until the 1960s most cocoa was grown under shade, and since the mid-1970s most cocoa has been grown unshaded. Bahia, southern Cameroon, and southwest Nigeria stand out as regions where complex and seemingly sustainable cocoa agroforests have evolved and been well conserved into the twenty-first century. Why has this happened, and what are their chances of surviving into the future? Are there lessons to be learned for the introduction of more sustainable cocoa-growing practices in other tropical regions?

Agroforest Estates in Bahia, Brazil

Bahia is exceptional in the cocoa world for a second reason: most of its cocoa is grown on large estates (Ramos 1976; Alger and Caldas 1992; Greenhill 1996). This makes it difficult to apply lessons from this case study directly to the smallholder farms that dominate the cocoa sector in most other countries. Nevertheless, it is instructive to see that the attitudes toward shade in this region, though insufficiently researched, oppose to some extent those of the Côte d'Ivoire.

Historical evidence suggests that in the nineteenth century cocoa growing in Bahia was associated with the familiar shifts in local production zones. Martius (cited by Monbeig 1937) writes:

> In 1820, the grains (beans) were spread over the hot banks of fine sand to dry, then taken down on pirogues to the maritime ports, Ilhéos, Belmonte, Caravellas etc., a journey of three to four days. This traffic has considerably diminished, firstly due to the construction of the railway, but also because of the progressive shift away from the river, which has led to the abandonment of the old fazendas; near Itabuna, all the way along the river, one often sees the now deserted buildings of these fazendas. (p. 210)

In the early twentieth century, a structure of large estates was built in Bahia (Mahony 1996), and the practices of cocoa growing on these estates were described by James (1942, cited by Leeds 1957) as follows:

> Plantation practices in Bahia are notably extensive and exploitative. . . . Once planted, the young trees are given almost no attention until they come of bearing age. . . . Thereafter, instead of clearing away the brush and weeds each year, this kind of work is done only every four or five years. When yields decline, the older plantations are abandoned and new ones are set out on virgin soils. . . . Here is speculative and destructive economy at its worst; one that is bringing temporary and unstable activity to Ilhéus and Salvador. . . . In short, the land, for the cacao zone capitalist class, is to be raped like a woman of easy life, rather than cherished like a wife. (p. 400)

Ironically, the structure of large estates where these "notably extensive and exploitative" practices were observed was also a key factor favoring the development of the cabruca system. As Alger and Caldas (1992) note, farmers on large estates tended to plant cocoa under native tree shade (i.e., in the cabruca system) and to invest in only part of their holdings, leaving more under forest, whereas small farmers were more likely to clearcut the forest on a larger part of their property and use bananas and other planted shade instead of native trees for their cocoa crop. Planting cocoa in the cabruca system took smaller investments per unit area than the clearcut system and was more amenable to the minimal management system practiced by absentee owners of large estates (Hill 1999). The cabruca system also gave them the flexibility to hire workers and intensify plantation management when the cocoa prices were high and to fire the workers when the prices fell. Under the sociopolitical and legal conditions of the late twentieth century, with improved workers' rights, and since the outbreak of the witches' broom fungus, which necessitates continuous efforts to keep the disease in check, these former advantages (from the perspective of estate owners) of the cabruca system have to some extent been lost.

In the late 1980s and 1990s, low cocoa prices coincided with the spread of the witches' broom disease in Bahia, encouraging some planters to (illegally) sell timber from their residual forests that had been reserved for future plantations and the cocoa agroforests themselves to compensate for low cocoa revenues. Moreover, some cocoa agroforests were abandoned, but others were converted into pasture (Alger and Chaldas 1992; Trevizan 1996; Johns 1999). In a November 1996 survey, 30 farmers declared an average loss of 70 percent of their labor force and a similar drop in production. All planters mentioned tree felling and timber selling by neighbors, and several planters anticipated a large-scale conversion of cocoa farms into other land uses, including pasture.

In 2003, however, the picture looks different. According to Brazilian agronomists and private cocoa pod counters and forecasters, the rate of tree felling and transformation of cabruca cocoa into pastures in the mid- to late 1990s, though higher than before, did not exceed 10 percent in the last 10 years (P. Petersen, pers. comm., 2003). Most of it occurred in drier parts of the region where the cocoa trees had also been affected by an increased frequency of droughts and inconsistency of rainfalls since 1982 (Carzola et al. 1995). These climatic events appear to have increased the sensibility of the Bahian cocoa planters to the ecological functions of shade trees, especially the retention of soil moisture and the microclimatic protection of the cocoa trees. In the mid-1990s, a study highlighted the importance of ecological functions of the shade canopy in the farmers' perception: protection from the sun and conservation of soil moisture and soil fertility (Johns 1999). In the same period, many farmers accused the extension service of having misled them in the 1980s by promoting shade removal in old farms and the establishment of new farms with little or no shade and expressed their worries about the climate change, especially more frequent drought, some giving drought the same importance as the witches' broom disease in reducing cocoa yields (F. Ruf, unpublished survey data, 1996). In 2003, protection of the cocoa trees from drought is the very first function of shade trees that Bahian cocoa farmers mention (P. Petersen, pers. comm.).

Why this emphasis on the ecological functions of shade trees is found in Bahia but not in some other regions that have also experienced droughts, such as the Côte d'Ivoire, is an open question in need of research. It may be related to shallow soils (P. Alvim, pers. comm., 1996), but this hypothesis must be tested. One way to explore how the experiences of Bahian cocoa growers can provide lessons for their counterparts in other cocoa-growing regions would be to promote the exchange of experiences between farmers from these regions and joint visits of their respective plantations.

Cocoa Agroforests in Southern Cameroon

Similar to the cabruca cocoa in Brazil, the 50-year-old cocoa agroforests under heavy shade formed by natural forest trees in southern Cameroon are among the best examples in Africa of seemingly permanent agriculture that preserved a forest environment and some of its biodiversity. Satellite imagery is unable to distinguish these cocoa agroforests from closed canopy forest. That complex agroforests have developed in this region as the predominant form of cocoa growing, in contrast to most of the Côte d'Ivoire, is best explained by the predominance of indigenous farmers among the cocoa growers of southern Cameroon. Other cocoa-producing regions in Cameroon, such as the M'Bam region and the southwest region, received more immigrants and followed more monocultural trends of cocoa growing, comparable to those in the Côte d'Ivoire (Losch et al. 1991). Furthermore, research and extension services in Cameroon favored shade both before and after independence, and cocoa farms established under forest tree shade in the 1950s have now turned into huge chocolate forests as the forest trees have also aged by 50 years.

A disadvantage of heavy shading in these cocoa agroforests is low yields, on the order of 300 kg of cocoa per hectare per year (Arditi et al. 1989; Losch et al. 1991). Where shading is too intense, it may also increase pod rot (*Phytophthora megakarya*), the most serious cocoa disease in this country, so reducing shade intensity may be a component of integrated disease management schemes (Berry 2001). However, it is difficult to regulate the shade provided by such giant trees, except by cutting them down and making planks (Kaiser 1987; Ruf and Zadi 1998). The heavy, permanent shade of the forest trees also raises its own type of replanting difficulties: farmers trying to regenerate their cocoa farms by underplanting the agroforests with cocoa seedlings often find that in the dense shade the trees become tall and thin in their search for light and form their pods 2–3 m high, where they are difficult to harvest (and diseased pods are difficult to remove). Therefore, farmers may choose to establish new cocoa plots at another site, outside the plantation, where the shade intensity is easier to regulate. Ideally, this would be a previously planted area under secondary forest, as in the old rotational system, but in practice farmers often use primary forest if available.

Production shifts at village and regional scales provide historical evidence of this process. Between 1960 and 1963, the East province and the department of Dja et Lobo, where cocoa is grown mainly in agroforests, produced a yearly average of 8,000 and 9,000 metric tons of cocoa, respectively. Until 1984, annual production had fallen to 5,300 metric tons per year, while cocoa production in the southwest province jumped from 7,300 to 27,000 metric tons per year and that in the M'Bam department from 6,200 to 10,900 metric tons per year (Losch et al. 1991). In the latter regions, cocoa is produced mainly by migrants in lightly shaded systems. Production shifts also occurred within the departments: in the Nyong et M'foumou department, where cocoa is produced in ancient chocolate forests, the main cocoa production centers in the 1970s were the road from Akonolinga to Yaoundé and Endom in the south; by the late 1980s, cocoa production had moved to other districts such as Nwane Soo and Ayos Fang Biloun in the north of the department (F. Ruf, unpublished survey data, 1990).

Available data suggest that noncocoa revenues from these agroforests usually are insufficient to compensate the farmers for reduced cocoa revenues at times of low cocoa prices. Therefore, when the cocoa price collapsed in the 1990s, farmers resorted to new forest clearing oriented toward food crop production as a survival strategy (J. Gockowski, pers. comm., 1998). However, during this time of economic crisis cocoa farmers close to the urban market of Yaoundé were successful at diversifying their farms by planting mandarin orange trees, often in places where cocoa trees had died and were difficult to replant (Aulong et al. 1999; Gockowski and Dury 1999), and this could indicate a way to further commercially oriented diversification of the cocoa farms in other regions. In the 1980s, farmers in the Nyong et M'foumou department mentioned the tree *Voacanga africana* (Obahtoan) as a source of an exportable medicinal product, but this export trade has ceased, probably because of a lack of certification and nonconformity to the legislation of the European market (Arditi et al. 1989). An interesting species is the African plum tree (*Dacryodes edulis*), which according to a survey of 300 farmers in southern Cameroon was planted by 83 percent of the respondents in their cocoa farms (Sonwa et al. 2000).

In conclusion, although cocoa agroforests have successfully conserved part of the forest environment of southern Cameroon, they have not been able to sustain farm revenues at times of crisis, although examples of successful economic diversification of cocoa farms are emerging. Replanting problems in agroforests are different from those in no-shade systems, but they do exist and may have contributed, together with immigrations into the southwest and

M'Bam regions, to historic shifts in cocoa production at department and province levels. As long as the demographic pressure remains low in southern Cameroon and migrants do not enter the region through a potential land market, the cocoa agroforests have a good chance to survive. However, the model may be difficult to reproduce if demographic pressure increases rapidly, unless cocoa and noncocoa revenues from these agroforests can be increased.

Conclusions

Historically, cocoa has been an important source of tropical deforestation, and it is still a nonnegligible one today. At the same time, it is a crop on which many conservationists and natural resource managers base their hopes for an agriculture that not only provides a living for tropical farmers but also helps to conserve a degree of biodiversity in tropical forest landscapes. A critical question is whether agroforestry practices can help stabilize cocoa growing systems and prevent the further move of this crop to new forest frontiers while providing sustainable income to successive generations of tropical farmers.

In all three countries discussed in this chapter—the Côte d'Ivoire, Brazil, and Cameroon—there are or were traditions of growing cocoa in agroforests. Obviously, this fact did not prevent deforestation by cocoa farmers, but it helped to slow down the process, at least in certain parts of these countries and during certain periods, by extending the useful life of the cocoa tree and, critically, providing a basis for the replanting of cocoa after a period of forest fallow and thereby for more permanent cocoa systems. This basis was seriously compromised when cocoa was cultivated in strict monoculture using more destructive techniques of forest conversion, as in the case of the Côte d'Ivoire. What is the future of the existing agroforests, and what are the chances of such techniques being adopted in regions where cocoa is grown in monoculture? Only preliminary answers can be given to these questions.

Complex cocoa agroforests have evolved under specific technological, economic, social, and historical contexts. When these contexts change, as through immigration, such traditional systems may become unstable. As we have seen in the case study of the Côte d'Ivoire, several such changes may occur simultaneously. An important factor that has historically favored the development of complex agroforests worldwide was the need to reduce labor costs when establishing new plantations against a background of a low level of technology and abundant land. By maintaining a large part of the forest trees, farmers saved time for forest clearing and weed control. As land became less abundant through immigration and more effective techniques of forest clearing became available, important premises of complex cocoa agroforestry were lost. An important factor in the move of the Ivorian and part of the Ghanaian cocoa economies toward zero-shade systems and monocultures was the introduction of new cocoa varieties that needed less shade and had a more rapid initial development, thereby reducing the attractiveness of extensive and labor-saving agroforestry practices during the long investment phase. Furthermore, the use of insecticides and later of mineral fertilizers gave an immediate advantage to the monoculture system, and once these innovations had been adopted there was even less incentive to adopt (or keep) agroforestry practices.

Despite the advantages of cocoa agroforests in terms of longevity of the tree crops and ease of replanting, the chances of the traditional agroforestry practices using primary forest trees as shade and forest fallows to facilitate replanting probably are low in most regions, especially where demographic pressure increases, unless strong incentives develop or are created for the maintenance or adoption of such systems (e.g., markets for the timber from forest fallows, certification schemes). In many regions, the most promising option to promote cocoa agroforestry probably is not so much to try to save the old forest trees in the remaining agroforests as to rebuild a new agroforestry tradition based on the planting of valuable timber or other useful trees together with cocoa.

The chance that these new agroforests will be adopted will increase with their ability to provide higher and more stable income to farmers from both cocoa and noncocoa products. Cocoa revenues from agroforests may increase with the introduction of clonal materials resistant to major cocoa diseases, as has recently started in Bahia (witches' broom disease) and West Africa (*Phytophthora* pod rot). Stable revenues cannot be obtained under conditions of fluctuating cocoa prices without markets for noncocoa products, like those in the Yaoundé area in southern Cameroon, leading to the diversification of cocoa farms with fruit trees. In all tropical countries, the development of urban markets will encourage this type of diversification on the plot and farm scale.

Specifically in the Côte d'Ivoire, a hope for cocoa agroforestry is the opening of the timber market to farmers in 1999, which legalizes the commercialization of timber and makes it less subject to informal taxation. Once the information about their new rights has reached the cocoa smallholders, which is not the case yet, it should increase the attractiveness of managing trees in cocoa farms. Whether this potential can be used to help establish more diversified and potentially more sustainable cocoa systems will be highly instructive for other African cocoa growing regions. The key factor in cocoa sustainability is not necessarily longevity of the tree crop but successful replanting. Replanting is costly, so it is important that at the end of the cocoa trees' life cycle there is capital that can be used to finance replanting, and this could be provided by the trees. This idea is nothing new: "As cocoa plantations should not live more than some twenty years, native people should be encouraged to intercrop cocoa with other trees, every 15 meters, for instance with oil palms, colas and avocados which provide them with valuable produce when the main crop disappears" (Vuillet 1925, 5).

Whereas intercropping with oil palms and fruit trees can help diversify the revenues from productive cocoa groves, timber trees may accumulate the capital needed for replanting during the life of a plantation. Imagine a couple of Côte d'Ivoire migrants who started clearing forest and planting cocoa in 1968 when they were in their twenties. Between 1989 and 1993, at the height of the cocoa crisis, suppose they sold a few iroko (*Milicia excelsa*) or frake (*Terminalia superba*) trees, or perhaps even a sipo (*Entandrophragma utile*). They could use the timber money to replant their cocoa grove, which was facilitated by still some large trees and enough trees of intermediate size to provide the necessary shade. They could retain and tend these trees in the plantation as retirement capital and as an inheritance for their sons and daughters to finance the next replanting. Perhaps they even planted some tree seedlings that they had collected in a nearby forest or received from the extension service. Their maintenance in the cocoa farm, along with the cocoa crop, would be almost cost-free.

Ways to trigger such investments in sustainability through contracts and institutional arrangements between farmers and forestry services or wood-processing companies should be explored. They need to be backed up by measures to protect remaining forests, signaling to the farmers that further shifts to the forest frontier are not an option. Whether the diversification of cocoa farms relies on timber, fruits, medicinal products, ecotourism, carbon credits, or payments for conserving biodiversity in the buffer zone of a forest park, what is most needed to make cocoa agroforests more sustainable is access to reliable and diversified markets for their products and services.

Acknowledgments

Most of the data used in writing this chapter, especially from the Côte d'Ivoire, were collected in a research project granted to Coopération Internationale en Recherche Agronomique pour le Développement (CIRAD) by the Fonds Industriel pour la Recherche Cacaoyère from sales of the cocoa bean stock. The CIRAD enumerator team in the Côte d'Ivoire, especially Allagba Konan and Abelle Kla, deserve special thanks for the quality of their fieldwork. Comments by A. M. N. Izac, J. Gockowski, and A. Angelsen greatly helped to improve the quality of this chapter.

References

Alger, K., and M. Caldas. 1992. *The crisis of the cocoa economy and the future of Bahian Atlantic Forest.* Los Angeles: Latin American Studies Association Conference.

Alves, M. C. 1990. *The role of cacao in the conservation of the Atlantic Forest of southern Bahia, Brazil.* Master's thesis, University of Florida, Gainesville.

Amanor, K. S., and M. K. Diderutuah. 2001. *Share contracts in the oil palm and citrus belt of Ghana.* Paris: HED/GRET.

Arditi, C., F. Jarrige, and F. Ruf. 1989. *Le Département du Nyong et M'Foumou dans la crise du marché international.* Montpellier, France: CIRAD.

Aulong, S., S. Dury, and L. Temple. 1999. *Les conditions d'extension de l'agrumiculture dans le centre du Cameroun.* Working paper. Montpellier, France: CIRAD.

Berry, D. 2001. Rational chemical control and cultural techniques. Pages 152–192 in D. Mariau (ed.), *Diseases of tropical tree crops.* Montpellier, France: CIRAD.

Bureau pour le Développement de la Production Agricole. 1963. *Etude pour la reconversion des cultures de caféier dans la République de Côte d'Ivoire.* Résumé du Rapport de Synthèse, Paris.

Carzola, I. M., L. P. Dos Santos Filho, and A. Gasparetto. 1995. Cocoa harvest shortfalls in Bahia, Brazil: long and short term factors. Pages 175–187 in F. Ruf and P. S. Siswoputranto (eds.), *Cocoa cycles. The economics of cocoa supply.* Cambridge, UK: Woodhead Publishing.

Chauveau, J. P., and E. Léonard. 1996. Côte d'Ivoire's pioneer fronts: historical and political determinants of the spread of cocoa cultivation. Pages 176–194 in W. G. Clarence-Smith (eds.), *Cocoa pioneer fronts since 1800.* London: Macmillan.

Cope, F. W. 1976. Cacao. Pages 285–289 in N. W. Simmonds (ed.), *Evolution of crop plants.* London: Longman.

Dean, W. 1995. *With broadax and firebrand: the destruction of the Brazilian Atlantic Forest.* Berkeley: University of California Press.

Delawarde, J. B. 1935. *Les défricheurs et les petits colons de la Martinique au XVIIe siècle.* Paris: René Buffault.

Delerue, F. 2003. *Aperçu sur les stratégies agroforestières des planteurs de cacao en Côte d'Ivoire. Premiers résultats et hypothèses.* FIRC Project, Report No. 2. Montpellier, France: CIRAD.

de Rouw, A. 1987. Tree management as part of two farming systems in the wet forest zone (Ivory Coast). *Acta Oecologica Oecologia Applicata* 8:39–51.

Direction des Grands Travaux. 1992. *Etude du bilan des superficies de forêt dense en Côte d'Ivoire.* Abidjan, Côte d'Ivoire: Direction des Grands Travaux.

Ekanade, O. 1985. The effects of cocoa cultivation on some physical properties of soil in south-western Nigeria. *International Tree Crops Journal* 3:113–124.

FAO. 1957. *Cacao. Tendances actuelles de la production, des prix et de la consommation.* Série des Monographies de Produits 27. Rome: Food and Agriculture Organization of the United Nations.

Gockowski, J., and S. Dury. 1999. *The economics of cocoa-fruit agroforests in southern Cameroon.* Working paper, IITA and CIRAD, Yaoundé, Cameroon, and Montpellier, France.

Gouyon, A. 1995. Government politics and smallholder changes in sustainable tree crop development in the tropics: a comparison of rubber and cocoa. Pages 291–313 in F. Ruf and P. S. Siswoputranto (eds.), *Cocoa cycles. The economics of cocoa supply.* Cambridge, UK: Woodhead Publishing.

Greenhill, R. G. 1996. A cocoa pioneer front, 1890–1914: planters, merchants and government policy in Bahia. Pages 86–104 in W. G. Clarence-Smith (ed.), *Cocoa pioneer fronts since 1800. The role of smallholders, planters and merchants.* London: Macmillan.

Hill, P. 1999. *Cacau acabou: crisis and change in the Bahian cocoa economy.* Unpublished document.

Johns, N. D. 1999. Conservation in Brazil's chocolate forest: the unlikely persistence of the traditional cocoa agroecosystem. *Environmental Management* 23:31–47.

Kaiser, M. 1987. *Rapport d'enquête sur la culture du cacao au Cameroun.* Paris: EHESS.

Knapp, A. W. 1923. *The cocoa and chocolate industry. The tree, the bean, the beverage.* London: Sir Isaac Pitman & Sons.

Leeds, A. 1957. *Economic cycles in Brazil: the persistence of a total culture-pattern: cacao and other cases.* Ph.D. thesis, Columbia University, New York.

Legrand, A. 1999. *La filière cacao en Côte d'Ivoire dans le contexte de la liberalisation: evolution de la concurrence, des prix et de la qualité.* Montpellier, France: CNEARC and CIRAD.

Léonard, E., and M. Oswald. 1996. Une agriculture forestière sans forêt. Transformation de l'environnement et mise en place de systèmes stables en Côte d'Ivoire forestière. *Nature Science et Sociétés* 4:3–12.

Losch, B. 1995. Cocoa production in Cameroon: a comparative analysis with the experience of Côte d'Ivoire. Pages 161–177 in F. Ruf and P. S. Siswoputranto (eds.), *Cocoa cycles. The economics of cocoa supply.* Cambridge, UK: Woodhead Publishing.

Losch, B., J. L. Fusiller, and P. Dupraz. 1991. *Stratégie des producteurs en zone caféière et cacaoyère du Cameroun. Quelles adaptations à la crise?* Collection Documents Systèmes Agraires No. 12. Montpellier, France: CIRAD-DSA.

MacLeod, M. 1973. *Spanish Central America. A socioeconomic history, 1520–1720.* Berkeley: University of California Press.

Mahony, M. A. 1996. *The world cocoa made: society, politics and history in southern Bahia, Brazil, 1822–1919.* Ph.D. dissertation, Yale University, New Haven, CT.

Monbeig, P. 1937. Colonisation, peuplement et plantation de cacao dans le sud de l'Etat de Bahia. *Annales de Géographie* 46:278–299.

Navel, C. 1921. *Les principaux ennemis du cacaoyer aux Îles de San Thome et de Principe.* Paris: E. Larose.

N'Goran, K. 1998. Reflections on a durable cacao production system: the situation in the Ivory Coast, Africa. In *First International Workshop on Sustainable Cocoa Growing, 1998, Smithsonian Institute, Panama.* Online: http://nationalzoo.si.edu/ConservationAnd Science/MigratoryBirds/Research/Cacao.

Oswald, M. 1997. *Recomposition d'une société au travers de plusieurs crises: la société rurale Bété.* Thèse de 3e Cycle, Institut National Économique Paris-Grignon, Paris.

Preuss, P. 1901. *Cocoa, its cultivation and preparation.* Archives of Cocoa Research Vol. 3, edited by H. Topoxeus and G. Topoxeus, 1985, International Office of Cocoa and Chocolate, Brussels.

Ramos, H. S. 1976. Estructura agraria. In *Diagnostico socioeconomico da região cacaueira,* Vol. 14. Bahia, Brazil: Comissão Executiva do Plano da Lavoura Cacaueira (CEPLAC), Ilhéus-Itabuna.

Revue Générale de Botanique. 1924 (Vol. 36). Paris: Librairie Générale d'Enseignement.

Ruf, F. 1981. Le déterminisme des prix sur les systèmes de production en économie de plantation ivoirienne. *Cahiers du CIRES* 28–29:35–52.

Ruf, F. 1985. Politiques économiques et dévelopement agricole. Quelques points de repère dans les pays africains producteurs de cacao. In CIRAD 1985, "Etats, développement, paysans." Proceedings of the seminar organized in Montpellier by CIRAD, September 1985, CIRAD, Montpellier, pp. 62–71.

Ruf, F. 1987. Eléments pour une théorie sur l'agriculture des régions tropicales humides. I: De la forêt, rente différentielle au cacaoyer, capital travail. *L'Agronomie Tropicale* 42:218–232.

Ruf, F. 1992. *Après la forêt, quelle stabilisation de l'agriculture de plantation? Le cas du Département d'Abengourou, Côte d'Ivoire.* Unpublished report, CIRAD, Montpellier, France.

Ruf, F. 1995. *Booms et crises du cacao. Les vertiges de l'or brun.* Paris: Karthala.

Ruf, F. 1996. *Les booms cacao de la Côte d'Ivoire . . . et du Burkina Faso. L'Accélération des années 1980/90.* Montpellier and Paris: CIRAD and Ministère de la Coopération.

Ruf, F. 2001. Tree crops as deforestation and reforestation agents: the case of cocoa in Côte d'Ivoire and Sulawesi. Pages 291–315 in A. Angelsen and D. Kaimowitz (eds.), *Agricultural technologies and tropical deforestation.* Wallingford, UK: CABI Publishing.

Ruf, F., and P. S. Siswoputranto. 1995. *Cocoa cycles. The economics of cocoa supply.* Cambridge, UK: Woodhead Publishing.

Ruf, F., and Yoddang. 2001. Cocoa migrants from boom to bust. Pages 97–156 in F. Gérard and F. Ruf (eds.), *Agriculture in crisis: people, commodities and natural resources in Indonesia, 1996–2000.* Richmond, VA: Curzon Press; Montpellier, France: CIRAD.

Ruf, F., and H. Zadi. 1998. Cocoa: from deforestation to reforestation. In *First International Workshop on Sustainable Cocoa Growing, 1998, Smithsonian Institute, Panama.* Online: http://nationalzoo.si.edu/ConservationAndScience/MigratoryBirds/Research/Cacao.

Ruthenberg, H. 1980. *Farming systems in the tropics.* Oxford, UK: Clarendon.

Schroth, G., B. Pity, and K. N'Goran. 1992. *Stabilisation des Systèmes de production dans la région d'Abengourou. Aspects agro-écologiques.* Unpublished report, Centre Technique Forestier Tropical de Côte d'Ivoire, Abidjan, Côte d'Ivoire.

Schwarz, L. J. 1931. *Cocoa in the Ivory Coast.* Washington, DC: U.S. Department of Commerce.

Sonwa, D. J., S. F. Weise, A. A. Adesina, M. Tchatat, O. Ndoye, and B. A. Nkongmeneck. 2000. Dynamics of diversification of cocoa multistrata agroforestry systems in southern Cameroon. *International Institute of Tropical Agriculture Annual Report,* 1–3. Ibadan, Nigeria: International Institute of Tropical Agriculture.

Touzard, J. M. 1993. *Croissance et déclin cacaoyer dans l'Amerique Centrale coloniale: essai d'histoire économique à l'usage des réflexions prospectives contemporaines.* Collection Repères. Paris: CIRAD.

Trevizan, S. D. P. 1996. *Mudanças no sul da Bahia associadas à vassoura-de-bruxa do cacau.* Paper presented at XII International Conference on Cocoa, Salvador, Bahia, 17–23 November 1996.

van Hall, C. J. J. 1914. *Cocoa.* London: Macmillan.

Vuillet, J. 1925. Notes sur la culture du cacaoyer en Côte d'Ivoire. *L'Agronomie Coloniale* 91:1–10.

Chapter 7

Achieving Biodiversity Conservation Using Conservation Concessions to Complement Agroforestry

Eduard Niesten, Shelley Ratay, and Richard Rice

Throughout the tropics, cultivation of agricultural commodities drives habitat conversion and biodiversity loss (McNeely and Scherr 2001). A role clearly exists for targeted interventions to decelerate and mitigate the impacts of this process, and in many cases agroforestry presents an alternative that is preferable to clearcuts and monocultures. However, from a strict conservation perspective, agroforestry systems are a compromise rather than a solution (Terborgh and van Schaik 1997). Although agroforestry initiatives can create corridors or buffer zones in a patchwork of forest and production areas, they nevertheless impose a disturbance on the ecosystem; given a choice, biodiversity protection is better served by continuous intact habitat than by the fragmentation inherent in a patchwork (Laurance and Bierregaard 1997). Moreover, agroforestry systems may or may not be sustainable in the medium to long term and therefore offer uncertain outcomes even where adopted as a conservation strategy.

Agroforestry rests on the premise that forests and the natural resource base must generate income from a flow of products to benefit farmers. Confining income to that which can be generated from flows of physical output limits the scope for action by conservation interests and income opportunities for local stakeholders. The danger of exclusively linking income to production is particularly well illustrated in areas with deteriorating economic prospects for agricultural commodity production, such as coffee-producing regions under conditions of low coffee prices (Sanchez 2002). In such situations, the driver of continued habitat loss is not so much market incentives as a lack of viable exit options for farmers, as many farmers living at tropical forest margins have no choice but to rely on destructive agricultural practices for their survival. Increasing international willingness to pay for global biodiversity conservation creates another land use option that does not rely on physical extraction; instead, local stakeholders can be financially rewarded for reduced dependence on physical product flows from irreplaceable reservoirs of biodiversity, thereby advancing both conservation and alternative income opportunities.

One promising land use alternative that provides direct remuneration for conservation services may be found in the conservation concession approach developed by the Center for Applied Biodiversity Science (CABS) at Conservation International (CI), in collaboration with Hardner and Gullison Associates, LLC (Hardner and Rice 2002). Conservation concessions may be applicable in a variety of resource-based land use contexts (such as timber extraction, perennial crop production, and livestock grazing), but this chapter focuses on its potential uses as a complement to agroforestry efforts. After a generalized overview of the conservation concession model, this chapter presents two scenarios for applying the model. The first relates to coffee production in Colombia, describing how conservation concessions may be used to retire aging coffee farms challenged by adverse economic conditions. The second suggests ways in which the conservation concession approach can fortify agroforestry efforts, targeting cocoa production in Bahia, Brazil. Although these two scenarios do not address all the potential intricacies and variations of conservation concessions or agroforestry, they do suggest that these two tools together can generate substantial biodiversity benefits. The final section explores various considerations that may complicate practical implementation of the conservation concession approach.

Agroforestry and Conservation Concessions

Several contributors to this volume discuss ways in which agroforestry can be an improvement over other land uses from ecological, agricultural, and economic perspectives. Where habitat conversion already has occurred, the ecological benefits of agroforestry systems relative to land used for annual crops, cattle pasture, or monoculture plantations justify efforts to promote agroforestry. Insofar as habitat conversion is inevitable (or already has taken place), agroforestry systems can serve as corridors or buffer zones, with benefits including lower use of agrochemicals, reduced soil erosion, less nutrient leaching and watershed degradation, and, depending on selection of species included in the system, enhanced nitrogen fixation and carbon sequestration. Perhaps the greatest biodiversity benefit of certain agroforestry practices, under certain conditions, is the potential to stabilize the spatial proliferation of production systems, thus protecting forest borders (see Chapter 5, this volume).

Any degree to which agroforestry can halt or slow shifting agricultural frontiers offers a valuable contribution to biodiversity conservation. Agricultural commodity production drives deforestation as conventional techniques for a broad range of crops exhaust soil resources; under these techniques, declining productivity on aging plots leads growers to establish new plots in areas of intact habitat (Angelsen and Kaimowitz 2001; see also Chapter 6, this volume). However, agroforestry land uses also can induce deforestation in a similar dynamic, as when declining productivity of cocoa trees leads farmers to establish new plots in pristine forests (Petithuguenin 1995; Ruf 1995). Even if agroforestry supports greater biodiversity than other cultivation systems, the greatest ecological benefits from agroforestry are derived if production is concentrated on already cleared lands; if forests are being disturbed by the introduction of agroforestry, then alternatives that lead to more direct protection are warranted.

Conservation concessions offer one such alternative. A conservation concession directly compensates local stakeholders and relevant government bodies for providing expanded conservation services. National resource authorities and local resource users protect natural ecosystems in return for a steady stream of structured investments under a negotiated agreement between local stakeholders, the host government, and investors such as environmental conservation organizations or private companies seeking environmental offset opportunities. For instance, a conservation concession might replace cultivation altogether, or the agreement might designate certain areas for agroforestry and set aside others exclusively for conservation. Negotiated elements of the agreement include the size of payments, the duration of the payment, the investment portfolio where these payments will be directed, and norms and guidelines for monitoring and enforcing natural resource protection.

Components of a Conservation Concession Agreement

A conservation concession involves a long-term contract, typically lasting between 25 and 40 years, that provides periodic payments from an endowed fund in return for the conservation of a specified area. The payments in part reflect the opportunity cost of not exploiting the natural resources in the area, addressing issues of lost employment, government revenue, and foreign currency capture. Payments can be negotiated to reflect other factors, such as government administration and enforcement burdens surrounding the concession, but the basis of any payment should be the economic value of exploitation forgone by conserving the area. Because financial flows do not incorporate environmental costs and benefits, the economic value does not simply equal profits from exploitation. Financial valuations may be adjusted for benefits preserved by the concession, such as traditional uses and watershed protection, as well as the risk-free nature of payments. Consequently, the level of compensation emerges as an output of a negotiation process that includes these considerations.

The payment structure of conservation concession agreements may be extremely attractive relative to other land uses. Potential economic benefits from logging, agriculture, and agroforestry include employment, income, export earnings in foreign currency, and public tax revenues. However, for a broad range of resource-based products, economic prospects look less than promising. In recent years, prices for coffee and cocoa have reached all-time lows, although in 2002–2003 cocoa prices rebounded largely because of civil conflict in the world's largest supplier, the Côte d'Ivoire. Periods of low commodity prices can lead to desperate actions; for example, in 2000 major West African cocoa producers announced their intention to destroy 250,000 tons of cocoa, or about 8 percent of global supply, in an effort to raise prices (West African Market Report 2001). In some cases, agroforestry may reduce dependence on a single crop and thus provide a hedge against adverse price trends. Nevertheless, income will remain sensitive to international market conditions and capricious weather patterns. Government revenue streams will remain vulnerable to limited capacity to capture all taxes and fees. A conservation concession offers regularly scheduled, risk-free payments, denominated in stable foreign currency, for as long as the terms of the agreement are met. Ideally, the payments are guaranteed through the use of an endowed fund that covers the duration of the agreement.

Conservation concession benefits must outweigh returns from alternative uses of the target area. Where appropriate, this is accomplished by investing payments in economic activities that will provide alternative jobs and improve human welfare. Negotiated terms of a conservation concession can include a description of the portfolio of activities to which annual payments will be directed. Although the role of the biodiversity investor is not to strictly delineate host government public investment decisions, economic benefits from concession payments should accrue to those who might forgo jobs or other economic benefits because of conservation. The conservation investor can voluntarily supplement concession payments with health or education investments to benefit local stakeholders, particularly in remote communities that lie beyond the effective reach of government services. For instance, salary supplements may persuade teachers to serve in local schools in communities that face difficulties attracting qualified educators. Similarly, payments can take the form of subsidies for medical supplies that otherwise would be inordinately expensive. Such investments help structure appropriate compensation at the community level and generate trust and support for the concession among local stakeholders.

The conservation concession approach emphasizes compensation and social investments on one hand and appropriate enforcement and monitoring activities on the other. The key to the approach lies in decoupling income from habitat modification and natural resource extraction and instead linking economic benefits to successful conservation. Perhaps the most basic illustration of the approach comprises a conservation concession that pays local stakeholders to desist from forest clearing and remunerates them for monitoring and enforcing habitat protection. In July 2002 CI concluded such an agreement with the government of Guyana, covering an area of about 81,000 ha in southeastern Guyana. Under the terms of the 30-year lease, CI pays the government annual acreage fees and royalties equal to those payable by timber concessionaires, amounting to about US$30,000 and $11,000, respectively, and includes voluntary annual investments of $10,000 in development projects to benefit three communities living near the concession (*Guyana Chronicle* 2002).

Some Advantages of the Conservation Concession Approach

A conservation concession yields immediate, transparent conservation benefits that can be easily identified and measured in spatial terms, thereby demonstrating clear conservation benefits to potential biodiversity investors. Although international willingness to pay for conservation is substantial and increasing, a growing trend emphasizes outcome-based rather than process-based indicators of effectiveness of conservation funds; conservation investments must generate unambiguous, measurable results in terms of area and species protected (Porter and Kramer 2000). The concrete geographic basis of conservation concessions, in which conservation of a clearly defined area derives from a negotiated business transaction, directly responds to this trend.

The underlying objective of a conservation concession is long-term habitat maintenance. Nevertheless, from the perspective of a host government, the expiration of a concession's term presents an opportunity to reexamine the best use of the area in question. Renegotiation and extension of the agreement may present an attractive option: conservation concessions offer substantial, secure revenue for the host government and local stakeholders and are cost-effective from the perspective of the international conservation community. Most importantly, this mechanism enables conservation to pay for itself on a large scale, in a way that avoids many of the obstacles and complications facing other conservation approaches such as high-maintenance integrated conservation and development projects and elusive sustainable extraction models (Wells et al. 2000; Rice et al. 2001).

Much of the appeal of the conservation concession approach lies in its simplicity. However, the model must be tailored to specific circumstances that differ by country, climate, principal natural resource, population density, ownership regimes, and more. The basic structure of the approach conforms readily to large tracts of unpopulated, state-owned land of only marginal economic value. Greater numbers of local resource users, intensity of resource exploitation, and complexity of tenure rights can raise the opportunity cost of conservation and the transactional cost of negotiating and implementing conservation concessions. Although these considerations may influence the price of a particular concession, they do not necessarily preclude applying the approach in a given situation as long as effective tenure rights are secure. The basic principle of promoting conservation as a negotiated business transaction among resource users, governments, and international conservation investors holds promise in many contexts.

Conservation concessions and sustainable agroforestry programs can serve as complementary approaches to biodiversity conservation and income generation. First, when sustainable agroforestry faces obstacles, conservation concessions provide an exit option to farmers who otherwise have few alternatives. Second, given that agroforestry per se does not guarantee a stable forest frontier (see Chapter 5, this volume), conservation concessions applied in conjunction with sustainable agroforestry efforts can increase the probability of achieving a stable spatial equilibrium. Under such scenarios, conservation concession payments can be used to facilitate agroforestry efforts (rather than more destructive agricultural land uses) on land designated for production in return for commitments to fortify protection of remaining natural habitat. When applied to land set-asides and retirement strategies, a conservation concession may facilitate a transition to permanent protected status for the area in question. The following sections present two scenarios in which conservation concessions can enhance biodiversity protection within agroforestry land uses.

Retiring Coffee Farms in Colombia

Coffee cultivation accounts for nearly 12 million ha of land in nearly 80 countries throughout the tropics (FAO 2001). Nineteen of the 25 global biodiversity hotspots emphasized as conservation priorities by CI (see Chapter 1, this volume) are major coffee-growing regions (Myers et al. 2000). This overlap results from the fact that agroclimatic conditions ideal for growing coffee also support ecosystems of high conservation value. Sustainable coffee cultivation is an attempt to ameliorate the ensuing conflict between coffee cultivation and biodiversity, supported by a proliferation of certified "green" coffees in the marketplace. However, although shade coffee poses less threat to conservation than other potential land uses (see Chapter 9, this volume), it is not a substitute for natural habitat. Moreover, certified coffee accounts for less than 1 percent of global coffee markets, limiting the potential for sustainable agroforestry to effect change on a scale necessary to avert conflict between coffee and biodiversity at a global scale (Giovannucci 2001).

Colombia, a long-time coffee producer, has more than 750,000 ha under coffee, many of which coincide with the country's biodiversity hotspot regions (Figure 7.1). In the country's most intensive coffee-growing areas, surviving ecosystems host up to 15 percent of the earth's terrestrial biodiversity, making biodiversity conservation imperative (Mittermeier et al. 1997). Specifically, Colombia is home to the greatest variety of birds and amphibians in the world, including 1,815 bird species, of which at least 142 are endemic, and about 600 amphibian species, of which more than half are endemic. Furthermore, roughly one-third of the country's estimated 45,000 higher-order plant species are thought to be endemic, making Colombia second only to Brazil in this regard, and with a much smaller land area. Colombia also is the world leader in orchid diversity and ranks third in butterfly diversity. This extraordinary biological richness is threatened by the commitment of Colombia's National Federation of Coffee Growers to "producing specialty coffee all over the country, even in places where you would least imagine it" (Villelabeitia 2001).

In 2001 coffee export prices in constant U.S. dollars reached their lowest

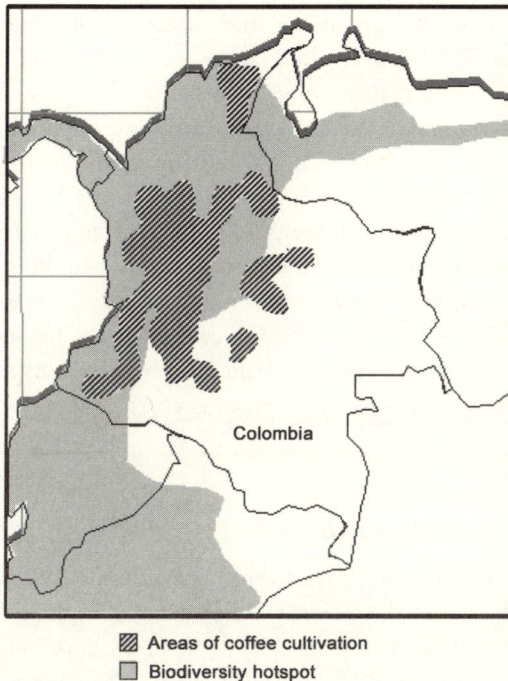

🖾 Areas of coffee cultivation
☐ Biodiversity hotspot

Figure 7.1. Coffee-growing regions and biodiversity hotspots in Colombia.

level since 1900, below even those recorded during the Great Depression of the 1930s (Esarey 2001). Between 1999 and 2001 alone, the international price of coffee beans plunged 60 percent, to just over $1 per kilogram (Miller 2001). At a time when global coffee prices are devastatingly low, Colombia's coffee farmers are at a particular disadvantage because of high labor costs (Sanchez 2002). A wealth of opportunities for achieving conservation presents itself in this context of widespread coffee farming, low market prices, high production costs, and high biodiversity levels. Coffee farmers, many of whom operate at a financial loss and remain in the coffee business for lack of alternatives, may be willing to retire their shade coffee farms for restoration to forest land or set aside intact forests as private reserves in exchange for compensation that would enable them to pursue alternatives. Colombia facilitates private reserve creation by providing tax incentives for landholders and legal recognition of the reserve status of their land (Gaviria 1997).

Under these conditions, arrangements to compensate growers for permanently conserving neighboring forests or retiring shade coffee farms through a combination of public and private financing may benefit farmers and conservationists alike. A strategy of acquiring or leasing low-yield coffee farms and retiring them, buttressed by compensation schemes for farm owners and workers, could catalyze private reserve creation and directly reward local stakeholders for conservation. Landowners and workers would benefit from secure income streams, and biodiversity conservation would no longer depend on finding an elusive balance between agriculture and conservation. Such arrangements could be negotiated with individual landowners holding large properties with particularly high conservation value. Another option, particularly in the context of agroforestry, might involve a community of smallholders jointly setting aside fragments of intact forest and retiring aged farms for conservation in exchange for access to pro-conservation coffee markets or other communitywide benefits such as education or health care. Arrangements of this sort could realize far-reaching social and conservation benefits as increasing numbers of struggling coffee farmers opt for an attractive land use alternative and greater connectivity is achieved between remnant habitat patches. The overlap between cultivated areas and the biodiversity hotspot illustrated in Figure 7.1 suggests great potential for creating corridors between larger protected areas through this strategy.

Set-Asides on Cocoa Farms in Bahia, Brazil

Cocoa farming plays a major role in the expansion of agricultural frontiers throughout the tropics (see Chapter 6, this volume). Globally, cocoa planting claims about 8 million ha, principally in biodiversity hotspot areas of West Africa, Brazil, and Indonesia. Stabilizing the area under cocoa cultivation has proved difficult because of economic considerations that dissuade private landowners from replanting and maintaining aged cocoa orchards; moreover, the cost advantages of planting cocoa in virgin soils make cocoa cultivators an ever-present threat to intact forests on the agricultural frontier (Ruf 1995).

The Atlantic Forest of Bahia, Brazil, is extraordinarily rich in biodiversity and one of the most threatened forests in the world (Figure 7.2; see also Chapter 17, this volume). A joint study conducted there by the New York Botanical Garden and the Brazilian government commission CEPLAC (Comissão Executiva do Plano da Lavoura Cacaueira) found the second-highest tree diversity in the world (450 species in 1 ha, of which 25 percent are found only in southern Bahia; Thomas et al. 1998). The region also features a wealth of endemic fauna: 80 percent of 22 primate species, 45 percent of 77 rodent species, and 37 percent of known marsupials exist only in the Atlantic Forest (Mittermeier et al. 1997). Only 5–7 percent of the original forest cover remains, and this is composed of numerous small fragments. Much of the region's forest comprises small patches on private lands, separated by areas dedicated to agriculture, ranching, and other economic activities. Fragmentation poses a severe threat to biodiversity in the region because small, isolated patches of forest cannot support genetically viable populations of endemic species (Bierregaard et al. 1992).

Cabruca cocoa farms that maintain a portion of canopy vegetation connect many of Bahia's natural forest fragments. The Brazilian cocoa sector has been suffering since 1989, when world cocoa prices dropped because of surplus

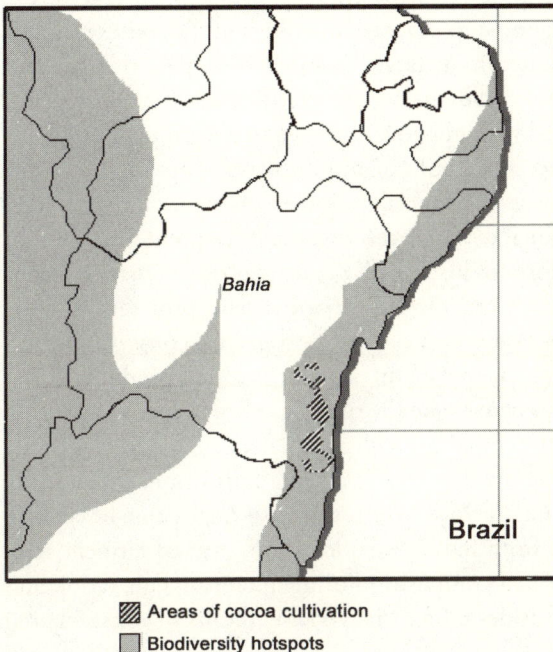

Bahia

Brazil

▨ Areas of cocoa cultivation
▧ Biodiversity hotspots

Figure 7.2. Cocoa-growing regions and biodiversity hotspots in Bahia, Brazil.

production by low-cost producers such as the Côte d'Ivoire. Furthermore, witches' broom disease has devastated crops and greatly increased production costs for Brazilian farmers because controlling the disease requires substantial labor investments (FAS 1999). As profits plummeted, Bahian cocoa output shrank by 60–70 percent. Some landowners have turned to destructive and economically dubious alternatives such as cattle ranching or robusta coffee farming. Others have abandoned their farms or even encouraged land occupation by subsistence farmers, hoping to accelerate deals and leverage more generous compensation packages under the agrarian reform program. Small-scale subsistence farmers in turn survive on annual crops and use slash-and-burn techniques that exacerbate forest destruction, fragmentation, and isolation. Finally, many landowners are seeking to completely divest or consolidate properties to raise capital for investment in remaining farms (Alger 1998).

One element of strategies to conserve Bahia's biodiversity involves a conservation corridor in which cocoa agroforestry systems serve as biological links between protected natural forest fragments (CABS and IESB 2000). However, as with shade coffee farms, even the most diversified cocoa farms do not harbor the same biodiversity level as natural forest (see Chapter 10, this volume). Moreover, because of constant disturbance of the understory through weeding and other activities, the canopy maintained by cabruca may be transitory, limiting long-term biodiversity benefits. Therefore, a vital component of conservation strategy consists of efforts to preserve remaining forest fragments and encourage forest restoration where feasible. Brazilian law requires that landowners maintain at least 20 percent of their land in natural forest. Although this law often is ignored in practice, members of the southern Bahia farmer cooperative for organic products are required to respect this law and maintain set-asides as reserves to obtain organic certification. The prospect of organic premiums, combined with nongovernment and government programs to train farmers, provide access to markets, and facilitate adoption of organic cultivation methods, serves as a financial incentive to maintain set-asides. However, just like conventional cocoa prices, organic price premiums are subject to market trends. As a form of insurance against market fluctuations, conservation concession payments can directly reward farmers for setting aside portions of their land, such that compensation is based on conservation services rather than the indirect channel of markets for certified products.

Conservation concessions could take the organic cabruca cocoa program a step further by offering farmers additional compensation for setting aside more than the minimum requirement of 20 percent. This would enhance the attractiveness of the program to those who otherwise might not be able to forgo the potential revenue from resource exploitation, expand farmers' freedom to determine their optimal composition of income from cultivation and conservation, and make set-asides a financially viable means of consolidating conservation corridors. The conservation concession model thus can complement sustainable agroforestry efforts in Bahia.

The two scenarios just sketched out suggest that the search for solutions in areas where agroforestry and biodiversity may collide can benefit from tools to finance set-asides and retirement of cultivated plots as a conservation service. Particularly in areas where economic prospects for cultivation are dim, such tools can prove a welcome source of relief to local stakeholders and are cost-effective from a conservation perspective. In essence, set-asides and retirement involve direct compensation for choosing conservation rather than cultivation. The conservation concession approach offers a conceptual model for designing such compensation mechanisms.

Conservation Concessions in Practice: Issues and Considerations

Biodiversity is under imminent threat from a wide range of activities throughout the world, and it would be unreasonable to burden agroforestry, conservation concessions, or any other single tool with the expectation of addressing them all. The conservation concession approach should not compete with agroforestry but instead should serve as a complementary tool, particularly in areas where sustainable agroforestry is unfeasible for institutional or financial reasons. In areas where farming may be a sunset industry, conservation concessions can help the transition from cultivation to biodiversity protection by financing retirement of cultivated areas. Finally, in areas where agroforestry or other interventions in productive systems are deemed necessary, the two approaches can work hand in hand: a conservation concession can help to stabilize the agricultural frontier by designating set-asides for protection, in return for social investments in changes in agricultural practices in remaining areas.

Agroforestry remains a second-best option as a tool for biodiversity conservation. Though potentially less detrimental than other forms of land use, agroforestry systems entail an environmental disturbance that does not necessarily result in a stable spatial equilibrium. The robustness of agroforestry systems over time in the face of changing agroclimatic and economic conditions remains an open question. Market mechanisms cannot be relied on to induce adoption of sustainable agroforestry systems because discount rates are high and international markets for green-labeled products are limited. Moreover, the more financially attractive sustainable agroforestry becomes, the stronger the incentives to convert remaining natural habitat become (see Chapter 5, this volume). Fundamentally, agroforestry rewards farmers for increasing physical demands on ecosystems, perhaps less so than under other forms of land use, but certainly relative to conservation of natural habitat. By complementing sustainable agroforestry efforts with conservation concession payments, direct compensation for conservation services can buttress forest frontiers against expansion of cultivated areas and strengthen incentives to adopt environmentally compatible resource use in production areas.

Conservation concession is an effort to arrive at a mutually agreeable level of habitat protection by offering appropriate compensation. Ideally, then, such arrangements are self-enforcing; should any of the parties to the agreement fail to comply with the terms of the concession, payments cease, so the challenge is to design compensation schemes that provide all parties with a vested interest in compliance. Nevertheless, an emphasis on monitoring and enforcement must be maintained to verify and ensure compliance with the agreement, often necessitating capacity building. Given that alternative employment for local stakeholders often is essential to concession acceptability and success, capacity building can take the form of employing stakeholders as monitoring and enforcement agents, thus achieving multiple aims.

Another concern relates to the potential danger of simply displacing pressure from one habitat area to another. For example, a farmer participating in a conservation concession arrangement may invest resources freed by the inability to clear a particular area (and the compensation for doing so) in clearing another area that falls outside the agreement. However, with respect to retirement and set-asides of productive plots, application of the conservation concession approach in the first instance will target areas with a low opportunity cost of conservation, that is, areas with weak prospects for cultivation. In the Colombian scenario, payments offer farmers a way out of a desperate situation; conservation concessions in this context provide an exit option, not a supplement to ongoing activities. The poor economic performance of coffee is precisely what makes retirement through the conservation concession approach an attractive strategy to both conservationists and farmers.

The question of displacement might seem pertinent if farming could be redirected to entirely different parts of the country that offer more promising prospects for the same or a different crop. However, areas with a high opportunity cost of conservation, implying high profit potential from conversion, are under threat regardless of whether conservation concessions are applied elsewhere. Moreover, this dynamic is not unique to conservation concessions; any effort to protect habitat, by definition, implies that activities that might have taken place in the target area might now take place elsewhere. Only detailed stakeholder analysis can determine whether beneficiaries of a retirement program through conservation concessions are likely to seek alternative rural employment in the area, establish new plots elsewhere, migrate to urban areas, or pursue other possibilities. However, one strength of the conservation concession approach is the explicit attention to alternative employment opportunities as an investment target. As an integral component of the mechanism, such opportunities form part of the compensation for conservation services and temper the tendency to displace ecologically undesirable activities elsewhere.

Alternatively, establishing a firm frontier implied by an effective conservation concession may spur intensification on remaining land under production, with potentially harmful environmental consequences. In an agroforestry context, a conservation concession may, for example, induce a farmer to remove shade trees or apply more agrochemicals on production plots when prevented from expanding to new areas. In the Bahian scenario described earlier, the result would be greater protection for forest fragments but reduced connectivity between them, with ambiguous net benefits to biodiversity. This potential dynamic illustrates that interactions between sustainable agroforestry programs and conservation concessions must be carefully managed to ensure desirable outcomes. Stakeholder analysis, agreement design, and compensation negotiation must be structured within a long-term, systemwide perspective, with the boundaries of the system defined broadly enough to capture the full intertemporal and spatial impacts of both agroforestry efforts and conservation concessions.

Another concern arises when ownership is easily established over new forest areas, as in large stretches of the Amazon. When a precedent of paying private landowners for conservation services is set, an incentive may arise to claim areas simply for the sake of leveraging them as potential conservation concessions. However, because the areas in question did not attract ownership claims or investment in the absence of conservation payments, the implication is that the opportunity cost of conserving them would be very low. Thus, by clarifying ownership regimes and compensating new owners for the opportunity cost of conservation, this dynamic could facilitate long-term, cost-effective protection of large areas previously inaccessible because of ill-defined tenure. More importantly, such a process might compel governments to consider the benefits of negotiating long-term conservation concessions over such areas.

The conservation concession approach is particularly suitable for extensive areas controlled by a single entity, such as the government or large landowners. For areas not under government or private ownership, the practicability of the approach depends on the ownership structure that is in place. Agroforestry systems operate under a variety of ownership regimes, ranging from small, privately owned plots to large plantations to communal land distributed by traditional chiefs. The complexity of tenure arrangements itself influences the identification of stakeholders and the relationships between them, involving local, regional, and national governments, traditional authorities, landowners, land renters, sharecroppers, hired labor, farmer organizations, and others. This feature of agroforestry systems raises two factors to consider with respect to the conservation concession approach. First, to achieve meaningful conservation outcomes, a critical mass (in the number of participants and spatial configuration of plots) typically is needed for an effective corridor or buffer zone.

Second, the number of farmers and their degree of organization in cooperatives or collectives influence the costs of negotiating and transacting an agreement.

As a rule, transaction costs incurred in negotiating a conservation concession are lower, on a per hectare basis, for large areas with few stakeholders. The degree to which transaction costs increase in scenarios such as those considered earlier, with numerous farmers on small plots, remains to be seen. In any case, over the long term the total costs of a conservation concession are dominated by compensation payments. Because payments reflect opportunity costs, situations in which farmers in agroforestry systems are struggling to break even suggest highly cost-effective conservation opportunities in the form of set-asides or plot retirement. Additionally, conservation concessions in areas with marginal economic prospects may yield promising opportunities for co-financing and coordination with income support and poverty alleviation initiatives of governments and nongovernment development organizations.

Conclusions

The preceding discussion suggests that a range of issues, some predictable and others unanticipated, will accompany the implementation of conservation concessions in any given context. From a global perspective, the applicability of the conservation concession approach depends on cost considerations and conservation priorities. In some areas the approach is prohibitively expensive, and in others the opportunity cost of conservation may be low but biodiversity values limited. The configuration of stakeholders and property rights also is a crucial determinant of suitability of a target area for the approach. Ultimately, appropriate design of compensation, monitoring, and enforcement components of a conservation concession rests on a thorough analysis of stakeholder needs and interests, from local communities to regional authorities and national governments. However, the inherent flexibility of the framework presents an invitation for creative adaptation of the model to locally specific circumstances. In many situations in which enhanced protection entails engagement with local communities to explore alternative cultivation forms, the joint application of conservation concessions and sustainable agroforestry programs may yield a powerful answer to the economic, social, and institutional forces that threaten biodiversity around the globe.

References

Alger, K. 1998. The reproduction of the cocoa industry and biodiversity in southern Bahia, Brazil. In *Shade-grown cacao research papers, Smithsonian Migratory Bird Center.* Online: http://natzoo.si.edu/smbc/Research/Cacao/cacao.htm.

Angelsen, A., and D. Kaimowitz (eds.). 2001. *Agricultural technologies and tropical deforestation*. Wallingford, UK: CAB International.

Bierregaard, R. O., Jr., T. E. Lovejoy, V. Kapos, A. A. dos Santos, and R. W. Hutchings. 1992. The biological dynamics of tropical rainforest fragments. *BioScience* 42:859–866.

CABS (Center for Applied Biodiversity Science) and IESB (Instituto de Estudios Sócio-Ambientais do Sul da Bahia). 2000. *Designing sustainable landscapes: the Brazilian Atlantic Forest*. Washington, DC: Conservation International.

Esarey, S. 2001. *Farmers turn depressed farm commodities into gold*. Reuters World Environment News, May 29. Online: http://www.planetark.org.

FAO (Food and Agriculture Organization). 2001. *FAOSTAT agriculture database*. Online: www.fao.org (last accessed May 2, 2001).

FAS (Foreign Agricultural Service). 1999. *Cocoa update (October)*. Washington, DC: United States Department of Agriculture. Online: http://www.fas.usda.gov/htp2/tropical/1999/99-10/Oct99txt.htm.

Gaviria, D. 1997. Economic and financial instruments for sustainable forestry in Colombia. *Unasylva* 48: 32–35.

Giovannucci, D. 2001. *Sustainable coffee survey of the North American specialty coffee industry*. Report prepared for the Summit Foundation, The Nature Conservancy, North American Commission for Environmental Cooperation, Specialty Coffee Association of North America, and the World Bank. Online: www.fairtradestudents.org/sustainablecoffee.pdf.

Guyana Chronicle. 2002. Guyana sells forest for conservation purposes. Georgetown, Guyana, July 18.

Hardner, J., and R. Rice. 2002. Rethinking green consumerism. *Scientific American* 286:88–95.

Laurance, W. F., and R. O. Bierregaard Jr. (eds.). 1997. *Tropical forest remnants*. Chicago: University of Chicago Press.

McNeely, J., and S. Scherr. 2001. *Common ground, common future: how ecoagriculture can help feed the world and save wild biodiversity*. Gland, Switzerland: International Union for the Conservation of Nature and Natural Resources; Washington, DC: Future Harvest.

Miller, M. 2001. Prices for java, beans just don't jive. *The Seattle Times,* May 19.

Mittermeier, R. A., P. R. Gil, and C. G. Mittermeier (eds.). 1997. *Megadiversity: Earth's biologically wealthiest nations*. Mexico City: CEMEX, Agrupacion Sierra Madre.

Myers, N., R. A. Mittermeier, C. G. Mittermeier, G. A. B. da Fonseca, and J. Kent. 2000. Biodiversity hotspots for conservation priorities. *Nature* 403:853–858.

Petithuguenin, P. 1995. Regeneration of cocoa cropping systems: the Ivorian and Togolese experience. Pages 89–105 in F. Ruf and P. S. Siswoputranto (eds.), *Cocoa cycles: the economics of cocoa supply*. Cambridge, UK: Woodhead Publishing.

Porter, M., and M. Kramer. 2000. Philanthropy's new agenda: creating value. *Harvard Business Review* November–December: 121–130.

Rice, R. E., C. A. Sugal, S. M. Ratay, and G. A. B. da Fonseca. 2001. *Sustainable forest management: a review of conventional wisdom*. Advances in Applied Biodiversity Science No. 3. Washington, DC: Center for Applied Biodiversity Science/Conservation International.

Ruf, F. 1995. From forest rent to tree-capital: basic "laws" of cocoa supply. Pages 1–53 in F. Ruf and P. S. Siswoputranto (eds.), *Cocoa cycles: the economics of cocoa supply* Cambridge, UK: Woodhead Publishing.

Sanchez, M. 2002. Colombia's coffee research center braces for hard times. *Reuters World Environment News,* January 28. Online: http://www.planetark.org.

Terborgh, J., and C. P. van Schaik. 1997. Minimizing species loss: the imperative of protection. Pages 15–35 in R. Kramer, C. van Schaik, and J. Johnson (eds.), *Last stand: protected areas and the defense of tropical biodiversity.* New York: Oxford University Press.

Thomas, W. W., A. M. de Carvalho, A. M. Amorim, J. Garrison, and A. L. Arbeláez. 1998. Plant endemism in two forests in southern Bahia, Brazil. *Biodiversity and Conservation* 7:311–322.

Villelabeitia, I. 2001. Colombia seeks to double specialty coffee exports. Reuters, July 19.

Wells, M., S. Guggenheim, A. Kahn, and W. Wardojo. 2000. *Investing in biodiversity: a review of Indonesia's integrated conservation and development projects.* Directions in Development Series. Washington, DC: World Bank.

West Africa Market Report. 2001. West Africans remain committed to destruction plan. June. Online: http://www.otal.com/mkrjun01.htm#tog2.

PART III

The Biodiversity of Agroforestry Systems: Habitat, Biological Corridor, and Buffer for Protected Areas

This section of the book reviews the effects of some important and widespread agroforestry practices on the biodiversity of human-dominated landscapes in the tropics. Chapter 8 looks at the most widespread and one of the oldest agroforestry practices: shifting cultivation. Landscapes dominated by shifting cultivation cover vast areas in the tropics and are therefore of particular importance for biodiversity conservation in agricultural landscapes. Although previous studies have often emphasized the threats that this land use practice poses for tropical forests, this chapter shows that traditional shifting cultivation (as opposed to land uses often practiced by immigrants) can be surprisingly stable over time and maintain a high percentage of the landscape under tree cover. The chapter reviews the diversity of plants and animals in the patchwork of crop fields, fallows of variable age, secondary and primary forest that is characteristic of shifting cultivation landscapes, and the factors that determine the abundance and diversity of organisms present. It also identifies management options that might help increase the conservation value of this land use type.

Chapter 9 focuses on the conservation potential of coffee agroforestry systems, which have attracted interest and numerous scientific studies because they cover large areas of the tropics and often overlap with areas of high biodiversity. The history of coffee growing, which has led to a wide range of shade use and management practices, is summarized, and the diversity of floral and faunal groups in coffee plantations as affected by their degree of shading, shade species composition, and management is reviewed. Trade-offs and synergies between biodiversity conservation and production objectives are

151

identified, especially with respect to the large numbers of coffee pests and their natural enemies in complex coffee ecosystems.

The most rustic coffee plantations, where the tree crops are grown under a dense, highly diversified shade canopy, form part of a group of agroforestry systems that are collectively called complex agroforests or simply agroforests because of their diversity, structural complexity, and similarity to natural (secondary) forests. Chapter 10 reviews the conservation potential of such agroforests, which have evolved in several parts of the humid tropics, distinguishing between systems based on canopy tree crops, such as rubber and damar trees, and shade-tolerant understory tree crops, such as cocoa and tea. Because of their extensive management, these systems are characterized by a substantial amount of spontaneous vegetation and associated fauna. Because agroforests are able to provide a range of timber and nontimber forest products and maintain high plant and animal diversity, they are particularly valuable as landscape conservation tools.

The aforementioned agroforestry systems occur in tropical landscapes as patches or blocks of variable shape and size. However, many tropical landscapes that have been cleared of forest to establish fields or pastures do not contain such distinct agroforestry patches but may still retain trees in the form of live fences, windbreaks, or isolated trees, which may either be remnants of the original forest vegetation, planted, or retained from spontaneous regeneration. Although these agroforestry elements may cover only a very small percentage of the landscape, they may still offer some habitat and increase the permeability of the agricultural matrix for many organisms that are not adapted to wide, treeless areas, thereby increasing connectivity between islands of natural vegetation. Chapter 11 discusses the roles of live fences, isolated trees, and windbreaks as habitats, corridors, and stepping stones and identifies knowledge gaps that must be filled before the potential of these agroforestry elements for landscape-scale conservation strategies can be fully appreciated.

Chapter 12 focuses on a role of trees in agricultural landscapes that is even less obvious: their effect on gene flows between trees in forest fragments. In the past, the knowledge that most tropical trees are outcrossing led to the assumption that many trees that occur in small numbers in forest fragments or even individually as remnants in agricultural areas are "living dead," or unable to produce viable offspring because of the lack of nearby mating partners. However, recent studies show that agroforestry trees that occur in agricultural areas near forest fragments may contribute significantly to gene flow across landscapes, which occurs over larger distances than had been assumed. However, problems may arise when agroforestry trees come to dominate the pollen pool of certain species in adjacent forests, thereby narrowing the genetic base of subsequent generations. Although these aspects warrant further study, the chapter shows clearly the potential of tree management, especially of remnants and natural regeneration of forest trees in agricultural areas, for the conservation of viable tree populations in fragmented forest landscapes.

Chapter 8

The Biodiversity and Conservation Potential of Shifting Cultivation Landscapes

Bryan Finegan and Robert Nasi

The most commonly used definitions of shifting cultivation (or swidden agriculture) are based on the work of Conklin (1957) and define it as any agricultural system in which the fields are cleared and cultivated for periods shorter than those over which they are fallowed. A more dynamic approach emerged in more recent works, with McGrath (1987) defining shifting cultivation as "a strategy of resource management in which fields are shifted in order to exploit the energy and nutrient capital of the vegetation-soil complex of the future site (p. 221)." Watters (1971) summarizes the principal characteristics that define shifting cultivation as it is practiced in the tropics: the shift between fields rather than between crops on the same field, short (1- to 3-year) cropping periods alternating with longer fallow periods (4–60 years), cutting and burning of the fallow vegetation at the beginning of each cropping period, and the almost exclusive use of human energy in land management operations. It is the alternation on the same site of crops and fallow vegetation dominated by woody plants that permits the definition of shifting cultivation as an agroforestry land use.

Shifting cultivation creates unique landscapes composed of a dynamic patchwork of crop fields, fallows of various ages, secondary forest derived from fallows, and remnants of the original vegetation. Crop fields and old secondary forests are clearly defined communities (Finegan 1992; Smith et al. 1997), but scientists from different disciplines may see fallows in different ways. In a forestry or ecological context, fallow communities are seen as secondary vegetation within a framework of dynamic relationships between vegetation types, centered on primary or old-growth vegetation. The definition of "forest fallow systems" used by FAO (1998) typifies this forest-centered approach, referring to "complexes of woody vegetation deriving from the clearing of forest for

153

agriculture" (p. 8) and mosaics of "various reconstitution phases." However, we emphasize that fallows are primarily components of an agricultural land use system, and their ecological or forestry status as secondary vegetation or phases in the "reconstitution" of forest in this context is, indeed, secondary. Fallows are components of an integrated farming system in which multiple objectives for the livelihoods of the farmers have to be met. They exist for a number of ecological and socioeconomic reasons, among which are soil fertility restoration, erosion reduction, weed control, and opportunities to gather products for sustaining the household. As far as forest "reconstitution" is concerned, in many tropical landscapes fallows may never develop into a community resembling the original one of the site, even if they are not subject to further disturbance. We therefore follow Burgers et al. (2000) in defining fallow communities as the vegetation and associated fauna that occupy land that has been cleared for cultivation but is not currently so used, although the community may have multiple other uses, such as the provision of firewood or nontimber forest products. The vegetation component of the community normally is made up of plants that regenerate naturally when the land is left fallow (we follow authors such as Spencer [1966, cited by Watters 1971] and Smith et al. [1997, 2001] in not using the term *abandonment,* with its inappropriate connotations, in reference to the transition from crop to fallow). It also contains useful plants that are conserved by the farmer, whether planted or naturally regenerated (jungle rubber cultivation systems are a special case here; see Chapter 10, this volume), and remnants of agricultural crops and weeds.

Professional and popular attitudes toward shifting cultivation vary. It is one of the major agricultural systems used by humanity, is ancient in origin and in the context of certain levels of available technology, capital, and population densities, and is considered by many to be a sophisticated and sustainable land use (Nye and Greenland 1960; Watters 1971; Whitmore 1989). On the other hand, the fact that shifting cultivation leads to the replacement of natural communities by anthropogenic ones means it is often identified as a major cause of habitat destruction and biodiversity loss in the tropics (Myers 1980), although other opinions have recently emerged (Brown and Schreckenberg 1998). The growing consensus that tropical biodiversity conservation can no longer be centered solely on protected areas, but will entail action in all land use types across landscapes and regions (Western and Pearl 1989; Aide 2000), makes an analysis of shifting cultivation necessary in the context of the present book.

From many points of view, including that of biodiversity maintenance and generation, it is important not to confuse shifting cultivators who have lived for a long time in a region (established shifting cultivation *sensu* Conklin, called "traditional" in this chapter) and recent immigrant shifting cultivators (pioneer shifting cultivation *sensu* Conklin). Traditional shifting cultivation is itinerant, thus temporary on any given patch of land, and is strongly but not completely dependent on the existence of large areas of undisturbed forest. Land conversion as carried out by agricultural settlers or cattle ranchers in many modern agricultural frontiers is characterized by greater degrees of permanence, although similarities to traditional systems may be found even though modern colonizers may lack the knowledge of traditional shifting cultivators. However, forest lands converted to other use in the latter context are 20–50 times greater in area than those affected by itinerant slash-and-burn agriculture.

Well-established among the major agricultural functions of the fallow period are soil fertility recovery and pest and weed control. Most biophysical research on fallows has focused on these ecological functions (reviewed by Nye and Greenland 1960 and Whitmore 1989), although weeds established during the cropping period have also been shown to contribute to the reduction of nutrient losses (Lambert and Arnason 1989). The fact that, by definition, fallows are to be reconverted to cropland may have contributed to a lack of interest in fallows as systems for the study of forest succession or as potential routes for the restoration of tropical forest and the goods and services it provides, among which, of course, is biodiversity conservation. This lack of interest is illustrated by recent collections of papers on the restoration of tropical forests, which focus either on the "catalytic" role of forest plantations in the process or on forest restoration through natural succession on pastures (see Parrotta and Turnbull 1997 and Aide 2000, for example). Smith et al. (2001) provide one exception to this latter trend (others are reviewed later in this chapter). However, they emphasize that the development of land use options for increasing forest cover in heavily deforested tropical regions faces a major challenge in the need to combine conservation biological desirability and technical solvency with the interest of farmers.

In this chapter we seek to answer the following questions linked to the role of shifting cultivation systems, and the landscapes they create, in biodiversity generation and maintenance:

- What are the spatial characteristics of shifting cultivation landscapes, how do they change over time, and what are the relationships between landscape spatial characteristics and local and landscape biodiversity at the species level?
- What are the principal regeneration mechanisms operating in fallow vegetation, and what are their implications for plant biodiversity?
- How do the richness, diversity, and composition of individual floral and faunal communities change over the shifting cultivation cycle, and what are the mechanisms of these changes?
- Do such landscapes contribute to the conservation of organisms characteristic of the original forests of those landscapes, either because those organisms are members of the communities of the shifting agriculture mosaic,

because that mosaic meets at least some of their needs, or because the mosaic at least provides connectivity between remnant forest habitats?

- How might management maintain or increase the conservation value of individual components of the landscape and of the landscape as a whole?

The Spatial Characteristics and Ecological Dynamics of Shifting Cultivation

In the tropics, shifting cultivation landscapes characterize particular regions at any given time and sometimes particular time periods in the development of agricultural frontiers in any given region. Shifting cultivation is not an important feature of all tropical agricultural frontiers (Moran et al. 1994), and some authors (e.g., McKey 2001) assert that overall, the major phenomenon occurring in modern tropical agricultural frontiers is not the implementation of shifting cultivation systems but land conversion to uses that are intended to be more permanent. On the other hand, shifting cultivation is not limited to young agricultural frontiers or to remote forest areas almost devoid of inhabitants. In many places in the tropics, humid or dry, it is indeed a stable land use over time.

In the Amazon basin, extensive areas in which shifting cultivation is a principal land use are found in the west in lowland Peru (Dourrojeanni 1987) and in the east in Pará State, Brazil (Vieira et al. 1996, 2003). In West and Central Africa, shifting cultivation is the norm and is part of almost all agricultural systems, together with permanent crop fields and more sophisticated agroforests (de Rouw et al. 1990; Dounias 2001). Shifting cultivation is the most widespread form of land use in the nonirrigable parts of northern Thailand (Schmidt-Vogt 1999) and in neighboring Laos, Cambodia, and Vietnam. In Laos, up to 1 million people may be involved in shifting cultivation, which makes up 40 percent of the land area dedicated to the country's principal crop, rice, with 200,000 ha under cultivation in any given year (Fujisaka 1991; Roder and Maniphone 1998). It is also the favored land use in less populated regions of Indonesia, such as Kalimantan, and in the Philippines.

Fallows occupy much of the land area in swidden landscapes, and for this and other reasons much of the potential of these landscapes for forest biodiversity conservation depends on them. Analyzing a 137,800-ha municipality in eastern Pará State, Brazilian Amazon, by remote-sensing techniques, Alencar et al. (1994) found that around 50 percent of the total area was occupied by fallow vegetation in three distinct developmental stages, and only 8–11 percent of the area was occupied by cropland. In contrast to the enormous area of fallow, around 15 percent of the land was covered by residual primary forest, most of it in riparian strips. Vieira et al. (2003) carried out a similar study in another Pará State municipality, finding 18, 17, and 5 percent of the

47,700 ha studied in young (3- to 10-year), intermediate (ca. 20-year), and advanced (40- to 70-year) fallow and secondary vegetation, respectively. Taking into account the 12 percent of land area covered by primary forest, again mainly riparian, total forest cover exceeds 50 percent in this municipality if a broad definition of forest is adopted (discussed later in this chapter), despite more than a century of agricultural settlement. In northern Thailand, the Lawa people of Ban Tung cultivate around 30 ha per year for an available fallow area of 800 ha (Schmidt-Vogt 1999). In a drier savanna context, Petit (1999) showed that the proportion of village land under various types of fallow was 34.5 percent for the Senufo people of northern Côte d'Ivoire and 64 percent for the Musey of northern Cameroon.

Beyond the overriding importance of fallows, it is important to note that secondary forests or old fallows play a significant and often overlooked role in shifting cultivation systems. The proportion of new crop fields established in primary forests is only 5 percent for the ethnic groups of south Cameroon (Mvae, Ntumu), 13 percent for the Kenyah of Sarawak, 24 percent for the Palikur in French Guyana, and 31 percent for the Kantu of Borneo (Dounias 2001).

Overall, it is clear that shifting cultivation in general and fallows in particular are the defining characteristics of many tropical landscapes and are therefore of fundamental importance to prospects for biodiversity conservation in such landscapes.

Dynamic Processes at the Landscape Scale and the Dynamics of Agricultural Frontiers

The temporal dimension of the analysis of shifting cultivation and biodiversity at the landscape scale is as important as the spatial. Tropical agricultural landscapes are dynamic at different temporal and spatial scales, and these dynamics must be understood if the potential of shifting cultivation to generate and maintain biodiversity is to be understood. Land use dynamics under traditional shifting cultivation differ from those imposed by modern colonization processes. Modern deforestation is synonymous with the advance and development of agricultural frontiers, a process for which numerous authors have proposed models. For example, Henkel (1971, cited by Thiele 1993) proposed that originally forested land passes through four stages after the arrival of the first colonists. This model may be summarized as follows, drawing parallels with the model proposed by Richards (1996, cited by Smith et al. 2001). A pioneer fringe (Richards's early pioneer stage) advances into the forest, with deforested areas then evolving into the commercial core of the frontier, a stage in which farmers take advantage of improved infrastructure and access to markets; Richards calls this the stage of the emerging market economy. This commercial core may evolve into a zone of decay as agricultural productivity declines and farmers abandon land and may migrate back into the pioneer fringe; land is concentrated in the hands of fewer property owners. The zone of decay is subsequently revitalized as originally small plots are amalgamated for livestock production or higher-technology agriculture is introduced. The last two stages of Henkel's model are equivalent to Richards's closing frontier, which he characterizes in terms of low or zero availability of forest land for further colonization and further improvements of infrastructure and farmer integration with markets. These simple models contribute to the explanation of the spatial zonation of agricultural frontiers at any time and the changes that are observed over time at any given place.

Stages in the development of agricultural frontiers are accompanied by changes in the proportion of the landscape at some stage of the shifting cultivation cycle. Smith et al. (1999b) studied these changes in farming communities near the city of Pucallpa in the Peruvian Amazon, using data derived from interviews with farmers in settlements characterized as belonging to the emerging market economy (EM) and closing frontier (CF) stages of Richards's framework. Interpreting their results as points in the temporal development of a single frontier, they conclude that excluding cattle ranches, approximately 50 percent of the original primary forest cover of an area is lost between the pioneer and EM stages. Half the cleared land in EM is in fallow at a given time, and this proportion increases to about 60 percent in CF as residual primary forest cover plummets. Despite the marked decline of residual forest cover between the two stages, the net decline of overall forest cover as the frontier evolves is low because primary vegetation is replaced by secondary (including fallow). Forest cover in CF is still 40 percent, composed of 23 percent secondary and 17 percent primary. However, the rise of cattle ranching in CF means that the loss of original forest cover and the net overall reduction in forest cover including secondary vegetation are both greater when the whole landscape is considered. A final comment on Smith et al.'s analysis is that whether net change in total forest cover during the development of swidden landscapes is limited depends, of course, on the definition of *forest* that is used. If the definition includes young fallow vegetation, then net change is indeed limited. In the context of this chapter, however, it is vital to remember that fallow vegetation and secondary forest are very different, biodiversity-wise, from primary forest.

Agricultural frontiers are sometimes dynamic, as we have seen, but in some cases shifting cultivation is a stable land use in both traditional and modern contexts. It probably began in 3500 BC in the Maya Zone of Mesoamerica and is still practiced there today. This time period naturally involves a switch from traditional to modern forms of the practice, and the extent of land used and the size of the dependent population undoubtedly have fluctuated over time for many reasons, including the widely cited hypothesis that land degradation was a factor in the decline of Maya civilization (Lambert and Arnason 1989;

Hammond 2000; Harrison 2000). Shifting cultivation has persisted for a century or more on extremely poor oxisols in the Bragantina region of the Brazilian Amazon (Alencar et al. 1994; Smith et al. 1999a), and in vast regions of Central Africa and New Guinea it has been one of the major types of agriculture for centuries or millennia (Dounias 2001; Kocher-Schmid 2001). The ecological, social, economic, and political contexts under which shifting cultivation persists over the long term are an interesting area for research. De Wachter (1997) studied the traditional swidden cultivation system practiced by the Badjwe people at the periphery of the Dja Reserve in Cameroon. Modeling suggests that under this land use system, the current population density of 4.2 inhabitants per square kilometer of arable land could grow to up to 38 inhabitants per square kilometer and still maintain, after more than 50 years and within 5 km of roads, a landscape mosaic with 50 percent of the land under shifting cultivation, 43 percent under primary forests (including 26 percent of swamp forests), and 7 percent of pure *Raphia* palm stands. Under a scenario of intensification of cash crop–oriented activities (cocoa, plantain), the numbers fall to 18 inhabitants per square kilometer and about 35 years to maintain a similar mosaic within 5 km of the road.

The Shortening of Fallow Periods: Causes and Consequences

The length of fallow periods is a key factor in the degree to which biodiversity accumulates on individual patches in swidden landscapes and the degree to which landscapes as a whole maintain certain components of biodiversity; in many senses, the longer the fallow, the better. This simple rule leads us in an equally simple way to one of the major conflicts regarding biodiversity in shifting cultivation landscapes. In the classic Boserup model (Boserup 1965), agricultural gains were accomplished through land expansion, increased labor, and, critically from our point of view, shortening of fallow periods (but see Stone and Downum 1999 for counterexamples). Many recent studies report a tropics-wide tendency toward shorter fallows in shifting cultivation, and this is one of the best-documented aspects of the subject from both ecological and agronomic standpoints (Jones and O'Neill 1993; de Jong et al. 2001).

In recent discussions, Smith et al. (1997, 1999a, 1999b, 2001) evaluate the factors that underlie the shortening of fallow periods by farmers. They observe that farmers set fallow periods by weighing the costs and benefits associated with different fallow lengths. Longer fallows have potential benefits that include higher yields from the cropping period, lower labor costs for weed control during that period (weed control is one of the most important functions of the fallow), and greater opportunities for harvesting the products of unplanned biodiversity during the fallow. The incentive to increase fallow length may also arise because as the time during which land has been under the shifting cultivation cycle increases, the productivity of fallows and the speed with which soil fertility is recovered decrease, and the time needed to control certain weeds in the patch increases. Increases in fallow length may be observed under some circumstances (Smith et al. 2001). Among the costs associated with longer fallow lengths is the need to achieve a greater area of land under the shifting cultivation cycle to maintain a given quantity of agricultural produce and the attendant possibility that this greater land area must be obtained by the labor-intensive cutting of more primary forest. A critical moment arrives when it is no longer possible to cut more primary forest, so that land area under cultivation can be increased—as a response to declining crop yields or an increasing population to be supported from the land—only by shortening fallow length. The shortening of fallows is much more common than their lengthening. For example, Thiele (1993) reports that traditional fallow lengths in seasonal environments of lowland Bolivia were 6–12 years but by the 1980s had declined markedly, with a modal length of 4 years. Fallow lengths had declined to a range of 3–10 years in the 1990s in northeastern India from a traditional value of 60 years (Ramakrishnan 1992; Shankar Raman 1996). Traditional fallow lengths in the Taï forest region of the Côte d'Ivoire were 14–30 years and had been reduced to 6–10 years by the end of the twentieth century (de Rouw 1995).

Shortening fallow lengths can exacerbate the decline of crop yields and increase weed problems and therefore labor needs during cultivation (de Rouw 1995; Roder and Maniphone 1998; Smith et al. 1997, 2001). For example, short fallows in the Côte d'Ivoire led to a 72 percent increase in weed biomass during cropping periods (Becker and Johnson 2001). A logical response to these tendencies, from both farmers and the research and development community, has been to seek alternative technologies. Permanent agriculture—as sought by the government of Laos to replace shifting cultivation, which is considered to be environmentally undesirable (Fujisaka 1991)—is one alternative (see also Thiele 1993). Improved (planted) fallows are another main focus of attention (Fujisaka 1991; Buckles and Triomphe 1999; Szott et al. 1999). Improved fallows are intended to fulfill the agroecological function of the fallow in short time periods (perhaps 3 years or even less) and may themselves bring increased labor needs (Szott et al. 1999), although examples of their spontaneous development and wide adoption by farmers are documented (Buckles and Triomphe 1999).

To conclude, the relationship between fallow length and the floral and faunal characteristics of the community is one of the most important elements in the potential of shifting cultivation landscapes for biodiversity conservation, as discussed later in this chapter. The tendency toward shorter fallows and the development of techniques for fallow "improvement" represent severe limitations on this potential. In particular, improved fallows seem to be of little value for forest biodiversity conservation because of their often monospecific composition and short duration and will not be considered further. Besides its agronomic importance, fallow length clearly is an important issue in the management of shifting cultivation landscapes for biodiversity conservation objectives.

Mechanisms of Plant Regeneration in Fallows

In general terms, plants may regenerate in fallows from seed already present in the soil seed bank, from seed dispersed onto the site after it is left fallow, or by sprouting from cut but living plant parts, either above the ground or in the form of root suckers (Kammesheidt 1998). Regeneration of fallow vegetation immediately after the cropping period is predominantly vegetative. Practically all the trees, shrubs, vines, and large herbaceous perennials and a majority of the grasses regenerate from stumps, roots, or rhizomes. Factors that impede seed regeneration of trees and shrubs include burning and weeding during the cropping period and the short lives of seeds of some species. Burning at the beginning of the cropping period and frequent weeding also eliminate the seedlings of trees and shrubs, most of which have only recently germinated. All these factors impoverish the seed bank in the soil and reduce the contribution of seed regeneration to the reestablishment of tree cover in the fallow period (Denich n.d.; Vieira and Proctor 1998).

Few studies have directly addressed the importance of sprouting as a regeneration mechanism in tropical secondary vegetation (Kiyono and Hastaniah 1997; Kammesheidt 1998; Pacioreck et al. 2000; van Nieuwstadt et al. 2001). Although in structural and floristic terms overall successional pathways in fallows may strongly resemble those that follow other types of human land use, the importance of resprouts as a regeneration mechanism is arguably one of the defining ecological characteristics of fallow vegetation and secondary forest derived from it. Uhl (1987) has shown experimentally that repeated weeding during cultivation can reduce the density of sprouts in fallows. In real situations, however, the short periods of low-intensity cultivation that characterize shifting cultivation probably do not significantly reduce the regenerative capacity of tree stumps and root fragments, which resprout when land is left fallow.

Species individualism is an important aspect of understanding plant regeneration and therefore plant community composition and diversity in fallows. For example, species in communities subject to natural disturbance differ in the relative importance of resprouts and seed as postdisturbance regeneration mechanisms (Boucher et al. 1994). Experimental studies are strictly required to confirm whether tree species regenerate from sprouts, but descriptive work provides a clear pointer to the existence of patterns. Kammesheidt (1998) showed that 28 of the 58 tree species he recorded in small plots in fallows at his Paraguayan study site were regenerated from both mechanisms, whereas 7 were found only as resprouts and 23 were regenerated only from seed. Among typical genera of short-lived pioneer trees (*sensu* Finegan 1996) of the neotropics, *Vismia* spp. may resprout frequently, whereas *Cecropia* spp. show this regeneration mechanism much less frequently (Kammesheidt 1998; Mesquita et al. 2001). These differences probably explain the dominance of young fallows by *Vismia* spp. on Amazonian sites observed by Uhl (1987) and Vester and Cleef (1998), although the latter authors did not identify regeneration mechanisms. Work on slash-and-burn practices in subtropical Australia (Stocker 1981) and Indonesia (Riswan and Kartawinata 1991; Kiyono and Hastaniah 1997) has also shown variations in regeneration mechanisms between species. In shifting cultivation landscapes, the abundance and diversity of tree species with limited powers of regeneration by sprouting depend partially on the presence of seed trees and dispersal agents. Most *Macaranga* species, very common short-lived pioneers in Africa and Asia, seldom resprout and depend essentially on the soil seed bank and the seed rain for regeneration.

The relative importance of resprouts in the development of fallow vegetation may vary between agricultural frontiers of different ages or landscapes with different areas of remnant primary vegetation. Resprouts were the most important sources of woody regeneration in 2- to 5-year-old vegetation on low-fertility oxisols in Kammesheidt's (1998) Paraguayan subtropical moist forest study site, but their relative importance declined with age, and trees regenerated from seed were more important in 10- to 15-year-old stands. Results from 5- to 20-year-old fallows on similar soils in the Bragantina region of eastern Pará State in the Brazilian Amazon were in marked contrast. There, resprouts contributed the greatest proportion of both stems and species 5 cm or more in diameter at breast height (130 cm) throughout the range of stand ages studied (Vieira and Proctor 1998).

Possible reasons for such intersite differences in the importance of resprouts as a regeneration mechanism may be suggested. It seems likely that the dominance of fallows and secondary forest by resprouts over long time periods in Bragantina (Vieira and Proctor 1998) results from the much greater time since settlement there than in Paraguay (more than 100 years in comparison with 30 years; Kammesheidt 1998). Especially when farmers are shortening the length of fallow periods, it is reasonable to hypothesize that on-site seed production by many tree species must be reduced or nonexistent in shifting cultivation landscapes such as those of Bragantina simply because stems barely or never reach reproductive status. Possible reductions in the size and diversity of the seed rain caused by the preceding factors might be exacerbated by the loss of primary forest habitat and its function as a seed source (Denich 1991; Vieira et al. 1996). Additionally, if most remnant primary forest is riparian or swamp vegetation, some of the plant species that make it up probably have limited abilities to colonize dryland sites, especially in competition with the pioneers typical of such sites. Vieira et al. (1996) nevertheless recorded adventive species originating from riparian vegetation and swamp forest (*igapó*) in Bragantina fallows.

What are the characteristics of seed dispersal processes in shifting cultivation landscapes? Studies of forest succession on abandoned neotropical pastures have emphasized the importance of wind dispersal in the natural regeneration of forest cover in anthropogenic habitats (Janzen 1988; Finegan and Delgado 2000). Shifting cultivation fields generally are small (less than 2 ha) in comparison with pastures, however, and are set in landscape mosaics that may favor the persistence of populations of seed-dispersing vertebrates. Younger shifting cultivation landscapes also are generally characterized by the presence of scattered, often large trees preserved by the farmer. We therefore predict that the role of vertebrate dispersal in vegetational dynamic processes is likely to be greater in shifting cultivation landscapes than in those dominated by pastures (even though cattle may sometimes disperse seeds; see Chapter 19, this volume). This point, which is crucial with respect to biodiversity generation and maintenance among plants and interacting frugivorous and granivorous animals, has not attracted the attention of most workers concerned with fallows. In one exception to this trend, Ferguson (2001) found wind-dispersed *Lonchocarpus guatemalensis* and *Trichospermum* sp. among the most numerically important tree species in fallows in Guatemalan subtropical moist forest, although vertebrate-dispersed tree species such as *Spondias mombin* were also common. In the paleotropics, some of the most common pioneer species (e.g., *Macaranga* spp. and *Musanga cecropioides*) rely on zoochory, and their seeds are generally short-lived. In the coastal plains of Gabon (Nasi 1997; Fuhr 1999), where *Musanga cecropioides* is absent because of the poor, sandy soils, the three most common species colonizing old fields or savannas are *Aucoumea klaineana* (a wind-dispersed, large, long-lived pioneer), *Sacoglottis gabonensis* (an emergent of the mature forest with elephant-dispersed seeds), and *Xylopia aethiopica* (a small, short-lived pioneer whose seeds are bird dispersed). These results tend to support our prediction, although the wide range of life forms and dispersal strategies among the tree species of shifting cultivation landscapes renders difficult any generalization about the relative importance of dispersal mechanisms.

Besides emphasizing the importance of wind dispersal to forest regeneration in neotropical pastures, recent studies show that trees in them facilitate succession by providing habitat for seed-dispersing vertebrates, an interaction whose importance has long been recognized in many successional environments. As in the case of dispersal mechanisms, however, the limited work on seed dispersal in shifting cultivation landscapes does not provide a similarly firm basis for generalization. Ferguson (2001) found no evidence that the *Attalea cohune* palms common on his shifting cultivation sites performed such a function and suggested that the lack of facilitation by *A. cohune* resulted from the shifting cultivation landscape providing a generally favorable habitat for seed-dispersing vertebrates. Conversely, Carrière (1999) and Carrière et al. (2002) found that the practice of leaving remnant trees accelerates regeneration and produces a landscape that is more hospitable as a habitat for biodiversity in the Ntumu agricultural system in Cameroon.

To sum up, it is clear that a significant proportion of the individuals of woody plant species in any shifting cultivation landscape has regenerated by resprouting, and it seems highly likely that the agricultural cycle exerts strong selection for species capable of resprouting after cutting, burning, and weeding. Shifting cultivation landscapes may be generally more favorable to the maintenance of vertebrate-mediated seed dispersal processes than other anthropogenic habitats such as pastures. They nevertheless remain landscapes dominated by pioneer plant species. If the lack of seed sources means that resprouting is the main or even the only mechanism of regeneration of many forest-dependent tree species in shifting cultivation landscapes (Denich 1991; Vieira and Proctor 1998), then the chance that diverse forest with at least some of the characteristics of the original vegetation will ever be recovered on shifting cultivation land seems limited. The vigorous regeneration from root sprouts of exceptionally valuable multiple-use species such as *Platonia insignis* (see Shanley et al. 1998) appears to have contributed to some authors' optimism regarding the potential of secondary succession in shifting cultivation landscapes for production and forest restoration (e.g., Vieira et al. 1996). On the other hand, Denich (1991) suggests that fallows should not be viewed as a stage in the regeneration of primary forest but as a new, wholly anthropogenic vegetation type. If this is the case, it does not mean that farmers cannot manage fallows for certain products as they have always done (Unruh 1988), but it is a sobering idea in the context of biological conservation.

Successional Dynamics of Fallows and the Factors That Underlie Successional Change

The successional dynamics of fallows and the secondary forests sometimes derived from them have been studied in moist forest of Mesoamerica (Mexico, Guatemala, Panama), the Amazon basin, the Guianas, and the South American subtropics and show successional sequences broadly similar to those in Africa and Asia (Kenoyer 1929, cited by Richards 1976; Gómez-Pompa and Vásquez-Yanes 1981; Vieira et al. 1996; Kammesheidt 1998; Vester and Cleef 1998; Ferguson 2001; Peña-Claros 2001). Reviewers differ in the number of successional stages they identify within this overall framework of similarity, and the identification of stages in an essentially continuous process is largely for convenience. At a very general level, Finegan (1996) described three stages of neotropical lowland rainforest succession during the first century of the process. Initial dominance by pioneer herbs, shrubs, and climbers often is followed by stages dominated by short-lived and then long-lived pioneer tree species, with dominance by this latter group potentially lasting for several decades. Forest structural characteristics such as canopy height and basal area may reach values similar to those of primary forest in as little as two decades. Individuals that colonize the site quickly after its abandonment or assignment to fallow dominate the site for decades, and succession unfolds because, to a great extent, of the different life histories and degrees of shade tolerance of this set of species. Successional processes similar to these also occur in tropical and subtropical regions of Africa (de Namur and Guillaumet 1978; Donfack et al. 1995; King et al. 1997; Fuhr 1999) and Asia (Whitmore 1984; Riswan and Kartawinata 1991). Botanical species, not tree functional groups, vary between continents.

It is vitally important to go beyond basic general descriptions of successions and characterize and understand the great diversity of successional processes that are observed in reality. Tropical secondary successions occur as described in the previous paragraph only under optimal conditions. These conditions include little or no degradation of site conditions, a well-stocked seed bank, seed trees located within dispersal range and with functional seed dispersal processes, viable tree stumps and root systems for resprouting, and minimal additional disturbance to the site after succession begins. Common sense suggests that these conditions are most likely to be met on small areas, with light or no agricultural use of the land, embedded in matrices that include large proportions of primary or old secondary forest. It is unlikely that such conditions will be met in most of the situations in which secondary succession will occur in the tropics in the coming years. Successional processes on large areas such as many abandoned pastures of the mainland neotropics are likely to be more complex than in the optimal scenario and in successional communities of slower development, lower diversity, and lower productivity. This is especially likely when a large area is accompanied by a suite of site factors that represent barriers to succession initiation (Janzen 1988; Nepstad et al. 1991; Finegan and Delgado 2000). With respect to succession on shifting cultivation plots, it is important to emphasize that these are small habitat patches created by a land use system designed with the maintenance of site productivity for agriculture as a principal goal. This scenario is favorable, in principle, for the development of fallow vegetation—for secondary succession. On the other hand, these are habitat patches that experience a high frequency of drastic disturbance (even traditional fallow lengths should be considered short time periods in relation to the recovery of many forest characteristics) embedded in a landscape in which fallow vegetation of intermediate value as a seed source, probably is the most important single land use. In seasonal environments, fallow vegetation is also chronically prone to disturbance by uncontrolled fire (Smith et al. 2001), and in general, long-term declines in site productivity caused by repeated cropping cycles and ever-shorter fallows appear inevitable. All these factors, in different ways, represent limitations on the development of secondary vegetation. Limitations may be especially serious with respect to forest-dependent plant species. These have a low proportional abundance in secondary successions for many decades in any case (Finegan 1996), although under favorable conditions they may accumulate in the understories of secondary forest (Guariguata et al. 1997). Conditions in shifting cultivation landscapes are far from optimal, however, and are especially hostile for this group of species, which therefore become a priority for conservation action.

Discussion of factors that limit successional development leads us to important specific cases of departure from the model presented at the beginning of this section. Conditions in shifting cultivation patches may be so far from the optimal for successional regrowth that they lead to the colonization of the site by species that are capable of establishing dominance and inhibiting successional change for many years. Among these are the grass *Imperata cylindrica* (Whitmore 1984) and the shrub *Chromolaena odorata* (family Asteraceae; de Rouw 1995). The pantropical *C. odorata* often is the dominant plant in young fallows in Asia and West Africa (de Rouw 1995; Roder and Maniphone 1998; Kanmegne et al. 1999). De Rouw (1995) describes the conditions under which the species may become dominant in the Côte d'Ivoire, which we present as an example of how trends in shifting cultivation may lead to the replacement of "normal" fallow vegetation by persistent lower-diversity communities. Short fallows are the trigger. They are associated with greater densities of arable weeds and lower densities of "forest plants" at the end of the cropping period than is the case under traditional, longer fallows. This situation is in turn associated with slower canopy closure by the fallow vegetation after cultivation ceases, which permits the establishment of *C. odorata*. The tendency toward shorter fallows therefore is accompanied by a steady increase in land area covered by *C. odorata* thickets (de Rouw 1995). This herb then shades out other fallow species, becoming dominant. *C. odorata*-dominated fallows, by definition, are of lower diversity than those free of this species.

Biodiversity and Conservation in Shifting Cultivation Landscapes

The contribution of shifting cultivation landscapes to biodiversity conservation must be assessed in relation to two different human-made entities: crop fields and fallows. Great care must also be taken regarding the approach to assessment. Although biodiversity often is evaluated solely on the basis of species richness, a full evaluation entails consideration of the composition of the community and its ecological diversity (Finegan 1996). Biodiversity evaluations must also weight species, for example, according to whether they are widely distributed or endemic, pioneers or forest-dependent (Pielou 1995). Weighting is particularly important for agricultural communities and landscapes. This is because species richness may be high in such settings and could play an important role in the use value of a community or landscape to people and in the maintenance of ecological functions in it, as we will see. But if communities are composed largely of species adapted to human disturbance, their importance in the context of biodiversity conservation will be low, however diverse they are. Our review of biodiversity and conservation in shifting cultivation landscapes places particular emphasis on forest-dependent species of both plants and animals.

Biodiversity of Crop Fields in Shifting Cultivation

In all shifting cultivation systems, the field is made of several strata of vegetation. It occupies a three-dimensional space and at first sight gives an impression of vegetable chaos. The various intermixed crops encompass all possible life forms: grasses, shrubs, small and large trees, palms, and lianas. Crop mixtures vary between and within continents, regions, and ethnic groups, but major crops belong to a limited number of taxa not really different from the ones used in permanent agricultural systems: cassava (*Manihot esculenta*), maize (*Zea mays*), banana and plantain (*Musa* spp.), rice (*Oryza* spp.), sweet potato (*Ipomoea* spp.), taro (*Colocasia* spp., *Xanthosoma* spp.), and yam (*Dioscorea* spp.). Important features of shifting cultivation systems from the biodiversity point of view are as follows:

- Traditional shifting cultivation systems use numerous cultivars or landraces (Table 8.1) of each of these major crops; these landraces generally are different from the varieties used in intensive systems, making shifting cultivation a possible reservoir of genetic resources for essential crops.

Table 8.1. The landraces used in shifting cultivation.

Crop	Landraces	Ethnic Group	Country
Banana	28	Maring	Papua New Guinea
	17	Mvae	Cameroon
Cassava	61	Caboclo	French Guyana
	31	Wayapi	French Guyana
	46	Kuikuru	Brazil
	76	Makushi	Guyana
Rice	92	Harunôo	Philippines
	44	Kantu	Borneo
Sweet potato	17	Daribi	Papua New Guinea
Taro	69	Elia	Papua New Guinea
	20	Yafar	Papua New Guinea
Yam	32	Maring	Papua New Guinea
	80	Wusi	Vanuatu

Source: Modified from Dounias (2001).

Table 8.2. The species cultivated in shifting cultivation.

Number of Cultivated Species	Ethnic Group	Country
Asia		
50–200	Kenyah	Sarawak, Kalimantan
413	Harunôo	Philippines
66	Daribi	Papua New Guinea
144	Yopno	Papua New Guinea
America		
38–41	Yanomami	Brazil, Venezuela
38	Andoke	Colombia
71	Yekwana	Venezuela
Africa		
34	Badjwe	Cameroon
38	Mvae	Cameroon
37	Tikar	Cameroon
40	Ngbaka	Central African Republic

Source: Modified from Dounias (2001).

- A great number of secondary crops are grown together with the main food crops (Table 8.2).

Unlike wild biodiversity, the agricultural biodiversity of shifting cultivation crop fields has attracted little attention from ecologists. The result is that processes that create and maintain this diversity, although they may have been recorded and appreciated by agronomists and anthropologists, are poorly understood. This is particularly true for the vegetatively propagated crops that dominate many shifting cultivation systems, such as banana, cassava, plantains, and sweet potatoes. Several external forces currently threaten the maintenance of this local agrobiodiversity, including intensification, commercialization, technological change, and the loss of traditional knowledge. There seem to be two answers to these threats. One is that genetic erosion is inevitable and therefore ex situ conservation is the solution (Frankel 1995). The other is that for several reasons (including site adaptation, risk aversion, and culture) farmers are reluctant to abandon their landraces or agricultural practices for new varieties or new practices (Brush 1995; Louette et al. 1997). In the first view, once landraces have been conserved ex situ, traditional farmers have little importance; in the latter view, in situ conservation continues to make unique contributions even in a modernized world (McKey 2001). Fortunately, the "ex situ conservation only" paradigm is moving toward a new, more balanced, "ex situ and in situ" paradigm. This paradigm shift might increase the importance of traditional shifting cultivation systems in agrobiodiversity conservation strategies.

A small fraction of the plant diversity of the original forest may be present in crop fields because some forest trees are preserved by shifting cultivation farmers. Their wood may be too hard to be cut using traditional tools (e.g., *Samanea dinklagei* and *Lophira alata* in Africa, *Koompassia* spp. in Southeast Asia, *Dipteryx panamensis* in Mesoamerica). Their crown may produce a very light shade and therefore not hinder the crop growing beneath (e.g., *Piptadeniastrum africanum* in Africa). Another, more compelling reason is that these trees are the source of important products for purposes such as construction, food, and medicine, or they have strong cultural values. Among the trees spared for food, *Coula edulis* (nuts), *Garcinia kola* (nuts), *Irvingia gabonensis* (fruits), and *Ricinodendron heudelotii* (fruits) are examples of species common in forest fallows throughout humid lowland Africa. Another interesting example is *Sterculia rhinopetala*. The Ntumu in southern Cameroon preserve this species because the fruits are considered to attract small antelopes, a favorite game animal in the region (Carrière 1999). Among the forest tree species nurtured by the Krissa of Papua New Guinea, 9 percent are retained because when in flower or fruit they attract a range of game animals and birds and are consequently used by hunters as hides to ambush game (Kocher-Schmid 2001). In Southeast Asia, *Koompassia* spp. trees are preserved not only because of the hardness of their wood but also because they are a favorite place for bees to establish their hives (Mabberley 1987; Prebble et al. 1999).

Switching the focus to the neotropics, Unruh (1988) recorded a number of valuable primary forest plants conserved by shifting cultivation agriculturists on farms in Peruvian Amazonia. Brazil nut (*Bertholletia excelsa*) is a prime example of a primary forest tree species conserved because of its high value (backed up in Brazil by a law that prohibits the cutting of this species). The low density of *B. excelsa* trees in most Amazonian primary forest (Shanley et al. 1998) may mean that trees are not abundant on farms, although they are certainly present on many (A. da Silva Dias, pers. comm., 2001). Peña-Claros (2001) suggests that *B. excelsa* may be just as abundant in anthropogenic habitats as in primary forest in the southwest Amazon, or more so, because of a combination of tree conservation by farmers and successful regeneration caused by the light-demanding nature of this species. *B. excelsa* is suitable for planting in fallows and actively used agricultural habitats, although this practice is not common (Peña-Claros 2001).

The planting of trees by farmers seems related more to complex agroforests (see Chapter 10, this volume) than to classic shifting cultivation. Limits are not always clear, however, and the Krissa people in Papua New Guinea practice a form of shifting cultivation focusing on woody species, nurturing or planting trees (*Gnetum gnemon*) and palms (*Areca catechu*, *Cocos nucifera*, *Metroxylon sagu*) as major crops in their shifting cultivation gardens. A cursory count in a Krissa garden revealed 11 different trees, two palms and two bamboo species planted, and seven tree and two palms nurtured from the regrowth. The main function of gardening thus appears to be the propagation of useful trees rather than the immediate and direct provision of crops (Kocher-Schmid 2001), creating a blurred frontier between forest and gardens, wild and cultivated. Such an activity, when carried out for centuries as in this case, undoubtedly has profound implications for the structure, composition, and biodiversity of the forest. Jungle rubber (see Chapter 10, this volume) is a further example of a production system that straddles the limits between shifting cultivation and other types of agroforestry.

As we have emphasized, "weeds," which can be simply defined as all noncrop species present in crop fields (Lambert and Arnason 1989), rank with declining soil fertility as one of the main reasons why crop fields are left fallow in the tropics (Lambert and Arnason 1989; Hinvi et al. 1991; Thurston 1997). They contribute to the species richness but, by their very nature, not to the overall conservation value of shifting cultivation landscapes. However, they may play a role in maintaining ecosystem functions. Many common weeds are also potential sources of medicinal and food products, and to the forest ecologist they are the pioneer species of the succession that begins when land is left fallow.

As long ago as 1929, Kenoyer (cited by Richards 1976) listed some of the plant families typical of weed communities of crop fields and young fallows on Barro Colorado Island, Panama, citing Amaranthaceae, Euphorbiaceae, Asteraceae (Compositae), Solanaceae, and Fabaceae or Mimosoideae among dicots and Poaceae and Cyperaceae as characteristic monocot families. The representative nature of Kenoyer's families is illustrated in Table 8.3, which lists some typical herbaceous species of neotropical shifting cultivation landscapes on the basis of more recent literature. The list is illustrative, not exhaustive, and permits the identification of some important characteristics of the herbaceous weed flora of shifting cultivation landscapes. First, it is clear that, as is the case for many organisms associated with human disturbance, this flora is characterized by many species with wide geographic distributions, some of which are known to be consequences of human introduction and some of which are pantropical. By the criteria of geographic distribution typically used in priority-setting exercises for conservation (Bibby et al. 1992; Myers et al. 2001) and by their abundance in anthropogenic communities, the herbaceous weed flora associated with shifting cultivation is not of any particular conservation value. On the other hand, weed floras can be diverse (Croat 1978). Although weeds are better known as one of the main agroecological bases of the use of fallows, they may contribute to ecosystem nutrient retention in the cropping stage of some shifting cultivation systems (Lambert and Arnason 1989), and the local diversity of weed communities may be related to the quality and magnitude of this contribution to the maintenance of ecosystem function.

The definition of weeds as noncrop species, with its lack of explicitly neg-

Table 8.3. Examples of typical weeds of crop fields and young fallows in the neotropics.

Family	Genus or Species	Observations	Uses	Source
Amaranthaceae	*Amaranthus hybridus*	Pantropical	Genus: F	1,2
Amaranthaceae	*Iresine celosia*		Genus: M	1, 3, 8
Asteraceae	*Bidens pilosa*	Pantropical		1
Asteraceae	*Eupatorium cerasifolium*			4, 5
Cyperaceae	*Cyperus, Scleria*			
Euphorbiaceae	*Phyllanthus urinaria,* *Phyllanthus* spp.	Species introduced from Asia	Genus: F, M, P	3, 5, 7
Loganiaceae	*Spigelia anthelmia*		Species: M, P	1, 3, 6, 7, 8
Phytolaccaceae	*Phytolacca*		Genus: F	1,3
Poaceae	*Andropogon bicornis*			5
	Panicum laxum	America and Africa		4
	Paspalum decumbens			5
	P. conjugatum	Pantropical		1
Rubiaceae	*Borreria latifolia,* *B. verticillata*		*B. verticillata:* M	4, 7, 8

Abbreviations: F, food; M, medicine; P, poison.
Sources: 1, Rico-Bernal and Gómez-Pompa (1976); 2, Judd et al. (1999); 3, Mabberley (1987); 4, Uhl et al. (1981); 5, Uhl (1987); 6, House et al. (1995); 7, Grenand et al. (1987); 8, Croat (1978).

ative connotations, is appropriate because of the use value many such species have to rural people. Weeds, particularly herbaceous species, are potentially important sources of nontimber products in shifting cultivation landscapes (Table 8.3). Indeed, several studies have shown that disturbed neotropical plant communities contain more individuals of more useful species than undisturbed forests and produce a greater diversity of products that are more likely to be familiar to and used by rural people (Voeks 1996; Chazdon and Coe 1999). The regeneration of some valuable light-demanding tree species is an important element of the consideration of the weed community of crop fields and fallows. Examples of such species, which may be actively encouraged by farmers, are *Platonia insignis,* valuable for fruits (Shanley et al. 1998), the multiple use *Inga* spp. (Pennington and Fernandes 1998; Unruh 1988; B. Finegan, pers. obs., 2000), and timber trees such as the widely distributed *Jacaranda copaia* (Unruh 1988; Finegan 1996; Finegan, pers. obs., 2000) and the western Amazonian *Guazuma crinita* (Smith et al. 2001). A complete review of this subject is beyond the scope of this chapter, but the potentially high utilitarian value (Chazdon and Coe 1999) of the taxonomic and

functional diversity of weeds in crop fields and young fallows should not dis-
tract us from the fact that their value from the point of view of taxonomic biodiversity conservation is very low.

All consideration of the utilitarian or conservation value of noncrop plant
species in the context of shifting cultivation must eventually return to the fact
that a weed is, by definition, a plant that is undesirable to the farmer. The
degree to which weed control is attempted during the growth of a crop is
variable and may be low, although this activity seems bound to limit the accumulation of species diversity during the cropping phase, especially when
accompanied by burning after harvests during that phase (Watters 1971; Lambert and Arnason 1989; de Rouw 1995).

Plant Biodiversity of Fallows in Shifting Cultivation

In terms of taxonomic and functional composition of woody species, fallows
and the secondary forest arising from them share many characteristics with
secondary vegetation developing in other situations. A low representation of
forest-dependent plant species, other than those conserved by farmers for
some reason, is a universal characteristic of fallows and therefore a major limitation on their conservation value.

Among pioneer trees, the first ones to dominate fallows, numerous genera
and species are widely distributed in secondary lowland tropical moist vegetation. Examples for the neotropics (Finegan 1996) are *Cecropia, Croton, Helio-
carpus, Trema,* and *Trichospermum* with short-lived species and *Cordia,
Guazuma, Rollinia, Spondias,* and *Vochysia* with long-lived species. *Didy-
mopanax (Schefflera) morototoni, Jacaranda copaia,* and *Simarouba amara* are
additional examples of long-lived pioneer species. This is also true for the paleotropics. Examples of short-lived pioneer genera and species in Africa are
Anthocleista spp., *Harungana madagascariensis, Macaranga* spp., *Musanga cecropioides,* and *Solanum verbascifolium.* Long-lived pioneer species are represented
by *Albizia* spp., *Aucoumea klaineana, Milicia (Chlorophora)* spp., *Ricinodendron
heudelotii, Terminalia superba,* and *Triplochyton scleroxylon.* Several of these
species are major timber species in Africa, and Finegan (1992) and Chazdon
and Coe (1999) have pointed out the prevalence of desirable characteristics
among the woods of the neotropical long-lived pioneers. In Asia (Whitmore
1983, 1984; Riswan and Kartawinata 1991) we again find similarly widespread
genera of short-lived pioneers, such as *Melastoma* spp., *Macaranga* spp., and
Trema spp., accompanied by long-lived pioneer species such as *Anthocephalus
chinensis, Duabanga molucana, Endospermum* spp., and *Octomeles sumatrana.*

In general, the species richness and diversity of fallows are initially low but
higher than in crop fields, and they increase over time as the vegetation develops; several decades may elapse before these parameters approach values similar to those of primary forest in small sample plots (Lescure 1986; Saldarriaga

et al. 1988; Fuhr 1999; White 1994). Even in the secondary forest that is sometimes derived from fallow, the long period of vegetation dominance by long-lived pioneer tree species, with a low representation of forest-dependent species caused by factors discussed earlier, ensures that recovery of the compositional characteristics of mature forests probably will take centuries even if forest-dependent species are colonizing the site, which they may not be (White 1994; Finegan 1996; Fuhr 1999). Shade and root competition from pioneer trees undoubtedly slow the increase of species richness and diversity. In Africa, Leroy-Deval (1973) and Kahn (1978) have shown that *Macaranga huriifolia* (a short-lived pioneer species) and *Aucoumea klaineana* (a long-lived pioneer species) established root grafts early in succession, increasing the competitive power of the species and allowing the establishment of pure stands.

Despite low species richness in the early stages of fallow development, shifting cultivation landscapes may be quite diverse when the species of fallows of different ages are added up. For example, Christanty et al. (1986) reported that the kebun-talun system of Java contained 112 plant species, largely because of a long period of perennial production in a managed fallow. It is also obvious that species richness in fallows increases dramatically in comparison with that of crop fields. Hart (1980) and Ewel (1986) have suggested that such systems may be designed as analogs of natural forest systems in that they tend to mimic successional stages of the forest in structure and presumably in function. On the other hand, the increasing tendency to fallow invasion by highly dominant species such as *Chromolaena odorata* (discussed earlier) is a tendency toward further reductions in the plant species diversity of fallows.

From this subsection and those preceding, it is clear that the contribution of fallows to the recovery of local-scale (alpha) diversity of vegetation, and the compositional characteristics of primary forest, is small. This is because of the shortness of fallow periods and the dominance of the vegetation by resprouts and, in relation to regeneration from seed, by widespread short-lived pioneer tree species. Plant diversity of fallow landscapes may be high, on the other hand, although it does seem very unlikely that species numbers increase with area at the same rate as in primary vegetation. The number of forest-dependent species present in landscapes seems likely to depend principally on the area of remnant primary forest.

Animal Biodiversity of Fallows in Shifting Cultivation

Shifting cultivation landscapes are a spatially and temporally heterogeneous habitat for vertebrates, to a degree that is undoubtedly influenced by the scales at which different species perceive such environmental variation. As in the case of plant species and communities, research on vertebrates has focused on variations of species diversity and composition between different habitat types in the landscape. Researchers have sought to relate this variation to factors such as vegetation and landscape structure, special habitat features, resource avail-ability, and hunting pressure. Less information is available on vertebrates than on plants, however (Shankar Raman et al. 1998). Published studies vary in the ages of fallow or secondary forest studied and in the characteristics of the sur-rounding landscape in terms of factors such as the total area of primary forest. The important influence of the surrounding landscape is acknowledged by authors such as Shankar Raman et al. (1998) for the specific case of shifting cultivation and by Saunders et al. (1991) as a general principle, but is not assessed quantitatively in most studies of shifting cultivation. Nevertheless, the available information indicates that differences of vertebrate diversity and composition between fallow stands and landscapes on one hand and primary forest on the other often are not as clear-cut as they are for plants. This differ-ence results from factors such as the mobility of vertebrates and the provision of resources such as food in plant communities of widely varying structural and compositional characteristics.

Six-year-old fallows of 0.9–2.9 ha in a primary rainforest matrix in lowland Chiapas, Mexico, did not differ from primary forest with respect to the species richness of small and medium-sized mammals (Medellín and Equiha 1998). All the mammal species recorded in traps were found in both habitats, although differences of relative abundances between habitats were recorded. Two monkey species were among those described as obligate arboreals and absent from the fallows (Medellín and Equiha 1998). Shankar Raman (1996) found that two primarily canopy-dwelling squirrel species were absent from fallows less than 25 years old in a tropical moist forest landscape of northeast-ern India. Switching the focus specifically to primates, Cowlishaw and Dun-bar (2000) observe that the large areas of woody vegetation that persist in shifting cultivation landscapes may offer adequate habitat for many species of this group. Tropical studies often show no differences of richness and compo-sition between primates observed in fallow and secondary forest vegetation and primary forest, and some primates are more abundant in the anthro-pogenic vegetation than in "natural" communities in Asia and Africa (Cowlishaw and Dunbar 2000). Shankar Raman (1996) and Medellín and Equiha (1998) nevertheless show evidence that as in the case of birds (dis-cussed later in this chapter), the species composition of mammal assemblages using fallow vegetation will change over time, at least under some circum-stances.

At a more general level, studies in Africa (Wilkie and Finn 1990; Thomas 1991; Lahm 1993) have shown that shifting cultivation landscapes may sus-tain a high biomass of small to medium-sized mammals and sometimes large ones (buffalos, *Syncerus caffer;* chimpanzees, *Pan troglodytes;* gorillas, *Gorilla gorilla;* large antelopes; and bush pigs, *Potamoecherus porcus*) as well. Elephants (*Loxodonta africana*) are found in greater numbers in these landscapes than in areas with greater proportions of primary forest land (Barnes et al. 1991).

Work in shifting cultivation landscapes suggests some important patterns in the characteristics of bird communities. Terborgh and Weske (1969, Amazonian Peru), Johns (1991, Amazonian Brazil), Andrade and Rubio-Torgler (1994, Amazonian Colombia), Vieira et al. (1996, Amazonian Brazil), and Anderson (2001, Mesoamerica) have carried out studies of birds in neotropical shifting cultivation landscapes. Studies from Africa and Asia are less numerous but include Blankenspoor (1991, Liberia) and Shankar Raman et al. (1998, northeastern India). The numbers of habitat types included in comparisons and the landscape contexts in which shifting cultivation communities were analyzed vary between studies. The fields and fallows where Johns, Andrade and Rubio-Torgler, and Anderson worked were embedded in a primary forest matrix, for example, whereas Vieira et al. (1996) worked in the Bragantina landscape of Pará State, Brazil, where fallow and secondary forest is the dominant land cover type and only a single 200-ha fragment of dryland forest could be found for comparisons with fallows. Sampling methods also vary between studies and affect the comparability of the results. The scale of these studies should be considered the patch, with the exception of that of Anderson, which related the characteristics of the raptor community to landscape characteristics. In common with the studies of vegetation cited previously, all these authors used a chronosequence approach to sampling, working simultaneously in fallows of different ages and assuming that any differences between them represent patterns over time in a single habitat patch.

No single, clear-cut pattern of bird species richness and diversity emerges from these studies; we will consider compositional patterns later in this chapter. Neither Terborgh and Weske (1969) nor Andrade and Rubio-Torgler (1994) found between-habitat differences of species diversity per 100 individuals in understory mist net captures, although total numbers of species observed by all methods differed between habitats in the former study and were lowest in second growth and a cocoa plantation. Adding to this indication that bird species richness may be lower in fallow vegetation than in primary forest, Vieira et al. (1996) caught fewer species in 10-year and 20-year fallows than in their primary forest fragment. They also found more forest bird species in the 20-year fallow than in the 10-year one, and, similarly, bird species richness increased markedly between fallows of increasing age in both the African (Blankenspoor 1991) and the Asian (Shankar Raman et al. 1998) studies cited. In contrast, however, fallow was the *most* species-rich habitat at Johns's (1991) site. On a much larger spatial scale, the abundance and diversity of raptors observed by Anderson (2001) increased with the increasing structural heterogeneity contributed to the landscape by shifting cultivation, with primary forest being the least structurally diverse landscape. Such a relationship seems likely to be observed on many different scales in many groups of organisms as long as the number of disturbed-habitat species gained through human modification of part of the habitat is greater than the number of forest species lost for the same reason.

Compositional patterns in bird communities of shifting cultivation landscapes are perhaps clearer than those in species richness and diversity, especially in relation to composition by feeding guilds. Even when many species are common to all habitats in a landscape, as was the case in many of the studies cited here, variations in their relative abundances mean that compositional similarity at the community level tends to be greatest between sites with the most similar vegetation (e.g., primary forest and old fallows) and least between sites at opposite extremes of the disturbance gradient (primary forest and very young fallow; Blankenspoor 1991; Johns 1991; Andrade and Rubio-Torgler 1994; Vieira et al. 1996; Shankar Raman et al. 1998; Table 8.4). The degree of isolation of anthropogenic habitat patches undoubtedly also plays a role in determining compositional patterns in bird communities, although this is largely undocumented. As an example, however, Stiles and Skutch (1989) state that four of the five most abundant species recorded by Vieira et al. (1996) use shaded habitats, such as cocoa plantations and older secondary forest, when these are *adjacent* to primary forest. On the same theme, Shankar Raman et al. (1998) recognize the important role that the habitat surrounding patches of fallow and secondary forest may play in determining the abundance, composition, and diversity of the vertebrate assemblages observed in those patches. However, they point out that the bird assemblages of replicate patches of fallow vegetation in given age classes separated by several kilometers were more similar to each other than to those of the adjacent habitats. These results support the conclusion that vegetation composition and structure of a given patch play major roles in determining the characteristics of bird

Table 8.4. Compositional similarity (Horn's Index of Overlap) between birds observed in contrasting vegetation types in an Amazonian rainforest, Amazonas State, Brazil.

	Unlogged Primary Forest	Logged Primary Forest	35-ha Forest Fragment	Fallow	Crop Fields
Unlogged primary forest	—				
Logged primary forest	.75	—			
35-ha forest fragment	.35	.23	—		
Fallow	.44	.32	.54	—	
Crop fields	.24	.23	.18	.55	—

Source: Johns (1991).

Note: Higher values indicate greater similarity.

assemblages observed in that patch, irrespective of the habitat type that adjoins it.

Although abundance patterns over disturbance gradients are documented for individual species by some authors, patterns at the guild level probably are a more practical indicator of the main ecological processes affecting bird community characteristics. Several studies have shown that terrestrial forest birds, particularly insectivores, are less species rich and abundant in fallows and crop fields than in primary forest (Terborgh and Weske 1969; Johns 1991; Andrade and Rubio-Torgler 1994). Bark-gleaning and foliage-gleaning insectivores may also decrease in fallows, with a corresponding increase of sallying insectivores such as flycatchers (Tyrannidae) and insectivore frugivores such as some tanagers (Thraupidae; Blankenspoor 1991; Johns 1991; Andrade and Rubio-Torgler 1994; Shankar Raman et al. 1998). Changes similar to these have been observed in neotropical forest disturbed by logging (Thiollay 1992, 1997) or fragmentation (Bierregaard and Stouffer 1997) and appear likely to be a general response of bird communities to habitat modification and simplification. The appearance in young second-growth and crop fields of granivores, as well as doves and pigeons (Columbidae), which are generally absent from forest habitats, is an important but not universal change in shifting cultivation landscapes (Johns 1991; Andrade and Rubio-Torgler 1994; Shankar Raman et al. 1998).

Because raptors are a single guild, compositional differences between landscapes of different structural diversity must be examined in terms of individual species (Anderson 2001). Only one raptor species was noticeably more abundant in primary forest than in shifting cultivation and forest mosaic landscapes, although three species observed only once, or observed outside sample plots, may be considered forest dependent; this group included the harpy eagle (*Harpia harpyja;* Anderson 2001). Three species were more abundant in the more heterogeneous landscapes, which included crop fields and fallows (Anderson 2001). It was the appearance of these species, among other factors, that accounted for the greater abundance and diversity of raptors in the more structurally diverse landscapes.

What habitat factors contribute to the changes in species richness and composition of vertebrate assemblages that are observed when shifting cultivation habitats are compared with primary forest? We will emphasize habitat structure, food availability and foraging behavior, and microclimate. Our focus is on factors linked to the lower abundances or absence of forest species in agricultural habitats because of the conservation importance of these species, not on the reasons why disturbed-habitat species colonize landscapes because they are able to use crop fields and fallows.

It is well established that habitat structural diversity can influence animal species diversity (a recent summary is provided by Begon et al. 1996). Terborgh and Weske (1969) suggested variation in foliage height profiles as the explanation for between-patch type differences in total species richness; for example, the secondary vegetation in their study lacked foliage above 20 feet (6.1 m) and therefore lacked the species that use this component of the habitat in other patch types, such as primary forest. Other authors demonstrate differences in vegetation vertical structure between fallows of different ages and point out their likely relationship to changes in the species richness of vertebrate assemblages (Blankespoor 1991; Medellín and Equiha 1998; Shankar Raman et al. 1998). In general, woody secondary vegetation tends to have a more uniform structure than mature forests in terms of canopy height and the absence of treefall gaps. Nevertheless, some of the other ways in which habitat structural factors are linked to vertebrate community characteristics may be subtle. For example, Terborgh and Weske (1969) found that variation in foliage height profiles was insufficient to explain all the observed variation in bird community composition and diversity at their site, so they invoked and justified an additional set of special habitat quality factors. Examples of such factors are vine tangles, absent, along with their associated birds, from the second-growth site studied by these authors.

Habitat structure may also influence the characteristics of vertebrate communities at the landscape scale. Although fallow vegetation is less structurally diverse than primary forest at the stand level, the landscapes with shifting cultivation activity in Anderson's (2001) study area were *more* structurally diverse than those with only primary forest. Three raptor species of the landscapes modified by human activity are common in anthropogenic open habitats of Honduras, so that the greater structural diversity of these landscapes presumably underlies the greater richness, diversity, and abundance of raptors observed in them, in ways linked to the habitat preferences, foraging tactics, and preferred prey of the different bird species (Anderson 2001).

Overall, the information available justifies the simple conclusion that elements of the forest fauna that use habitat structural elements or types that are absent or uncommon in areas influenced by shifting cultivation probably will be less abundant in those areas than in forest. Conversely, habitat features associated with shifting cultivation will bring species adapted to those features into the community. These relationships will operate at different spatial scales depending on the characteristics of the vertebrate guild or species involved.

It is highly likely that spatial and temporal patterns in the availability of food influence the characteristics of vertebrate communities in shifting cultivation landscapes. Cowlishaw and Dunbar (2000) emphasize that *Musanga cecropioides,* a pioneer tree abundant in African shifting cultivation landscapes, produces fruit attractive to primates over much of the year and so contributes to the use of agricultural habitats by these animals. However, habitat use patterns by primates may vary over the year in relation to the fruiting phenologies of the different communities of the landscape (Fimbel 1994, cited by

Cowlishaw and Dunbar 2000). In the same vein, Shankar Raman (1996) interprets contrasting patterns of habitat use by northeast Indian forest primates in relation to feeding habits. For example, frugivorous hoolock gibbon (*Hylobates hoolock*) was observed largely in primary forest habitats and has a documented preference for fruits of certain primary forest tree species. Phayre's leaf-monkey (*Presbytis phayrei*), on the other hand, is a folivore preferring a group of early and midsuccessional trees and was not observed in primary forest.

Studies of neotropical birds typically identify several feeding guilds, some of which we have already mentioned. Guilds are delimited in relation to the preferred food type and the specific habitat component used during foraging (from broad categories such as "canopy" and "understory" to more specific ones such as "bark") and whether activity is diurnal or nocturnal. The most detailed classification of feeding guilds consulted by the present authors was that of Robinson and Terborgh (1990), who identify 22 guilds. Species in these guilds may additionally be considered generalists or specialists or grouped in relation to the size of food articles taken (Fleming et al. 1987). The studies of birds in shifting cultivation landscapes cited earlier mention several hypotheses linking the community- and species-level patterns found to food availability.

In some cases, the reduction or absence of preferred food sources in shifting cultivation landscapes, in comparison with forest, might be linked to the decline of forest bird populations. The small proportion of terrestrial forest insectivores that follows army ant swarms is a clear example of how trophic relationships may break down in human-influenced landscapes. Studies in fragmented forests emphasize that these birds need to track several ant swarms simultaneously, so they may disappear if forest fragments are of insufficient size for the necessary numbers of swarms (Stouffer and Bierregaard 1995). Although the decline of understory and terrestrial insectivores in disturbed habitats has already been emphasized, whether or not such causal relationships to ant swarms apply in shifting cultivation landscapes remains to be determined. For example, Johns (1991) believed that the frequency of army ant swarms did not differ between the habitat types he sampled. It is also the case that some terrestrial forest birds may experience physiological stress in the microclimates of second-growth vegetation, so that this factor, rather than pattern in food availability, could be responsible for responses of bird communities to habitat disturbance.

A variety of species or guilds of forest birds evidently use or prefer anthropogenic habitat patches in shifting cultivation landscapes when foraging. The insectivore-nectarivore guild, made up in the neotropics by hummingbirds (Trochilidae), normally maintains or increases its representation in fallow vegetation because of the high abundance there of both floral and invertebrate resources (Johns 1991; Andrade and Rubio-Torgler 1994). Johns (1991) reports toucans (Ramphastidae) and cotingas (Cotingidae) from fallows and crop fields and observes that their use of these habitats is related to generalist feeding habits. Birds of both families are principally frugivores, and many species take a wide range of fruit types that may include those of pioneer plants (Stiles and Skutch 1989). Species richness in families that include consumers of small berries, such as Pipridae (manakins) and Thraupidae (tanagers), may similarly be maintained or increased in fallows (Johns 1991; Andrade and Rubio-Torgler 1994). Such increases may be at least partially linked to increased availability of the small berries of species of plant families such as Rubiaceae, Melastomataceae, and Piperaceae, which can be both more abundant (Laska 1997; Guariguata and Dupuy 1997) and more fecund (Levey et al. 1994) in disturbed habitats than in the shaded forest understory. However, it is important to emphasize that patterns of habitat use by frugivorous birds in human-influenced landscapes do not always correlate well with patterns of fruit availability and therefore must be related to other factors (Restrepo et al. 1999).

Microclimatic variation between tropical forest environments with different degrees of disturbance, and its effects on populations and communities of plants at least, is well documented and has been the subject of several reviews over the years (e.g., Clark 1990). Shifting cultivation clearings are large in the context of tropical forest canopy gaps, and it is obvious that their microclimates will differ markedly from those of the forest understory and forest treefall gaps. The regeneration of vegetation can quickly buffer microclimatic change in large clearings, however (Fetcher et al. 1985), and it is unfortunate that there appear to be no published comparative studies of microclimate in different-aged fallows and primary forest. Do microclimatic differences between primary forest and fallow affect the distributions of some forest vertebrates over shifting cultivation landscapes in the same way as they affect plant regeneration patterns? Karr and Freemark (1983) demonstrated that habitat use patterns of many neotropical rainforest birds are partly related to spatial and temporal microclimatic gradients, suggesting that physiological stress, rather than a microclimatically mediated pattern in food availability, was the main causal factor. This suggestion apparently has been interpreted as a tested hypothesis by subsequent authors (Johns 1991; Andrade and Rubio-Torgler 1994).

Hunting, rather than habitat structure and quality, is likely to be the significant factor in determining some vertebrate community characteristics in many shifting cultivation landscapes (see Chapter 14, this volume). The apparently greater effect of shifting cultivation on primates indicated for neotropical sites, in comparison with those from Africa and Asia, may have been a consequence of historical or contemporary hunting rather than habitat factors (Cowlishaw and Dunbar 2000). The high biomass of small and medium-sized mammals in African shifting cultivation landscapes is sustained even under significant hunting pressure, but if large mammals are absent from the land, it is probably because they have been expelled or extirpated by human activities (Wilkie and Finn 1990; Thomas 1991; Lahm 1993).

In conclusion, differences in diversity and composition sometimes are found when the vertebrate assemblages observed in fallows and crop fields are compared with those of primary forest. It is likely that up to a point, anthropogenic modification of forest habitat increases the diversity of the associated fauna at many spatial scales, as in the case of Anderson's raptor community at the landscape scale. Diversity increases often come about because species of disturbed habitats enter areas after disturbance and should be considered neutral from the conservation standpoint. Changes in the characteristics of vertebrate assemblages are likely to be related to changes in habitat structure, spatial and temporal patterns in the availability of food, and possibly microclimatic variation. The type and degree of changes observed at the community level depend at least partly on the characteristics of the landscape in which observations take place.

Managing Shifting Cultivation Landscapes for Increased Biodiversity and Conservation Value

The patchwork of stages of the shifting cultivation cycle is the dominant single land use in many tropical landscapes. This land use, and the way it interacts with remaining areas of primary communities and other human land uses, therefore must become a main target of biodiversity-oriented research, development, and management in such landscapes. Likely objectives for biodiversity management in a shifting cultivation mosaic might be

- To maintain in the landscape as much biodiversity (human-made and wild) as is compatible with the satisfaction of other human needs in a sustainable way, though not necessarily to maximize biodiversity in each patch within the landscape
- To contribute to regional efforts to conserve forest-dependent plant and animal species

Such objectives might be integrated with planning and action for biological conservation in a context of good land management at still larger scales than the landscape, such as in the context of the management of buffer zones for protected areas.

We believe that the landscape scale must be the primary management focus for biodiversity conservation. Action at the landscape scale would be complemented by management at the scale of individual patches. This is because there does not seem to be much potential for achieving increases in the biodiversity conservation value of individual habitat patches in shifting cultivation landscapes. These habitat patches are small, and most of them are subjected to the drastic disturbances associated with the agricultural cycle; the shortening of fallows means that disturbance frequency is increasing. Farmers are reluctant to abandon one of their most environmentally undesirable management tools, fire, for alternative ways of accessing the nutrients contained in fallow biomass (Szott et al. 1999). In any case, the fallows must be cut. It may turn out to be easier to conserve forest vertebrates in shifting cultivation landscapes than to conserve forest plants because the mobility of vertebrates allows them to use different habitat patches in meeting their needs.

Landscape Management

Our summary of the characteristics of shifting cultivation landscapes (Box 8.1) points first to the application of simple basic principles to their management in the context of objectives such as those set out in this chapter. Two such principles stand out, neither of which is specific to this type of landscape (Hartley 2002):

- *Conserve as much of the remaining primary forest in landscapes as possible.* Justification of such a measure hardly seems necessary, but if it were needed, indications that the best single correlate of animal species diversity observed in some studies of forest plantations is the amount of "native vegetation" in the landscape (Hartley 2002) are more than adequate. In such a situation, primary forest remnants arguably become keystone habitat patches in the landscape, playing a similar role to that of keystone species (Meffe and Carroll 1994) in that they would have an effect on biodiversity in the landscape that is disproportionate to their relative area. Forest in each of the major physical environments of the landscape ideally would be included in that conserved, a "coarse-filter" approach (Hunter 1991; Noss 1996) today found in many precautionary frameworks for biodiversity conservation in human-impacted ecosystems such as forests managed for timber production (Finegan et al. 2001).
- *Maintain connectivity between patches of habitat that are essential for the maintenance of populations of forest-dependent organisms.* Whether or not connectivity (Meffe and Carroll 1994) exists depends on the species or group of organisms under consideration, and as we have seen, even young fallow vegetation may provide part of the habitat used by some mobile organisms in shifting cultivation landscapes. However, an important element of a precautionary approach to the provision of connectivity would be to try to ensure the physical continuity of areas of the most important and least extensive habitats in the landscape: mature forest and older secondary forest.

Building on ideas set out by Smith et al. (2001), it is perhaps self-evident that all the preceding ideas for landscape management for biodiversity would best be implemented during the early stages of agricultural frontier development, when significant areas of primary forest still remain. The magnitude of

---

Box 8.1. Main Features of Landscape-Scale Biodiversity in Shifting Cultivation Landscapes

- Some shifting cultivation landscapes are permanent, others temporary stages in the evolution of agricultural frontiers whose potential for biodiversity conservation therefore is transient.
- Shifting cultivation landscapes may be tens or even hundreds of thousands of hectares in extent.
- Proportions of shifting cultivation landscapes are delimited by sociopolitical criteria (e.g., Latin American municipalities, African village land), covered by fallow vegetation, 50 percent or more, with less than 20 percent covered by disturbed primary vegetation and less than 10 percent by cropland.
- Biodiversity in shifting cultivation landscapes therefore is strongly dependent on anthropogenic communities; fallows are most important in terms of area.
- Rapid turnover of community types on individual patches in the landscape is a defining characteristic of shifting cultivation landscapes, of fundamental ecological importance.
- Relative proportions of land assigned to different uses vary between stages in agricultural frontier evolution, with old-growth forest declining, fallows increasing, and, in more advanced stages of frontier development, increases in non–shifting cultivation land uses.
- Mature and perhaps old secondary forests are presumably keystone communities, crucial for organisms that are forest dependent in some way, including those that use many landscape patch types but need well-developed forest for at least some of the conditions and resources crucial to their survival.
- Some community-level characteristics (e.g., the proportions of plants regenerated from resprouts and those regenerated from seed or the presence of forest-dependent vertebrates in anthropogenic habitats) presumably depend on the structure and composition of the landscape surrounding the community.

---

the challenge of managing landscapes during those stages of landscape colonization is also self-evident.

Beyond action related specifically to primary and old secondary forest habitats, the maintenance of biodiversity in general and of forest-dependent species in particular could be evaluated and managed for in relation to overall landscape structure and diversity. Shifting cultivation landscapes are diverse at the scale of patch types in the landscape. As shown earlier, the increase in diversity at this scale that accompanies conversion of part of a forested area to shifting cultivation may in some circumstances be accompanied by increases in the species diversity of organisms with a coarse-grained perception of habitat, such as diurnal raptors, and undoubtedly has an important influence on species-level diversity in general. Trends in landscape structure and diversity that intuition tells us probably are detrimental to biodiversity and probably are associated with the trend toward shorter fallows would include greater dominance of the landscape by younger fallow habitats and increases in the mean areas of patches of anthropogenic vegetation. Specific aims for management to counter these trends might be to maintain or increase the area of older fallow vegetation and to maintain a high degree of interspersedness of different patch types. More detailed quantitative analyses of shifting cultivation landscapes than those available should provide further pointers to technically desirable management objectives (Metzger 2002).

The landscape-scale consequences of adopting improved (i.e., planted) or managed fallows would depend on how this change affects the relative area and spatial configuration of natural fallow vegetation and other patch types in the landscape, which is impossible to assess at present. It is clear that improved, planted fallows, which are short (often less than 3 years) and often feature a single planted species as a major component (Szott et al. 1999), in general are of lower diversity at the patch scale than natural fallows.

Finally, by analogy with ecological principles related to fragmented communities (Laurance et al. 2000; Metzger 2000), factors such as patch size and shape, the type of community or communities bordering a given patch, and distances to similar patch types must also influence biodiversity in any given patch. There appears to be no published information on this aspect of biodiversity and its dynamics in shifting cultivation landscapes.

Community-Level Management

Box 8.2 contains a summary of our review of aspects of the biodiversity of the communities that make up shifting cultivation landscapes, on which we base the following suggestions for biodiversity management at the level of individual communities or patch types within such landscapes. Box 8.2 makes clear that fallow vegetation and crop fields are anthropogenic communities whose characteristics are largely shaped by drastic, high-frequency disturbances. As such, they are inhospitable to forest-dependent plant species, and whether forest-dependent vertebrates are observed in them is likely to depend at least partly on the presence of older secondary or mature forest in the landscape. The anthropogenic nature and hostility to patch-scale biodiversity of forest species probably is even more marked in planted fallows.

Analysis and management at the scale of the individual patch arguably are most important from the point of view of plants because mobile animals and birds range over a variety of patches within the landscape. However, aspects of vegetation composition and structure of fallow and secondary forest could be

Box 8.2. Main Features of Community-Level Biodiversity in Shifting Cultivation Landscapes

- Cultivated plants belong mainly to a few species, but crop fields may harbor important agrobiodiversity at the genetic level whose creation and maintenance are little understood.
- Weed communities sometimes are quite diverse and potentially valuable for medicines and food and perhaps for some aspects of ecosystem function but have low intrinsic conservation value.
- Weeds such as *Chromolaena odorata* may become dominant and inhibit successional change for many years.
- Forest tree species are among the trees conserved by farmers but usually belong to the small subset of species that have some kind of value to farmers.
- The fallow plant community is dominated by species with high resprouting capacity and pioneers with precocious fruiting and functioning seed dispersal mechanisms.
- Although fallow plant communities change over time, their species richness and diversity, as well as their representation of forest-dependent species, remain low in comparison with original forest.
- A trend toward shorter fallows further limits potential for biodiversity recovery in patches.
- Fallow and secondary forest communities are not a stage in the recovery of original forest but are entirely new, anthropogenic vegetation types.
- The number of vertebrate species that make exclusive use of single habitat patches probably is limited because of the small size of such patches, so their community characteristics must be characterized and understood at the landscape scale (see Box 8.1).
- Crop fields, fallows, and secondary forest patches nevertheless are used by many forest vertebrates (with some more frequently observed in such patches than in mature forest) and species of disturbed habitats.
- The diversity of observed vertebrate assemblages or particular components of them may be the same as, lower than, or higher than in mature forest, depending on factors such as surrounding vegetation.
- The composition of vertebrate assemblages is more sharply differentiated between patch types than richness and diversity; declines in some bird guilds and increases in others are predictable, for example, when comparing mature forest with fallow and secondary vegetation.
- The diversity and composition of observed vertebrate assemblages often change over time.
- Compositional changes are linked to variation in factors such as habitat structure, food availability and foraging behavior, microclimate, and hunting pressure.
- Overall, the prevailing short fallows greatly limit the potential for biodiversity recovery, and especially mature forest attributes, in most patches in the landscape.

manipulated to increase the number of animal and bird species using particular habitat patches and therefore could increase the total habitat area in a landscape suitable for at least some of the needs of animal and bird species.

Options for biodiversity-conscious management of plant communities might concentrate on two related, specific objectives: to increase the length of fallow periods so that more species accumulate and (a point not touched upon in case studies) greater numbers of individuals reach reproductive maturity and to increase of the rate of accumulation of plant diversity so that more diversity accumulates for a given fallow length. The former objective might be achieved by promoting uses of fallows other than the normal ones of weed control and the recovery of soil fertility, ones that entail longer periods of vegetation development, another significant challenge in the context of general tendencies toward shortening fallows and adopting the planted fallow. A possible strategy here is fallow management for timber and nontimber forest products (Smith et al. 2001). The latter objective could be pursued by thinning to favor longer-lived or forest-dependent plant species over pioneers, focusing specifically on reducing the degree and duration of dominance of the vegetation by low-diversity assemblages of short- or long-lived pioneer species (Finegan 1996). The regeneration of the species to be favored is a basic premise here and cannot by any means be guaranteed. In an ideal world, managers would evaluate regeneration using techniques of silvicultural diagnosis, as Finegan and Delgado (2000) have suggested in the context of forest restoration through secondary succession on abandoned neotropical pastures, and may conclude that planting is necessary for biodiversity conservation objectives. In general, however, there is little or no experience in this type of silvicultural intervention in the neotropics, although advances have been documented for temperate zones (Smith et al. 2001).

Two main areas of action suggest themselves in relation to vertebrates: management of vegetation structure, composition, and microclimates; and management of hunting (Bennett and Robinson 2000; Robinson and Bennett 2000; see Chapter 14, this volume).

Data from other contexts (Hartley 2002) suggest that the conservation of more trees of the original forest than is usual, of a wider range of species, would make an important contribution to the animal and bird diversity of shifting cultivation communities. Different spatial configurations of conserved trees may vary in their effectiveness in this context (Hartley 2002). Given that shifting cultivation land is burned frequently, however, and most tropical forest tree species are highly vulnerable to death even from ground fires (Uhl and Kauffman 1990), trees probably would have to be conserved in strips between fields and burning carried out with care. In terms of resistance to fire, riparian forests also may have a special place in habitat management.

Tree conservation is another measure that would obviously be more effective if taken during the early phases of frontier development; it is important to take into account that this is a temporary measure because once they are gone, primary forest trees will not be replaced in the landscape.

Vegetation structure of individual fallow and secondary forest patches could usefully be diversified to promote vertebrate diversity. The conservation of forest trees would contribute to the diversification of vegetation structure, as would the favoring of some individuals of fast-growing pioneers regenerating from seed, all with the aim of broadening foliage height profiles and increasing vertical and horizontal structural diversity. Among more specific habitat features identified in this chapter as important to vertebrates and to be conserved are vine tangles, continuous tree canopy cover at all possible levels in habitat patches, and moribund or dead trees. Many pioneer plants provide food to frugivorous or omnivorous vertebrates, and by the very nature of these plants maintaining this function might take little management attention. However, management interventions designed to accelerate the increase of stand diversity or favor forest-dependent plant species in older vegetation could also increase the frequency and size of fruit crops if they were to involve canopy opening. Moves to diversify stand vertical structure could be integrated with the conservation of species providing fruits to vertebrates, taking into account that some vertebrate fruit consumers also have foraging height preferences. Any manipulation of habitat characteristics would need to take the avoidance of a return to early-successional microclimates and their associated species as a basic rule.

Conclusions

Debate on shifting cultivation tends to become polarized, with the practice characterized as either the fate or the future of tropical forests and their biodiversity, over significant areas of the tropics. However, polarization is based on fundamental misunderstandings of the nature of shifting cultivation and its effects on biodiversity, including gross simplification of an agricultural production system that in fact consists of a variety of practices applied under a variety of conditions. Effects on biodiversity will be very different from place to place because of variation in agricultural systems, sociocultural organizations, external drivers, and site ecological conditions. Before any sensible conclusion can be reached regarding shifting cultivation and biological conservation, this variety must be analyzed. This chapter is an attempt to begin such an analysis.

Claims by conservationists or foresters that shifting cultivation in general is a major cause of deforestation or forest degradation, and hence of biodiversity loss, are based on an improper analysis of the logic, practice, and impact of the range of food production systems included under the rubric of shifting cultivation. Too often systems dominated by new migrants clearing land by fire are assimilated with ancient traditional shifting cultivation systems. Not enough attention has been paid to the fact that nowhere is shifting cultivation the only food production system in the agriculture practiced by traditional forest people. It is always complemented by hunting-and-gathering activities, homegardens, and often complex agroforests. Overall, as this chapter has emphasized, shifting cultivation often may create landscapes that maintain high levels of biodiversity in general, in which some components of forest biodiversity probably can be conserved, especially vertebrates.

In assessing the possible contribution shifting cultivation landscapes can make to the conservation of tropical biodiversity, it is vital to distinguish between biodiversity in general and forest-dependent biodiversity in particular. We have tried to emphasize this distinction throughout this chapter. It is a basic tenet of modern approaches to agriculture and natural resource management that biodiversity is a good thing. But in the humid tropics, natural forest communities and the species that depend on them are the priority for conservation action. In this context, biodiversity often is seen too simplistically by many observers, including social scientists (McKey 2001). Anthropogenic communities—crop fields and fallows—may support, or contribute to supporting, much biodiversity at the species and genetic levels. A great part of tropical diversity was in place long before human influence on the characteristics of tropical forests became important, however. Tropical forests are the most biodiverse terrestrial ecosystems for reasons that are not yet fully understood, and the drive to understand the creation and maintenance of their diversity is a significant element of tropical forest research (Huston 1994; Hubbell 2001). The "forest-dependence" of much biodiversity should be self-evident, as should the vulnerability of this biodiversity to the high-frequency drastic disturbance inherent in shifting cultivation and to modern tendencies toward shorter fallows and consequently greater areas of crop fields and young fallow vegetation in landscapes.

In conclusion, it is clear that shifting cultivation systems can play a positive role in biodiversity conservation and especially—although we have not emphasized this comparison—a much more positive one than any modern intensive agricultural system. Modern tropical landscapes are being increasingly shaped by people, and shifting cultivation therefore is a relatively biodiversity-friendly land use in the face of this reality. However, its contribution to biological conservation will be important only if shifting cultivation landscapes do not become merely transient stages of frontier development, as is often the case already. Population growth, economic policies, and government relocation programs are important forces that may increase the transience of shifting cultivation landscapes. In addition, over much of the humid tropics, fallow periods are shortening, and the relative extents of crop fields and young fallow vegetation are increasing dramatically. These trends have dismaying implications for the continued agricultural productivity of shifting cultivation systems and for biodiversity maintenance in the landscape as a whole. Planted fallows are an important response to the crisis of agricultural productivity, but unless their adoption is accompanied by measures to maintain or increase cover of old secondary and primary forest in the landscape, they are not beneficial from the biodiversity point of view. Management of shifting cultivation landscapes for biodiversity, along lines such as those we have described and emphasizing forest-dependent species, could increase their contribution to biological conservation, but clearly it faces major implementation challenges.

References

Aide, T. M. 2000. Clues for tropical forest restoration. *Restoration Ecology* 8:327.

Alencar, A. A. C., I. C. G. Vieira, D. C. Nepstad, and P. Lefebvre. 1994. Análise multitemporal do uso do solo e mudança da cobertura vegetal em antiga área agrícola da Amazônia Oriental. In *Anais do VIII Simpósio Internacional de Sensoramento Remoto*, Santander, Bahia, Brazil.

Anderson, D. L. 2001. Landscape heterogeneity and diurnal raptor diversity in Honduras: the role of indigenous shifting cultivation. *Biotropica* 33:511–519.

Andrade, G. I., and H. Rubio-Torgler. 1994. Sustainable use of the tropical rain forest: evidence from the avifauna in a shifting-cultivation habitat mosaic in the Colombian Amazon. *Conservation Biology* 8:545–554.

Barnes, R. F. W., K. L. Barnes, M. P. T. Alers, and A. Blom. 1991. Man determines the distribution of elephants in the rain forests of northeastern Gabon. *African Journal of Ecology* 29:54–63.

Becker, M., and D. E. Johnson. 2001. Cropping intensity effects on upland rice yield and sustainability in West Africa. *Nutrient Cycling in Agroecosystems* 59:107–117.

Begon, M., J. L. Harper, and C. R. Townsend. 1996. *Ecology*. Oxford, UK: Blackwell Science.

Bennett, E., and J. Robinson. 2000. *Hunting of wildlife in tropical forests: implications for biodiversity and forest peoples*. Biodiversity Series, Impact Studies No. 76. Washington, DC: The World Bank.

Bibby, C. J., N. J. Collar, M. J. Crosby, M. F. Heath, C. Imboden, T. H. Johnson, A. J. Long, A. J. Stattersfield, and S. J. Thirgood. 1992. *Putting biodiversity on the map: priority areas for global conservation*. Cambridge, UK: International Council for Bird Preservation.

Bierregaard, R. O., Jr., and P. C. Stouffer. 1997. Understorey birds and dynamic habitat mosaics in Amazonian rainforests. Pages 138–155 in W. F. Laurance and R. O. Bierregaard Jr. (eds.), *Tropical forest remnants: ecology, management, and conservation of fragmented communities*. Chicago: University of Chicago Press.

Blankenspoor, G. W. 1991. Slash-and-burn shifting agriculture and bird communities in Liberia, West Africa. *Biological Conservation* 57:47–71.

Boserup, E. 1965. *The conditions of agricultural growth: the economics of agrarian change under population pressure.* Chicago: Aldine.

Boucher, D. H., J. H. Vandermeer, M. A. Mallona, N. Zamora, and I. Perfecto. 1994. Resistance and resilience in a directly regenerating rainforest: Nicaraguan trees of the Vochysiaceae after Hurricane Joan. *Forest Ecology and Management* 68:127–136.

Brown, D., and K. Schreckenberg. 1998. *Shifting cultivation as an agent of deforestation: assessing the evidence.* ODI Natural Resources Perspectives. Online: http://www.one world.org/odi/nrp/29.html.

Brush, S. B. 1995. In situ conservation of landraces in centers of crop diversity. *Crop Science* 35:346–354.

Buckles, D., and B. Triomphe. 1999. Adoption of mucuna in the farming systems of northern Honduras. *Agroforestry Systems* 47:67–91.

Burgers, P., K. Hairiah, and M. Cairns. 2000. *Indigenous fallow management. Lecture Note 4.* Bogor, Indonesia: International Center for Research in Agroforestry, South East Asian Research Programme.

Carrière, S. 1999. *"Les orphelins de la forêt". Influence des pratiques agricoles ancestrales des Ntumu sur le maintien et l'évolution du couvert forestier tropical du sud Cameroun.* Ph.D. thesis, Université des Sciences et Techniques du Languedoc, Montpellier, France.

Carrière, S., M. André, P. Letourmy, I. Olivier, and D. McKey. 2002. Seed rain beneath remnant trees in a slash-and-burn agricultural system in southern Cameroon. *Journal of Tropical Ecology* 18:353–374.

Chazdon, R. L., and F. G. Coe. 1999. Ethnobotany of woody species in second-growth, old-growth and selectively logged forests of northeastern Costa Rica. *Conservation Biology* 13:1312–1322.

Christanty, L., O. E. Abdoellah, G. G. Marten, and J. Iskandar. 1986. Traditional agroforestry in West Java: the pekarangan (homegarden) and kebun-talun (annual-perennial rotation) cropping systems. Pages 132–158 in G. G. Marten (ed.), *Traditional agriculture in Southeast Asia. A human ecology perspective.* Boulder, CO: Westview.

Clark, D. B. 1990. The role of disturbance in the regeneration of neotropical moist forests. Pages 291–315 in K. S. Bawa and M. Hadley (eds.), *Reproductive ecology of tropical forest plants.* Man and the Biosphere Series Vol. 7. Paris: UNESCO; Carnforth, UK: Parthenon.

Conklin, H. C. 1957. *Hanunoo agriculture: a report on an integral system of shifting cultivation in the Philippines.* Forestry Development Paper No. 12. Rome: Food and Agriculture Organization of the United Nations.

Cowlishaw, G., and R. Dunbar. 2000. *Primate conservation biology.* Chicago: University of Chicago Press.

Croat, T. 1978. *Flora of Barro Colorado Island.* Stanford, CA: Stanford University Press.

de Jong, W., U. Chokkalingam, and G. A. D. Perera. 2001. The evolution of swidden fallow secondary forests in Asia. *Journal of Tropical Forest Science* 13:800–815.

de Namur, D. C., and J. L. Guillaumet. 1978. Grands traits de la reconstitution dans le sud-ouest ivoirien. *Cahier ORSTOM, Série Biologie* 13:197–201.

Denich, M. 1991. *Estudo da importância de uma vegetação secundária nova para o incremento da produtividade do sistema de produção na Amazônia oriental Brasileira.* Belém, Brazil: EMBRAPA CPATU; Eschborn, Germany: GTZ.

Denich, M. (n.d.). *Regeneration of secondary vegetation in the agricultural landscape. Secondary forests and fallow vegetation in the eastern Amazon region: function and management.*

A German-Brazilian research project carried out by DAT, University of Göttingen, and CPATU/EMBRAPA, Belém-Pará. Online: http://www.gwdg.de/~jwiesen/iatpages/mdenich5.htm.

de Rouw, A. 1995. The fallow period as a weed-break in shifting cultivation (tropical wet forests). *Agriculture, Ecosystems and Environment* 54:31–43.

de Rouw, A., H. C. Vellema, and W. A. Blokhuis. 1990. *Land unit survey of the Tai region, south-west Côte d'Ivoire.* Tropenbos Technical Series No. 7. Ede, the Netherlands: The Tropenbos Foundation.

de Wachter, P. 1997. Economie et impact spatial de l'agriculture itinérante Badjoué (Sud-Cameroun). In D. V. Joiris and D. de Laveleye (eds.), *Les peuples des forêts tropicales: systèmes traditionnels et développement rural en Afrique équatoriale, Grande Amazonie et Asie du Sud-est. Civilisations* (Special Issue) 44(1–2).

Donfack, P., C. Floret, and R. Pontanier. 1995. Secondary succession in abandoned fields of dry tropical northern Cameroun. *Journal of Vegetation Science* 6:1–10.

Dounias, E. 2001. La diversité des agricultures itinérantes sur brûlis. Pages 65–106 in S. Bahuchet (ed.), *Les peuples des forêts tropicales aujourd'hui*, Vol. II: *Une approche thématique.* Paris: Programme Avenir des Peuples des Forêts Tropicales (APFT).

Dourrojeanni, M. 1987. Aprovechamiento del barbecho forestal en áreas de agricultura migratoria en la Amazonía Peruana. *Revista Forestal del Perú* 14:15–61.

Ewel, J. J. 1986. Designing agricultural ecosystems for the humid tropics. *Annual Review of Ecology and Systematics* 17:245–271.

FAO. 1998. FRA 2000 Terms and Definitions. Forest Resources Assessment Programme, Working Paper 1. Online: http://www.fao.org/forestry/fo/fra/docs/FRA.

Ferguson, G. B. 2001. *Post-agricultural tropical forest succession: patterns, processes and implications for conservation and restoration.* Ph.D. thesis, University of Michigan, Ann Arbor.

Fetcher, N., S. F. Oberbauer, and B. R. Strain. 1985. Vegetation effects on microclimate in lowland tropical forest in Costa Rica. *International Journal of Biometeorology* 29:145–155.

Finegan, B. 1992. The management potential of neotropical secondary lowland rain forest. *Forest Ecology and Management* 47:295–321.

Finegan, B. 1996. Pattern and process in neotropical secondary rain forests: the first 100 years of succession. *Trends in Ecology and Evolution* 11:119–124.

Finegan, B., and D. Delgado. 2000. Structural and floristic heterogeneity in a 30-year-old Costa Rican rain forest restored on pasture through natural secondary succession. *Restoration Ecology* 8:380–393.

Finegan, B., W. Palacios, N. Zamora, and D. Delgado. 2001. Ecosystem-level forest biodiversity and sustainability assessments for forest management. Pages 341–378 in R. J. Raison, A. G. Brown, and D. W. Flinn (eds.), *Criteria and indicators for sustainable forest management.* Wallingford, UK: CAB International.

Fleming, T. H., R. Breitwisch, and G. H. Whitesides. 1987. Patterns of tropical vertebrate frugivore diversity. *Annual Review of Ecology and Systematics* 18:91–109.

Frankel, O. H. 1995. Landraces in transit: the threat perceived. *Diversity* 11:14–15.

Fuhr, M. 1999. *Structure et dynamique de la forêt côtière au Gabon. Implications pour une succession secondaire dérivant de la forêt dominante à Okoumé (Aucoumea klaineana Pierre).* Ph.D. thesis, Université de Montpellier II, Montpellier, France.

Fujisaka, S. 1991. A diagnostic survey of shifting cultivation in northern Laos: targeting research to improve sustainability and productivity. *Agroforestry Systems* 13:95–109.

Gómez-Pompa, A., and C. Vásquez-Yanes. 1981. Successional studies of a rain forest in

Mexico. Pages 246–266 in D. C. West, H. H. Shugart, and D. B. Botkin (eds.), *Forest succession: concepts and application*. Berlin: Springer-Verlag.

Grenand, P., C. Moretti, and H. Jacquemin. 1987. *Pharmacopées traditionelles en Guyane: Créoles, Palikur, Wayapi*. Editions de l'ORSTOM, Collection Mémoires No. 108. Paris: ORSTOM.

Guariguata, M. R., R. L. Chazdon, J. S. Denslow, and J. M. Dupuy. 1997. Structure and floristics of secondary and old-growth forest stands in lowland Costa Rica. *Plant Ecology* 132:107–120.

Guariguata, M. R., and J. M. Dupuy. 1997. Forest regeneration in abandoned logging roads in lowland Costa Rica. *Biotropica* 29:15–28.

Hammond, N. 2000. The origins of Maya civilization: the beginnings of village life. Pages 35–47 in N. Grube (ed.), *Maya: divine kings of the rain forest*. Cologne, Germany: Könemann.

Harrison, P. D. 2000. Maya agriculture. Pages 71–79 in N. Grube (ed.), *Maya: divine kings of the rain forest*. Cologne, Germany: Könemann.

Hart, R. D. 1980. A natural ecosystem analog approach to the design of a successional crop system for tropical forest environments. *Biotropica* 12:73–82.

Hartley, M. J. 2002. Rationale and methods for conserving biodiversity in plantation forests. *Forest Ecology and Management* 155:81–95.

Hinvi, J. C., J. K. Totongnon, C. Dahin, and P. Vissoh. 1991. *Les systèmes traditionnels de culture face à la dégradation de l'environnement: cas du département de l'Atlantique. Résultats d'Enquêtes et d'Experimentation de 1986 à 1990*. Calavi, Benin: Regional Action Centre for Rural Development, Atlantique (mimeo).

House, P. R., S. Lagos-Witte, L. Ochoa, C. Torres, T. Mejía, and M. Rivas. 1995. *Plantas medicinales comunes de Honduras*. Tegucigalpa: Universidad Nacional Autónoma de Honduras.

Hubbell, S. P. 2001. *The unified neutral theory of biodiversity and biogeography*. Princeton, NJ: Princeton University Press.

Hunter, M. L., Jr. 1991. Coping with ignorance: the coarse-filter strategy for maintaining biodiversity. Pages 266–281 in K. A. Kohm (ed.), *Balancing on the brink of extinction: the Endangered Species Act and lessons for the future*. Washington, DC: Island Press.

Huston, M. A. 1994. *Biological diversity: the coexistence of species on changing landscapes*. Cambridge, UK: Cambridge University Press.

Janzen, D. H. 1988. Management of habitat fragments in a tropical dry forest: growth. *Annals of the Missouri Botanical Garden* 75:105–116.

Johns, A. D. 1991. Responses of Amazon rain forest birds to habitat modification. *Journal of Tropical Ecology* 7:417–437.

Jones, D. W., and R. V. O'Neill. 1993. Human-environmental influences and interactions in shifting agriculture. Pages 297–309 in T. R. Lakshmanan and P. Nijkamp (eds.), *Structure and change in the space economy*. Berlin: Springer-Verlag.

Judd, W. S., C. S. Campbell, E. A. Kellogg, and P. F. Stevens. 1999. *Plant systematics, a phylogenetic approach*. Sunderland, MA: Sinauer.

Kahn, F. 1978. Évolution structurale d'un peuplement de *Macaranga hurifolia*. *Cahiers ORSTOM, sér. Biologie* 13:223–238.

Kammesheidt, L. 1998. The role of tree sprouts in the restoration of stand structure and species diversity in tropical moist forest after slash-and-burn agriculture in eastern Paraguay. *Plant Ecology* 139:155–165.

Kanmegne, J., B. Duguma, J. Henrot, and N. O. Isirimah. 1999. Soil fertility enhancement by planted tree-fallow species in the humid lowlands of Cameroon. *Agroforestry Systems* 46:239–249.

Karr, J. R., and K. E. Freemark. 1983. Habitat selection and environmental gradients: dynamics in the "stable" tropics. *Ecology* 64:1481–1494.

King, J., J. B. Moutsinga, and G. Dufoulon. 1997. Conversion of anthropogenic savanna to production forest through fire-protection of the forest-savanna edge in Gabon, Central Africa. *Forest Ecology and Management* 94:233–247.

Kiyono, Y., and Hastaniah. 1997. *Slash-and-burn agriculture and succeeding vegetation in East-Kalimantan*. Special publication no. 6. Samarinda, Indonesia: Tropical Rain Forest Research Center.

Kocher-Schmid, C. 2001. Overview. Pages 5–28 in C. Kocher-Schmid and R. Ellen (eds.), *Tropical rainforest people today*, Vol. 5: *Pacific region: Melanesia*. Brussels: Future of the Rainforest People Program.

Lahm, S. A. 1993. Utilization of forest resources and local variation of wildlife populations in northeastern Gabon. Pages 213–226 in C. M. Hladik, A. Hladik, O. Linares, H. Pagézy, A. Semple, and M. Hadley (eds.), *Tropical forest, people and food: biocultural interactions and applications to development*. Paris: Parthenon/UNESCO.

Lambert, J. D. H., and J. T. Arnason. 1989. Nutrient mobility in a shifting cultivation system in Belize, C.A. Pages 160–168 in S. R. Gliessman (ed.), *Agroecology*. Ecol. Studies 78. New York: Springer-Verlag.

Laska, M. S. 1997. Structure of understory shrub assemblages in adjacent secondary and old growth tropical wet forest, Costa Rica. *Biotropica* 29:29–37.

Laurance, W., H. L. Vasconcelos, and T. E. Lovejoy. 2000. Forest loss and fragmentation in the Amazon: implications for wildlife conservation. *Oryx* 34:39–45.

Leroy-Deval, J. 1973. Les anastomoses racinaires chez l'Okoumé (*Aucoumea klaineana*). *Comptes Rendus de l'Académie des Sciences de Paris*, série D, 276:2425–2428.

Lescure, J.-P. 1986. *La reconstitution du couvert végétal après agriculture sur brûlis chez les Wayãpi du haut Oyapock (Guyane Française)*. Ph.D. thesis, Université de Paris VI.

Levey, D. J., T. C. Moermond, and J. S. Denslow. 1994. Frugivory at La Selva: an overview. Pages 282–294 in L. A. McDade, K. S. Bawa, H. A. Hespenheide, and G. S. Hartshorn (eds.), *La Selva: ecology and natural history of a neotropical rainforest*. Chicago: University of Chicago Press.

Louette, D., A. Charrier, and J. Berthaud. 1997. In situ conservation of maize in Mexico: genetic diversity and maize seed management in a traditional community. *Economic Botany* 51:20–38.

Mabberley, D. J. 1987. *Tropical rain forest ecology*. 2nd edition. London: Chapman & Hall.

McGrath, D. G. 1987. The role of biomass in shifting cultivation. *Human Ecology* 15:221–242.

McKey, D. 2001. Tropical forest peoples and biodiversity. Pages 12–32 in *Rapport final*, Vol. II: *Un approche thématique*. Paris: Projet Avenir des Peuples des Forêts Tropicales (APFT).

Medellín, R. A., and M. Equiha. 1998. Mammal species richness and habitat use in rainforest and abandoned agricultural fields in Chiapas, Mexico. *Journal of Applied Ecology* 35:13–23.

Meffe, G. K., and C. R. Carroll. 1994. *Principles of conservation biology*. Sunderland, MA: Sinauer.

Mesquita, R., P. Delamonica, and W. F. Laurance. 2001. Effect of surrounding vegetation on edge-related tree mortality in Amazonian forest fragments. *Biological Conservation* 91:129–134.

Metzger, J. P. 2000. Tree functional group richness and landscape structure in a Brazilian tropical fragmented landscape. *Ecological Applications* 10:1147–1161.

Metzger, J. P. 2002. Landscape dynamics and equilibrium in areas of slash-and-burn agriculture with short and long fallow period (Bragantina region, NE Brazilian Amazon). *Landscape Ecology* 17:419–432.

Moran, E. F., E. Brondizio, P. Mausel, and Y. Wu. 1994. Integrating Amazonian vegetation, land use and satellite data. *BioScience* 44:329–338.

Myers, N. 1980. *Conversion of tropical moist forests.* Washington, DC: National Academy of Sciences.

Myers, N., R. Mittermeier, C. G. Mittermeier, G. A. B. da Fonseca, and J. Kent. 2001. Biodiversity hotspots for conservation priorities. *Nature* 403:853–858.

Nasi, R. 1997. Les peuplements d'Okoumé au Gabon: dynamique et croissance en zone côtière. *Bois et Forêts des Tropiques* 251:5–25.

Nepstad, D., C. Uhl, and E. A. S. Serrão. 1991. Recuperation of a degraded Amazonian landscape: forest recovery and agricultural restoration. *Ambio* 20:248–255.

Noss, R. F. 1996. Ecosystems as conservation targets. *Trends in Ecology and Evolution* 11:351.

Nye, P. H., and D. J. Greenland. 1960. *The soil under shifting cultivation.* Farnham Royal, UK: Commonwealth Agricultural Bureau.

Paciorek, C. J., R. Condit, S. P. Hubbell, and R. B. Foster. 2000. The demographics of resprouting in tree and shrub species of a moist tropical forest. *Journal of Ecology* 88:765–777.

Parrotta, J. A., and J. W. Turnbull (eds.). 1997. Catalyzing native forest regeneration on degraded tropical lands. Special Issue, *Forest Ecology and Management* 99.

Peña-Claros, M. 2001. *Secondary forest succession: processes affecting the regeneration of Bolivian tree species.* PROMAB Scientific Series 3. Riberalta, Bolivia: PROMAB.

Pennington, T. A., and E. C. M. Fernandes (eds.). 1998. *The utilization of the genus* Inga *(Fabacae).* Kew, UK: The Royal Botanic Gardens; Nairobi, Kenya: ICRAF.

Petit, S. 1999. *Structure, dynamique et fonctionnement des parcs agroforestiers traditionnels. Cas de Dolekaha, nord Côte d'Ivoire et Hollom, nord Cameroun.* Ph.D. thesis, Université de Paris I, CIRAD, Montpellier, France.

Pielou, E. C. 1995. Biodiversity versus old-style diversity: measuring biodiversity for conservation. Pages 5–18 in T. J. B. Boyle and B. Boontawee (eds.), *Measuring and monitoring biodiversity in tropical and temperate forests.* Bogor, Indonesia: CIFOR.

Prebble, C., A. Ella, and W. Subansenee. 1999. *ITTO: making the most of NWFP.* Tropical Forest Update 9(1). Online: http://www.itto.or.jp/newsletter/v9n1/04.html.

Ramakrishnan, P. S. 1992. *Shifting agriculture and sustainable development: an interdisciplinary study from north-eastern India.* Man and the Biosphere Series Vol. 10. Paris: UNESCO; Carnforth, UK: Parthenon.

Restrepo, C., N. Gómez, and S. Heredia. 1999. Anthropogenic edges, treefall gaps and fruit-frugivore interactions in a neotropical montane forest. *Ecology* 80:668–665.

Richards, P. W. 1976. *The tropical rain forest: an ecological study.* Cambridge, UK: Cambridge University Press.

Rico-Bernal, M., and A. Gómez-Pompa. 1976. Estudio de las primeras etapas sucesionales de una selva alta perennifolia en Veracruz, México. Pages 112–202 in A. Gómez-Pompa, S. del Amo, C. Vásquez-Yanes, and A. Butanda (eds.), *Regeneración de selvas: investigaciones sobre la regeneración de selvas altas en Veracruz, México.* Mexico City: CECSA.

Riswan, S., and K. Kartawinata. 1991. Regeneration after disturbance in a lowland mixed dipterocarp forest in East Kalimantan, Indonesia. Pages 295–301 in A. Gomez-Pompa, T. C. Whitmore, and M. Hadley (eds.), *Rainforest regeneration and management.* Man and the Biosphere Series Vol. 6. Paris: UNESCO.

Robinson, J., and E. Bennett (eds.). 2000. *Hunting for sustainability in tropical forests.* New York: Columbia University Press.

Robinson, S. K., and J. Terborgh. 1990. Bird communities of the Cocha Cashu Biological Station in Amazonian Peru. Pages 199–216 in A. Gentry (ed.), *Four neotropical forests.* New Haven, CT: Yale University Press.

Roder, W., and S. Maniphone. 1998. Shrubby legumes for fallow improvement in northern Laos: establishment, fallow biomass, weeds, rice yield and soil properties. *Agroforestry Systems* 39:291–303.

Saldarriaga, J. C., D. C. West, M. L. Tharp, and C. Uhl. 1988. Long-term chronosequence of forest succession in the upper Rio Negro of Colombia and Venezuela. *Journal of Ecology* 76:938–958.

Saunders, D. A., R. J. Hobbs, and C. R. Margules. 1991. Biological consequences of ecosystem fragmentation: a review. *Conservation Biology* 5:18–32.

Schmidt-Vogt, D. 1999. *Swidden farming and fallow vegetation in northern Thailand.* Geoecological Research 8. Stuttgart, Germany: Franz Steiner.

Shankar Raman, T. R. 1996. Impact of shifting cultivation on diurnal squirrels and primates in Mizoram, northeast India: a preliminary study. *Current Science* 70:747–750.

Shankar Raman, T. R., G. S. Rawat, and A. J. T. Johnsingh. 1998. Recovery of tropical rainforest avifauna in relation to vegetation succession following shifting cultivation in Mizoram, north-east India. *Journal of Applied Ecology* 35:214–231.

Shanley, P., M. Cymerys, and J. Galvão. 1998. *Frutíferas da Mata na vida Amazônica.* Belém, Brazil: Editora Supercores.

Smith, J., S. Ferreira, P. van de Kop, C. Palheta, and C. Sabogal. 1999a. *The persistence of secondary forest cover on slash-and-burn farms in the Amazon: implications for improving slash-and-burn agriculture.* Bogor, Indonesia: CIFOR.

Smith, J., B. Finegan, C. Sabogal, M. S. G. Ferreira, G. Siles, P. van de Kop, and A. Díaz Barba. 2001. Management of secondary forests in colonist swidden agriculture in Perú, Brazil and Nicaragua. Pages 263–278 in M. Palo, J. Uusivuori, and G. Mery (eds.), *World forests, markets and policies.* World Forests Vol. III. Dordrecht, the Netherlands: Kluwer.

Smith, J., C. Sabogal, W. de Jong, and D. Kaimowitz. 1997. *Bosques secundarios como recurso para el desarrollo rural y la conservación ambiental en los trópicos de América Latina.* Occasional Paper no. 13. Bogor, Indonesia: CIFOR.

Smith, J., P. van de Kop, K. Restegui, I. Lombardi, C. Sabogal, and A. Díaz. 1999b. Dynamics of secondary forests in slash-and-burn farming: interactions among land use types in the Peruvian Amazon. *Agriculture, Ecosystems and Environment* 76:85–98.

Stiles, F. G., and A. F. Skutch. 1989. *A guide to the birds of Costa Rica.* New York: Cornell University Press.

Stocker, G. C. 1981. Regeneration of a North Queensland rain forest following felling and burning. *Biotropica* 13:86–92.

Stone, G. D., and C. E. Downum. 1999. Non-Boserupian ecology and agricultural risk: ethnic politics and land control in the arid southwest. *American Anthropologist* 101:113–128.

Stouffer, P. C., and R. O. Bierregaard Jr. 1995. Use of Amazonian forest fragments by understorey insectivorous birds. *Ecology* 76:2429–2445.

Szott, L. T., C. A. Palm, and R. J. Buresh. 1999. Ecosystem fertility and fallow function in the humid and subhumid tropics. *Agroforestry Systems* 47:163–196.

Terborgh, J., and J. S. Weske. 1969. Colonization of secondary habitats by Peruvian birds. *Ecology* 50:765–782.

Thiele, G. 1993. The dynamics of farm development in the Amazon: the *barbecho* crisis model. *Agricultural Systems* 42:179–197.

Thiollay, J. M. 1992. Influence of selective logging on bird species diversity in a Guianan rain forest. *Conservation Biology* 6:47–63.

Thiollay, J. M. 1997. Disturbance, selective logging and bird diversity. A neotropical forest study. *Biodiversity and Conservation* 6:1155–1173.

Thomas, S. C. 1991. Population densities and patterns of habitat use among anthropoid primates of the Irturi forest (Zaïre). *Biotropica* 23:68–83.

Thurston, H. D. 1997. *Slash/mulch systems: sustainable methods for tropical agriculture.* Boulder, CO: Westview.

Uhl, C. 1987. Factors controlling succession following slash and burn agriculture in Amazonia. *Journal of Ecology* 75:377–407.

Uhl, C., K. Clark, H. Clark, and P. Murphy. 1981. Early plant succession after cutting and burning in the Upper Rio Negro region of the Amazon basin. *Journal of Ecology* 69:631–649.

Uhl, C., and J. B. Kauffman. 1990. Deforestation, fire susceptibility and potential tree responses to fire in the eastern Amazon. *Ecology* 71:437–499.

Unruh, J. 1988. Ecological aspects of site recovery under swidden-fallow management in the Peruvian Amazon. *Agroforestry Systems* 7:161–184.

van Nieuwstadt, M. G. L., D. Sheil, and K. Kartawinata. 2001. The ecological consequences of logging in the burned forests of East Kalimantan, Indonesia. *Conservation Biology* 15:1183–1186.

Vester, H., and A. M. Cleef. 1998. Tree architecture and secondary tropical rain forest development. *Flora* 193:75–97.

Vieira, I., and J. Proctor. 1998. Dinamica de sementes e regeneracao vegetativa em florestas successionais da Amazonia oriental. Pages 89–97 in M. R. Guariguata and B. Finegan (eds.), *Ecology and management of tropical secondary rain forest: science, people and policy.* Turrialba, Costa Rica: CATIE.

Vieira, I. C. G., R. P. Salomao, N. A. Rosa, D. C. Nepstad, and J. C. Roma. 1996. O renascimento da floresta no rastro da agricultura. *Ciencia Hoje* 20:38–44.

Vieira, I. C. G., A. S. de Almeida, E. A. Davidson, T. A. Stone, C. J. R. de Carvalho, and J. B. Guerrero. 2003. Classifying successional forests using Landsat spectral properties and ecological characteristics in eastern Amazonia. *Remote Sensing of the Environment* 87: 470–481.

Voeks, R. A. 1996. Tropical forest healers and habitat preference. *Economic Botany* 50:354–373.

Watters, R. F. 1971. *La agricultura migratoria en América Latina.* Cuadernos de Fomento Forestal no. 17. Rome: Food and Agriculture Organization of the United Nations.

Western, D., and M. Pearl (eds.). 1989. *Conservation for the twenty-first century.* New York: Oxford University Press.

White, J. L. T. 1994. Biomass of rain forest mammals in the Lope reserve (Gabon). *Journal of Animal Ecology* 63:199–212.

Whitmore, T. C. 1983. Review article: secondary succession from seed in tropical rain forests. *Forestry Abstracts* 44:767–779.

Whitmore, T. C. 1984. *Tropical rain forests of the Far East.* Oxford, UK: Oxford University Press.

Whitmore, T. C. 1989. Tropical forest nutrients, where do we stand? Pages 1–14 in J. Proctor (ed.), *Mineral nutrients in tropical forest and savanna ecosystems.* Special Publications Series of the British Ecological Society Number 9. Oxford, UK: Blackwell Scientific.

Wilkie, D. S., and J. T. Finn. 1990. Slash-burn cultivation and mammal abundance in the Ituri Forest, Zaïre. *Biotropica* 22:90–99.

Chapter 9

Biodiversity Conservation in Neotropical Coffee (*Coffea arabica*) Plantations

Eduardo Somarriba, Celia A. Harvey, Mario Samper,
François Anthony, Jorge González, Charles Staver, and Robert A. Rice

The unprecedented high rate of destruction of natural forests and other natural ecosystems has led scientists to focus on biodiversity conservation in managed landscapes and agroecosystems. Agroforestry systems, renowned for their high tree species richness and complex vegetation structure, stand out as promising biodiversity conservation tools. Well-known examples include shaded coffee (*Coffea* spp.) and cocoa (*Theobroma cacao*) plantations, homegardens, rubber and fruit tree agroforests (see Chapter 10, this volume), grazed dry scrubs and forests, and long fallows (see Chapter 8, this volume).

In recent years, shaded coffee plantations have been singled out for their ability to harbor diverse and abundant wildlife. This strong interest in shaded coffee plantations, in part, reflects the following facts:

- Shaded coffee plantations that have a diverse and structurally complex tree component have a high potential to retain biodiversity and may play critical roles in regional conservation efforts (Perfecto et al. 1996).
- Coffee is of paramount economic importance in more than 50 countries, providing economic support to 20–25 million people and covering 11 million ha of land, so the potential exists to influence biodiversity conservation over large areas (however, coffee cultivation can also be a cause of deforestation; see Nestel 1995 for Mexico).
- In most areas where coffee is grown, the landscape has been so severely deforested and transformed that the only remaining tree cover is that in the coffee plantations; for example, in El Salvador most of the so-called forest cover is actually shade-grown coffee (E. Somarriba, pers. obs., 2002).
- Coffee is grown mostly in regions that are highly biologically diverse, such as Mexico, Ecuador, Peru, Brazil, Colombia, the Côte d'Ivoire, Tanzania, the Western Ghats of India, Sri Lanka, Papua New Guinea, and New Caledo-

nia. For example, Colombia not only is one of the most important producers of coffee but also has the world's richest diversity of birds and amphibians (Botero and Baker 2001; see also Chapter 7, this volume). In individual countries, coffee production areas sometimes overlap with priority areas for conservation that include high numbers of species or endemics. In Mexico, for example, 14 of the 155 conservation priority regions are in or near traditional coffee-growing areas (Llorente-Bosuquets et al. 1996, cited in Moguel and Toledo 1999). Consequently, activities that promote biodiversity conservation in coffee plantations could have impact at both national and regional scales.

• Shaded coffee plantations in the neotropics also play key roles as habitat for migrating birds and therefore have important effects on conservation of biodiversity at supraregional scales.

In this chapter we begin with a historical account of the use of shade in coffee plantations, followed by a review of the literature on vegetation structural types of coffee plantations; plant diversity in the shade canopies and the ground cover, including the genetic diversity of the coffee crop itself; the diversity of other vegetation including the shade canopy and ground cover plants; and the diversity of fauna and microorganisms (including coffee pests, diseases, and their natural enemies) that use the coffee ecosystem as temporary or permanent habitat. Emphasis is given to neotropical coffee plantations, with the exceptions of the historical account, which is global, and the review of pests, pathogens, and their complexes of natural enemies, which have been studied mostly in India.

Shade or No Shade: The Structure of Coffee Agroecosystems

Whether coffee should be grown under a shade canopy has been debated for as long as coffee has been cultivated. Several reviews cover the advantages and disadvantages of shade in coffee (Willey 1975; Beer et al. 1998). As a result of historical processes, pedoclimatic differences between coffee-growing regions, and socioeconomic factors, a wide variety of structural types of coffee agroecosystems, with different levels of biodiversity, have evolved in different parts of the tropics.

Historical Perspective

Coffee (*Coffea arabica*) was discovered in Ethiopia in about AD 850 and was cultivated in the Arabian colony of Harar, an Ethiopian province. It then spread to Mecca, whence it was taken home by pilgrims to other parts of the

Islamic world (Smith 1985). The agroecological needs of the Typica cultivar (the main variety cultivated during the coffee expansion) are characteristic of the Ethiopian hillsides where it originated, at 6°–9° north and 1,300–2,000 m altitude: moderate temperatures (lower and upper extremes of 4° and 31°C, respectively, and means of 20° to 25°C, with hot days and cool nights), 1,500–1,800 mm of annual precipitation with a well-defined dry season of 4–5 months, and a photoperiod of 10.5 to 15 hours per day. In their wild state, the Ethiopian coffee plants grew under a canopy of natural or modified forests on hillsides and along riverbanks (Haile-Mariam 1973). When coffee was introduced into Yemen (fourteenth and fifteenth centuries), in the extreme south of the Arabian Peninsula, with a drier climate and sandier soils than in Ethiopia, it had to be cultivated under shade (Roque 1988). The consumption and cultivation of coffee expanded south through the humid tropics of Asia in the sixteenth and seventeenth centuries, following the expansion of the Islamic culture. Coffee was cultivated below shade in homegardens and thinned forests. In the eighteenth and nineteenth centuries, first the Dutch and then the British promoted intensive coffee production under full sun in India and Ceylon (Smith 1985; Clarence-Smith 1998; Kurian 1998; Tharakan 1998).

In the Indonesian archipelago, coffee cultivation expanded under the colonial Dutch regime from the end of the seventeenth century onward, with coffee being cultivated under shade on small peasant farms. In the mid-eighteenth century, there were three types of coffee systems: the colonial model in high areas where forests were cleared and lines of coffee bushes and shade trees were planted, coffee plantations planted as hedges, and coffee plantations below natural forest. The last two systems were preferred by small farmers because they allowed the simultaneous production of food crops (Fernando 1998).

In the Americas, coffee growing began in the Caribbean in the nineteenth century on plantations of various sizes and with distinct degrees of production intensity, using slave labor. In Saint Domingue (now Haiti), the prevalent plantation type was that of an intensively cultivated plantation, without shade, but with trees planted in field borders or in widely spaced lines throughout the plantations as windbreaks. This Antillean model of coffee plantation (with little or no shade, intensive cultivation with high labor inputs, and wet processing of coffee beans, which greatly increased the cup quality of the coffee) was introduced to Cuba by French emigrants after the Haitian revolution and the abolition of slavery at the end of the eighteenth century (Laborie 1797). Seeds and coffee technology were exported from Cuba to the rest of the Spanish territories in Central and South America. In Spanish Puerto Rico, coffee was cultivated under a planted and managed canopy of *Inga* spp., where densities varied with the altitude of the plantation (Díaz-Hernández 1983; Picó 1983).

In the British Caribbean, which was more specialized in sugar production, the coffee plantations evolved from intensive production (without nutrient replenishment and high soil erosion rates) using slave labor to less intensive systems with shade and other crops grown between the coffee plants (Lowndes 1807; Smith 1998).

In South America, the type of coffee production system varied depending on whether it was located on the Atlantic slope or the Pacific Andean slope. For example, in the Guyanas in the eighteenth century and the beginning of the nineteenth century, coffee was planted in full sun and was associated with intensive land use and slave labor. Both the genetic material and the practice of cultivating coffee under full sun appear to have been transferred to Suriname and to the north of Brazil, then to the state of Rio de Janeiro, and then to São Paulo (Cardoso and Pérez 1977). On the peasant farms and in the large *haciendas* of the Andes in Colombia and Venezuela, bananas (*Musa* spp.) were used as temporary shade, whereas legume trees (*Inga laurina* and *Erythrina fusca*) provided permanent shade. The use of coffee polycultures was a common practice on small peasant farms, with shade densities being lower in high, cool zones than in warm, dry zones (Izard 1973; Ardao 1984; Rios de Hernández 1988).

In Central America, the use of shade in coffee plantations varied from one coffee region to the next and even, within the same region, from one period to the next (Duque 1938; Hearst 1929; Cardoso and Pérez 1977; Samper 1994). For example, between the end of the eighteenth century and 1870, Costa Rica followed the French Antillean model of cultivating coffee in full sun or minimal shade. As productivity declined (because of plantation aging and more severe pest infestation), shade was introduced into the plantations to the extent that in the last third of the nineteenth century, the use of shade was common throughout the country. On the dry and hot Pacific coast of Central America (i.e., El Salvador and Nicaragua), coffee was always grown under a shade canopy (Samper 1994).

The technical intensification of coffee production accelerated in the twentieth century in several parts of the world at varying rates and with marked differences between countries and between different types of farms and farmers. After the mid-twentieth century, new, shorter coffee varieties were introduced in a number of regions; smaller coffee bushes permitted higher planting densities, increasing self-shading and reducing the need for shade trees. These new varieties produced higher yields but also necessitated a greater use of agrochemicals. Shade management was simplified in many areas and reduced to the use of only a few species, mainly fast-growing leguminous trees (notably various species of *Inga, Erythrina,* and *Gliricidia*) that rapidly resprouted after crown pruning, fixed nitrogen, and could be propagated and managed easily. Shade was eliminated altogether, and later reintroduced, in several coffee-growing regions.

Factors That Determine the Use of Shade in Coffee Plantations

This historical analysis indicates that the observed variations in shade design and management of coffee plantations worldwide result from combinations of three major factors:

- *Local climate and extreme environmental conditions that limit coffee yields:* Shading is needed in dry and hot areas, windy places (e.g., *Croton reflexi-folius* windbreaks in El Salvador; Escalante 2000), frost-prone areas (Caramori et al. 1996; Baggio et al. 1997), and sites affected by acid rain from nearby volcanoes (Bonilla 1999). The use of shade is unnecessary in cool, humid, and cloudy highlands. For instance, little shade is used in coffee-growing areas of Costa Rica at more than 1,200 m altitude and 2,500–3,500 mm year^{-1} of rainfall (E. Somarriba, pers. obs., 2002).

- *The compromise between expected coffee yields and plantation longevity:* High coffee yields can be attained in lightly shaded or open sun coffee plantations and with intensive use of agrochemicals. However, plantation longevity is reduced, and the whole plantation must be renovated more frequently (e.g., every 12–15 years in open sun plantations compared with 15–20 years in shaded plantations). Plantation renovation is an expensive task, and the farmer must face 2–4 years without coffee production while coffee plants are still young. The use of shade increases plantation longevity and reduces the need for expensive agrochemicals at the expense of lower coffee yields. Each farmer chooses a place along the continuum between these two extremes.

- *Plantation size and the need for production diversification:* Big and wealthy farms that specialize in coffee production use very simple shade canopies with one service shade species (such as *Inga* or *Erythrina* spp.) that is planted, pollarded, and thinned according to the needs of the coffee plants. In contrast, small coffee farmers commonly opt for a diversified, polycultural system with a diverse shade canopy including several species of fruit, timber, firewood, and other types for home consumption or sale. In these polycultural systems shade cannot be regulated to satisfy the needs of only the coffee plants (e.g., pruning fruit trees to enhance fruit production may not be the best way to regulate shade for the coffee beneath), and this may reduce coffee yields.

Structural Types of Coffee Plantations

Structurally, coffee plantations vary along a continuum from very simple to very complex (Figure 9.1). Schematically, the following structural types can be distinguished:

- Open sun monocultures (with no shade canopy).

Open sun monoculture

Coffee with *Inga* spp.

Coffee with tree rows

Coffee under remnant forest

Coffee with *Cordia alliodora* and Banana

Coffee with *Cordia alliodora* and *Inga* spp.

Multistrata polyculture

Rustic system

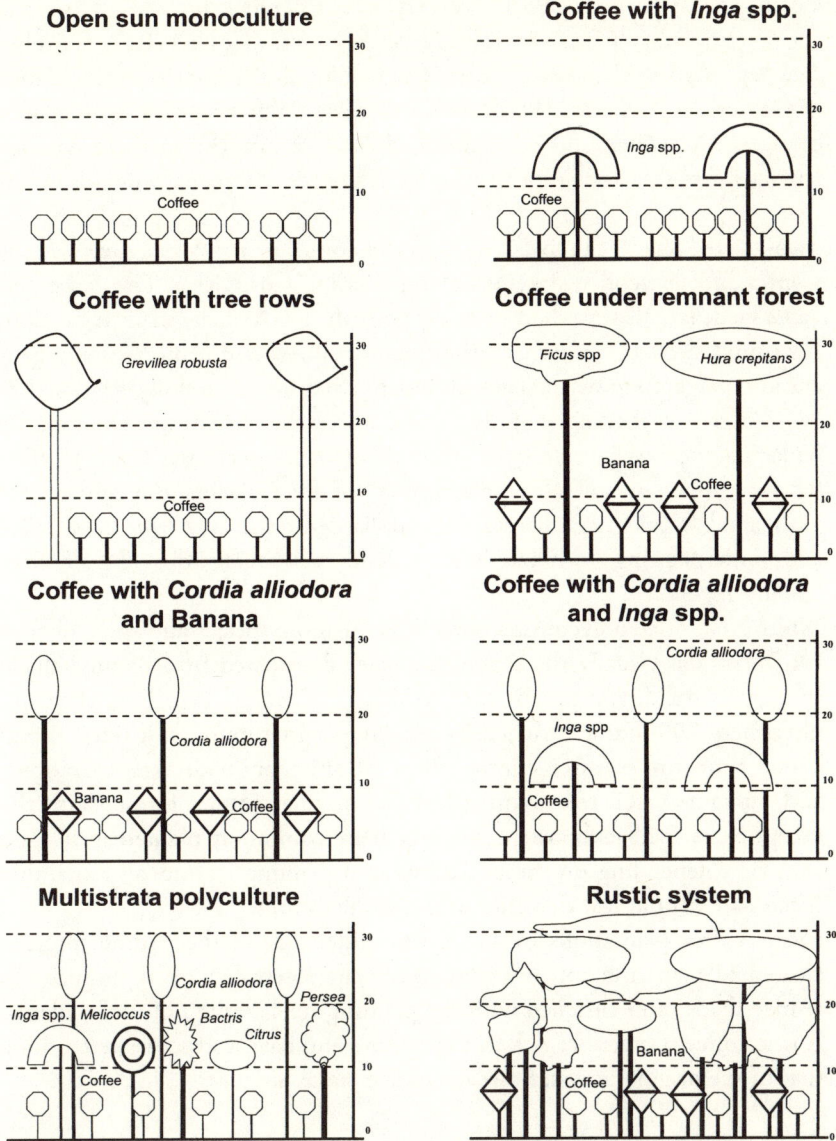

Figure 9.1. Idealized vertical structures and botanical compositions in coffee plantations.

- Coffee plantations with lateral shading from linear tree plantings in field borders and along roads that block wind, avoid excessive shading or facilitate air movement and reduce pathogen infestation in humid, cloudy sites.
- Monolayered shade canopies: coffee plantations with one shade stratum and, typically, only one shade species, be this a service tree that is grown for

shading and soil improvement (several species of the genus *Inga* [Lawrence 1995], *Erythrina, Gliricidia, Albizia,* or *Ficus* spp. [Cook 1901]), a timber tree (e.g., *Cordia alliodora* in Costa Rica or *Grevillea robusta* in the highlands of Guatemala [Villatoro 1986]), or a second commercial crop (e.g., bananas, oranges, *Macadamia* spp., cinnamon, clove, or avocados). In these systems, species richness is low, vertical structure is simple, and management is intensive.

- Two-layered shade canopies, such as the common *Erythrina poeppigiana– Cordia alliodora* coffee systems in midelevation Costa Rica. The *E. poeppigiana* shade layer is kept short by heavy pollarding to facilitate shade regulation; the timber layer (*C. alliodora* either planted or selected from natural regeneration) is left to grow unchecked, but tree density is carefully regulated to avoid excessive shading. Popular variations of this system are obtained by replacing the service tree with bananas or other perennial crops (coffee-banana-timber or coffee-oranges-timber) or by replacing the timber tree with a tall service legume tree (nonpollarded *Inga* or *Erythrina* species) and underplanting with bananas or other perennial crops (e.g., coffee-banana-*Inga*).

- Multistory coffee polycultures with three or more species and three or four vertical strata. Usually the shade canopy is dominated by a planted shade species (a "backbone species" such as *Inga* or *Erythrina* spp. *sensu* Rice and Greenberg 2000) and enriched by planting a mixture of fruit trees, useful palms, timber trees (often selected from the abundant natural regeneration), and, in some cases, trees remnant of the original natural forest. With the exception of the backbone species, which are commonly planted at 50–300 trees ha^{-1} depending on the pollarding and pruning regime, all remaining species are kept at low densities in the shade canopy.

- Rustic coffee plantations in which the understory of the natural forest is cleared to plant the coffee bushes while the forest canopy is thinned (to reduce shade) and enriched with the planting (or favoring) of useful plants. Rustic coffee systems are rich in tree species and have a structure resembling that of the original forest; however, coffee yields are low.

Examples of the aforementioned coffee shade systems have been described for Costa Rica (Lagemann and Heuveldop 1983; Espinoza 1985; Salazar 1985), Nicaragua (Rice 1991), Colombia (Chamorro et al. 1994), Venezuela (Escalante et al. 1987), Ecuador (Mussak and Laarman 1989), Guatemala (Villatoro 1986), Mexico (Jimenez-Avila 1979; Granados and Vera 1995; Moguel and Toledo 1999), Puerto Rico (Weaver and Birdsey 1986), Uganda (Odoul and Aluma 1990), Ethiopia (Teketay and Tegineh 1991), Indonesia (Michon et al. 1986; Godoy and Bennett 1989), Kenya (Njoroge and Kimemia 1993), India (Rao 1975; Awatramani 1977; Reddy et al. 1982;

Bheemaiah and Shariff 1989; Korikanthimath et al. 1994; Reddy and Rao 1999), and Papua New Guinea (Bourke 1985).

Biodiversity in Coffee Agroecosystems

The biodiversity of coffee systems can be divided into the genetic diversity of the coffee crop itself; the diversity of other vegetation including the shade canopy and ground cover; and the diversity of fauna and microorganisms that use the coffee ecosystem as temporary or permanent habitat, including coffee pests and diseases and their biological control organisms.

Genetic Diversity of Coffee

Coffees originated in Africa. They are classified into two genera of the Rubiaceae family, *Coffea* and *Psilanthus,* with each genus being divided into two subgenera (Charrier and Berthaud 1985). More than 80 taxa have been identified in the subgenus *Coffea,* and recent collections of several new taxa in Cameroon (Anthony et al. 1985) and Congo (de Namur et al. 1987) indicate that the inventory is not yet complete. Commercial coffee production relies mainly on two species, *Coffea arabica* (66 percent of world production) and *Coffea canephora* (34 percent). Better cup quality is associated with *C. arabica,* which has its primary center of diversity in the highlands of East Africa; *C. canephora* has its primary center of diversity in the lowlands of the Congo River basin. *C. arabica* is the only self-fertile, tetraploid species (2n = 4x = 44); other *Coffea* species are diploid (2n = 2x = 22) and generally self-incompatible (Charrier and Berthaud 1985).

Genetic diversity in existing *C. arabica* plantations worldwide is very low because of intense reductions of both genetic diversity and polymorphism during domestication, a process favored by its self-fertility. Most commercial cultivars currently grown (Caturra, Catuai, and Mondo novo) were selected from two narrow genetic base populations, spread in the early eighteenth century and known as Typica and Bourbon cultivars (Anthony et al. 2001). Both cultivars have weak polymorphism (Anthony et al. 2002) and are highly susceptible to several major diseases, especially coffee rust (*Hemileia vastatrix*). Fortunately, genes from other diploid species (*Coffea* and some *Psilanthus*) can be transferred into *C. arabica* by controlled hybridization (Couturon et al. 1998), and this has become a priority for the genetic improvement of commercial coffee (Carvalho 1988; Lashermes et al. 2000). Many modern coffee plantations are based on the extensive use of a few introgression lines selected from natural interspecific hybrids: the Timor hybrid in Latin America (*C. arabica* x *C. canephora*) and (*C. arabica* x *C. liberica*) in India. Selected lines include Costa Rica 95 and IHCAFE 90 in Central America, Variedad Colombia in Colombia, IAPAR 59 and Icatu in Brazil, Riuru 11 in Kenya, and Sln 12 in India.

Plant Species Richness and Botanical Composition of Shade Canopies

Plant species richness in coffee shade canopies varies widely between countries, between coffee regions within a country, and between farms in a region. In a series of studies conducted in Central America,[1] the estimated total plant species richness varied between 19 and 49 species for Costa Rica and between 92 and 136 species for El Salvador (Figure 9.2 and Table 9.1; Llanderal 1998; Bonilla 1999; Escalante 2000; Zuñiga 2000). Average tree densities varied between 198 and 488 stems ha[-1], and Shannon diversity indices ranged from 1.57 to 3.08. In Venezuela (625-m^2 plots) 19 species were recorded in 20 coffee farms (Escalante et al. 1987), and in Puriscal, Costa Rica, 82 species were recorded in 117 coffee farms (Espinoza 1985). A total of 261 tree species (including 23 endangered species) and 32 fern species have been reported in Salvadorian coffee shade canopies (Monro et al. 2001, 2002). In Sumatra, Indonesia, coffee shade may include 10–15 additional crops (Godoy and Bennett 1989).

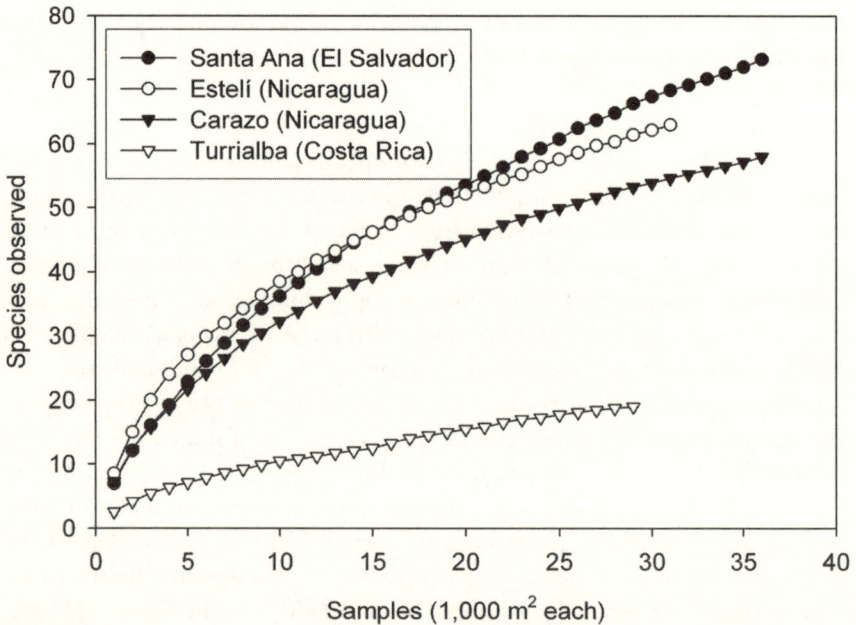

Figure 9.2. Species-area accumulation curves for shade canopies in selected coffee zones in Central America.

Table 9.1. Plant species parameters in the shade canopy of *Coffea arabica* plantations in Central America.

	Turrialba, Costa Rica	Carazo, Nicaragua	Estelí, Nicaragua	Santa Ana, El Salvador
Number of 1,000-m² plots sampled	29	36	31	40
Total plant species richness observed	19	36	63	77
Maximum expected richness (see Colwell 1997)	19–49	68–94	74–129	92–136
Modal plant richness per farm (minimum–maximum)	2 (1–8)	7 (1–12)	10 (2–17)	3 (2–14)
Stems ha⁻¹ (standard deviation)	386 (184)	472 (386)	488 (477)	198 (46)
Shannon diversity index	1.57	2.06	2.85	3.08
Simpson diversity index	3.0	3.471	8.82	9.87
Alpha diversity index (Fisher)	3.25	11.61	13.4	21.0

Source: E. Somarriba, unpublished data, 2002.

Farmers tend to keep species richness at the farm level low to facilitate shade regulation. For instance, most farms in Costa Rica and El Salvador have only 2 or 3 species in the shade canopy; corresponding figures for two Nicaraguan sites are 7 or 10 species per farm (Table 9.1). The species used in coffee shade canopies differ between countries. For instance, bananas (several species and varieties of *Musa*, 60–240 stems per hectare) abound in Nicaraguan coffee plantations but less so in El Salvador and Costa Rica (10–37 stems per hectare). *E. poeppigiana* and several *Inga* species dominate the shade canopies in Costa Rica and El Salvador, respectively (Table 9.2). Most species had fewer than five trees per hectare. Similar results have been reported for Mexico (Marten and Sancholuz 1981).

A total of 25 *Inga* species are used regularly in the shade strata of neotropical coffee plantations; in some countries (e.g., Honduras and El Salvador) most shade canopies include a mixture of three to six *Inga* species. *Inga* species support nectarivorous birds and provide fruit, firewood, and ecological services such as water and nutrient maintenance (Wadsworth 1945; Gutierrez and Soto 1976; Jimenez-Avila 1979; Espinoza 1985; Lawrence 1995; Lawrence and Zuñiga 1996).

Table 9.2. Average density and frequency of occurrence of the 10 most common shade species in selected Central American coffee zones.

Site	Species	Local Name	Stems ha^{-1}	Frequency
Carazo, Nicaragua	Musa AAB	Platano	241	55
	Gliricidia sepium	Madreado	53	83
	Citrus sinensis	Naranja	50	38
	Cordia alliodora	Laurel	21	38
	Simarouba glauca	Acetuno	12	30
	Cedrela odorata	Cedro	10	44
	Persea americana	Aguacate	9	41
	Mangifera indica	Mango	8	27
	Solanum erianthum	Lavaplato	6	16
	Amphipterygium adstringens	Copel	5	19
Estelí, Nicaragua	Musa AA	Guineo negro	125	16
	Musa AAA	Datil	65	25
	Musa AAB	Platano	61	35
	Inga oerstediana	Guama	33	67
	MusaAAA (Gros Michel enano)	Guineo caribe	30	9
	Sapium glandulosum	Lechoso	17	58
	Inga vera	Guama negra	12	35
	Persea caerulea	Aguacate colorado	12	48
	Cinnamomum costaricanum	Aguacate canelo	10	38
	Ocotea helicterifolia	Aguacate pachon	9	19
Santa Ana, El Salvador	Inga punctata	Pepeto peludo	51	77
	Inga vera	Guama negra	27	67
	Inga ruiziana	Pepeto negro	16	40
	Musa AAA	Minimo	10	22
	Inga minutula	Nacaspilo	9	20
	Cordia alliodora	Laurel	7	35
	Eugenia jambos	Manzana rosa	6	17
	Unknown	Unknown	6	7
	Persea americana	Aguacate	6	27
	Mangifera indica	Mango	5	25
Turrialba, Costa Rica	Erythrina poeppigiana	Poró	209	82
	Cordia alliodora	Laurel	55	44
	Musa AAA	Minimo	37	24
	Musa AAB	Platano	33	27
	Macadamia integrifolia	Macadamia	20	17
	Theobroma cacao	Cacao	11	3
	Citrus sinensis	Naranja	4	13
	Carica papaya	Papaya	4	3
	Eucalyptus deglupta	Eucalipto	3	3
	Bactris gasipaes	Pejibaye	3	6

Useful Plant Diversity in Coffee Shade Canopies

Farmers (especially smallholders) have long managed coffee shade canopies to diversify production, cope with unexpected family needs and pest outbreaks, buffer themselves against persistent low coffee prices, and reduce both weed competition and the need for expensive inorganic fertilizers. A rich literature is available on the design and management of useful plants in coffee shade canopies (e.g., from India, Kenya, and Central and South America). For instance, species valued only for shade represent 54 percent of all stems in the shade canopy in Costa Rica but less than 12 percent in El Salvador and Nicaragua. Bananas are very important in Nicaragua (50–57 percent of all stems) but not so in El Salvador (where they represent only 5 percent of all stems); timber production is equally important in all Central American countries, whereas firewood is of no relevance in some areas of Costa Rica (Table 9.3). Firewood is the most commonly mentioned reason for planting *Inga* spp. as shade in Salvadorian coffee plantations (Lawrence and Zuñiga 1996), and timber trees are perceived as a savings account that can be used when coffee prices are low or when unexpected family needs arise.

Products from the shade canopy may be important sources of income to small coffee holders. For example, in Peru and Guatemala products from the shade canopy may account for 28 percent and 19 percent, respectively, of the total value emerging from the coffee plantation (R. Rice, unpublished data, 2002). Firewood for family use (52 percent) and fruit sold (19 percent) or consumed by the family (15 percent) accounted for much of the total value obtained from the coffee plantations; firewood for sale (8 percent) and lumber for family use (5 percent) or sale (1 percent) are less important. Most of the fruit production is lost (53 percent), 28 percent is sold, and 19 percent is consumed by the family (R. Rice, unpublished data, 2002). These figures may

Table 9.3. Relative abundance (stems per use group in percent of the total number of stems at a site) in percentage and number of useful plant species (in parentheses) in the shade canopy of Central American coffee plantations.

Use	Carazo, Nicaragua	Estelí, Nicaragua	Santa Ana, El Salvador	Turrialba, Costa Rica
Only shade	4 (12)	11 (21)	6 (18)	54 (1)
Citrus spp.	10 (2)	<1 (1)	3 (2)	1 (4)
Minor fruits	5 (15)	2 (6)	12 (13)	10 (8)
Firewood	13 (6)	17 (14)	56 (14)	0 (0)
Timber	14 (19)	9 (16)	16 (29)	15 (3)
Musa spp.	50 (1)	57 (4)	5 (1)	18 (2)
Other uses[a]	4 (3)	4 (1)	2 (2)	2 (1)

Source: E. Somarriba, unpublished data, 2002.

[a]Posts, ornamental, or medicinal.

vary between regions. For instance, in small coffee holdings in a suboptimal dry area in Venezuela, fruit production from the shade trees accounted for 55–60 percent of total gross revenues from the coffee plantations, timber 3 percent of total gross revenues, and coffee the remaining 37–42 percent (Escalante et al. 1987). In Puriscal, Costa Rica, sales from oranges and other fruits from coffee shade canopies account for 5–11 percent of total sales from the coffee plot (Lagemann and Heuveldop 1983). Aside from these products with easily quantifiable market values, shade trees and other plants in the canopy have other, less easily assessed values. The use of plant parts in traditional or home remedies is a common cultural practice found in shaded coffee systems throughout the tropics. For example, bark-cloth is made from *Ficus natalensis* (a common coffee shade tree species) in Uganda, and plant parts are used in ceremony, ritual, or as adornment in other places.

The Ground Cover of Coffee Systems

Coffee plantations commonly have 20–90 plant species in the ground cover plant layer. The diversity of this layer in a given field depends on the prior land use and within-field variability in soil, drainage, tree canopy distribution, species composition, and ground cover management practices. Goldman and Kigel (1986) listed 24 species in a 600-m^2 shaded coffee field in Mexico grouped in two distinct weed associations: one in microhabitats with more than 75 percent canopy coverage and the other in microhabitats with less than 25 percent canopy cover. Weed species richness was greater in sunnier places. This response of weed associations to different shade, soil, and management history was also recorded in a study of ground cover by plant growth habit in five coffee fields in northern Nicaragua: perennial broadleafs varied between fields from 2 to 26 percent of the between-row area, grasses from 12 to 47 percent, and sedges from 0 to 9 percent (Staver 1999).

In a comparison of ground cover plant families in coffee systems in open sun and with single- and multiple-species shade (Nestel and Altieri 1992), weed families were similar in the three systems but varied in their abundance. Asteraceae were more common in open sun coffee, whereas Commelinaceae were more common in shaded coffee. Similar results have been reported for Mexico (Jimenez-Avila 1979). In Andhra Pradesh, India, 74 dicotyledonous, 8 monocotyledonous, and 1 fern species were recorded in the ground layer of coffee plantations; extensive differences were observed depending on the degree of shading and season. Acanthaceae, Amaranthaceae, Asteraceae, Cucurbitaceae, Fabaceae, Gentianaceae, Malvaceae, Poaceae, and Rubiaceae were well represented in the weed complexes (Reddy and Reddy 1980). At least 28 main weed species have been reported for small coffee holdings in Papua New Guinea (de Silva and Tisdell 1990), and in Venezuela, prevalent weed complexes in coffee plantations included some 17 species of Poaceae, 7 Cyperaceae, and 42 other broadleaf species (Garcia 1988). In Puriscal, Costa Rica, 84 species were recorded in the weed complex in four small coffee plantations (Mora-Delgado and Acosta 2001).

Although the management of ground cover as habitat for beneficial organisms has been successfully included in pest management strategies in citrus and nut tree plantations (Bugg and Waddington 1994), most coffee pests are highly specific to the coffee plant itself or to other tree species (Staver et al. 2001), so the potential for direct interactions with ground cover species is reduced. Plant parasitic nematodes of coffee such as the root-knot nematode *Meloidogyne* may have alternative hosts in the ground cover layer.

Farther up the food web, parasitoids of the coffee berry borer (*Hypothenemus hampei*) feed on flower nectar for survival while they search for berry borer larvae, suggesting that the ground cover could be managed to promote flowering during critical periods of the parasitoid life cycle. A dipterous larva has been identified that feeds on rust spores on the weed species *Alternanthera pubiflora* and also on coffee rust spores (C. Staver, pers. obs., 2002). These are indicative of other interactions that may occur but remain to be studied.

Vertebrates in Coffee Ecosystems

A significant amount of research has been devoted to studying the fauna in coffee plantations; indeed, coffee agroecosystems probably are the best studied of all agroforestry systems in terms of their biodiversity. A wide variety of animals use or visit shaded coffee plantations, including birds, bats and other mammals, insects, and reptiles. Many of the animals that use shaded coffee plantations depend heavily on the tree component and other flora as food resources, nesting, mating, and foraging sites, shelter, or habitat; the monospecific coffee layer itself, with its low structural complexity, provides few resources and is of only limited habitat value for a few species (Perfecto et al. 1996). In addition to the trees themselves, the occasionally diverse communities of epiphytes on tree trunks and branches may offer a wide variety of microhabitats for both plants and animals.

A subset of mammal species may take refuge in coffee plantations, although many of these species depend on other habitats for their survival. For example, studies in Mexico reported a total of 24 large mammal species, including three types of cats (Gallina et al. 1996), in shaded coffee systems. In Costa Rica, 15 mammal species were found in a shaded coffee plantation (J. Gonzalez, unpublished data, 2000). The mammals recorded in shaded coffee plantations include agoutis (*Dasyprocta punctata*), anteaters (*Tamandua mexicana*), bats (various species), coatis (*Nasua narica*), coyotes (*Canis latrans*), howler monkeys

(*Alouatta palliata*), kinkajous (*Potus flavus*), margays (*Leopardus wiedii*), mice (several species), opossums (*Didelphis* spp.), pumas (*Puma concolor*), raccoons (*Procyon lotor*), and squirrels (*Sciurus* spp.) (Estrada et al. 1993; Gallina et al. 1996; Gonzalez 1999a). Many of these mammal species are adapted to an arboreal life, seeking shelter or building nests in the shade trees and feeding on the flowers, leaves, and fruits of the shade canopy, and would not be present if the shade canopy were absent. Although most of the species are generalists, the presence of a few endangered mammalian species in shaded coffee plantations in Mexico, such as the tamandua anteater (*Tamandua mexicanus*), river otter (*Lutra longicaudis*), Mexican porcupine (*Sphiggurus mexicanus*), and margay (*Leopardus wiedii*), suggests that traditional coffee systems could play an important role in the conservation of forest species threatened by deforestation and habitat loss (Gallina et al. 1996). However, it is important to note that many of the mobile animals in coffee plantations are likely to forage over large areas, including forest patches, and their survival may be more closely linked to the availability of forest habitat in the region than to the availability of the shaded coffee plantation itself. This factor is important when considering biodiversity baseline studies involving specific land use types.

Shaded coffee plantations are widely known to harbor a diverse and abundant avifauna. For example, studies in the coffee-growing region of Colombia have reported 170 bird species using shaded coffee plantations, representing roughly 10 percent of the known bird species in this country (Botero and Baker 2001). The high avifaunal richness in shaded coffee plantations includes a mixture of bird species characteristic of open and second-growth habitats, forest generalists and forest edge species, that belong primarily to frugivorous, insectivorous, and nectarivorous guilds (Greenberg et al. 1997; Moguel and Toledo 1999). In general, the bird diversity in shaded coffee plantations is less than that in the original forests and is distinct in terms of its species composition: only rarely are specialized forest bird species (such as those associated with forest understory) found in coffee plantations in Central and South America (Wunderle and Latta 1996; Greenberg et al. 1997). For example, in a study of bird populations in Central Guatemala, Greenberg et al. (1997) found that forest habitats had the highest number of species (87–122), followed by shaded coffee plantations (73) and sun coffee plantations (65). However, in other areas, studies have shown that the bird diversity in coffee plantations can be similar to (or even greater than) that of intact forest, as was found in a comparison of bird diversity in rustic and *Inga*-shaded coffee plantations in eastern Chiapas, Mexico (Greenberg et al. 1997), and in rustic coffee systems in Mexico (Moguel and Toledo, 1999) and northern Panama (Roberts et al. 2000a, 2000b).

In the New World, shaded coffee plantations are also critical habitats for large numbers of migratory birds that arrive from the north at the beginning of the dry season and overwinter in the coffee plantations, feeding on fruits, nectar, and insects. Shaded coffee plantations are the most widely used of all agricultural habitats in the tropics, and the abundance of migratory birds in shade coffee sometimes is even higher than that in primary forest (Greenberg et al. 1997). Among the common migrants that visit coffee plantations are fly-catchers, wood warblers, tanagers, and orioles. The migrant birds appear to fare well in shaded coffee plantations (and better than resident bird populations), with survival rates of many migrant bird species comparable to those in natural habitats (Wunderle and Latta 2000). This ability of migrant birds to survive in shade coffee plantations is thought to reflect their more flexible habitat needs compared with residents who breed in the area (Perfecto et al. 1996).

Arthropods in Coffee Ecosystems

The arthropod fauna in shaded coffee plantations can also be large, with the abundance of leaf litter, fallen twigs, trees, and weeds in coffee plantations providing habitat for both ground-dwelling and arboreal species (Nestel et al. 1993, 1994). A total of 609 morphospecies of arthropods was found in shaded coffee systems in Mexico (Ibarra-Nuñez 1990), 78 families of arthropods (mostly Diptera, Coleoptera, Hymenoptera, and Collembola) were collected from coffee plantations in Chiapas (Moron and Lopez-Mendez 1985), 322 insect species were found in a Costa Rican coffee plantation (Gonzalez 1999b), more than 30 ant species were registered in a traditional, shaded coffee agroecosystem in Costa Rica (Perfecto and Vandemeer 1994; Barbera 2001), 50 species of Pimplinae wasps have been recorded in Salvadorian coffee plantations (The Natural History Museum 2002), and 168 butterfly species were found in a shaded coffee plantation in Colombia (Botero and Baker 2001). A recent study reported 130 species of Auchenorrhyncha (Homoptera), mostly Cicadellidae (82 species), in very simple coffee–*Erythrina poeppigiana* or coffee–*E. poeppigiana–Cordia alliodora* systems in Turrialba, Costa Rica (Rojas et al. 2001a, 2001b). Perhaps most surprisingly, 30 ant species, 103 other Hymenopterans, and 126 beetle species were collected from the canopy of a single *E. poeppigiana* tree in a coffee plantation in Costa Rica. In another *E. poeppigiana* tree, located less than 200 m away, 30 ant species, 103 other Hymenopteran species, and 126 beetle species were found. Species overlap between these two trees was only 14 percent for beetles and 18 percent for ants, suggesting that shaded coffee plantations may have very high levels of arthropod diversity; however, the species composition often is distinct from that of natural forest (Perfecto et al. 1996).

Coffee agronomists have documented the varied arrays of natural enemies, notably several species of Hymenoptera, Diptera, and Fungi, that control the population of every known insect pest of coffee. Borers are notorious pests in coffee. In robusta coffee, the shot-hole borers *Xylosandrus* spp. (Coleoptera, Scolytidae) are considered serious pests; they are controlled by several species of Hymenoptera and Coleoptera (Dhanam et al. 1992a; Sreedharan et al. 1992). The white stem-borer *Xylotrechus quadripes* (Coleoptera, Cerambycidae), a serious pest of arabica coffee, is controlled by a complex of 11 hymenopteran parasites in Vietnam (Le Pelley 1968) and by *Allorhogas pallidiceps* (Hymenoptera, Braconidae) in India (Prakasan et al. 1986). The coffee berry borer (*Hypothenemus hampei*) is economically the most important of several species of this genus that feed and live mostly on *Coffea* spp. (Johanneson and Mansingh 1984). Several fungi have been reported as natural enemies of *H. hampei* (Kumar et al. 1994; Balakrishnan et al. 1995); one of these, *Beauveria bassiana,* is known to also attack several other species of coffee borers, including *Xylosandrus compactus* and *Xylotrechus quadripes* (Balakrishnan et al. 1994).

Common shade trees are also attacked by borers (Dhanam et al. 1992b). In several countries, *Erythrina* spp. are attacked by *Terastia meticulosalis* (Lepidoptera, Pyralidae), which is in turn controlled by *Aparkeles leptoura* in India and *Bracon terestiae* in Congo (Samuel and Bhat 1988). Other Lepidopteran borers (Hepialidae) attacking robusta coffee and several shade tree species include *Sahyadrassus malabaricus* in India (Balakrishnan et al. 1988) and *Phassus damor* in Indonesia (Le Pelley 1968). *Grevillea robusta,* one of the most common shade trees in most coffee-growing regions of India, is attacked by a whole complex of borers (Sreedharan et al. 1991).

More than 60 species of scale insects and mealybugs (Homoptera, Coccoidea) have been reported in coffee plantations, but only 20 are known as pests (Le Pelley 1968). In India, some 17 coccoid species have been recorded on coffee (Chacko 1979). In Guatemala, seven scale species are commonly found in coffee plantations; some are also found on the weed complex or on shade trees (Garcia et al. 1995). Scales and mealybug populations are controlled by several Hymenoptera, Diptera, and Coleoptera species. For instance, the white-tailed mealybug (*Ferrisia virgata,* Homoptera, Pseudococcidae), a cosmopolitan and highly polyphagous pest of robusta coffee, is controlled by eight parasitoid species and 24 predator species in India (Balakrishnan et al. 1991). Coccids and pseudococcids are associated with ants all over the world. About 27 ant species have been recorded in association with homoptera attacking coffee (Venkataramaiah and Rehman 1989).

Leaf-miners are occasionally important among the various pests damaging coffee foliage. In India, several species of minor importance are controlled by a complex of hymenopteran parasites (Balakrishnan et al. 1986). Thrips, such as *Retithrips syriacus* (Thysanoptera, Thripidae), have been reported as minor pests of coffee in Tanzania, Kenya, Uganda, and India (Kumar et al. 1984). Sixteen termite species attack coffee bushes and their shade trees around the world (Kashyap et al. 1984). In India, five termite species have been observed attacking coffee and the shade trees *Grevillea robusta* and *Erythrina lithosperma* (Gowda et al. 1995).

Nematodes and Mycorrhizal Fungi in Coffee Ecosystems

Of the many nematode species that attack coffee, the 11 species of *Meloidogyne*, 4 species of *Pratylenchus, Radopholus similis,* and *Rotylenchulus reniformis,* and 3 species of *Hemicriconemoides* are considered to be of economic importance; nonpathogenic nematodes found in coffee soils include the genera *Criconemoides, Helicotylenchus, Nothocriconema, Rotylenchus, Scutellonema,* and *Xiphinema* (Kumar and Samuel 1990).

Vesicular arbuscular mycorrhizae in arabica and robusta coffee roots have received some research attention because of their impact on phosphorus uptake. In India, *Glomus macrocarpum, Sclerocystis rubiformis, Gigaspora gigantea,* and *Gigaspora heterogama* have been identified (Rangeshawaran et al. 1990). In Colombia, 20 species of vesicular arbuscular mycorrhizae have been identified in coffee plantations, including six species of *Acaulospora,* nine species of *Glomus, Gigaspora* sp., *Scuttellospora* sp., *Entrophospora colombiana,* and *Sclerocystis sinuosa* (Bolaños et al. 2000).

Benefits and Costs of High Faunal Diversity in Coffee Ecosystems

Although the presence of a diverse and abundant fauna in shaded coffee plantations is clearly beneficial from a conservation viewpoint, it is not clear what the benefits or costs are to the farmer who owns and manages the plantation (but see Gobbi 2000; Bray et al. 2002). Potential benefits include the fulfillment of ecological services such as pollination (Roubik 2002), seed dispersal, soil regeneration, and pest regulation. Although there are few data on the potential importance of these ecological services, several studies suggest that the presence of parasites and predators in shaded coffee plantations could help control insect pest outbreaks. For example, almost 25 percent of all arthropods collected and 42 percent of the species found in a species-rich shade plantation near Tapachula, Chiapas, Mexico, were predators and parasites (e.g., spiders, ants). Similarly, 83 percent of the 322 insect species found in Costa Rican coffee plantations were species with potential as biological control agents (Gonzalez 1999b). Some of the mammals found in shaded coffee plantations might also control pest populations by feeding on insects or small rodents (Gallina et al. 1996). From a farmer's point of view, the drawbacks of having a rich animal community include the potential for damage to occur to nearby crops and the potential for harm to domestic animals or humans from large mammals and snakes. Unfortunately, no studies have considered these potential impacts (see also Chapter 13, this volume).

Factors Affecting Biodiversity in Coffee Plantations

The ability of coffee plantations to harbor wildlife depends on a variety of factors, including the diversity and density of trees, the presence of wild plants in the understory, plantation management (especially the use of agrochemicals), and the composition and structure of the surrounding landscape.

The floristic and structural complexity of shaded coffee plantations is of utmost importance in ensuring the sustainability of biodiversity, given that positive relationships have been found between vegetational complexity and insect diversity, mammal diversity, and birds (Perfecto and Snelling 1995; Greenberg et al. 1997; Wunderle and Latta 1998). Consequently, coffee plantations that have a structurally and floristically diverse tall tree layer generally host a higher animal abundance and species richness than plantations of low stature and low diversity. For example, studies in Mexico found up to 184 bird species in shaded coffee plantations (Moguel and Toledo 1999), compared with only 6–12 bird species in nonshaded plantations (Martinez and Peters 1996). More than 30 species of terrestrial ants have been encountered in the traditional, shaded coffee agroecosystem of the Central Valley in Costa Rica, whereas only 6 species were found in the modern, unshaded systems (Perfecto and Vandemeer 1996). This pattern of higher diversity in diverse, shade-grown coffee plantations than in sun-grown coffee is consistent among bird, insect, and mammal populations (Perfecto et al. 1996; Moguel and Toledo 1999).

Differences in the fauna present in different types of coffee plantations may also reflect differences in food availability in individual coffee plantations. For example, coffee plantations shaded with *Inga* trees are likely to have high numbers of nectarivorous and omnivorous bird species because *Inga* flowers are important sources of nectar (Greenberg et al. 1997; Wunderle and Latta 2000); in addition, the wide variety of insects found on *Inga* leaves (including grasshoppers, katydids, lepidopteran larvae, beetles, and spiders) makes them key foraging sites for insectivorous species (Koptur 1983; Wunderle and Latta 2000). In contrast, rustic coffee plantations that contain high densities of fruit-producing trees are likely to support high numbers of frugivorous bird species (Greenberg et al. 1997). The number of butterflies in coffee plantations is similarly influenced by the abundance and diversity of wild plant species that provide nectar and larval food sources.

The presence of fauna in coffee plantations may also be related to management aspects such as the thinning or pollarding of shade trees and the use of pesticides. It is to be expected that coffee plantations that are more intensively managed or use greater quantities of pesticides would have lower populations and lower species richness (Perfecto and Vandermeer 1994; Perfecto et al. 1997). However, no exact data are available about these relationships, although the comparisons of biodiversity in sun-grown coffee systems (which typically have high agrochemical inputs) with shade-grown coffee (with lower agrochemical inputs) tend to support this idea. It is similarly likely that the regular and drastic pollarding of shade trees could reduce the plantation's ability to support animals.

Finally, the abundance and diversity of fauna in coffee plantations are likely to be influenced by the overall landscape in which the coffee plantation is located, particularly the patch size and frequency of forests; the presence of live fences, windbreaks, dispersed trees, and other remnant vegetation (see Chapter 11, this volume); the connectivity of the coffee plantation to nearby forests; and the overall degree of disturbance and degradation in the landscape. Many of the animals visiting or using coffee plantations probably move throughout the landscape and depend, at least in part, on resources and habitats outside the coffee plantation (Wunderle and Latta 2000). Consequently, the abundance and distribution of other habitats in the landscape and their proximity to the coffee plantations are important. For example, traditional coffee fields adjacent to a tropical forest had a total of 184 bird species, whereas a similar coffee plantation isolated from forest remnants had only 82 species (Martinez and Peters 1996), suggesting the importance of forest habitat to the persistence of many bird species. Where shaded coffee plantations abut natural forests, they may also buffer the natural forests from outside influences and thereby increase the habitat area for some wildlife species. At the same time, the presence of trees in coffee plantations may help increase the overall landscape connectivity, thereby facilitating animal movement to and from isolated forest patches in the coffee farming matrix. Additional research is needed to elucidate the role of shaded coffee plantations in conserving biodiversity at the landscape scale.

Improving Biodiversity Conservation in Coffee Plantations

The role of coffee agroforestry systems in conserving biodiversity can be enhanced by designing and managing the landscape in which the coffee plantation occurs and by increasing the floristic and structural diversity of the shade canopy. Shaded coffee plantations that have a diverse and structurally complex tree component have a high potential to retain biodiversity and may play critical roles in regional conservation efforts. For instance, coffee plantations that are certified as bird-friendly must

- Maintain a minimum of 10 tree species in the shade layer (preferably native and evergreen species) and a shade cover of at least 40 percent throughout the year, preferably integrating a mixture of tree species and creating several strata

- Allow epiphytes and parasitic plants to grow on the canopy trees
- Retain dead limbs and snags in the plantation to provide additional resources for birds
- Allow the tree canopy to attain at least 12 m in height
- Ensure that coffee fields are bordered with a living fence or border strip of trees and shrubs or natural second-growth vegetation (Smithsonian Migratory Bird Center 1999).

In commercial polyculture and specialized shade systems where the shade canopy is planted by the farmers, a variety of nondeciduous *Inga* species (rather than *Erythrina* and *Gliricidia,* which are deciduous part of the year) have been suggested as the dominant shade species to ensure that flowers and fruits are available for longer periods of time and that the coffee plantation has shade in the dry season when canopy cover for both migrant and resident birds is most critical (Smithsonian Migratory Bird Center 1999). There are hopes that the higher price of these certified products will entice farmers to conserve shaded coffee plantations and the species in them.

Biodiversity in coffee plantations can be further increased; for instance, species richness in the ground cover plant layer can be promoted by creating a mosaic of shade conditions and applying selective management. Native species suitable for shading coffee and producing useful products could be identified, propagated, and introduced into coffee plantations (Linkimer 2001; Yépez 2001). Useful diversity can also be increased by exchanging management information and perhaps germplasm between coffee-growing areas. For instance, very few timber species are used in Central American coffee plantations despite the high potential to use many other native species. It is worth noting that earlier attempts to diversify coffee production systems have failed because no market or processing facilities were concurrently developed with the planting of useful species in the coffee plantations (Godoy and Bennett 1989). Special efforts should be made to diversify big farms because many small coffee farms already maintain high levels of useful diversity.

Conclusions

The concern about the environmental impacts of agriculture, the global loss of forests and biodiversity, the valorization of environmental services (e.g., water, soil conservation, carbon sequestration to mitigate global warming, and aesthetic and recreational aspects), and the fall of international coffee prices caused by overproduction have changed the value of coffee plantations grown under diverse shade canopies. In shaded coffee plantations production costs can be reduced, coffee grain and cup quality can be increased (Guyot et al. 1996), plantation longevity can be increased, and new markets with preferential prices can be accessed (i.e., organic markets, markets for environmentally friendly products). The time is right for designing and managing shaded coffee plantations that are both productive and of high value for maintaining and conserving biodiversity.

Acknowledgments

Andy Gillison and Götz Schroth provided valuable comments to improve the quality of this chapter. Partial funding to Eduardo Somarriba was provided by the CASCA Project (INCO, ICA4-2000-10327) and to Celia Harvey by the European Community Fifth Framework Programme "Confirming the Role of Community Research" (INCO, ICA4-CT-2001-10099).

Endnote

1. A total of 136 coffee farms were studied in four coffee-growing regions in Costa Rica, Nicaragua, and El Salvador. Study farms were selected randomly from lists of farmers in local coffee associations and processing plants. A standard diagnostic method was used in all cases. Land use sketches were prepared for each farm, depicting all coffee plots (shade types, age, variety). One temporary 1,000-m² plot was established in the largest and most representative coffee plot of the farm. All shade plants were identified and counted. Only trees greater than 10 cm diameter at breast height were considered; palms and fully developed banana stems were also counted. Results in Tables 9.1 and 9.2 come from these studies.

References

Anthony, F., B. Bertrand, O. Quiros, P. Lashermes, J. Berthaud, and A. Charrier. 2001. Genetic diversity of wild coffee (*Coffea arabica* L.) using molecular markers. *Euphytica* 118:53–65.

Anthony, F., M. C. Combes, C. Astorga, B. Bertrand, G. Graziosi, and P. Lashermes. 2002. The origin of cultivated *Coffea arabica* L. varieties revealed by AFLP and SSR markers. *Theoretical Applied Genetics* 104:894–900.

Anthony, F., E. Couturon, and C. de Namur. 1985. Les caféiers sauvages du Cameroun. Résultats d'une mission de prospection effectuée par l'ORSTOM en 1983. Pages 495–505 in *Proceedings of the 11th International Scientific Colloquium on Coffee, Lomé.* Vevey, Switzerland: ASIC.

Ardao, A. 1984. *El café y las ciudades en los andes venezolanos (1879–1939).* Caracas, Venezuela: Academia Nacional de la Historia.

Awatramani, N. A. 1977. Multistoreyed cropping patterns with coffee for maximizing production. *Indian Coffee* 41:253–254.

Baggio, A. J., P. H. Caramori, A. Androcioli Filho, and L. Montoya. 1997. Productivity of southern Brazil coffee plantations shaded by different stockings of *Grevillea robusta. Agroforestry Systems* 37:111–120.

Balakrishnan, M. M., P. K. V. Kumar, and T. S. Govindarajan. 1986. Role of natural enemies in the suppression of the leaf miner *Tropicomyia* sp. (Diptera; Agromyzzidae) on coffee. *Journal of Coffee Research* 16:89–93.

Balakrishnan, M. M., P. K. V. Kumar, and C. B. Prakasan. 1988. Record of *Sahyadrassus malabaricus* Moore (Lepidoptera, Hepialidae) on coffee. *Journal of Coffee Research* 18:120–125.

Balakrishnan, M. M., K. Sreedharan, and B. Krishnamoorthy. 1994. Occurrence of the entomopathogenic fungus *Beauveria bassiana* on certain coffee pests in India. *Journal of Coffee Research* 24:33–35.

Balakrishnan, M. M., K. Sreedharan, Venkatesha, and P. K. Bhat. 1991. Observations on *Ferrisia virgata* Ckll. (Homoptera, Pseudococcidae) and its natural enemies on coffee, with new records of predators and host plants. *Journal of Coffee Research* 21:11–19.

Balakrishnan, M. M., V. A. Vijayan, K. Sreedharan, and P. K. Bhat. 1995. New fungal associates of the coffee berry borer *Hypothenemus hampei*. *Journal of Coffee Research* 24:33–35.

Barbera, N. 2001. *Diversidad de especies de hormigas en sistemas agroforestales contrastantes en café en Turrialba, Costa Rica.* M.S. thesis, CATIE, Turrialba, Costa Rica.

Beer, J., R. Muschler, D. Kass, and E. Somarriba. 1998. Shade management in coffee and cacao plantations. *Agroforestry Systems* 38:139–164.

Bheemaiah, M. M., and M. Shariff. 1989. Multiple cropping in coffee. *Indian Coffee* 53:9–13.

Bolaños, M. M., C. A. Rivilla, and S. Suarez. 2000. Identificación de micorrizas arbusculares en suelos de la zona cafetera colombiana *Cenicafé* 51:245–262.

Bonilla, G. 1999. *Tipologías cafetaleras en el Pacífico de Nicaragua.* M.S. thesis, CATIE, Turrialba, Costa Rica.

Botero, J. E., and P. S. Baker. 2001. Coffee and biodiversity, a producer-country perspective. Pages 94–103 in P. S. Baker (ed.), *Coffee futures: a source book of some critical issues confronting the coffee industry.* Wallingford, UK: CAB International.

Bourke, R. M. 1985. Food, coffee and casuarina: an agroforestry system from the Papua New Guinea highlands. *Agroforestry Systems* 2:273–279.

Bray, D. B., J. L. Plaza-Sánchez, and E. Contreras-Murphy. 2002. Social dimensions of organic coffee production in Mexico: lessons for eco-labeling initiatives. *Society and Natural Resources* 15:429–446.

Bugg, R., and C. Waddington. 1994. Using cover crops to manage arthropod pests of orchards: a review. *Agriculture, Ecosystems and Environment* 50:11–28.

Caramori, P. H., A. Androcioli Filho, and A. C. Leal. 1996. Coffee with *Mimosa scabrella* Benth for frost protection in southern Brazil. *Agroforestry Systems* 33:205–214.

Cardoso, C. F. S., and H. Pérez. 1977. *Centroamérica y la economía occidental (1520–1930).* San José, Costa Rica: Editorial Universidad de Costa Rica.

Carvalho, A. 1988. Principles and practice of coffee plant breeding for productivity and quality factors: *Coffea arabica*. Pages 129–165 in R. J. Clarke and R. Macrae (eds.), *Coffee*, Vol. 4: *Agronomy.* London: Elsevier.

Chacko, M. J. 1979. Natural enemies of coffee pests. *Plant Protection Bulletin* 31:63–68.

Chamorro, G., A. Gallo, and R. López. 1994. Evaluación económica del sistema agroforestal café asociado con nogal. *Cenicafé* 45:164–170.

Charrier, A., and J. Berthaud. 1985. Botanical classification of coffee. Pages 13–47 in M. N. Clifford and K. C. Wilson (eds.), *Botany, biochemistry and production of beans and beverage.* New South Wales, Australia: Croom Helm Ltd.

Clarence-Smith, W. G. 1998. The coffee crisis in Asia, Africa and the Pacific, c. 1870 to c. 1914. In *Conference on Coffee Production and Economic Development, c. 1700 to c. 1960.* September 10–12, 1998, Oxford, UK.

Colwell, R. K. 1997. *EstimateS: statistical estimation of species richness and shared species from samples*. Version 5. User's guide and application. Online: http://viceroy.eeb.uconn.edu/estimates.

Cook, O. F. 1901. *Shade in coffee culture*. Bulletin 25. Washington, DC: U.S. Department of Agriculture.

Couturon, E., P. Lashermes, and A. Charrier. 1998. First intergeneric hybrids (*Psilanthus ebracteolatus* Hiern x *Coffea arabica* L.) in coffee trees. *Canadian Journal of Botany* 76:542–546.

de Namur, C., E. Couturon, P. Sita, and F. Anthony. 1987. Résultats d'une mission de prospection des caféiers sauvages du Congo. Pages 397–404 in *Proceedings of the 12th International Scientific Colloquium on Coffee, Montreux*. Vevey, Switzerland: ASIC.

de Silva, N. T. M. H., and C. A. Tisdell. 1990. Evaluating techniques for weed control in coffee in Papua New Guinea. *International Tree Crops Journal* 6:31–49.

Dhanam, M., T. Raju, and P. K. Bhat. 1992a. *Tetrastichus* sp. nr. xyleboroum Domenichini (Hymenoptera, Eulophidae), a natural enemy of *Xylosandrus compactus* (Eichhoff). *Journal of Coffee Research* 22:127–128.

Dhanam, M., M. Ramachandran, and P. K. Bhat. 1992b. A note on *Xylotrechus suscutellatus* Chevrolat attacking *Trema orientalis*. *Journal of Coffee Research* 22:129–130.

Díaz-Hernández, L. E. 1983. *Castañer: una hacienda cafetalera en Puerto Rico (1868–1930)*. Rio Piedras, Puerto Rico: Editorial Edil.

Duque, J. P. 1938. Costa Rica, Nicaragua, El Salvador y Guatemala: informe del jefe del Departamento Técnico sobre su viaje de estudio a algunos países cafeteros de la América Central. *Revista Cafetera de Colombia* 7:2295–2460.

Escalante, E., A. Aguilar, and R. Lugo. 1987. Identificación, evaluación y distribución espacial de especies utilizadas como sombra en sistemas tradicionales de café (*Coffea arabica*) en dos zonas del estado de Trujillo, Venezuela. *Venezuela Forestal* 3:50–62.

Escalante, M. 2000. *Diseño y manejo de cafetales del occidente de El Salvador*. M.S. thesis, CATIE, Turrialba, Costa Rica.

Espinoza, L. 1985. *Untersuchungen uber die Bedeutung der Baumkomponente bei agroforstwirtschaftlichem Kaffeeanbau an Beispielen aus Costa Rica*. Ph.D. dissertation, University of Göttingen, Germany.

Estrada, A., E. Coates-Estrada, and D. Merrit Jr. 1993. Bat species richness and abundance in tropical rainforest fragments and in agricultural habitats at Los Tuxtlas, Mexico. *Ecography* 16:309–318.

Fernando, R. 1998. The effects of coffee cultivation on peasantry in Java, 1830–1970. In *Conference on Coffee Production and Economic Development, c. 1700 to c. 1960*. September 10–12, 1998, Oxford, UK.

Gallina, S., S. Mandujano, and A. Gonzalez-Romero. 1996. Conservation of mammalian biodiversity in coffee plantations of Central Veracruz, Mexico. *Agroforestry Systems* 33:13–27.

Garcia, A., B. Decazy, and C. Alauzet. 1995. Bioecología de los pseudococcidae, parásitos del cafeto en Guatemala. In *XV Simposio sobre Caficultura Latinoamerican*, Vol. 2. San José, Costa Rica: IICA-PROMECAFE.

Garcia, N. 1988. *Cafetales y café*. Caracas, Venezuela: Ministerio de Agricultura y Cria.

Gobbi, J. A. 2000. Is biodiversity-friendly coffee financially viable? An analysis of five different coffee production systems in western El Salvador. *Ecological Economics* 33:267–281.

Godoy, R., and C. Bennett. 1989. Diversification among coffee smallholders in the highlands of South Sumatra, Indonesia. *Human Ecology* 16:397–420.

Goldman, A., and J. Kigel. 1986. Dynamics of the weed community in coffee plantations grown under shade trees: effect of clearing. *Israel Journal of Botany* 35:121–131.

Gonzalez, J. A. 1999a. Diversidad y abundancia de aves en cafetales con y sin sombra, Heredia, Costa Rica. *Ciencias Ambientales* 17:70–81.

Gonzalez, J. A. 1999b. *Impacto económico-ecológico del manejo de los cafetales con y sin sombra para la conservación de aves e insectos en la III zona cafetalera de Heredia, Costa Rica.* Master's thesis, Universidad Nacional, Heredia, Costa Rica.

Gowda, D. K. S., M. G. Venkatesha, and P. K. Bhat. 1995. Preliminary observations on the incidence of termites on coffee and its shade trees. *Journal of Coffee Research* 25:30–34.

Granados, D., and J. Vera. 1995. El sistema agroforestal cafetalero en Córdoba, Veracruz. *Revista Chapingo de Ciencias Forestales* 1:97–108.

Greenberg, R., R. Bichier, and J. Sterling. 1997. Bird populations in rustic and planted shade coffee plantations of Eastern Chiapas, Mexico. *Biotropica* 29:501–514.

Gutierrez, G., and B. Soto. 1976. Arboles usados como sombra en café y cacao. *Revista Cafetalera (Guatemala)* 159:27–32.

Guyot, B., D. Guelule, J. C. Manez, J. J. Perriot, J. Giron, and L. Villain. 1996. Influence de l'altitude et de l'ombrage sur la qualité des cafés Arabica. *Plantations, Récherche et Développement* 20:272–280.

Haile-Mariam, T. 1973. *The production, marketing, and economic impact of coffee in Ethiopia.* Ph.D. dissertation, University of Stanford, CA.

Hearst, H. L. 1929. *The coffee industry of Central America.* M.S. thesis, Geography, University of Chicago.

Ibarra-Nuñez, G. 1990. Los artrópodos asociados a cafetos en un cafetal mixto del Soconusco, Chipas, Mexico. *Folia Entomológico Mexicana* 79:207–231.

Izard, M. 1973. *El café en la economía Venezolana del XIX.* Valencia, Spain: Vadell Hermanos.

Jimenez-Avila, E. 1979. Estudios ecológicos del agroecosistema cafetalero. I. Estructura de los cafetales de una finca cafetalera de Coatepe, Mexico. *Biotica* 4:1–12.

Johanneson, N. E., and A. Mansingh. 1984. Host-pest relationships of the genus *Hypothenemus* (Coleoptera, Scolytidae) with special reference to the coffee berry borer *H. hampei.* *Journal of Coffee Research* 14:43–56.

Kashyap, R. K., A. N. Verma, and J. P. Bhanot. 1984. Termites of plantation crops, their damage and control. *Journal of Plantation Crops* 12:1–10.

Koptur, S. 1983. *Inga.* Pages 259–261 in D. H. Janzen (ed.), *Costa Rica natural history.* Chicago: University of Chicago Press.

Korikanthimath, V. S., R. Hedge, M. N. Menugopal, K. Sivaraman, and B. Krishnamurthy. 1994. Multistoreyed cropping system with coffee, clove and pepper. *Indian Coffee* 58:3–5.

Kumar, A. C., and S. D. Samuel. 1990. Nematodes attacking coffee and their management: a review. *Journal of Coffee Research* 20:1–27.

Kumar, P. K. V., M. M. Balakrishnan, and T. S. Govindarajan. 1984. *Retithrips syriacus* (Mayet): a new record on coffee from south India. *Journal of Coffee Research* 14:131–132.

Kumar, P. K. V., C. B. Prakasan, and C. K. Vijayalakshmi. 1994. Record of entomopathogenic fungi on *Hypothenemus hampei* Ferrari from South India. *Journal of Coffee Research* 24:119–120.

Kurian, R. 1998. Land, labor and capital in the coffee plantations in nineteenth century South-West India. In *Conference on Coffee Production and Economic Development, c. 1700 to c. 1960.* September 10–12, 1998, Oxford, UK.

Laborie, P. J. 1797. *The coffee planter of Saint Domingo*. London: Cabell and Davis.

Lagemann, J., and J. Heuveldop. 1983. Characterization and evaluation of agroforestry systems: the case of Acosta-Puriscal, Costa Rica. *Agroforestry Systems* 1:101–115.

Lashermes, P., M. C. Mombes, P. Topart, G. Graziosi, B. Bertrand, and F. Anthony. 2000. Molecular breeding in coffee (*Coffea arabica* L.). Pages 134–146 in T. Sera, C. R. Soccol, A. Pandey, and S. Roussos (eds.), *Coffee biotechnology and quality*. Dordrecht, the Netherlands: Kluwer.

Lawrence, A. 1995. Farmer knowledge and use of *Inga* species. Pages 142–151 in D. O. Evans and L. R. Szott (eds.), *Nitrogen fixing trees for acid soils*. Morrilton, AR: Winrock International and Nitrogen Fixing Tree Association.

Lawrence, A., and R. A. Zuñiga. 1996. *The role of farmer's knowledge in agroforestry development: a case study from Honduras and El Salvador*. AERDD Working Paper 96-5. Reading, UK: University of Reading.

Le Pelley, R. H. 1968. *Pests of coffee*. London: Longman.

Linkimer, M. 2001. *Árboles nativos para diversificar cafetales en la zona Atlántica de Costa Rica*. M.S. thesis, CATIE, Turrialba, Costa Rica.

Llanderal, T. 1998. *Diversidad del dosel de sombra en cafetales de Turrialba, Costa Rica*. M.S. thesis, CATIE, Turrialba, Costa Rica.

Llorente-Bousquets, J., A. L. Martinez, I. Vargas-Fernandez, and J. Soberón-Mainero. 1996. Papilionoidea (Lepidoptera). Pages 531–548 in J. Llorente-Bousquets, A. García-Aldrete, and E. Bonzalez-Soriano (eds.), *Biodiversidad, taxonomía y biogeografía de artrópodos de México*. Mexico City: Universidad Nacional Autónoma de México.

Lowndes, J. 1807. *The coffee planter, or an essay on the cultivation and manufacturing of the article of West-India produce*. London: C. Lowndes.

Marten, G. G., and L. A. Sancholuz. 1981. Estudio ecológico de las zonas cafetaleras de Veracruz, Puebla, Hidalgo y Tamaulipas. Evaluación estadística de los muestreos. *Biótica* 6:7–32.

Martinez, E., and W. Peters. 1996. La caficultura biológica: la finca Irlanda como estudio de caso de un diseño agroecológico. Pages 159–183 in J. Trujillo, F. de Leon-González, R. Calderon, and P. Torres-Lima (eds.), *Ecología aplicada a la agricultura: temas selectos de México*. Mexico City: Universidad Autónoma Metropolitana.

Michon, G., F. Mary, and J. Bompard. 1986. Multistoried agroforestry garden systems in West Sumatra, Indonesia. *Agroforestry Systems* 3:315–338.

Moguel, R., and V. M. Toledo. 1999. Biodiversity conservation in traditional coffee systems of Mexico. *Conservation Biology* 13:11–21.

Monro, A., D. Alexander, J. Reyes, M. Renderos, and N. Ventura. 2002. *Árboles de los Cafetales de El Salvador*. London: The Natural History Museum.

Monro, A., J. Monterrosa, N. Ventura, D. Godfrey, D. Alexander, and M. C. Peña. 2001. *Helechos de los cafetales de El Salvador*. London: The Natural History Museum.

Mora-Delgado, J., and L. Acosta. 2001. Uso, clasificación y manejo de la vegetación asociada al cultivo de café (*Coffea arabica*) desde la percepción campesina en Costa Rica. *Agroforestería en las Américas* 8:20–27.

Moron, M. A., and J. A. Lopez-Mendez. 1985. Análisis de la entomofauna necrófila de un cafetal de Soconusco, Chiapas, Mexico. *Folia Entomológica Mexicana* 63:47–59.

Mussak, M. F., and J. G. Laarman. 1989. Farmers' production of timber trees in the cacao-coffee region of coastal Ecuador. *Agroforestry Systems* 9:155–170.

The Natural History Museum. 2002. *Guía para la identificación de las pimplinae de cafetales bajo sombra de El Salvador (Hymenoptera: Ichneumonidae)*. London: The Natural History Museum.

Nestel, D. 1995. Coffee in Mexico: international market, agricultural landscape and ecology. *Ecological Economics* 15:165–178.

Nestel, D., and M. Altieri. 1992. The weed community of Mexican coffee agroecosystems: effect of management upon plant biomass and species composition. *Acta Ecologica* 13:715–726.

Nestel, D., F. Dickschen, and M. A. Altieri. 1993. Diversity patterns of soil macro Coleoptera in Mexican shaded and unshaded coffee agroecosystems: an indication of habitat perturbation. *Biodiversity and Conservation* 2:70–78.

Nestel, D., F. Dickschen, and M. A. Altieri. 1994. Seasonal and spatial population loads of a tropical insect: the case of the coffee leaf-miner in Mexico. *Ecological Entomology* 19:159–167.

Njoroge, J. M., and J. K. Kimemia. 1993. Current intercropping observations and future trends in arabica coffee, Kenya. *Outlook on Agriculture* 22:43–48.

Odoul, P. A., and J. R. W. Aluma. 1990. The banana (*Musa* spp.)–*Coffea robusta:* traditional agroforestry system of Uganda. *Agroforestry Systems* 11:213–226.

Perfecto, I., R. Rice, R. Greenberg, and M. E. van Der Voorst. 1996. Shade coffee: a disappearing refuge for biodiversity. *BioScience* 46:598–608.

Perfecto, I., and R. Snelling. 1995. Biodiversity and the transformation of a tropical agroecosystem: ants in coffee plantations. *Ecological Applications* 5:1084–1097.

Perfecto, I., and J. Vandermeer. 1994. Understanding biodiversity loss in agroecosystems: reduction of ant diversity resulting from transformation of the coffee agroecosystem in Costa Rica. *Entomology (Trends in Agricultural Sciences)* 2:7–13.

Perfecto, I., and J. Vandermeer. 1996. Microclimatic changes and the indirect loss of ant diversity in a tropical agroecosystem. *Oecologia* 108:577–582.

Perfecto, I., J. Vandermeer, P. Hanson, and V. Cartin. 1997. Arthropod biodiversity loss and the transformation of a tropical agroecosystem. *Biodiversity and Conservation* 6:935–945.

Picó, F. 1983. *Libertad y servidumbre en el Puerto Rico del siglo XIX.* Rio Piedra, Puerto Rico: Ediciones Huracán.

Prakasan, C. B., K. Sreedharan, and P. K. Bhat. 1986. New record of a parasite of coffee white stem-borer *Xylotrechus quadripes* Chevr. from India. *Journal of Coffee Research* 16:38–40.

Rangeshwaran, R., W. K. Rao, and P. K. Ramaiah. 1990. Vesicular arbuscular mycorrhiza in coffee. *Journal of Coffee Research* 20:55–68.

Rao, H. H. 1975. Diversification in coffee. *Indian Coffee* 39:16–18.

Reddy, A. G. S., and I. V. A. Rao. 1999. Coffee as an intercrop to coconut in the plains of Karnataka. *Indian Coffee* 63:3–6.

Reddy, A. G. S., and D. S. Reddy. 1980. Coffee weeds in Andhra. *Indian Coffee* 44:79–82.

Reddy, A. G. S. M., K. V. V. S. N. Raju, and K. C. Naidu. 1982. Coffee shade in Andhra. *Indian Coffee* 56:337–341.

Rice, R. 1991. Observaciones sobre la transición en el sector cafetalero en centroamérica. *Agroecologia Neotropical* 2:1–6.

Rice, R. A., and R. Greenberg. 2000. Cacao cultivation and the conservation of biological diversity. *Ambio* 29:167–173.

Rios de Hernandez, J. 1988. *La hacienda Venezolana.* Caracas, Venezuela: Tropykos.

Roberts, D. L., R. J. Cooper, and L. J. Petit. 2000a. Flock characteristics of ant-following birds in premontane moist forest and coffee agroecosystems. *Ecological Applications* 10:1414–1425.

Roberts, D. L., R. J. Cooper, and L. J. Petit. 2000b. Use of premontane moist forest and shade coffee agroecosystems by army ants in Western Panama. *Conservation Biology* 14:192–199.

Rojas, J., C. Godoy, P. Hanson, and L. Hilje. 2001a. A survey of homopteran species (Auchenorrhyncha) from coffee shrubs and poró and laurel trees in shaded coffee plantations, in Turrialba, Costa Rica. *Revista de Biología Tropical* 49:1057–1065.

Rojas, J., C. Godoy, P. Hanson, C. Kleinn, and L. Hilje. 2001b. Hopper (Homoptera: Auchenorrhyncha) diversity in shaded coffee systems of Turrialba, Costa Rica. *Agroforestry Systems* 53:171–177.

Roque, J. de la. 1988. Voyage à l'Arabie hereuse. Pages 238–239 in G. Wrigley (ed.), *Coffee*. New York: Longman.

Roubik, D. W. 2002. The value of bees to the coffee harvest. *Nature* 417:708.

Salazar, R. 1985. Producción de leña y biomasa de *Inga densiflora* Benth en San Ramón, Costa Rica. *Silvoenergía* (Costa Rica) 3:1–4.

Samper, M. 1994. Café, trabajo y sociedad en Centroamérica (1870–1930): una historia común y divergente. Pages 11–110 in *Historia general de Centroamérica*, Vol. 4, San José, Costa Rica: Flasco.

Samuel, S. D., and P. K. Bhat. 1988. A note on *Terastia meticulosalis* Guenee attacking dadad (*Erythrina* spp.) on Pulney Hills. *Journal of Coffee Research* 18:52–53.

Smith, R. F. 1985. A history of coffee. Pages 1–12 in M. N. Clifford and K. C. Wilson (eds.), *Botany, biochemistry and production of beans and beverage*. New South Wales, Australia: Croom Helm Ltd.

Smith, S. D. 1998. Sugar's poor relation: coffee planting in the British West Indies, 1720–1833. *Slavery and Abolition* 19:68–89.

Smithsonian Migratory Bird Center. 1999. *El cultivo de café con sombra: criterios para cultivar un café "amistoso con las aves."* Online: http://web2.si.edu/smbc/coffee/cofcrisp.htm.

Sreedharan, K., M. M. Balakrishnan, and P. K. Bhat. 1992. *Callimerus* sp. (Coleoptera, Cleridae), a predator of the shot-hole borer, *Xylosandrus compactus* (Eichh.). *Journal of Coffee Research* 22:139–142.

Sreedharan, K., M. M. Balakrishnan, S. D. Samuel, and P. K. Bhat. 1991. A note on the association of wood boring beetles and a fungus with the death of silver oak trees on coffee plantations. *Journal of Coffee Research* 21:145–148.

Staver, C. 1999. Managing ground cover heterogeneity in perennial crops under trees: from replicated plots to farmer practice. Pages 67–96 in L. E. Buck, J. Lassoie, and E. C. M. Fernandes (eds.), *Agroforestry in sustainable agricultural systems*. New York: CRC Press.

Staver, C., F. Guharay, D. Monterroso, and R. Muschler. 2001. Designing pest-suppressive multi-strata perennial crop systems: shade-grown coffee in Central America as a case study. *Agroforestry Systems* 53:151–170.

Teketay, D., and A. Tegineh. 1991. Shade trees of coffee in Harerge, eastern Ethiopia. *International Tree Crops Journal* 7:17–27.

Tharakan, P. K. M. 1998. Coffee, tea or pepper? Factors affecting choice of crops of agro-entrepreneurs in nineteenth century South-West India. In *Conference on Coffee Production and Economic Development, c. 1700 to c. 1960*. September 10–12, 1998, Oxford, UK.

Venkataramaiah, G. H., and P. A. Rehman. 1989. Ants associated with mealybugs of coffee. *Indian Coffee* 33:13–14.

Villatoro, R. M. 1986. *Caracterización del sistema agroforestal café-especies arbóreas en la cuenca del Río Achiguate, Guatemala.* Thesis, Universidad San Carlos, Guatemala.

Wadsworth, F. H. 1945. Forestry in the coffee region of Puerto Rico. *Caribbean Forester* 6:71–84.

Weaver, P. L., and A. Birdsey. 1986. Tree succession and management opportunities in coffee shade stands. *Turrialba* 36:47–58.

Willey, R. W. 1975. The use of shade in coffee, cocoa and tea. *Horticultural Abstracts* 45:791–798.

Wunderle, J., and S. Latta. 1996. Avian abundance in sun and shade coffee plantations and remnant pine forest in the Cordillera Central, Dominican Republic. *Ornitología Neotropical* 7:19–34.

Wunderle, J., and S. Latta. 1998. Avian resource use in Dominican shade coffee plantations. *Wilson Bulletin* 110:271–281.

Wunderle, J., and S. Latta. 2000. Winter site fidelity of nearctic migrants in shade coffee plantations of different sizes in the Dominican Republic. *The Auk* 117:596–614.

Yépez, C. 2001. *Selección de árboles para sombra en cafetales diversificados de Chiapas.* M.S. thesis, CATIE, Turrialba, Costa Rica.

Zuñiga, C. 2000. *Tipologías cafetaleras y desarrollo de enfermedades en los cafetales de la Reserva Natural Miraflor-Moropotente, Estelí, Nicaragua.* M.S. thesis, CATIE, Turrialba, Costa Rica.

Chapter 10

Complex Agroforests: Their Structure, Diversity, and Potential Role in Landscape Conservation

Götz Schroth, Celia A. Harvey, and Grégoire Vincent

Complex agroforests (or simply agroforests) are understood here as a special type of agroforestry system characterized by a forest-like structure and significant plant diversity in which useful tree and tree crop species attain substantially greater density, compared with the natural forest, through planting, selection, and management of useful species from spontaneous regeneration. Agroforests are the most forest-like in their structure and appearance of all agroforestry systems, and some of them may be easily mistaken for natural forest if seen from a distance (see Figure I.3 in the Introduction).

Agroforests occur in all tropical regions and can be based on many different tree crop species. The principal tree crops in agroforests typically are either shade-tolerant subcanopy species, such as cocoa (*Theobroma cacao*), tea (*Camellia sinensis*), and coffee (*Coffea* spp.; see also Chapter 9, this volume), or canopy species, such as rubber (*Hevea brasiliensis*), damar (*Shorea javanica*, a resin-producing dipterocarp species), or durian (*Durio zibethinus*, a highly valued fruit species of Southeast Asia). Species from both groups often are associated in the same system. Of course, the tree crops that are grown in complex agroforests can also be cultivated in other systems of varying complexity and diversity, ranging from monocultures, such as unshaded and clean-weeded plantations, through simple associations of a few tree crops and shade trees (simple multistrata systems), to complex agroforests and extractively used natural forests (Figure 10.1).

Agroforests based on subcanopy species usually are established by selectively clearing natural forest and underplanting it with tree crops. In contrast, agroforests based on canopy trees often depart from a clear-felled, slash-and-burn plot into which tree crops are planted together with food crops; these agroforests develop their forest structure through the association of different

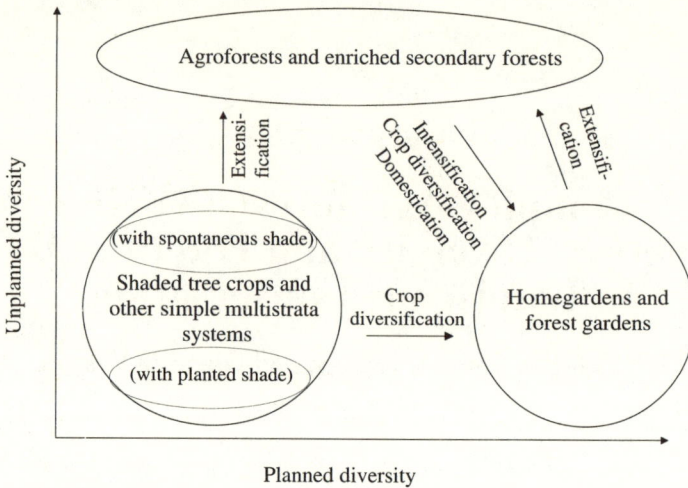

Figure 10.1. Relationships between different types of multistrata agroforestry systems on schematic gradients of planned and unplanned diversity and the types of management through which one system type may be converted into another. Extensification may include temporary abandonment.

tree crop species and selective tolerance of spontaneous regrowth, especially of useful species. Complex agroforests thus are the result of an increase in the density of valuable species that are already present and introduction of new valuable species into the framework of disturbed primary or regenerating secondary forests.

Agroforests can be distinguished from extractively used natural forests by a substantially higher density of useful tree species and the concomitant restructuring of the original vegetation (H. de Foresta, pers. comm., 2001). However, the limits between "pure" extractivism, as practiced in natural forest, and complex agroforest cultivation are flexible. The typically high plant diversity of tropical forests implies that most species, including those sought by extractivists, usually occur at low densities, with the exception of oligarchic forests that develop under specific conditions and may have a high value for extractivism (Peters et al. 1989). Efforts to increase the density and development of these useful plant species in natural forest therefore are a typical component of extractivism. For example, in the Amazon forest, extractively used natural rubber groves traditionally have been enriched through planting of rubber seeds or seedlings to counteract their decline (Dean 1987), and regeneration management also seems to be an essential feature of the formation of Brazil nut (*Bertholletia excelsa*) groves (Pereira 2000). Enrichment of natural forest with useful tree species, especially palms, in groves near campsites was the earliest form of agriculture in some forest regions, such as the northwestern Amazon (Politis 2001) and Indonesia (Michon and de Foresta 1999), suggesting that this type of agroforest historically has preceded slash-and-burn agriculture in some regions.

Some tree-dominated homegardens share with agroforests a high diversity of plant species and multilayered structure (see Figure I.4 in the Introduction). Homegardens typically are of small size, close to a homestead, and intensively managed. Their multilayered structure and high diversity result from the association of many useful trees, palms, shrubs, and herbaceous species that occupy different canopy positions. Their high diversity of planted and domesticated species (planned diversity) and usually intensive management contrast with the high diversity of wild species (unplanned diversity), resulting from extensive management or even temporary abandonment, in typical agroforests. The distinction between homegardens and agroforests is not clear-cut, however; in fact, homegardens (or forest gardens if more distant from the home) can be seen as a potential endpoint of an intensification and domestication trajectory of agroforests (Figure 10.1).

With their dominant tree cover, high plant diversity, structural complexity, and extensive management, complex agroforests often are of special interest for the conservation management of tropical forest landscapes. Not only may they harbor important on-site diversity of plant and animal species (Michon and de Foresta 1999), but they also often border forest areas, buffering them from the more intensively used agricultural surroundings and providing effective wildlife corridors (see Chapters 2 and 3, this volume). In this chapter, we review the actual and potential contribution of complex agroforests to tropical biodiversity conservation, including information on their geographic distribution, spatiotemporal association with other land uses, and structural attributes and species composition. We focus on cocoa and rubber agroforests, for which most information is available. Coffee-based agroforests are discussed in Chapter 9, this volume.

Role of Complex Agroforests in Tropical Landscapes and Farming Systems

Agroforests are important land uses in several parts of the tropics, occupying large areas of land and providing important sources of income for local people. Because the significance of agroforests for the ecology of tropical landscapes and the livelihoods of their inhabitants has been recognized only recently, and agroforests often are difficult to distinguish from secondary forest on aerial photographs and satellite images, quantitative data on their extent are available only for a few regions.

The jungle rubber systems of the lowlands of Sumatra and Borneo, in which rubber trees are grown in a secondary forest environment, cover 2.5–3 million ha, whereas fruit-dominated forest gardens cover hundreds of thousands of hectares and damar gardens cover about 50,000 ha on Sumatra. These agro-forests provide about 80 percent of the rubber, 95 percent of the fruits, and 80 percent of the dipterocarp resins produced in Indonesia and substantial amounts of bamboo, rattan, firewood, and medicinal plants (Michon and de Foresta 1999; de Foresta and Boer 2000). Rubber-based agroforests have been estimated to provide income for at least 5 million people in Indonesia (Gouyon et al. 1993). However, these extensive land use systems come under increasing pressure from more intensive land use forms; for example, rubber agroforests in lowland Sumatra are being replaced by monoculture plantations of oil palm (*Elaeis guineensis*) and rubber (Gouyon et al. 1993). Cocoa agro-forests cover 300,000 to 400,000 ha in southern Cameroon, where some 400,000 households depend on them for their income and food (Sonwa et al. 2001). Somarriba et al. (Chapter 9, this volume, citing Villagren and Boan-erges 1989) report that in El Salvador about 80 percent of the remaining for-est cover is actually shade-grown coffee. The distribution of cocoa agroforests in Bahia, Brazil, is discussed later in this chapter (see also Figure 10.8). In other parts of the world, agroforests may also constitute major land uses, but data are not readily available on their abundance and distribution.

The earliest forms of agroforests, at least in certain regions, apparently were patches where the forest was enriched with useful tree species by transfer of seeds and vegetative material from surrounding areas to places close to camp-sites, creating islands of increased productivity in a forested landscape (Politis 2001). Today, agroforests usually are a component of more thoroughly human-modified landscapes, in which they are associated with annual crop-ping systems such as lowland rice or upland slash-and-burn plots, home-gardens, pastures, fallows, perennial crop plantations, and often remnants of primary and secondary forest.

Agroforests often occupy a transitional place in landscapes between inten-sively used agricultural land and natural forest. Figure 10.2 shows a transect from a river valley to the mountains at the border of the Kerinci-Seblat National Park in Sumatra, Indonesia (Murniati et al. 2001). The river valley is occupied by human settlements and irrigated rice fields, and complex agro-forests form the transition to shrubland and community ("nagari") forest on the mountain slopes, which merge into the national park forest at higher ele-vations. These agroforests are composed of an upper story dominated by rub-ber, durian, jengkol (*Archidendron* [syn. *Pithecellobium*] *jiringa*), and coconut (*Cocos nucifera*) trees and a lower story of coffee (*Coffea canephora*) and cinna-mon (*Cinnamomum burmanii*) trees. As is often the case in Southeast Asia, where lowland rice is the staple food crop, tree-based systems occupy slope positions that are not suitable for rice cultivation (see also Figure I.3 in the Introduction). This gradient from rice cultivation and villages in the valley, through agroforests on the slopes, to forest on the mountaintops has also been described in the Lake Maninjau region in western Sumatra (Michon et al.

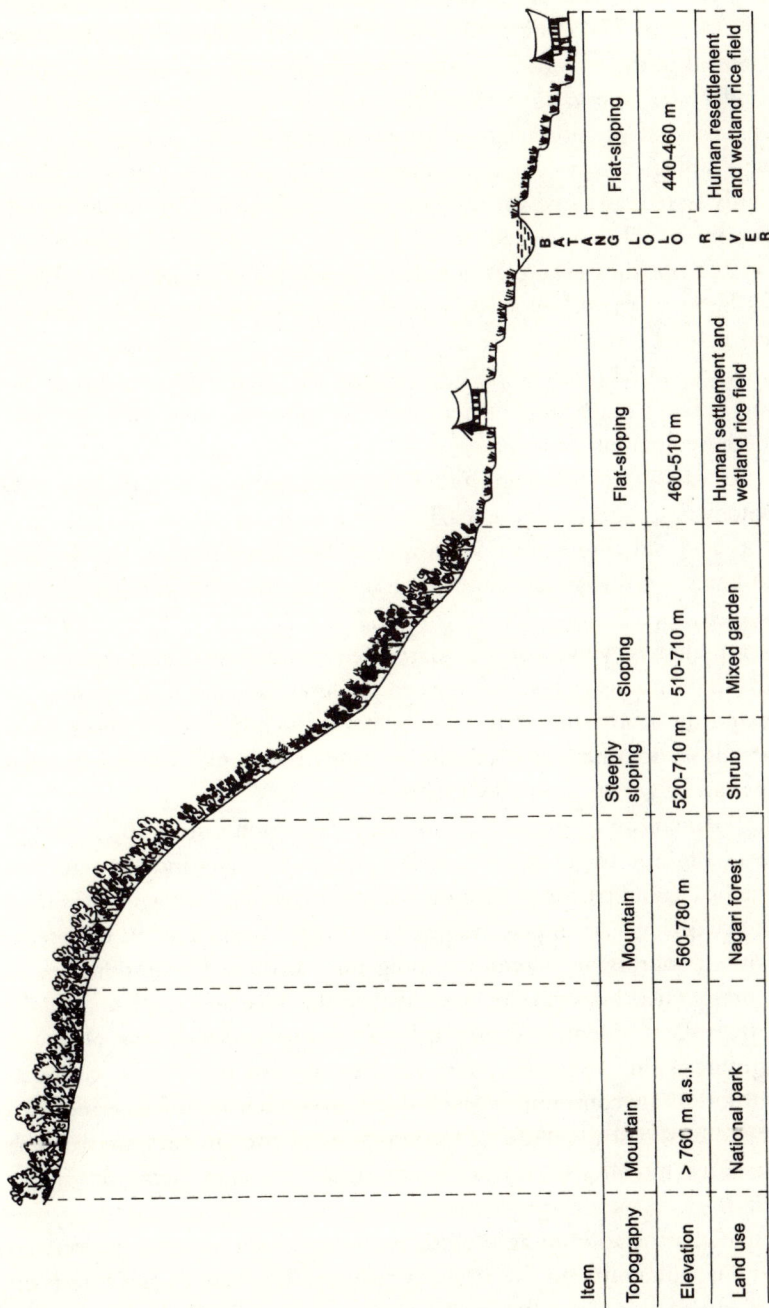

Item					BATANG LOLO RIVER	
Topography	Mountain	Mountain	Steeply sloping	Sloping	Flat-sloping	Flat-sloping
Elevation	> 760 m a.s.l.	560-780 m	520-710 m	510-710 m	460-510 m	440-460 m
Land use	National park	Nagari forest	Shrub	Mixed garden	Human settlement and wetland rice field	Human resettlement and wetland rice field

Figure 10.2. Transect of a landscape at the border of Kerinci-Seblat National Park in Sumatra (from Murniati et al. 2001, with permission).

1986). In this volcanic region, the permanent soil cover by the agroforests on the mountain slopes is particularly important for protecting the villages from landslides, to which these soils are particularly susceptible. In three villages, agroforests covered 22–63 percent of the agricultural area and accounted for 26–80 percent of the income from agricultural produce (Michon et al. 1986).

A similar land use gradient has been described in mountain villages in northern Thailand where miang tea for chewing is traditionally grown in an extensive "jungle tea" system (Preechapanya 1996; Preechapanya et al. in press). This system is based on the enrichment of the hill evergreen forests with tea trees, which may be associated with other crops and sometimes cattle. As in the previous examples, a belt of jungle tea occupies the midslope positions between the valleys, where settlements and homegardens are situated, and natural forest grows at higher elevations. The transition from the agroforests to natural forest is gradual because tea often is planted inside the forest and because some jungle tea areas revert to forest when they are abandoned because of labor shortages.

In other cases agroforests and other land uses occur in a more heterogeneous patchwork pattern. Figure 10.3 shows a land use mosaic of slash-and-burn plots and old (more than 50 years) rubber agroforests in the Tapajós National Forest on the margins and slope of a plateau a few kilometers from the Tapajós River, a major southern tributary of the Amazon. This small-scale land use mosaic has evolved on the plateau edge, which is characterized by a narrow band of humus-rich soils, between the ferralitic soils of the plateau on one side and the sandy soils of the riverbank on the other side. Rubber agroforests are also a dominant land use on the sandy riverbanks, where few other tree crops can be grown profitably (Schroth et al. 2003).

In the Talamancan region of Costa Rica, cocoa agroforests occur as part of a patchwork of forest fragments, small plantations of banana and plantain, rice and maize fields, and pastures, with individual cocoa plots averaging less than 2 ha and occupying a small percentage of the overall landscape but representing important sources of income for indigenous people. The importance of the agroforests varies between the lowlands and the hillsides (less than 300 m). Land in the valleys is dedicated primarily to bananas and plantains, although some agroforests and remnant forest patches are also present. In contrast, cocoa agroforests are common on the hillsides and often are interspersed with secondary forests, pastures, areas of forest regrowth, and annual crop fields. A similar land use mosaic with cocoa agroforests also exists in Bahia, Brazil (see Figure 10.8).

The spatial association of agroforests with other land use types reflects on one hand their suitability for the respective site conditions (e.g., agroforests on slopes and lowland rice on valley bottoms) and on the other their complementary role in the prevailing farming systems (Dove 1993). In farming systems composed of slash-and-burn agriculture and agroforests, the former provides

Figure 10.3. Slash-and-burn plot surrounded by old rubber agroforests in the buffer zone of the Tapajós National Park, Brazilian Amazon (from Schroth and Sinclair 2003, with permission).

most of the staple food such as rice, maize, or cassava, whereas the latter provides cash income (e.g., coffee, cocoa, rubber), fruits, timber, firewood, and medicinal products. The relative importance of the agroforests in providing such products depends on their composition, the availability of other sources of timber and nontimber forest products such as fallows and natural forest, and household needs. For example, in the Tapajós region, leaves of the understory palm *Attalea spectabilis* (curuá) that are used for roofs are an important byproduct of certain rubber agroforests, and the agroforests that are more distant from the villages are also used for hunting by many farmers (Schroth et al. 2003). Trees with medicinal properties may be retained or specifically planted, such as Brazil nut (whose bark has medicinal properties), *Himatanthus sucuuba,* a common tree in rubber agroforests on sandy soils at the Tapajós (Schroth et al. 2003), or *Alstonia boonei* and *Voacanga africana* in cocoa agroforests in Cameroon (Sonwa et al. 2000b).

Agroforests can also be a significant source of basic food such as plantains and palm fruits, such as peach palm (*Bactris gasipaes*) in Latin America and oil palm (*Elaeis guineensis*) in West Africa. They also serve as investments; for example, in Sumatra timber or cinnamon trees occasionally may be harvested as a source of larger sums of cash (Michon and de Foresta 1999; Burgers and William 2000). Especially in heavily overlogged landscapes, agroforests may contain valuable trees of species that have become rare in surrounding forests. Although it is formally illegal to cut them, rare trees such as *Dalbergia nigra* and *Cedrela odorata* from cocoa agroforests in Bahia, Brazil, have recently been sold for timber because of low cocoa prices (Johns 1999), and the same has occurred in the Côte d'Ivoire (see Chapter 6, this volume). In Talamanca, Costa Rica, many farmers retain *Cordia alliodora* trees in their cocoa agroforests to provide timber for house construction.

The association between agroforests and annual cropping systems is not only spatial but very often also temporal. In many cases, the establishment of an agroforest follows a successional process, starting with food crops and short-lived perennials, which are gradually replaced by longer-living trees. In this process, the food crops fulfill various ecological and economic functions: they sustain the farmer in the first years when the tree crops are not yet producing, give an early return to the investments for clearing and weeding the plot, cover the soil, use the nutrients released by burning and clearing, and reduce weed growth. In the Indonesian jungle rubber systems, the main staple food crop, rice, has the advantage of not being susceptible to fungal root rots, thereby helping to sanitize the land (Berry 2001). Food crops may also provide temporary shade for sensitive tree crop seedlings, as is the case with bananas and taro (*Xanthosoma* sp., *Colocasia* sp.) in West African and Central American cocoa and coffee plantations.

Cyclic and Permanent Agroforests

Although most agroforests are initially established after clearcutting or selective clearing of the original forest vegetation, some are, in principle, permanent systems that are continuously renovated in a small-scale pattern of replanting and spontaneous regeneration, whereas others undergo cycles of distinct management phases, including periodic replanting, which often involves a slash-and-burn phase. In cyclic systems, replanting often is preceded by a period of extensive management or abandonment, when the productivity of the aging agroforest has become low but the land is not urgently needed or resources for replanting (especially labor) are not available. Such extensively managed or abandoned agroforests gradually turn into secondary forests, which may eventually be clearcut or underplanted with tree crops again. There are thus parallels between cyclic agroforests with tree crops and shifting cultivation systems based on annual crops in that phases of establishment, more intensive management, and extensive management or fallowing alternate on the same site (although in the case of agroforests this may occur over several decades) and form mosaics of different land use phases within a landscape. Annual and semiperennial food crops are conveniently integrated into the renovation phase of agroforests, and in some cases the present, cyclic agroforest system has evolved from a shifting cultivation system by integrating tree crops into the traditional fallow.

Cyclic Agroforests

In some cyclic agroforests, the agroforest phase is a type of long-term, economically enriched fallow that alternates with slash-and-burn phases on the same piece of land. This type of system is especially developed in Indonesia, with the jungle rubber system as its most important representative (Dove 1993; Gouyon et al. 1993). There are also cyclic agroforests based on rattan (*Calamus* sp.) in Indonesia (Michon and de Foresta 1999).

The jungle rubber system was developed by farmers in Sumatra and Borneo who integrated rubber trees into their fallow rotations after the introduction of rubber around 1910 (Dove 1993; Gouyon et al. 1993; de Jong 2001). Rubber seedlings are planted in slash-and-burn fields shortly after rice is planted (initially, it seems that rubber seeds were used; H. de Foresta, pers. comm., 2002). They develop together with food crops and forest regrowth and can be tapped from an age of about 10 years onward, about 3 years later than rubber trees in weeded plantations. Under the influence of intensive competition, the rubber tree density usually falls from about 1,500–2,000 seedlings per hectare at planting time to 500–600 when tapping begins. With year-round tapping, high pressure from fungal diseases of the tapping panel and root rots, their density then progressively decreases in subsequent decades, although decaying trees are also replaced to some extent by spontaneous regeneration. Once the number of productive trees in a plot no longer provides sufficient latex yield, the plot may be abandoned for a variable period of time before it is eventually clearcut, burned, and replanted to start a new cycle (Gouyon et al. 1993). In a study reported by Gouyon et al. (1993), the rubber tree density fell to about 200 trees after 40 years, at which point tapping became unprofitable; however, there is a wide variability in the age of replanting, and in some cases the cycle may extend up to 80 years (Michon and de Foresta 1999). The renovation period, during which there is no income from the rubber trees, is especially difficult for poor farmers who own only a small rubber area, so farmers may extend the productive life of their agroforests by transplanting rubber seedlings into gaps that have developed through mortality of older trees, a technique locally known as *sisipan* (Wibawa et al. 1999). A complete transition from a cyclic to a permanent system with this method is hindered by the light-demanding nature of rubber trees and consequently slow growth of the seedlings under the canopy of an existing stand (Vincent et al. in press), so ultimately most stands are renovated by slashing and burning.

Another long-term form of rubber agroforests has been described from the region of the lower Tapajós river in the central Amazon (Schroth et al. 2003). These agroforests are also established through a slash-and-burn phase after which rubber seeds or seedlings are planted into the first annual crop, which is usually cassava (*Manihot esculenta*), and develop into agroforests when after some years the farmers restrict weeding to circulation paths between the rubber trees and allow forest regrowth to develop in the remaining areas. These systems resemble in many respects the Indonesian jungle rubber; however, they are not intended as rotational but rather seen by many farmers as being permanent, although their real lifetime may be reduced by the frequent dry season fires. Indeed, 50-year-old trees often are still in good condition, and some trees have remained productive for three generations of rubber tappers and may be almost a century old (Schroth et al. 2003). The often better health of old rubber trees at the Tapajós compared with those in Indonesian jungle rubber systems results from interruption of the tapping for 4–6 months or more per year during the dry season, a tradition of abandoning the groves at times of low rubber prices (which means that few if any agroforests have been tapped for their entire life), the lower pressure from fungal root rots, and a specific tapping technique that protects the trees from fungal infections of the tapping panel in a moist forest environment (Schroth et al. 2003). Little is known about the regeneration of these systems, but it seems likely that they should be considered long-term cyclic rather than permanent systems, for the aforementioned reasons. Unfortunately, no comparative studies of biomass, structural complexity, and diversity of Amazonian and Indonesian rubber agroforests of different age are available to quantify the effects of these regional and management differences.

Benzoin (*Styrax paralleloneurum*) agroforests are a peculiar case of cyclical canopy tree–based agroforests that do not go through a slash-and-burn phase. They are found in the highlands of North Sumatra, where silvicultural practices and associated levels of plant diversity in this system have been studied by García-Fernández et al. (2003). Establishment starts with the clearing of small to medium-sized trees in secondary or primary forest and subsequent planting of about 400 benzoin seedlings per hectare in the understory. Two years later, big canopy trees are girdled to reduce shade. Tapping of the benzoin trees starts at an age of 7 years, and by year 12 all trees are tapped. After about 50 years of tapping a plot usually is abandoned because of declining production and gradually reverts to forest. These benzoin agroforests are encountered only in quite recent settlements (four to eight generations) in northern Sumatra, and the system itself probably is less than 150 years old (Katz et al. 2002). Early Dutch colonial reports mention the domestication of benzoin in the Palembang area in southern Sumatra and the lowlands of northern Sumatra. At that time, benzoin (probably *Styrax benzoin,* a different species known to be less shade tolerant) was established after land clearing and rice cultivation in a similar way to rubber trees today in rubber agroforests. In the lowlands of Sumatra, benzoin cultivation was largely replaced by rubber in the 1920s and 1930s, but *S. benzoin* is still sporadically cultivated in mixture with rubber and sometimes cocoa in North Sumatra between 500 and 800 m above sea level. At higher elevations, some benzoin (*S. paralleloneurum*) gardens were also replaced by coffee- and cinnamon-based systems in the 1970s because benzoin resin prices were low (Katz et al. 2002). The case of benzoin gardens in Sumatra further illustrates the dynamic nature of agroforests that are based on internationally traded commodities and may be abandoned or converted into other land uses as a response to fluctuations in international market prices.

Cyclic agroforests were also the traditional form of cocoa growing by indigenous forest farmers in the Côte d'Ivoire (see Chapter 6, this volume). Cocoa trees were planted together with food crops under thinned forest and were cultivated in an extensive manner under the dense shade of forest remnant trees for some 35 years, after which the groves were abandoned and reverted into forest where after some years cocoa could be conveniently replanted with the same method. This system was basically a form of shifting cultivation adapted to the needs of a perennial crop that was easiest to replant in a forest environment. After the 1960s, it was increasingly replaced in the Côte d'Ivoire by more intensive, often almost monocultural methods, although shaded cocoa agroforests can still be found in this country today. In other countries, more permanent forms of cocoa growing in complex agroforests have developed and are still being practiced in parts of Cameroon, Nigeria, and Bahia, Brazil (see Chapter 6, this volume).

Permanent Agroforests

As discussed earlier, some cyclic agroforest types can attain ages of many decades before they are renovated, especially if pressure on the land is low and if resources for replanting are not available (e.g., at times of low commodity prices). Therefore, the distinction between permanent and cyclic agroforest types cannot be based on their age alone but must take into consideration the method of regeneration, which in permanent agroforests is a continuous, small-scale process based on either planting or natural regeneration in gaps rather than a distinct (though rare) replanting campaign that entails disturbance of the system on a large (plot) scale.

A good example of permanent agroforests is the damar gardens of southern Sumatra, which are also established in slash-and-burn fields by planting damar seedlings along with fruit trees into a rice crop interplanted with coffee and pepper. These agroforests are regenerated on a continuous basis largely from natural regeneration without disruption of the forest canopy (Michon and de Foresta 1999). A further example from Sumatra illustrates how farmers may choose between different options for growing their crops and the factors that may lead to the development of permanent agroforests. In the region of the Kerinci-Seblat National Park, farmers grow cinnamon in a variety of agroforestry systems, some of which are simple cyclic systems and may not appropriately be classified as agroforests and others of which are permanent agroforests. The dominant system is a cyclic one that starts with annual crops that are grown for about 2 years between new or resprouting (coppiced) coffee bushes. After the annual cropping phase, cinnamon trees are introduced in the system, whose growth allows two or three coffee harvests before their canopy has closed. During this phase, the plots are invaded by spontaneous vegetation that consists mostly of common weeds and not forest species. The cinnamon trees usually are harvested in 6–12 years, and the cycle starts again (Burgers and William 2000). Farmers with sufficient land usually go for a longer growth period of cinnamon trees (up to 25 years), because bark quality increases with tree age, and therefore manage a set of plots of different ages (Aumeeruddy and Sansonnens 1994).

Beside this simple cyclic system, permanent cinnamon agroforests occur in the immediate surroundings of Lake Kerinci, where irrigated rice fields are scarce and the availability of arable hill lands is limited by the steep terrain. In such agroforests many indigenous fruit trees and timber species are cultivated in association with cinnamon and coffee or rubber. Their structural complexity and associated plant diversity vary but are substantially higher than in the cyclic system: a hundred useful woody and herbaceous plant species have been identified in a single village (Aumeeruddy and Sansonnens 1994). Such permanent agroforests are assumed to have evolved in response to land scarcity; below a certain threshold area of arable land, reducing the rotation cycle or the plot size is no longer viable, and permanent agroforests seem a better option. In these systems, the lower productivity of the cinnamon under shade is compensated by the produce of the associated fruit and timber trees (Aumeeruddy and Sansonnens 1994).

Structure and Species Composition of Agroforests

Although all agroforests, by definition, include several vegetation strata and a number of planted or spontaneous tree and other plant species, their structure and plant species composition nevertheless differ substantially from one type of agroforest to another, and these in turn determine their value as habitat and biological corridors for various types of wildlife and their ability to physically protect and buffer forest boundaries. In this section we discuss factors that influence vegetation structure and composition of floral and faunal communities for a variety of agroforests, focusing on the types of crop and tree species planted, the method of establishment and subsequent management, and (to a lesser extent) their position in the landscape. An important distinction in agroforests with respect to vegetation structure and management is between those based on canopy trees, such as rubber, damar, and durian, and those based on understory tree crops, such as cocoa, coffee, and tea.

Structure and Floral Composition of Agroforests Based on Canopy Trees

Studies in Sumatra show that the vegetation structure of old jungle rubber is very similar to that of secondary forest, with a closed canopy 20–25 m high that is dominated by rubber trees and a dense understory of shrubs and small trees, including many seedlings of the canopy trees. In these agroforests, the rubber trees occupy the place of pioneer trees (Figure 10.4; Gouyon et al. 1993). Tree species richness in productive rubber agroforests can be as high as 70 species per hectare for trees more than 10 cm in stem diameter at breast height (dbh, 130 cm; Table 10.1). This is probably less than the species richness of secondary forests of similar age because of the dominance of rubber trees in the agroforests. Several studies show that tree species richness of rubber agroforests is negatively correlated with the density of rubber trees

Table 10.1. Structure and diversity of trees more than 10 cm in stem diameter at breast height in natural forest, damar agroforests, and rubber agroforests in Sumatra.

	Natural Forest		Damar Agroforests			Rubber Agroforests		
	NF1[a]	NF2[b]	DAF1[c]	DAF2[x]	DAF3[c]	RAF1[d]	RAF2[d]	RAF3[e]
Approximate age (yr)			>70	>70	>70	12–40	20–60?	35
Trees per hectare	648	495	360	296	294	450	468	496
Damar or rubber trees per hectare	—	—	222	162	97	270	184	220
Basal area (m² ha⁻¹)	30.7	25.4	41.0	23.8	21.6	20.6	21.7	14.0
Tree species per hectare	216	155	37	29	32	63	71	73
Shannon diversity index*	5.0	4.5	1.9	1.9	2.4	2.1	2.9	2.9
Simpson diversity index*	108	57.1	2.6	3.1	6.2	2.7	5.8	5.1
Fisher's alpha*	114	76.6	10.3	8.0	9.1	19.9	23.3	23.1
Simpson concentration index	0.01	0.02	0.39	0.33	0.16	0.37	0.17	0.20

Fisher's Alpha, Shannon, and Simpson diversity indices are different measures of species diversity combining species richness and species relative abundance, with higher values indicating higher diversity. Fisher's Alpha is more robust for small sample sizes than the other two indices. The Simpson concentration index is the probability that two randomly chosen individuals belong to the same species; higher values of this index indicate the dominance of the few planted species (e.g., rubber trees) in the overall tree population of an agroforest.

[a]Average for 6 ha (U. Rosalina, pers. comm., 2000).

[b]Average for 2 ha (Supriyanto, pers. comm., 2000).

[c]Average for 1 ha (Vincent et al. 2002).

[d]Hectare values obtained by combining results from five 0.2-ha plots (Hardiwinoto et al. 1999).

[e]1-ha plot (G. Vincent, unpublished data, 1999).

Figure 10.4. Structural profile of a 40- to 45-year-old jungle rubber plot in Muara Buat, Sumatra, based on a 50- by 20-m survey plot. Only trees more than 10 cm in diameter at breast height are shown (from Gouyon et al. 1993, with permission). All trees in the transect that are not listed in the following are rubber trees (*Hevea brasiliensis*): 1, *Artocarpus* sp.; 2, *Peronema canescens*; 5, *P. canescens*; 6, ditto; 7, *Pternandra echinata*; 8, *P.canescens*; 10, *Xerospermum* ?; 12, *Ficus* sp.; 14, Rattan "semambu"; 15, *Styrax benzoin*; 16, *Milletia atropurpurea*; 17, ditto; 18, ditto; 19, *Macaranga triloba*; 21, *S. benzoin*; 22, *P. canescens*; 24, Rattan "semambu"; 28, *P. canescens*; 29, ditto; 30, *M. atropurpurea*; 31, *Canarium patentinervium*; 32, *Artocarpus* sp.; 33, *S. benzoin*; 34, *Drypetes longifolia* ?; 36, *Payena* cf. *acuminata*; 37, Lauracea ?; 42, ? "Bintung"; 46, *Eugenia* ?; 48, *Pternandra echinata*; 49, ? "Ntango"; 50, Rattan "Manau tebu"; 51, ? "Jangkang"; 52, *P. echinata*; 53, *Eugenia* sp. "Kayu klat"; 56, *Xerospermum noronianum* ?; 57, *Polyalthia hypoleuca*; 58, *Eugenia* sp. ? "Kayu klat"; 59, *M. atropurpurea*; 60, *P. echinata*; 62, *Porterandia* ?; 66, *S. benzoin*; 67, Fagacea ?; 69, Fagacea ?; 70, *S. benzoin*.

(Lawrence 1996; Hardiwinoto et al. 1999), reflecting competition between the dominant rubber trees and other species for light and soil resources. However, rubber dominance diminishes as the plots grow older and rubber trees progressively die, allowing other plant species to establish and grow into the upper strata (Vincent et al. in press). The trend of increasing species richness with increasing age and decreasing density of rubber trees has also been documented for the understory of rubber agroforests (Carrier 2002; Girault 2002).

Thus, whereas the conservation value of rubber agroforests increases with their age, rubber production decreases, and this makes it more likely that the plot will be converted into a less diversified land use (e.g., an oil palm plantation) or rejuvenated through a slash-and-burn phase (Gouyon et al. 1993). This situation is to some extent representative for other cyclical agroforest types such as benzoin agroforests, where a strong positive correlation between plant diversity and plot age and a clear trade-off between agroforest productivity (itself tightly correlated to density of benzoin trees) and plant diversity have been shown (García-Fernández et al. 2003).

These insights have motivated research into the possibilities of increasing the productive life of rubber agroforests by maintaining a sufficient number of productive rubber trees per hectare. Although this would prolong the dominance of the systems by the rubber trees, their conservation potential would still increase because only old agroforests present the range of niches that allows significant colonization by late-successional plant species. Also, as disturbance frequency decreases, a larger percentage of the landscape would be maintained under old agroforest cover, so landscape connectivity for forest-dependent species would increase (Vincent et al. 2003).

Other canopy tree–based agroforests that, though smaller in extent than the rubber agroforests, hold conservation potential are the damar and durian agroforests of Sumatra (Michon and de Foresta 1999). In a comparison of the structure and composition of rubber, damar, and durian agroforests with primary forest in Sumatra, Thiollay (1995) found that damar agroforests were structurally the most similar to primary forest (Figure 10.5), with an often continuous canopy of 35–45 m height composed of at least 39 tree species of more than 20 cm dbh. Of the 245–500 trees per hectare, 56–80 percent were damars, which were associated mostly with fruit trees. The canopy of rubber agroforests was lower (20–30 m) and their understory denser than that of the damar agroforests. Of the 750 trees per hectare, 65 percent were rubber trees (about 50 percent for the plots in Table 10.1). Durian agroforests had a more open canopy at 30–45 m height composed of durian and other fruit and timber trees (350 trees per hectare) and a lower stratum of smaller trees such as clove (*Syzygium aromaticum*), cinnamon, nutmeg (*Myristica fragrans*), and coffee.

The data summarized in Table 10.1 show that although damar agroforests are structurally similar to primary forest, they have a much lower tree species

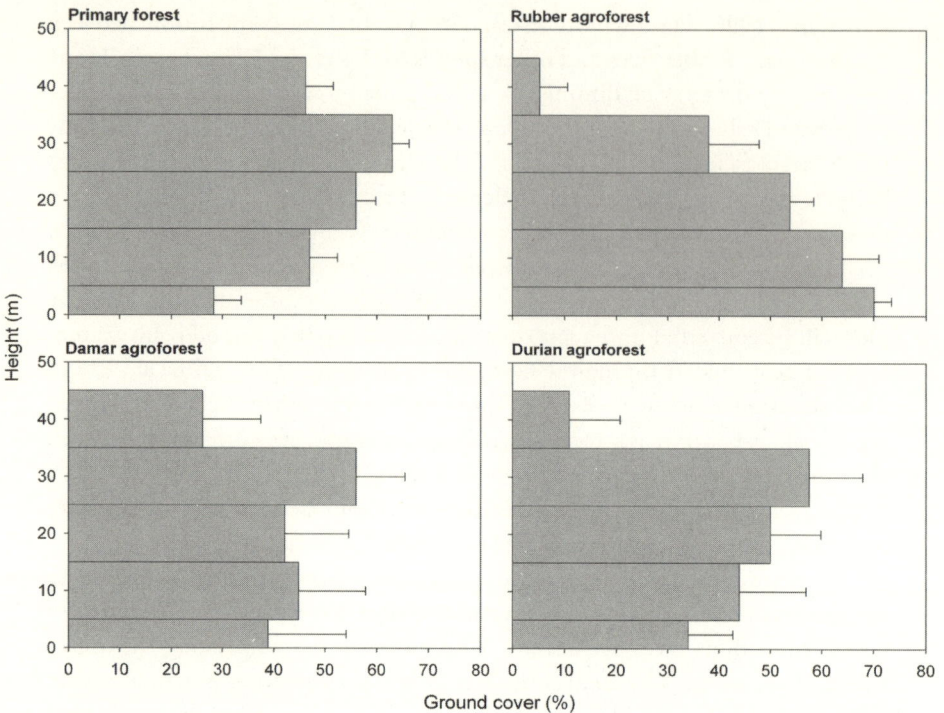

Figure 10.5. Height profiles of mean foliage cover in primary forest and agroforests based on rubber, damar, and durian in Sumatra (redrawn from Thiollay 1995, with permission).

richness. Their species richness also seems to be significantly lower than that of rubber agroforests despite the higher basal area and greater age of the damar forests. This is because damar agroforests tend to be more intensively managed than rubber agroforests and other species, especially fruit trees, are systematically interplanted with the damar trees. Langsat (*Lansium domesticum,* in the midcanopy), durian, and petai (*Parkia speciosa,* in the upper canopy) are among the most prized and commonly planted tree species. Selection among the spontaneously regenerating vegetation and clearing of unwanted species is also more systematic in damar than in rubber agroforests. This more intensive management of the damar agroforests probably is related to more limited access to land on the coastal strip in southwestern Sumatra, where these agroforests are common, compared with central Sumatra and Kalimantan, where rubber agroforestry is practiced. Another factor could be the perceived permanent nature of the damar agroforests (as opposed to the cyclic nature of the rubber agroforests), which encourages efforts to maintain late-arriving or -producing species of economic value at the expense of spontaneous vegetation.

Rubber groves in the Tapajós region of the central Amazon often are characterized by decreasing management (especially weeding) intensity with increasing distance from the homesteads, and mainly those occurring at greater distance from the villages are appropriately characterized as agroforests. In contrast, groves that are situated close to the villages often are more intensively maintained for fear of snakes and fire and perhaps a general preference for well-maintained groves (Schroth et al. 2003). In an inventory of eight agroforests of 23 to more than 50 years of age (G. Schroth, unpublished data, 2002), rubber tree density (more than 10 cm dbh) ranged from 100 to 700 trees per hectare and that of other trees ranged from 225 to 875 stems per hectare (more than 5 cm dbh; 0–575 stems per hectare for dbh more than 10 cm). Five of the eight plots contained no planted tree species other than rubber trees, and the others were close to (previous) homesteads and contained some planted fruit trees in the midstory (orange, *Citrus sinensis;* cupuaçu, *Theobroma grandiflorum;* mango, *Mangifera indica*). Despite a trend toward decreasing rubber tree densities with increasing age, groves of 50 years or older still had a high density of rubber trees (100–425 per hectare), many of which were of large size and good health. In accord with the aforementioned observations from Southeast Asian jungle rubber, the density of rubber trees was significantly negatively related to the density of other large trees (more than 10 cm dbh; $r^2 = 0.69$) and to the number of tree species present (more than 5 cm dbh; $r^2 = 0.70$), which ranged from 3 to 27 for the 400-m^2 plots. Tree species found in the groves included primary forest canopy species (Parrotta et al. 1995) such as mututí (*Pterocarpus amazonum*), quaruba verdadeira (*Vochysia maxima*), breu sucuruba (*Trattinnickia rhoifolia*), mirindiba doce (*Glycydendron amazonicum*), fava folha fina (*Pseudopiptadenia psilostachya*), fava barbatimão (*Stryphnodendron pulcherrimum*), tauarí (*Couratari guinanensis*), and cumarú (*Dipteryx odorata*). Of the two groves with the highest tree species richness (23 and 27 species, respectively), one apparently had been abandoned for some time at an early age, which had caused high mortality of the rubber trees so that other species could develop, and the other directly neighbored primary forest as a seed source.

Faunal Communities of Canopy Tree–Based Agroforests

Much less information is available on the faunal communities of canopy tree–based agroforests than about their floral composition and structure. In a study in the buffer zone of the Gunung Palung National Park in Kalimantan, Salafsky (1993) found that primates most commonly encountered in agroforests were species that are adapted to disturbed forest, such as leaf monkeys (*Presbytis rubicunda*) and gibbons (*Hylobates agilis*), rather than taxa that prefer primary forest, such as orangutan (*Pongo pygmaeus*), or open agricultural areas, such as macaques (*Macaca* spp.). Michon and de Foresta (1995) reported the presence of seven primate species (macaques, leaf monkeys, gibbons, and siamang, *Hylobates syndactylus*) in rubber and damar agroforests and five species in durian agroforests in Sumatra and noted that their density was similar to that in primary forest. The same authors mention the presence of highly endangered wildlife such as rhinoceros (*Dicerorhinus sumatrensis*) and tiger (*Panthera tigris*) in damar agroforests, suggesting that these systems may at least serve as corridors and temporary habitat for these species (Michon and de Foresta 1999).

Thiollay (1995) found lower bird species richness and diversity in damar, rubber, and durian agroforests than in primary forest and low coefficients of similarity between the agroforest and forest communities in Sumatra (Figure 10.6). However, the agroforests had much higher species richness than monoculture plantations of tree crops such as rubber, oil palm, and coconut palm in the same regions. In fact, in the monocultures so few species were found in preliminary investigations that no detailed data were collected. Of 216 bird species identified, 102 (47 percent) were present in forest but absent or significantly less common in the agroforests, and 43 species (20 percent) were present only in agroforests or significantly more common in agroforests than in forest. Seventy-one species showed no clear trend between the habitats. Large frugivore and insectivore species and terrestrial interior forest specialists were least common in the agroforests, whereas small frugivores, foliage insectivores, and nectarivores, species often associated with gaps, were more common in agroforests than in forest. Among the three agroforest types, the bird communities of the rubber agroforests were most similar to those of the primary forest, and those of the durian gardens were the most different and contained the

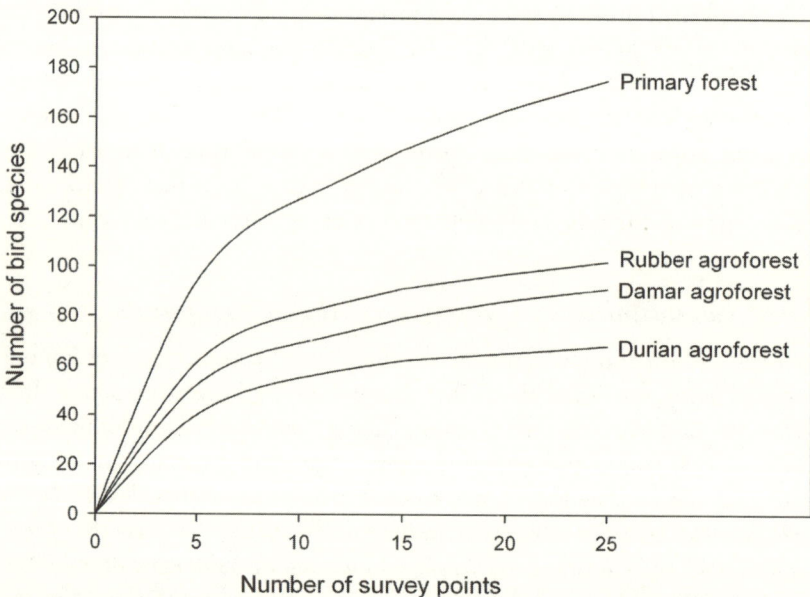

Figure 10.6. Estimated accumulation curves of bird species in primary forest and agroforests based on rubber, damar, and durian in Sumatra (redrawn from Thiollay 1995, with permission).

fewest forest specialists. The relative abundance of bird species associated with open woodlands and cultivated areas increased from forest to damar and durian agroforests, with rubber agroforests being intermediate. The lower species richness in the agroforests was explained by lower tree height, structural complexity, and variety of food resources and by hunting, human disturbance, and competition from dominant bird and mammal species.

Although no quantitative information is available on the faunal communities of Amazonian rubber agroforests, it is very likely that these are also significantly affected by hunting. From 51 owners of rubber agroforests interviewed in the buffer zone of the Tapajós National Forest, 50 percent hunted in their agroforests and another 6 percent did not hunt themselves but knew that others did (Schroth et al. 2003). Frequently mentioned game species included armadillo (*Dasypus* sp.), paca (*Agouti paca*), brocket deer (*Mazama* sp.), agouti (*Dasyprocta agouti*), and collared peccaries (*Tayassu tajacu*), which are seasonally attracted by fallen fruits (e.g., mango, taperebá [*Spondias mombin*]) and rubber seeds. At the beginning of the dry season, macaws (*Ara chloroptera*) also visit the agroforests to feed on unripe rubber seeds and are hunted for food with slingshots when sitting in the tree canopies (G. Schroth, pers. obs., 2002). Although hunting in the agroforests certainly reduces their value as faunal habitat, it may also help keep farmers from hunting in the forest itself (see Chapters 13 and 14, this volume), and the net cost or benefit of rubber agroforestry along forest boundaries for forest fauna warrants investigation.

In summary, these studies show that the forest-like character and the high tree diversity of agroforests based on canopy tree crops provide significant habitat and resources for wildlife, though less than those of the original forest. Most of these agroforests possess a closed canopy, which may be strongly dominated by a single species, and a dense understory and midstory of regeneration of the canopy trees and other spontaneous regrowth. If smaller fruit trees are grown in the understory, the canopy may need to be kept more open to permit sufficient light entry into the lower strata. Although the canopies of most of the agroforests described earlier were dominated by a single tree crop species, they also contained relevant numbers of spontaneous primary and secondary forest trees that may have developed from seeds or rootstocks of the previous forest vegetation and had been tolerated or, in the case of useful species, even favored during weeding. The presence of many primary forest species suggests that the microclimate in the agroforest understory is favorable to their regeneration and that abandoned agroforests may with time return to vegetation communities resembling primary forest.

Although these agroforests also offer habitat for many fauna species, it is clear that the faunal communities even of the extensive Indonesian agroforests differ substantially from those of undisturbed forest and may often be more similar to those of disturbed forests with many gap-associated species. No data are available for Amazonian rubber agroforests, but their faunal communities are likely to be influenced by the small size of many agroforests, which are typically embedded in a mosaic of slash-and-burn plots and fallows, their proximity to primary forest, and high hunting pressure.

Structure and Floral Composition of Agroforests Based on Understory Tree Crops

Agroforests based on understory crops differ from those based on canopy trees by having a more open canopy, which is managed to allow sufficient light transmission, and an understory and midstory dominated by planted tree crops rather than spontaneous regeneration. An example of such agroforests are the cabruca cocoa plantations of Bahia, Brazil. According to Johns (1999), the traditional practice of establishing cabruca plantations was to remove about one-third of the original forest canopy trees, often some of the largest, and underplant them with cocoa trees. This resulted in a shade canopy of about 50–60 percent, considered necessary by most farmers for maintaining a humid microclimate, conserving soil fertility, reducing weed growth and insect attack, and conserving pollinator species (Figure 10.7). In contrast, Alves (1990) reported a more drastic alteration of the forest canopy by cabruca farmers, who on average removed 90 percent of the trees (with densities of 65 trees per hectare in cabruca, compared with 742 trees per hectare in forest), the entire midlayer (replacing it with cocoa trees), and 83 percent of the forest herb layer. Moreover, most vines were removed from the cabruca systems, and natural regeneration of the canopy trees often was replaced by planted legume or other useful trees.

Nevertheless, a survey of 61 cabruca farms throughout southern Bahia carried out in 1964 found a density of 76 shade trees per hectare belonging to 171 species overall (Alvim and Pereira 1965). Many cabruca agroforests in Bahia may thus be important genetic reservoirs that today still contain valuable hardwood species that are otherwise severely logged from natural forests in the region (Johns 1999). However, this potential is threatened by the suppression of shade tree regeneration during the normal management of the systems, which involves periodic slashing of the undergrowth or even chemical weeding (Alves 1990).

Underplanting of selectively opened forest is also a traditional practice of cocoa and coffee planting in parts of the West African rainforest zone. However, the degree of forest clearing differs significantly between regions and ethnic groups, with important consequences for the structure and, certainly, the diversity of the resulting agroforest communities (see also Chapter 6, this volume). De Rouw (1987) compared the practices of establishing coffee and cocoa plantations in the southwestern Côte d'Ivoire between the local Oubi and immigrant Baoulé farmers from the savanna zone. When clearing a forest plot, the native Oubi leave two or three large forest trees per hectare, because of their hard wood or large buttresses, and also a variable number of smaller

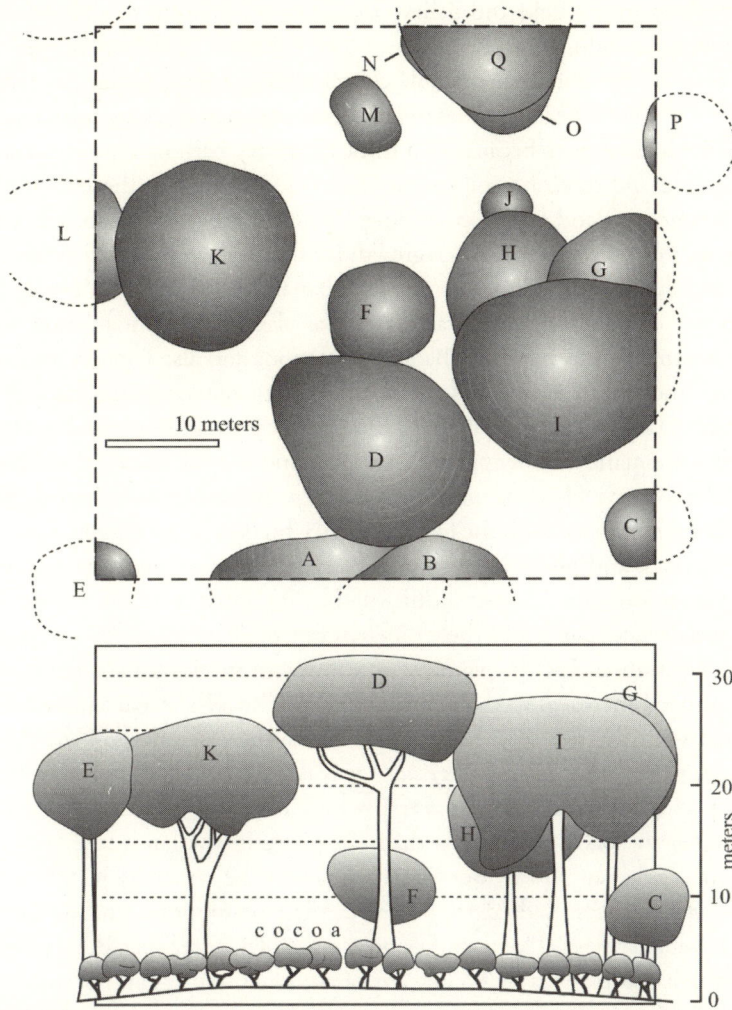

Figure 10.7. Plan view (*top*) and lateral view (*bottom*) of the shade canopy of a typical cabruca plot in Bahia, Brazil (from Johns 1999, with permission). A, Bacumixá (*Sideroxylon vastum*); B, Jitaí (*Apuleia* sp.); C, Inga (*Inga edulis*); D and M, Pau sange (*Pterocarpus violacens*); F, I, and L, Biriba (*Eschweilera speciosa*); G, Jatobá (*Hymenea stignocarpum*); H, Carobuçu (*Jacaranda mimosaefolia*); J, Fidalgo (*Aegiphila sellowiana*); K, Gameleira branca (*Ficus doliaria*); N, Jacaranda branca (*Swartzia macrostachya*); Q, Buranhém (*Pradosia lactescens*); O and P, Jaqueira (*Artocarpus integrifolia*—not native to Atlantic rainforest).

trees that are not harmful to the crops, are difficult to fell, produce useful seeds, or have religious functions. This results in stands of up to 19 forest trees of more than 15 m height per hectare. After burning the debris of the other vegetation, they sow rice and later plant coffee and cocoa seedlings when the food crop is maturing. The Oubi appreciate the shade of the retained trees when working in the field and believe that they will help to reestablish forest after a plot is abandoned and conserve a forest climate in the landscape.

In contrast, the immigrant Baoulé cut all smaller trees and burn the debris around the stems of the larger trees to kill them. No forest trees are tolerated, but some individuals survive because of a thick or watery bark or lack of firewood. After burning and intensive soil preparation to form mounts, the Baoulé plant yam (*Dioscorea* sp.) and tree crop seedlings. Whereas the native Oubi select suitable species from the spontaneous regrowth for shading the young tree crops, the Baoulé mainly use planted food crops (taro, bananas) and some fruit trees. As a consequence of these different practices, Oubi plantations contain more forest trees of intermediate height than Baoulé plantations and also contain some very large trees (more than 40 m) that are absent from plantations established by the immigrants (Table 10.2). The different strategies of plantation establishment and more intensive management applied to the plantations result in earlier and higher per-hectare cocoa yields of the immigrant farmers, but lower native tree densities and diversity, compared with the forest people, who usually have more forest area at their disposal and prefer larger, less intensively managed, and therefore more diverse plantations (see Chapter 6, this volume, for a discussion of the historical and socioeconomic context of these different practices).

The selection and active planting of useful tree species for shade may lead to pronounced increases in the density of certain species in cocoa plantations compared with the rest of the landscape. In southern Cameroon the density of African plum trees (*Dacryodes edulis*) was ten times higher and that of the timber species limba (*Terminalia superba*), and iroko (*Milicia excelsa*) was three times higher in cocoa plantations than elsewhere in the landscape (van Dijk 1999). In a survey of 300 farmers in 21 villages in the same region, 93 percent of the cocoa farmers planted fruit trees in their plantations, and 81 percent planted timber species. The most frequently used fruit tree species was African plum, which was planted by 83 percent of all cocoa farmers and occurred at a density of 17 trees per hectare in the cocoa plantations, and the most commonly used timber species were limba, obeche (*Triplochiton scleroxylon*), iroko, fuma (*Ceiba pentandra*), and *Ficus mucuso* (Sonwa et al. 2000a).

The replacement of native forest trees by useful species is also evident from

Table 10.2. Tree density per hectare in 10-year-old coffee and cocoa plantations of native Oubi and immigrant Baoulé farmers in the southwestern Côte d'Ivoire.

	Oubi (native)	Baoulé (immigrant)
Coffee and cocoa trees	600–1,000	600–1,300
Forest trees <40 m height	8–20	5–8
Forest trees >40 m height	2–3	0

Source: After de Rouw (1987).

floristic inventories of cocoa agroforests in Talamanca, Costa Rica (Guiracocha et al. 2001). In a comparison of five cocoa agroforests and five forest plots of 0.1 hectare each, total species richness of trees (more than 10 cm dbh) was higher in forests (80 species) than agroforests (35 species). As expected, tree densities were also significantly higher in forest than in agroforests (432 vs. 234 trees per hectare), although diameter distributions and canopy height were similar. The habitats differed in their floristic composition, sharing only seven species. Whereas the natural forest was dominated by the palms *Socratea exorrhiza* and *Iriartea deltoidea* and the dicot tree species *Pentaclethra macroloba* and *Goetalsia meiantha,* the agroforests were dominated by an important timber species, *Cordia alliodora* (Guiracocha et al. 2001).

Faunal Communities of Agroforests Based on Understory Tree Crops

Little information is available on the faunal communities of shaded cocoa ecosystems. In a comparison of cabruca and forest plots in Bahia, Brazil, Alves (1990) found that faunal groups that depend on the understory, such as specialized understory bird species, and large frugivores and large terrestrial mammals were underrepresented or missing from the cocoa ecosystems. Bird family richness in cabruca cocoa and forest was positively correlated with vegetation variables describing the height, density, and cover of the herb layer, midstory density, canopy cover, and structural complexity of the vegetation, that is, variables that are substantially altered by the establishment and management of the cocoa plantations.

In contrast, in a study of bird communities in abandoned and managed cocoa agroforests and natural forest in Talamanca, Costa Rica, Reitsma et al. (2001) found a 17 percent higher abundance and also higher species richness in cocoa than in forest (130, 131, and 144 species in forest, abandoned cocoa, and managed cocoa, respectively). However, although the cocoa plantations obviously offered habitat for a large number of forest-dependent species, there were fewer forest specialist species and more agricultural generalist species in cocoa plantations than in forest. As in Alves's Brazilian study, specialist species found in forest but not cocoa plots were mostly understory insectivores, which are not adapted to an understory dominated by cocoa trees. The number of forest specialist bird species per observation point increased with the density and diversity of canopy tree species, suggesting that the conservation value of the cocoa ecosystems could be further increased by appropriate shade management. In contrast, distance to forest was not significantly related to the number of forest specialists observed, which the authors attributed to the complex mosaic of habitat types in this region.

Mammal communities in cocoa agroforests may also be quite diverse. A study in Talamanca, Costa Rica, found that the species richness and relative abundance of large mammals (as registered from mammal tracks) were similar in cocoa agroforests and adjacent primary forest (Guiracocha et al. 2001). In a survey of five plots each of agroforests and forest, a total of 10 mammal species were found in each habitat, and overall mammal abundances were similar for all species with the exception of agoutis (*Dasyprocta punctata*), which were more abundant in forest. Local people mentioned a total of 22 animal species commonly seen in cocoa agroforests, as opposed to 27 species in forest. The presence of two endangered cat species, jaguarundi (*Felis yagouaroundi*) and puma (*Felis concolor*), in the cocoa agroforests also attests to their potential conservation value. The high densities and richness of mammals in these agroforests probably reflect their forest-like structure but also the abundance of forest cover in the areas surrounding these small agroforest patches, in combination with the ability of many mammal species to move through different habitat types within a landscape. Additional studies are needed to determine the extent to which these mammals depend on the presence of natural forest, that is, whether they are able to survive and reproduce in landscapes with agroforests but few or no forest remnants. Preliminary data from Talamanca suggest that cocoa agro-forests that are situated in an agricultural environment have a much poorer fauna, in terms of both quality and quantity, than agroforests that are embedded in a forested landscape (Gaudrain 2002). Moreover, the ability of the Talamancan cocoa agro-forests to conserve mammal populations depends heavily on hunting regulations and local people's attitudes toward conservation. Currently, almost all large mammal and bird species visiting these agroforests are subject to hunting, so that their conserva-tion potential is severely reduced (C. A. Harvey, pers. obs., 2002).

The understory of cocoa groves as related to their age, stand structure, and management is an important factor influencing small mammal communities. In a study of cocoa groves shaded principally by the native oil palm (*Elaeis guineensis*) in Sierra Leone, Barnett et al. (2000) found that biomass and diver-sity of small mammals were positively correlated with the density of the under-story vegetation, which was highest in either the very young groves or in the old, degrading cocoa groves and lowest in a plot where the ground vegetation had recently been slashed (Table 10.3). In a 10-year-old grove where ground cover was sparse, 50 percent of the rodents were caught on 4 percent of the area with dense ground cover, suggesting that areas excluded from manage-ment within agricultural systems could benefit certain species. The fauna of the cocoa groves comprised both savanna species, which also characterized the rodent fauna of agricultural fields, and forest species, although the latter were edge and gap specialists rather than forest interior species.

These results corroborated earlier findings by Jeffrey (1977) on the effect of forest conversion into cocoa plantations on rodent communities in Ghana. Comparing primary forest with new (1–2 years cleared and planted with food crops), immature (6–8 years cleared and planted with cocoa trees), and mature (about 20 years under cocoa) cocoa farms, she found that trap success (a proxy of small mammal density) initially increased when forest was cleared for plan-

Table 10.3. Small mammal communities in cocoa groves and agricultural fields in Sierra Leone.

Type	Cocoa Grove	Cocoa Grove	Cocoa Grove	Cocoa Grove	Elephant Grass Field	Rice Field
Age (years)	3	3	10		40	
Understory	Dense	Recently cleared	Sparse	Moderate		
Species	7	2	5	3	6	5
Biomass per 100 trap nights	207.1	8.4	45.9	542.5	54.9	556.2

Source: After Barnett et al. (2000).

tation establishment but then decreased again in mature cocoa, which typically has little live ground cover. In the same sequence, rodent species richness increased from 6 species in forest to 8 and 10 species in the new clearings and immature cocoa, respectively, and then fell back to 6 species in mature cocoa. Despite the overall increase in species, of the six species captured in forest, two (*Hybomys trivirgatus, Malacomys edwardsi*) practically disappeared after clearing, and one (*Hylomyscus stella*) was less abundant. All three species were found on land left fallow for 2–8 years, indicating again the important role that areas set aside for extensive (or no) management within agricultural landscapes could play for certain species that are not adapted to managed habitats.

These studies show that the transformation of native forest into shaded cocoa plantations involves a substantial modification of the original ecosystem, especially the opening of the canopy, replacement of most of the midstory by tree crops, enrichment of upper and midcanopy with a selection of fruit and timber tree species, and the suppression of ground vegetation by litter and shade of the tree crops and by mechanical or, in some cases, chemical weeding (Rice and Greenberg 2000). These alterations of the forest ecosystem necessarily affect faunal communities, depending on the intensity of management, the availability of nearby intact forest as population source, and hunting pressure, among other factors. Although no direct comparisons seem to be available, one would expect that the conditions encountered by fauna in a complex agroforest based on canopy tree crops such as rubber or damar are more similar to those in natural forest than those encountered in a shaded cocoa or coffee plantation. However, the degree to which different groups of fauna are affected by these structural and compositional differences warrants further study.

The Potential Role of Complex Agroforests in Landscape Conservation Strategies

Besides offering habitat for a substantial number of plant and animal species, including many forest-dependent species, complex agroforests can make an important contribution to the conservation of regional biodiversity by enhancing landscape connectivity, reducing edge effects, and improving local microclimates. Whereas the focus of the previous section was on the biodiversity of the agroforests themselves, the present section discusses the ability of complex agroforests to support biodiversity conservation in natural forest. By definition, these functions become ineffective with the disappearance of forest from a landscape, although agroforests may continue to play an important role

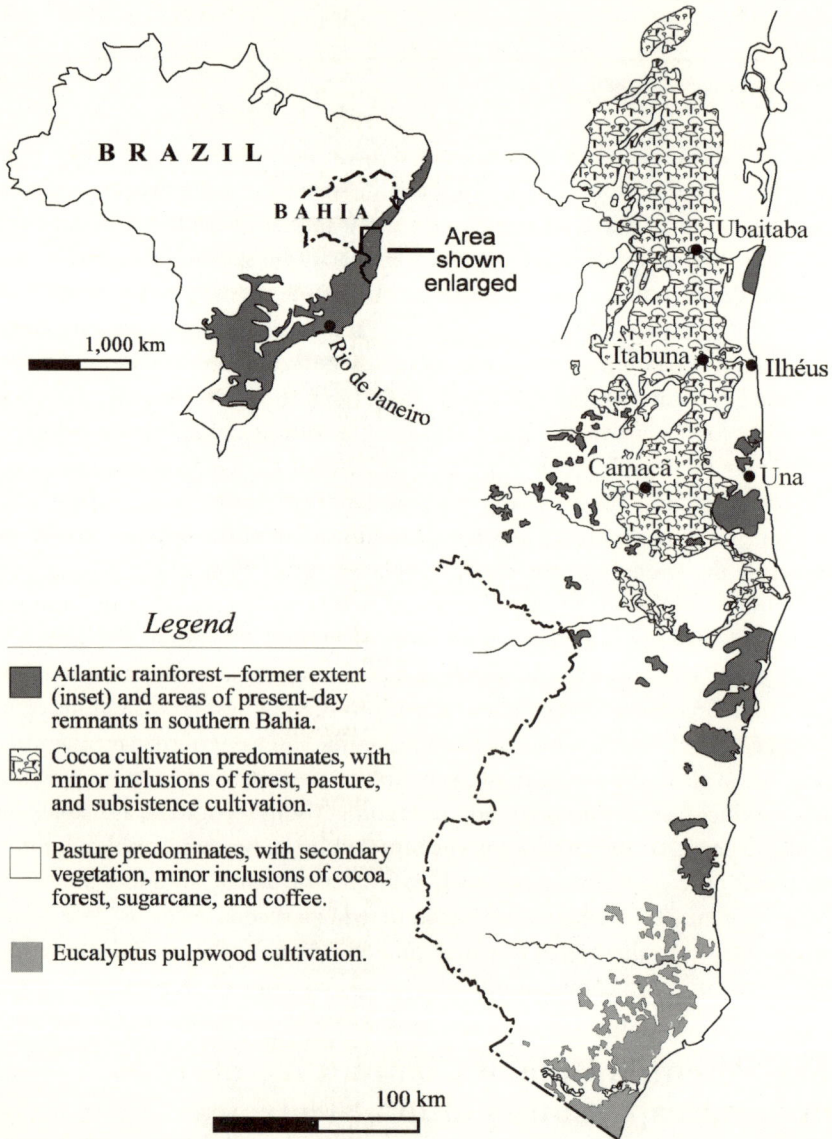

Figure 10.8. Distribution of cabruca cocoa, forest, and other land uses in southern Bahia, Brazil. The inset map shows the former extent of the Atlantic rainforest (from Johns 1999, with permission).

for the survival of species that are not strictly dependent on natural forest and may become the last refugia for partly forest-dependent species.

As has been shown, historical land use patterns in several tropical regions have led to landscapes where complex agroforests form the transition between agricultural or pasture land and natural forest (Figure 10.2). In other cases, remnant forests have been reduced to small patches within an agricultural matrix, where agroforests are interspersed with other agricultural land uses such as pasture and annual and perennial crops. Examples include the cocoa-growing region of Bahia, Brazil, where the few remaining fragments of Atlantic rainforest are interspersed with large areas dominated by pasture or cabruca cocoa (Figure 10.8). Cabruca cocoa has recently attracted much attention because of its ability to harbor some rare fauna such as the endangered golden-headed lion tamarin (*Leontopithecus chrysomelis*) and the recently discovered pink-legged graveteiro (*Acrobatornis fonsecai;* Rice and Greenberg 2000). There is evidence that these rare species depend on forest patches and use the cocoa systems only as secondary habitat (Alves 1990), suggesting that an important role of these agroforests within a land use mosaic may be to increase the available area for forest fauna and to provide wildlife corridors between forest fragments that otherwise would be separated by pastures or other open agricultural areas (see Chapter 3, this volume).

Similarly, observations of large endangered mammal species such as rhinoceros, tiger, and siamang in Sumatran damar gardens (Michon and de Foresta 1999) and tracks of endangered cat species on cocoa farms in Talamanca (Guiracocha et al. 2001) illustrate the value of these agroforests as secondary habitat of forest fauna and their potential for use in buffer zones and landscape corridors. The continuous tree cover of the agroforests is likely to facilitate such animal movements between forest and agroforest habitats. Consequently, cocoa agroforests are being promoted as buffer zones and forest corridors in Talamanca, which forms part of the Mesoamerican Biological Corridor (Rice and Greenberg 2000; Reitsma et al. 2001).

Because of their tall stature and forest-like structure, agroforests may also reduce edge effects that occur when forest borders on open agricultural fields or pasture, decreasing mortality of forest trees that are not adapted to the drier microclimate and gusty winds to which they are exposed near open forest edges (see Chapter 2, this volume). High tree mortality near edges, reinforced by the effects of vine invasions and fire, may lead to the shrinking and final collapse of isolated forest fragments and reserves in agricultural landscapes (Gascon et al. 2000). In long-term observations on forest fragment dynamics in the central Amazon, edge-related tree mortality was lower where forest edges bordered on tall secondary regrowth than where they bordered on open cattle pasture (Mesquita et al. 1999). Because complex agroforests, especially those based on canopy tree crops, are structurally similar to secondary forests, the same protective effects are to be expected if forest edges are buffered by agroforests rather than being exposed to open pastures or agricultural fields. Therefore, strategically placed agroforests may help maintain the integrity of forest borders and remnant forest patches in landscapes composed of agricultural and forest habitats. As shown before, traditional agroforests separate forest from more intensively used agricultural land in several tropical landscapes.

One of the most important threats to tropical forests bordering inhabited areas is fires that may be set in agricultural and pasture areas for land clearing or other management purposes but may encroach on forested areas. Most rainforest trees are very susceptible even to low-intensity fire, and increased leaf fall and tree mortality after fire tend to increase the likelihood of subsequent fires in a positive feedback cycle (Cochrane 2001; see also Chapter 2, this volume). Where agroforests form the borders of the forest, such risks are reduced because farmers take care to keep fire under control to avoid losing valuable tree crops. This may especially be the case if agroforests based on fire-sensitive tree crops, such as rubber, and slash-and-burn plots form a small-scale mosaic, as in parts of the Tapajós region in the Amazon (Figure 10.3). In this fire-sensitive region, farmers may prepare narrow firebreaks by removing the litter from the soil to stop low-intensity surface fires from passing into their agroforests (G. Schroth, pers. obs., 2002). According to Preechapanya (1996), farmers in northern Thailand prevent forest fires and seldom practice shifting cultivation in watersheds that are used for growing jungle tea. They suggest that jungle tea might have acted as a buffer zone and has prevented shifting cultivation from spreading into the forest. In areas without jungle tea, the forest has often been lost to shifting cultivation and fire. However, recent reductions in management intensity caused by out-migration of young men as well as reductions in use of the systems for cattle grazing seem to have led to more frequent fires, and the buffer function may progressively be lost (Preechapanya et al. in press). Griffith (2000) noted that when wildfires raged through the Petén region of Guatemala in 1998 and affected the Maya Biosphere Reserve, some agroforest farms in the buffer zone of the park may have served as critical refuges for forest fauna because they were actively protected from fire by their owners and retained some of the only intact vegetation in the area.

In buffer zones of officially protected forests, agroforests may also help to reduce multiple pressures from the surrounding population on forest resources (Michon et al. 1986). In the region bordering the Kerinci-Seblat National Park in western Sumatra that is depicted in Figure 10.2, the villagers establish agroforests with a variety of fruit and other useful trees on the transition between village and park land by progressively clearing community forest and, to a lesser extent, park forest. Villagers may possess rice fields, mixed gardens, or both, and may collect a variety of products from the park forest such as timber, fuelwood, rattan, incense, palm fiber, game, and fish. Murniati et al. (2001) showed that the highest dependency on forest resources was found for farmers owning only rice fields, and a drastically lower dependency was found for farmers that owned both rice fields and agroforests and that obtained their food from the former and cash, timber, and fuelwood from the latter. These households also had the most even distribution of labor through the year and little spare time for collecting forest products. Poor farmers generally were the most dependent on forest resources, suggesting that diversification of the farming systems combined with income generation through the increased use of agroforests would reduce pressure on the national park (Murniati et al. 2001). The gathering of wood and other forest products in agroforests rather than more distant forests may also substantially reduce transport time and effort, especially for women, which can thus be devoted to other household tasks (Sonwa et al. 2001).

Conclusions

Complex agroforests are the most forest-like of all agroforestry systems and the ones that hold the highest potential for contributing to biodiversity conservation in tropical forest regions. From his study of Sumatran agroforests, Thiollay (1995) concluded that "traditional agroforests are one of the best possible compromises between the conservation of biodiversity and the economic and sustainable use of natural resources" (p. 346). Similar statements can be made for traditional agroforests throughout the tropics. Although no direct comparisons of the species richness of agroforests based on canopy or understory tree crops are available, we suggest that the former may have an even greater potential to host forest-dependent understory species than the latter because of their closed canopy and greater tolerance of spontaneous regeneration in the understory and midstory. In contrast, agroforests based on understory tree crops may have greater potential for the conservation of canopy trees and organisms depending on diverse forest canopies. The conservation value of multistrata systems composed of both canopy and understory tree crops, such as cocoa shaded by rubber or coconut trees, is certainly much lower because both strata are intensively managed. The species richness of flora and fauna and especially the presence of many forest-dependent species in complex agroforests clearly justify efforts to conserve and promote these traditional agroecosystems, both in buffer zones of protected areas and in largely deforested regions, where complex agroforests may offer some of the last habitats for forest-dependent flora and fauna and greatly enhance landscape connectivity.

However, from the presented data it is also evident that despite their exceptionally high biodiversity for agricultural systems and the occasional presence of threatened fauna species, complex agroforests are poor substitutes for natural forest because many forest-dependent species are missing or underrepresented (Alves 1990; Thiollay 1995; Beukema and van Noordwijk in press). The extent to which complex agroforests are needed as partial substitutes for natural forest in landscape conservation strategies obviously depends on the availability of intact forest. For example, in parts of the Sumatran lowlands there is very little primary forest left, and secondary forests older than about 20 years are mostly jungle rubber (Gouyon et al. 1993). In this locale then, landscape-scale biodiversity conservation depends critically on the habitat quality of these agroforests for the native fauna and flora.

The situation may be very different in the buffer zone of a major forest reserve, such as in the Amazon, which still hosts vast areas of primary forest. Here, the role of an agroforest as habitat for wild species may be much less important for regional biodiversity conservation than its role in the protection of forest borders, especially from fire, and connection of forest remnants in increasingly fragmented landscapes. Many situations are intermediate, such as the small-scale mosaics of forest patches and agricultural areas, interspersed with complex agroforests, in Central America and Bahia, Brazil, and here complex agroforests may play an important role both as habitats in their own right and for the protection and linkage of remnant forests. In short, where there is still a lot of forest left, the most important role of agroforests may be to protect the forest, which, in turn, conserves biodiversity; where intact forest is scarce, complex agroforests often are the last available habitats for forest-dependent fauna and flora. The presence of many primary forest species in complex agroforests also suggests that they are a good starting point for forest regeneration if for reasons such as low productivity or consistently low commodity prices they have been abandoned—unless, of course, they are converted to another, less diversified land use.

Many traditional agroforests are threatened by increasing pressures to intensify or modernize land use in the tropics, such as the jungle rubber systems in the Sumatran lowlands, which are increasingly being lost to expanding oil palm and monoculture rubber plantations (Michon and de Foresta 1995). Others have survived difficult times but face better prospects today, such as the Amazonian rubber agroforests, most of which were abandoned and some converted into slash-and-burn fields or pastures 10–20 years ago but which now enjoy increased product prices and government support (Schroth et al. 2003), and cocoa ecosystems, which also experienced price slumps in the late 1980s and early 1990s but experience higher commodity prices today (see Chapter 6, this volume).

It is worth mentioning that complex agroforests often have survived only because farmers resisted pressure from government programs to modernize (and simplify) their traditional systems. Examples include Brazilian cabruca farmers who refused to reduce the shade canopies of their cocoa farms, which would have increased cocoa yields but made them more dependent on agrochemical inputs (Johns 1999). Similarly, jungle tea farmers in northern Thailand believed that the natural vegetation cover is more effective in controlling erosion and runoff than terraces and resisted extension efforts to change their traditional, diversified agroforests into terraced tea monocultures (Preechapanya 1996; Preechapanya et al. in press).

However, if complex agroforests are to play a role in future tropical landscapes and contribute to biodiversity conservation, they need to be profitable and make significant contributions to the livelihoods of their owners. A full discussion of the agronomic issues related to the intensification of low-productivity agroforests is not possible in this chapter, but a few relevant research and development approaches deserve mention (see also Chapter 6, this volume). Increased product diversification of agroforests through the association of more valuable fruit and timber tree species with the main tree crops is a way to buffer smallholder farmers against fluctuating commodity prices and ecological risks, such as the diseases that threaten cocoa agroforests in several tropical regions (Lodoen 1998). If applied to agroforests that contain a large amount of spontaneous vegetation, this strategy corresponds to an increased domestication of the system by replacing unplanned with planned diversity, in a trajectory that may ultimately lead to homegarden-like systems. The degree to which faunal diversity is affected by this type of domestication warrants further study. The increased management of agroforests for timber production is another important option, especially in regions such as Indonesia, Central America, and parts of West Africa where natural forests are becoming scarce (Michon and de Foresta 1995; see also Chapter 6, this volume) but also in the Amazon, where the exploitation of natural forests and marketing of its products are increasingly subject to environmental regulation.

Most importantly, more efforts are needed to increase the profitability of the economic backbone species of complex agroforests. This may include the selection of disease-resistant cocoa varieties (Lodoen 1998) supported by more intensive management to control diseases, the integration of more productive planting material into rubber agroforests (Williams et al. 2001), and management practices that increase the health and longevity of tree crops such as rubber without compromising, or even while increasing, their yields (Schroth et al. 2003; Schroth et al. in press). It is a major research challenge to develop intensification methods for tropical agroforests that increase their profitability without losing too much of their biodiversity benefits.

Acknowledgments

Thoughtful and detailed comments from Hubert de Foresta and Norman D. Johns have greatly helped to improve the content and readability of the manuscript. G.S. received support from the Center of Applied Biodiversity Science at Conservation International while working on this chapter.

References

Alves, M. C. 1990. *The role of cacao plantations in the conservation of the Atlantic Forest of southern Bahia, Brazil.* Master's thesis, University of Florida, Gainesville.

Alvim, P. de T., and C. P. Pereira. 1965. Sombra e espacamento nas plantações de cacau no Estado da Bahia. Pages 18–19 in *Relatorio anual 1964.* Itabuna, Brazil: Centro de Pesquisas do Cacau, CEPLAC.

Aumeeruddy, Y., and B. Sansonnens. 1994. Shifting from simple to complex agroforestry systems: an example for buffer zone management from Kerinci (Sumatra, Indonesia). *Agroforestry Systems* 28:113–141.

Barnett, A. A., N. Read, J. Scurlock, C. Low, H. Norris, and R. Shapley. 2000. Ecology of rodent communities in agricultural habitats in eastern Sierra Leone: cocoa groves as forest refugia. *Tropical Ecology* 41:127–142.

Berry, D. 2001. Rational chemical control and cultural techniques. Pages 153–192 in D. Mariau (ed.), *Diseases of tropical tree crops.* Montpellier, France: CIRAD.

Beukema, H., and M. van Noordwijk. In press. Terrestrial pteridophytes as indicators of a forest-like environment in rubber production systems in the lowlands of Jambi, Sumatra. *Agriculture, Ecosystems and Environment.*

Burgers, P., and D. William. 2000. Options for sustainable agriculture in the forest margins? Indigenous strategies. *ILEIA Newsletter* 16:10–11.

Carrier, J. 2002. *Structure and understorey composition of rubber agroforests in Jambi, Sumatra: methodology assessment and preliminary results.* Montpellier, France: ENGREF.

Cochrane, M. A. 2001. Synergistic interactions between habitat fragmentation and fire in evergreen tropical forests. *Conservation Biology* 15:1515–1521.

Dean, W. 1987. *Brazil and the struggle for rubber.* Cambridge, UK: Cambridge University Press.

de Foresta, H., and E. Boer. 2000. *Shorea javanica* Koord. and Valeton. Pages 105–109 in E. Boer and A. B. Ella (eds.), *Plant resources of South-East Asia,* No. 18: *Plants producing exudates.* Leyden, the Netherlands: Backhuys.

de Jong, W. 2001. The impact of rubber on the forest landscape in Borneo. Pages 367–381 in A. Angelsen and D. Kaimowitz (eds.), *Agricultural technologies and tropical deforestation.* Wallingford, UK: CAB International.

de Rouw, A. 1987. Tree management as part of two farming systems in the wet forest zone (Ivory Coast). *Acta Oecologica Oecologia Applicata* 8:39–51.

Dove, M. R. 1993. Smallholder rubber and swidden agriculture in Borneo: a sustainable adaptation to the ecology and economy of the tropical forest. *Economic Botany* 47:136–147.

García-Fernández, C., M. A. Pérez, and M. Ruiz. 2003. Benzoin gardens in North Sumatra, Indonesia: effects of management on tree diversity. *Conservation Biology* 17:829–836.

Gascon, C., G. B. Williamson, and G. A. B. da Fonseca. 2000. Receding forest edges and vanishing reserves. *Science* 288:1356–1358.

Gaudrain, C. 2002. *Chasse et diversité faunistique vue par les Indiens Bribri, en paysage fragmenté, Talamanca, Costa Rica.* Montpellier, France: ENGREF-IRD-CATIE.

Girault, R. 2002. *Effet de la distance à la forêt naturelle sur la diversité et la composition floristique du sous-bois des agroforêts de la province de Jambi (Sumatra, Indonesie).* Orleans, France: Université d'Orleans.

Gouyon, A., H. de Foresta, and P. Levang. 1993. Does "jungle rubber" deserve its name? An analysis of rubber agroforestry systems in southeastern Sumatra. *Agroforestry Systems* 22:181–206.

Griffith, D. M. 2000. Agroforestry: a refuge for tropical biodiversity after fire. *Conservation Biology* 14:325–326.

Guiracocha, G., C. A. Harvey, E. Somarriba, U. Krauss, and E. Carillo. 2001. Conservación de la biodiversidad en sistemas agroforestales con cacao y banano en Talamanca, Costa Rica. *Agroforestería en las Américas* 8:7–11.

Hardiwinoto, S., D. T. Adriyanti, H. B. Suwarno, D. Aris, M. Wahyudi, and S. M. Sambas. 1999. *Stand structure and species composition of rubber agroforests in tropical ecosystems of Jambi, Sumatra.* Bogor and Yogjakarta, Indonesia: ICRAF-SEA and Faculty of Forestry, Gadjah Mada University.

Jeffrey, S. M. 1977. Rodent ecology and land use in western Ghana. *Journal of Applied Ecology* 14:741–755.

Johns, N. D. 1999. Conservation in Brazil's chocolate forest: the unlikely persistence of the traditional cocoa agroecosystem. *Environmental Management* 23:31–47.

Katz, E., C. García, and M. Goloubinoff. 2002. Sumatra benzoin (*Styrax* spp.). Pages 182–190 in A. Guillen, S. A. Laird, P. Shanley, and A. R. Pierce (eds.), *Tapping the green market: certification and management of non-timber forest products.* London: Earthscan.

Lawrence, D. C. 1996. Trade-offs between rubber production and maintenance of diversity: the structure of rubber gardens in West Kalimantan, Indonesia. *Agroforestry Systems* 34:83–100.

Lodoen, D. 1998. Cameroon cocoa agroforests: planting hope for smallholder farmers. *Agroforestry Today* 10:3–6.

Mesquita, R. C. G., P. Delamônica, and W. F. Laurance. 1999. Effect of surrounding vegetation on edge-related tree mortality in Amazonian forest fragments. *Biological Conservation* 91:129–134.

Michon, G., and H. de Foresta. 1995. The Indonesian agro-forest model. Pages 90–106 in P. Halladay and D. A. Gilmour (eds.), *Conserving biodiversity outside protected areas: the role of traditional agroecosystems.* Gland, Switzerland: IUCN.

Michon, G., and H. de Foresta. 1999. Agro-forests: incorporating a forest vision in agroforestry. Pages 381–406 in L. E. Buck, J. P. Lassoie, and E. C. M. Fernandes (eds.), *Agroforestry in sustainable agricultural systems.* Boca Raton, FL: Lewis.

Michon, G., F. Mary, and J. Bompard. 1986. Multistoreyed agroforestry garden systems in West Sumatra, Indonesia. *Agroforestry Systems* 4:315–338.

Murniati, D., P. Garrity, and A. N. Gintings. 2001. The contribution of agroforestry systems to reducing farmers' dependence on the resources of adjacent national parks: a case study from Sumatra, Indonesia. *Agroforestry Systems* 52:171–184.

Parrotta, J. A., J. K. Francis, and R. R. de Almeida. 1995. *Trees of the Tapajós, a photographic field guide.* Rio Pedras, Puerto Rico: International Institute of Tropical Forestry, USDA Forest Service.

Pereira, H. S. 2000. Castanhais nativos: um caso de domesticação incidental de uma éspecie dominante do dossel de floresta tropical. Pages 353–356 in *III Congresso Brasileiro de Sistemas Agroflorestais.* Manaus, Brazil: Embrapa.

Peters, C. M., M. J. Balick, F. Kahn, and A. B. Anderson. 1989. Oligarchic forests of economic plants in Amazonia: utilization and conservation of an important tropical resource. *Conservation Biology* 3:341–349.

Politis, G. 2001. Foragers of the Amazon: the last survivors or the first to succeed? Pages 26–49 in C. McEwan, C. Barreto, and E. Neves (eds.), *Unknown Amazon.* London: The British Museum Press.

Preechapanya, P. 1996. Indigenous ecological knowledge about the sustainability of tea gardens in the hill evergreen forest of northern Thailand. Ph.D. Thesis, School of Agricultural and Forest Sciences, University of Wales, Bangor, UK.

Preechapanya, P., J. R. Healey, M. Jones and F. L. Sinclair. In press. Retention of forest biodiversity in multistrata tea gardens in northern Thailand. *Agroforestry Systems.*

Reitsma, R., J. D. Parrish, and W. McLarney. 2001. The role of cacao plantations in maintaining forest avian diversity in southeastern Costa Rica. *Agroforestry Systems* 53:185–193.

Rice, R. A., and R. Greenberg. 2000. Cacao cultivation and the conservation of biological diversity. *Ambio* 29:167–173.

Salafsky, N. 1993. Mammalian use of a buffer zone agroforestry system bordering Gunung Palung National Park, West Kalimantan, Indonesia. *Conservation Biology* 4:928–933.

Schroth, G., P. Coutinho, V. H. F. Moraes, and A. K. M. Albernaz. 2003. Rubber agroforests at the Tapajós River, Brazilian Amazon: environmentally benign land use systems in an old forest frontier region. *Agriculture, Ecosystems and Environment* 97:151–165.

Schroth, G., V. H. F. Moraes, and M. S. S. da Mota. In press. Increasing the profitability of traditional, planted rubber agroforests at the Tapajós River, Brazilian Amazon. *Agriculture, Ecosystems and Environment.*

Schroth, G., and F. L. Sinclair (eds.). 2003. *Trees, crops and soil fertility: concepts and research methods.* Wallingford, UK: CAB International.

Sonwa, D. J., S. F. Weise, A. A. Adesina, M. Tchatat, O. Ndoye, and B. A. Nkongmeneck. 2000a. *Dynamics of diversification of cocoa multistrata agroforestry systems in southern Cameroon.* International Institute of Tropical Agriculture Annual Report, 1–3. Ibadan, Nigeria: International Institute of Tropical Agriculture.

Sonwa, D. J., S. F. Weise, M. Tchatat, B. A. Nkongmeneck, A. A. Adesina, O. Ndoye, and J. J. Gockowski. 2000b. Les agroforêts cacao: espace intégrant développement de la cacaoculture, gestion et conservation des ressources forestières au Sud-Cameroun. *Second Pan African Symposium on the Sustainable Use of Natural Resources in Africa,* Ouagadougou, Burkina Faso, July 2000, 1–12.

Sonwa, D. J., S. F. Weise, M. Tchatat, B. A. Nkongmeneck, A. A. Adesina, O. Ndoye, and J. J. Gockowski. 2001. *The role of cocoa agroforests in rural and community forestry in southern Cameroon.* London: Overseas Development Institute.

Thiollay, J.-M. 1995. The role of traditional agroforests in the conservation of rain forest bird diversity in Sumatra. *Conservation Biology* 9:335–353.

van Dijk, J. F. W. 1999. *Non-timber forest products in the Bipindi-Akom Region, Cameroon. A socio-economic and ecological assessment.* Kribi: The Tropenbos-Cameroon Programme.

Villagren, G., and A. Boanerges. 1989. Situación actual del sector forestal en El Salvador. In R. Hernández, M. Juárez, and M. Zambrana (eds.), *Incentivos para la reforestación en El Salvador.* San Salvador, El Salvador: PROCAFE.

Vincent, G., F. Azhima, L. Joshi, and J. R. Healey. In press. Are permanent rubber agroforests an alternative to rotational rubber cultivation? An agro-ecological perspective. *Agroforestry Systems.*

Vincent, G., H. de Foresta, and R. Mulia. 2002. Predictors of tree growth in a Dipterocarp-based agroforest: a critical assessment. *Forest Ecology and Management* 161:39–52.

Wibawa, G., S. Hendratno, A. Gunawan, C. Anwar, Supriadi, A. Budiman, and M. van Noordwijk. 1999. Permanent rubber agroforest, based on gap replanting, as farmer strategy in Jambi, Indonesia. In F. Jiménez and J. Beer (eds.), *International Symposium on Multi-strata Agroforestry Systems with Perennial Crops, 22–27 Feb 1999, Extended Abstracts* (separate document). Turrialba, Costa Rica: CATIE.

Williams, S. E., M. van Noordwijk, E. Penot, J. R. Healey, F. L. Sinclair, and G. Wibawa. 2001. On-farm evaluation of the establishment of clonal rubber in multistrata agroforests in Jambi, Indonesia. *Agroforestry Systems* 53:227–237.

Chapter 11

Live Fences, Isolated Trees, and Windbreaks: Tools for Conserving Biodiversity in Fragmented Tropical Landscapes

Celia A. Harvey, Nigel I. J. Tucker, and Alejandro Estrada

At first glance, many deforested tropical landscapes appear to be simple mosaics of forest patches, interspersed with pastures and crop fields. However, closer examination reveals that many of the agricultural areas retain abundant and conspicuous tree cover, whether as individual isolated trees, live fences, windbreaks, or clusters of trees. Some of these trees are relicts of the original forest that were left standing when the area was cleared; others have regenerated naturally or been planted by farmers. Often, the isolated trees, live fences, and windbreaks form part of agroforestry systems that the farmers manage to obtain a wide array of goods and services. Although this on-farm tree cover is often overlooked or ignored in surveys of land use (FAO 2000; Kleinn 2000), analyses of forest fragmentation patterns, and conservation efforts, it may be critical to maintaining biodiversity in the fragmented landscapes that characterize many tropical regions (Guevara et al. 1998; Gascon et al. 1999; Harvey et al. 2000).

The presence of live fences, isolated trees, windbreaks, and other agroforestry elements in deforested regions could help conserve biodiversity by serving as habitats, corridors, or stepping stones for plant and animal species, adding structural and floristic complexity to the agricultural landscape and enhancing landscape connectivity. Whereas the importance of these agroforestry elements for conservation efforts has been studied in great detail in temperate regions (Forman and Baudry 1984; Baudry 1988; Capel 1988; Burel 1996), little attention has been focused on their ability to help conserve species in deforested regions in the tropics. Until recently, even the ample literature on the effects of forest fragmentation on the survival of plant and animal populations

261

in the tropics has largely ignored the ability of the surrounding agricultural matrix to support species diversity and enhance species persistence.

In this chapter, we examine the potential role of three common agroforestry elements—live fences, windbreaks, and isolated trees—in helping to retain plant and animal species and maintain the continuity of species populations and ecological processes in fragmented tropical landscapes. We focus on these elements because they are conspicuous in many regions of the tropics, are easily integrated into farm practices, and appear to hold potential for conservation efforts. We first characterize the abundance of live fences, isolated trees, and windbreaks in tropical regions and how farmers manage them. Next, we present information on the floristic and structural diversity that they represent and the fauna associated with them, focusing on the potential role of the agroforestry elements as habitats, food resources, stepping stones, and corridors (for their role in conserving genetic diversity and enhancing gene flow across fragmented landscapes see Chapter 12, this volume). Finally, we identify key gaps in our knowledge about their role in conservation efforts. Our focus on examples from Central America reflects the greater availability of information in this region; however, where possible we include examples from other tropical regions.

Throughout this chapter, *live fences* refers to narrow lines of trees or shrub species planted on farm boundaries or between pastures, fields, or animal enclosures whose primary purpose is to control the movement of animals or people (Westley 1990; Budowski and Russo 1993). Live fences usually are composed of a single row of trees or shrubs that are closely planted at uniform distances and may support barbed wire (Sauer 1979; Westley 1990), although sometimes they arise from natural regeneration underneath fence lines. *Windbreaks* refers to linear plantings of trees and shrubs (usually several rows wide) and linear strips of remnant vegetation whose primary function is to protect crops, livestock, and homes from wind damage (Finch 1988; Wight 1988). Although we focus on windbreaks, many of the generalizations about the relationships between windbreak structure and species composition and biodiversity conservation also hold for hedges. The term *isolated trees* refers to trees that are scattered in pastures, in fields, or around homes, occur in varying densities and spatial arrangements, and have variable origins (e.g., relicts of the original forest, naturally regenerated, or planted by farmers; Harvey and Haber 1999).

Importance of Live Fences, Windbreaks, and Isolated Trees in Tropical Regions

With the exception of commercial crops grown in large expanses (e.g., sugarcane, pineapple, and banana), most tropical agricultural landscapes contain at least some trees, although the density, diversity, and spatial arrangement vary greatly between sites. The use of live fences to delineate crop fields, pastures, and farm boundaries is common in Central America (Lagemann and Heuveldop

1983; Paap 1993), Mexico (Guevara et al. 1997), South America (Murgueitio and Calle 1999; Cajas-Giron and Sinclair 2001), Africa (Westley 1990), and several Caribbean countries (Budowski and Russo 1993), and early accounts of live fences and hedges exist from Africa, India, Australia, New Zealand, Peru, Cuba, Nigeria, and Costa Rica (Budowski and Russo 1993). In these countries, live fences occur in a wide range of conditions, from sea level to well above 1,500 m and from dry to humid environments (Budowski 1987). Isolated trees are also conspicuous features of many fragmented tropical landscapes, occurring in pastures in Central America (Guevara et al. 1998; Harvey and Haber 1999; Souza de Abreu et al. 2000), South America (Majer and Delabie 1999; Cajas-Giron and Sinclair 2001), and Australia (Crome et al. 1994; Fischer and Lindenmayer 2002a), African parklands (Gijsbers et al. 1994; Boffa 2000), and Central American milpas (small areas planted with maize, beans, or sorghum; Wilken 1977; Hellin et al. 1999). Windbreaks are commonly found in areas affected by heavy winds, extending along field borders, and in pastures and fields, creating complex networks of trees (Wilken 1977; Wight 1988).

Although an individual tree, live fence, or windbreak is likely to have little impact on landscape structure and be insignificant to conservation efforts, the presence of several agroforestry elements in the agricultural landscape may greatly enhance tree cover and structural heterogeneity and provide complementary habitats and resources to the remaining forest remnants, thereby contributing to biodiversity maintenance. In addition, by connecting forest

Table 11.1. Large-scale assessments of the occurrence of isolated trees, live fences, and windbreaks in tropical landscapes.

Agroforestry Element	Country	Area or Farms Surveyed	Prevalence in the Study Area	Reference
Live fences in pastures	La Fortuna, Costa Rica	35 cattle farms	85% of all cattle farms	Souza de Abreu et al. 2000
Live fences in farms	Guaitil, Costa Rica Tabarcia, Costa Rica	51 farms 41 farms	61% of farms 76% of farms	Marmillod 1989
Dispersed trees in pastures	Guaitil, Costa Rica Tabarcia, Costa Rica	51 farms 41 farms	93% of farms 79% of farms	Marmillod 1989
Dispersed trees in pastures	Veracruz, Mexico	5,509 ha	3.3% of the study area	Guevara et al. 1998
Dispersed trees in pastures	Moropotente, Nicaragua	278 km²	24.8% of the study area	Corrêa do Carmo et al. 2001
Dispersed trees in cattle farms	Litoral, Golfo de Morrosquillo, Sabana, and Valle de Sinu, Colombia	54 farms	26–69% of the farm area	Cajas-Giron and Sinclair, 2001

patches and other patches of remnant vegetation and forming complex, integrated networks of trees across agricultural landscapes, live fences may reduce the isolation between suitable habitats and influence animal movement patterns (Estrada et al. 1993, 1998; Guevara et al. 1998).

The prevalence of these agroforestry elements in many regions suggests that they may have a significant impact on conservation efforts (Table 11.1, previous page). For example, in Central and South American landscapes, 60–95 percent of the cattle farms have live fences and 25–93 percent of the farms have scattered, isolated trees in pastures (Table 11.1). In a study in Veracruz, Mexico, isolated trees covered approximately 3.3 percent of the total area in a 5,509-ha landscape and created a fragmented, discontinuous canopy that nevertheless enhanced biotic connectivity (Guevara et al. 1998).

Farmer Management and Use of Live Fences, Windbreaks, and Isolated Trees

In a particular region, the abundance and distribution of live fences, windbreaks, and isolated trees reflect the history of deforestation and land use as well as the management of farm tree resources (Browder 1996; Arnold and Dewees 1998; Janzi et al. 1999). When farmers clear forests to create agricultural lands, they often retain some forest patches, strips of trees along rivers or streams, and remnant forest trees as sources of future products and services, although in some tropical regions such as the Mata Atlantica of Brazil and parts of the Wet Tropics of northeastern Australia farmers have extensively cleared the land and left little tree cover.

Isolated trees typically are retained in pastures and agricultural areas because of their value as sources of timber, fenceposts, firewood, and fruits, as shade and forage for cattle, and as sources of organic matter for improving soil fertility or because their cutting is prohibited by law (Pezo and Ibrahim 1988; Marmillod 1989; Harvey and Haber 1999; Cajas-Giron and Sinclair 2001). They may also be retained or planted to beautify the farm landscape and increase its economic value (Wight 1988; Bird et al. 1992). Windbreaks are maintained or planted primarily to provide wind protection and prevent soil erosion, although they may provide additional functions and services (Baldwin 1988; Drone 1988; Wight 1988). In contrast, live fences usually are established to delineate borders with adjacent properties, divide pastures into smaller sections for cattle rotation, and prevent animals and humans from trespassing.

When choosing which trees to retain on their farms, farmers generally select healthy trees that have valuable timber or firewood, provide fruits for humans, or serve as cattle forage (Paap 1993; Barrance et al. 2003). Farmers may also carefully determine the distribution of trees within the farm, as is the case in Honduras where maize farmers tend to limit trees to field edges to minimize shading of associated crops (Barrance et al. 2003). This contrasts with tree distributions in pastures, where trees often are widely scattered across the entire pasture to offer shade and supplementary fodder to cattle while they are grazing (see also Chapter 19, this volume). Farmers may protect individual trees by clearing around the stem when they are saplings while weeding fields and pastures. To minimize competition between the trees and agricultural crops or pastures, farmers not only regulate tree densities and arrangements but also prune the lower branches of trees to reduce shade, taking care not to affect tree development (Kowal 2000; Barrance et al. 2003). Thus, tree management by farmers is likely to influence the potential of the land to conserve biodiversity.

Floristic and Structural Diversity of Live Fences, Isolated Trees, and Windbreaks

The value of individual agroforestry elements for conservation depends, to a large degree, on their floristic composition and structural diversity. In general, the greater the floristic and structural diversity, the greater the ability of the agroforestry element to provide habitat and resources for wildlife. Here we review information on the floristic and structural diversity documented in live fences, isolated trees, and windbreaks.

Floristic and Structural Diversity of Live Fences

When planted by farmers, live fences tend to be simple linear plantings of trees (usually of only a single species) that are evenly spaced and periodically pollarded and trimmed (Sauer 1979; Budowski 1987). Although numerous tree species may be used, a few species account for most live fences. For example, although more than 100 species are used in live fences in Costa Rica, only 8 species account for 95 percent of the posts (Budowski and Russo 1993). In the humid zones of Central America, northern South America, and several Caribbean countries, live fences generally consist of *Erythrina* spp. and *Gliricidia sepium*, whereas in dry areas they usually consist of *Bursera simaruba*, *Spondias purpurea*, and *Leucaena leucocephala* (Budowski 1987). Over time, some of the planted live fences are colonized by other plant species whose seeds are dispersed to the site by birds or other animals (Molano et al. 2002). However, because of the small area below the live fences, the open, exposed conditions, and the frequent disturbance by cattle and humans, only a limited number of plant species establish.

In contrast to planted fences, those that arise naturally underneath existing fences (from seeds dispersed to the site by animals or wind) or are relicts of the original vegetation harbor a greater diversity of life forms and plant species. In a survey of the flora in the understory of 19 naturally regenerated live fences in Piedemonte Llanero of Colombia, for example, a total of 247 plant species were found, most of which were bird-dispersed species (Molano et al. 2002). The high abundance of fruiting plants in naturally regenerated live fences makes them particularly attractive to birds, primates, and other frugivores (Molano et al. 2002; Luck and Daily 2003). As live fences age and become more structurally complex, the density and species richness of plants in the understory may change, reflecting a combination of ecological factors (seed input, seed banks, and regeneration dynamics), biophysical conditions, and management (e.g., pollarding and herbiciding), but only rarely do forest plants establish in these exposed areas.

Regardless of whether the live fences are species poor or floristically diverse, their presence enhances the structural diversity of the landscape, interrupting the monotony of pastures and crop fields and adding vertical and horizontal complexity. Live fences that contain a mixture of plant species with varying canopy physiognomies and with some fully grown trees clearly offer greater structural diversity than those that are uniform rows of a single tree species or those that are regularly reduced to large, leafless stumps by pruning. Often the live fences form complex rectilinear networks that follow field boundaries and topographic features, providing some degree of biotic connectivity (Estrada et al. 2000). For example, a study of cattle farms in La Fortuna, Costa Rica, found that there was an average of 0.16–0.19 km of live fence per hectare of pasture and that individual farms may include up to 52 km of live fences within their boundaries (Souza de Abreu et al. 2000), clearly influencing tree cover and connectivity within the farm.

Floristic and Structural Diversity of Isolated Trees

In contrast to planted live fences, isolated trees may represent a higher floristic and structural diversity depending on the tree origin (relict, regenerated, or planted), density, distribution within the landscape, and management by farmers. Although the floristic diversity represented by isolated trees is highly variable (Table 11.2), in some regions these trees may represent a significant portion of the original tree species present in the forest. For example, isolated trees in pastures of Monteverde, Costa Rica, represented 60 percent of the species present in the study area (Harvey and Haber 1999), whereas isolated trees in pastures in Veracruz, Mexico, represented 33 percent of the total rainforest tree flora, albeit at greatly reduced densities (Guevara et al. 1998; Table 11.2). In the traditional agricultural systems where farmers pollard or cut trees to provide mulch for crop production, tree diversity within the system can be quite high (Table 11.2) because many trees survive despite being pollarded and resprout in subsequent years (Wilken 1977; Hellin et al. 1999; Garcia Rodriguez et al. 2001; Barrance et al. 2003). However, in other regions where deforestation has been more complete and there are few isolated trees, the floristic diversity may be minimal: for example, in Rondônia, in the southwest

Table 11.2. Density and species richness of isolated trees in pastures and fields.

Habitat	Area Surveyed	Total Number of Tree Species	Total Number of Trees Surveyed	Mean Tree Density	Most Abundant Species	Reference
PASTURES						
Pastures in Veracruz, Mexico	81.4 ha	57 (61% were primary forest species)	265	0.3–39 trees ha^{-1}	*Nectandra ambigens* *Ficus yoponensis* *Brosimum alicastrum* *Ampelocera hottlei* *Bursera simaruba* *Spondias radlkoferi* *Zanthoxylum kellermanii*	Guevara and Laborde 1993; Guevara et al. 1994
Pastures in Veracruz, Mexico	173 ha (= 30 pastures)	98	735	3.3 trees ha^{-1} (range 0.4–11.9)[a]	*Bursera simaruba* *Zanthoxylum kellermanii* *Nectandra ambigens* *Ficus yoponensis* *Poulsenia armata* *Spondia radlkoferi* *Sapium nitidum* *Pouteria sapota* *Brosimum alicastrum*	Guevara et al. 1998
Pastures in Boaco, Nicaragua	40 ha	108	1,695	42.3 trees ha^{-1}	*Bursera simaruba* *Cordia alliodora* *Guazuma ulmifolia* *Tabebuia rosea* *Byrsonima crassifolia*	Zamora et al. 2002

(continues)

Table 11.2. *Continued*

Habitat	Area Surveyed	Total Number of Tree Species	Total Number of Trees Surveyed	Mean Tree Density	Most Abundant Species	Reference
Pastures in La Fortuna, Costa Rica	42.4 ha (9.6 ha in mixed systems, 4.2 ha in specialized milk systems, and 28.7 ha in dual-purpose systems)	NA	NA	12.46 trees ha^{-1} in mixed systems 22.1 trees ha^{-1} in dairy cattle systems 20.5 trees ha^{-1} in double-purpose systems	*Cordia alliodora*[b] *Cedrela odorata* *Terminalia oblonga*	Souza de Abreu et al. 2000
Pastures in Monteverde, Costa Rica	237 ha	190	5,583	25 trees ha^{-1}	*Psidium guajava* *Sapium glandulosum* *Acnistus arborescens* *Ocotea whitei* *Eugenia guatemalensis*	Harvey and Haber 1999
Pastures in Litoral, Golfo de Morrosquillo, Sabana, and Valle de Sinu, Colombia	(Based on interviews, not field surveys)	96[c]	(Based on interviews, not field surveys)	Ranges from a mean of 2.6 ± 0.68 (*SE*) trees ha^{-1} (on beef farms) to 53.1 ± 24.3 (*SE*) trees ha^{-1} on dual-purpose farms[c]	*Tabebuia rosea* *Albizia caribea* *Sterculia apelata*	Cajas-Giron and Sinclair 2001

CROP FIELDS						
Dispersed trees in milpas in El Salvador	2.5 ha	37	482		192 trees ha^{-1}	*Cordia alliodora* *Lysiloma auritum* *Genipa americana* *Tabebuia rosea*	Garcia Rodriguez et al. 2001
Trees in milpas of southern Honduras	NA[c]	41 (5 are exotic)	NA		13–139 stems ha^{-1} (of stems >2 m high)	*Cordia alliodora* *Swietenia humilis* *Lysiloma* spp. *Enterolobium cyclocarpum* *Albizia saman*	Barrance et al. 2003
Quezungal system in Lempira Sur, Honduras	2,250 m^2	20	300	coppiced tree	419 trees ha^{-1} (range: 190–666 trees/ha^{-1}) and 919 sprouts ha^{-1} (range: 400–1,421 sprouts ha^{-1})	*Cordia alliodora* *Gliricidia sepium* *Thounidium decandrum* *Lonchocarpus* sp. *Baubinia* sp.	Kowal 2000

[a]Based on aerial photo interpretation.

[b]Only timber species were identified in the survey.

[c]Calculations of species richness and tree densities are based on information provided by farmers, not tree inventories.

of the Brazilian Amazon, 10-year-old pastures retained only 20 of the 326 plant species present in the original forest and only 6 of the 196 tree species in the current forest (Fujisaka et al. 1998).

In some areas, isolated trees from a single species or group of species may dominate the landscape (Table 11.2). For example, in Moropotente, Nicaragua, pastures are dominated by *Acacia penatulata,* which occurs at mean densities of 240 trees per hectare (Nieto et al. 2001; see Chapter 19, this volume). In Costa Rica, some lowland pastures are dominated by *Cordia alliodora,* whereas some highland pastures are dominated by *Alnus acuminata* (Combe 1981; Lagemann and Heuveldop 1983), both of which are important timber species. Maize plots in Ilobasco, El Salvador, are dominated by *Cordia alliodora* trees that occur at mean densities of 86 trees per hectare (Garcia Rodriguez et al. 2001). The parklands of Burkina Faso, West Africa, are dominated by *Vitellaria paradoxa* (the shea butter tree), *Parkia biglobosa* (which produces edible fruits), and a few other species such as *Faidherbia albida* (Gijsbers et al. 1994). In contrast, Guevara et al. (1998) note that the species composition of isolated trees varies widely among pastures in Veracruz, Mexico, with no single species or group of species being dominant.

Isolated trees may further enhance the floristic diversity retained in the landscape by harboring diverse epiphyte communities, particularly if the trees are relicts of the original forest. For example, a study in Veracruz, Mexico, found a total of 35 orchid species occurring on isolated pasture trees, compared with 51 orchid species in forest fragments and 25 on shade trees in coffee plantations (Williams-Linera et al. 1995). Another study in the same region found that isolated trees retained 37 percent (58 species) of the vascular epiphytic and hemiepiphytic forest flora, despite the distinct microclimatic conditions in pastures compared with forests (Hietz-Seifert et al. 1996). Although epiphyte abundance was lower on isolated trees than on counterpart trees in adjacent forests, the epiphytic species richness per tree was similar in both habitats, suggesting that the isolated relict trees may be suitable habitats for epiphytes, at least in the short term after deforestation; however, whether these trees will maintain epiphytes in the long-term is not known. In contrast to remnant trees, trees that are planted in pastures tend to lack epiphytes or have poorly developed communities, probably because of the limited colonization in pasture habitats (Hietz-Seifert et al. 1996) compounded by the unsuitability of the pasture microclimate.

Another way in which isolated trees can increase floristic diversity is by serving as nuclei for forest regeneration. Many of the birds that visit isolated trees regurgitate or defecate seeds while perched in the trees, thereby dispersing seeds from forest patches into agricultural areas and enhancing both the abundance and species richness of seed input (Guevara and Laborde 1993; Galindo-Gonzalez et al. 2000; Holl et al. 2000). For example, seeds of 25 species of trees and shrubs were collected under isolated trees in pastures in the

Caribbean lowlands of Costa Rica (Slocum and Horvitz 2000). Similarly, seeds of a total of 107 plant species were deposited under isolated trees in pastures of Veracruz, Mexico, of which 56 species were dispersed by vertebrate frugivores (Guevara and Laborde 1993). In general, the seed rain arriving under isolated trees consists of mostly small-seeded pioneer species dispersed by frugivorous birds and bats (Guevara and Laborde 1993; Tucker and Murphy 1997; Galindo-Gonzalez et al. 2000). The amount and type of seeds arriving in pastures under isolated trees appear to depend on the type of fruit produced (fleshy or dry), tree height, the distance to adjacent forest, and possibly tree canopy architecture (Toh et al. 1999; Slocum and Horvitz 2000). Large, fruiting trees may attract more birds because they provide both feeding sites and good perch sites for spotting predators (Slocum and Horvitz 2000).

The modified microclimatic conditions (reduced solar irradiation and reduced temperature and humidity fluctuations) below tree crowns may be more favorable, and the soils may have better physical structure and water infiltration (Guevara et al. 1992; Belsky 1994), than open pastures, resulting in higher seed germination and plant establishment. In a study of vegetation under isolated trees in neotropical pastures in Veracruz, Mexico, the mean species richness of regenerating plants per quadrant was significantly higher under isolated tree canopies than at the canopy perimeter and in open pastures; a total of 193 species (109 woody and 84 herbaceous) were present under 50 isolated trees (Guevara et al. 1992). Similarly, in a subtropical rainforest site in southern Queensland, Australia, 48 canopy tree species were found regenerating underneath the crowns of tall residual trees, with the number of species increasing with tree height and crown area (Toh et al. 1999). By enhancing seed input and providing safe sites for tree establishment, the presence of isolated trees in pastures creates a positive feedback loop that, under appropriate management regimes, results in the growth of more trees and creates more perch and feeding sites for seed-dispersing animals (Slocum and Horvitz 2000). However, the long-term benefit of this enhanced forest regeneration will be realized only if the area is later abandoned and allowed to regenerate.

Floristic and Structural Diversity of Windbreaks

Planted windbreaks generally consist of a limited number of species carefully selected for their rapid growth, ability to provide adequate wind protection, and suitability for a given climatic zone. For example, windbreaks in the highlands of Costa Rica tend to consist of primarily exotic species such as *Cupressus lusitanica, Alnus jorullensis, Casuarina equisetifolia,* and *Croton niveus* (Combe 1981; Harvey et al. 2000). In Mexico, windbreaks are dominated by *Cupressus* sp. on the Pacific coast, *Tamarix* sp. and *Casuarina* sp. in the semi-arid areas, *Casuarina* sp. in the Golfo, and *Erythrina* sp. in the highlands of

Chiapas (Wilken 1977). Windbreaks are also common features of African countries, with the genera *Eucalyptus, Senna, Leucaena, Prosopis, Casuarina, Azadirachta,* and *Acacia* being used in dry areas (Krishnamurthy and Avila 1999). In tropical Australia, windbreaks are generally composed of *Eucalyptus* spp., hoop pine (*Araucaria cunninghamiana*), and the exotic conifer Caribbean pine (*Pinus caribea* var. *hondurensis;* Chapter 18, this volume). Unfortunately, many of these common windbreak species offer little in terms of resources for wildlife (Crome et al. 1994).

Despite the fact that the floristic diversity of planted windbreaks usually is quite limited, they can potentially facilitate natural regeneration in their understories by serving as perching and seed deposition sites for birds and other animals and providing a modified microclimate that enhances the establishment of some forest trees. A study in Monteverde, Costa Rica, found that windbreaks (consisting of *Montanoa guatemalensis, Cupressus lusitanica, Casuarina equisetifolia,* and *Croton niveus*) received 40 times as many tree seeds and more than twice as many species of seeds as adjacent pastures due to increased bird visitation, indicating the potential for windbreaks as foci for regeneration (Harvey 2000b). Surveys of the understories of windbreaks found a total of 91 tree species (including primary and secondary forest species) occurring as seedlings, just 5–6 years after the windbreaks were established (Harvey 2000a). Interestingly, windbreaks connected to forests had significantly higher numbers of tree species and higher densities of tree seedlings than those that were isolated from forests by 20–50 m (Harvey 2000a). This pattern probably reflects the greater activity of frugivorous birds in connected windbreaks (DeRosier 1995; Tucker 2001). Planted windbreaks consisting of *Eucalyptus camaldulensis, Tecoma stans,* and *Leucaena leucocephala* in León, Nicaragua, similarly appeared to serve as habitats for plant regeneration, although the density and species richness of trees (33 species) in windbreak understories were low, probably because of the frequent use of fire in adjacent agricultural lands (Alvarado et al. 2001). Although it is not clear how many of the regenerating seedlings will survive and grow into mature trees, there is at least a strong potential for the windbreaks to be colonized by native species. To a large degree, the fate of the seedlings depends on windbreak management practices, especially the exclusion of cattle (Capel 1988; Johnson and Beck 1988).

Fauna Associated with Live Fences, Isolated Trees, and Windbreaks

A variety of animal species may take advantage of agroforestry elements in fragmented landscapes, using them as habitats, foraging sites, corridors, or stepping stones to cross open areas. Here we review the available information on fauna using live fences, isolated trees, and windbreaks and identify factors that influence the value of these agroforestry elements for fauna conservation.

Fauna Associated with Live Fences

Live fences in tropical landscapes provide perching sites, cover, and foraging sites for some animals, including birds, bats, beetles, and nonflying mammals (Table 11.3). For example, a total of 98 bird species (representing 54 percent of the bird species detected in adjacent forest fragments) were detected in a 6-km-long live fence consisting of *Bursera simaruba* and *Gliricidia sepium* (with a few naturally regenerated species) in Veracruz, Mexico (Estrada et al. 1997). Similarly, in naturally regenerated live fences in Colombia, a total of 105 bird species of 45 families were found, with older, more structurally complex live fences having more bird species and more birds typical of forest borders and secondary growth (Molano et al. 2002). Although live fences often are dominated by bird species typical of edge or open habitats, a few forest interior resident species, including some that rarely leave the forest, also visit them (Estrada et al. 2000). The visiting bird community includes granivores, frugivores, and insectivores that use the fences as perches and foraging sites

Table 11.3. Summary of studies of fauna in live fences.

Type of Live Fence	Organisms	Number of Species	Guilds Represented	Reference
6-km live fence of *Bursera simaruba* and *Gliricidia sepium* in pastures of Veracruz, Mexico	Birds	98 species (58 resident, 40 migratory)	50% of birds were frugivores	Estrada et al. 2000; Estrada et al. 1997
	Bats	12 spp.	32% insectivores	Estrada et al. 2001
	Dung and carrion beetles	14 spp.	64% insectivores, 31% frugivores, 4% nectivores, 1% sanguinivores (represents 47% of the forest dung beetle species)	Estrada et al. 1998
Four live fences in pastures of Veracruz, Mexico	Nonflying mammals	11 spp.	36% scansorial, 27% arboreal, and 37% terrestrial	Estrada et al. 1994
Naturally regenerated live fences in El Pidemonte Llanero of Colombia	Birds	105 spp.	34% of avifauna in area	Molano et al. 2002
	Lizards	6 spp.	—	

(Estrada et al. 2000; Molano et al. 2002). Neotropical migrant bird species are also commonly sighted in live fences (Rappole 1995; Estrada et al. 1997).

The presence of live fences in fragmented landscapes may also benefit some bat species (Limpens and Kapteyn 1989; Verboom and Huitema 1997; Estrada and Coates-Estrada 2001). Twelve bat species were detected in *Bursera simaruba* and *Gliricidia sepium* live fences in Veracruz, Mexico, representing 37 percent of the bat species detected in the area and including some bat species that were not found in adjacent forests (Estrada and Coates-Estrada 2001). Another study in the same fragmented landscape showed that bat activity rates were significantly higher in the live fences than in pastures (Jimenez 2001). The presence of many bat species and the intense activity recorded in the live fence sites could reflect the enhanced insect distribution in these habitats (Lewis 1969; Dix and Leatherman 1988; Estrada and Coates-Estrada 2001).

Dung and carrion beetles may also use live fences. A total of 14 dung beetle species (representing 47 percent of the species detected in the forest samples) was found in live fences of *Bursera simaruba* and *Gliricidia sepium* in Veracruz, Mexico (Estrada et al. 1998). Although the live fences had lower species richness than the forests, shaded plantations (cocoa, coffee, and mixed plantations), and forest edges, they appeared to have more dung beetle species than open pastures. Dung beetles use live fences for perching, protection from predators, and foraging for dung and rotting fruit. By relocating and burying dung in the live fences, beetles may reduce the risk of predation of seeds contained in the dung (Estrada and Coates-Estrada 1991). Such secondary dispersal may aid in the establishment of other plant species under the shadow of the live fence trees.

Retaining live fences in agricultural landscapes could also help conserve nonflying mammals. In a study in Los Tuxtlas, Estrada et al. (1994) found a total of 11 species of nonflying mammals in four live fence sites, accounting for 29 percent of the nonflying mammal species detected in 35 forest fragments in the same landscape. The presence of these mammals suggests that individuals of these species are dispersing across the open landscape using live fences. Although such use may be primarily by small mammals, occasionally larger mammals, such as howler monkeys (average mass 6.5 kg), travel through old live fences (with trees with diameters at breast height greater than 25 cm) to reach forest fragments or fruiting trees (A. Estrada, pers. obs.).

Although live fences may be used as temporary or permanent habitats by many vertebrate species, the risks of reproducing in such exposed sites may be higher than those in remaining forested areas. In a study of artificial nest predation in Los Tuxtlas, Veracruz, Estrada et al. (2002) found that predation pressure (as measured by the mean number of nests surviving after 9 days) was higher in live fences than in the forest fragment interior but lower than that in forest-pasture edges or remnant corridors. Live fences could potentially be high-risk habitats for some animals because many crepuscular (e.g., bat falcon), nocturnal (owls), and diurnal raptors (e.g., hawks, falcons, and eagles) tend to use them to monitor open areas for potential prey (Estrada et al. 2000; Estrada and Coates-Estrada 2001).

Fauna Associated with Isolated Trees

Like live fences, isolated trees may provide habitats, perching and foraging sites, and stepping stones for a variety of animal species, particularly birds. For example, a study in Veracruz, Mexico, recorded 73 bird species visiting four isolated fig trees (*Ficus yoponensis* and *F. aurea*) in pastures (Guevara and Laborde 1993), and isolated trees in Costa Rican pastures were visited by at least 27 frugivorous bird species (Holl et al. 2000). Some of the frugivorous birds are resident species that nest in pastures, whereas other birds nest elsewhere and use the trees as perching or feeding sites (Guevara and Laborde 1993; Slocum and Horvitz 2000). Similarly, isolated *Eucalyptus* trees in sheep paddock of New South Wales, Australia, appear to be important for a large range of bird taxa, with 31 bird species observed using paddock trees (Fischer and Lindenmayer 2002a, 2002b). Although many of these birds are open-country birds, several birds considered to be woodland species were also observed visiting the trees (e.g., striated pardalote [*Paradalotus striatus*], scarlet robin [*Petroica multicolor*], grey shrike-thrush [*Colluricinla harmonica*], and crested shrike-tit [*Falcunculus frontatus*]; Fischer and Lindenmayer 2002a, 2002b).

Comparisons of pastures with and without isolated trees generally show higher bird diversity in the former. For example, recently abandoned pastures in Belize that contained isolated shrubs and trees had a total of 39 bird species, compared with only 15 species in actively grazed pastures with minimal tree cover (Saab and Petit 1994). These observations suggest that isolated trees or clumps of trees may provide important complementary habitats to woodland patches for birds and therefore warrant attention in conservation strategies.

Migrant birds may also benefit from the presence of scattered trees in active and recently abandoned pastures and fields (Greenberg et al. 1997; see also Chapter 19, this volume). For example, Lynch (1989a, 1989b) reported a total of 17 nearctic migrants occurring in habitats with isolated trees in the Yucatan Peninsula, Mexico, and noted that even small clumps of trees and bushes often sheltered forest migrants, such as the American redstart (*Setophaga ruticilla*), magnolia warbler (*Dendroica magnolia*), black-throated green warbler (*Dendroica virens*), and hooded warbler (*Wilsonia pusilla*), although typically at lower densities than were observed in the forest. In the Caribbean lowlands of Costa Rica, at least 15 migrant bird species use open areas, but 13 of these species are found almost exclusively foraging in trees that are remnants of the original closed forest or in hedgerows or wooded stream edges (Powell et al. 1989).

Isolated trees may also be important habitats for bats and other mammals. In Veracruz, Mexico, 20 bat species visited isolated trees in pastures, with frugivorous bats representing 83.1 percent of the total captures (Galindo-Gonzalez et al. 2000). The bats appeared to visit isolated trees year-round, even in periods when the fig trees were not fruiting. A wide variety of other animals may also benefit from the presence of isolated trees in pastures. For example, Slocum and Horvitz (2000) observed white-faced capuchins (*Cebus capucinus*) and howler monkeys (*Allouatta palliata*) feeding on isolated *Ficus* trees in pastures close to forest edges.

Arboreal and ground ant communities may also benefit from the presence of isolated trees in pastures. A detailed study of ant communities in isolated trees in pastures of the Atlantic rainforest region of Bahia, Brazil, reported 63 ant species on pasture trees and suggested that isolated trees play an important role in conserving elements of the original rainforest ant fauna (Majer and Delabie 1999). The richness of the ground-foraging ant community near isolated trees was nearly equivalent to that of the mature forest, and the species richness of arboreal ants was positively correlated with tree height, crown diameter, and epiphyte load. A similar study of isolated trees in both active and fallow fields of cassava, maize, yam, and groundnuts in Ghana found that ant species richness per tree, total ant species richness, and beetle abundance were higher near trees than in the open and noted that large trees had a greater effect on ant species than small trees, although not all ant species responded similarly to the presence of trees (Dunn 2000). Interestingly, although isolated trees had a strong local effect on species richness and abundance, the presence of isolated trees had little effect at the scale of whole fields.

The high abundance and diversity of birds, bats, and other animals using isolated trees in pastures and fields result partially from the high availability of fruits: many of the remnant trees produce large quantities of fruits in open areas (because of higher light availability), and many of the common pioneer tree species that regenerate naturally in pastures and fields are fruit-bearing species (Lynch 1989a). For example, 94 percent of the isolated tree species found in the pastures of Monteverde, Costa Rica, produce fruits that are dispersed by birds, bats, or other mammals (Harvey and Haber 1999), and 55 percent of the isolated trees in pastures of Chiapas, Mexico, are fleshy-fruited, presumably attracting birds and other animals (Otero-Arnaiz et al. 1999). Certain tree species, such as figs, may be particularly important food sources because they attract a wide variety of birds, bats, and other mammals (Guevara and Laborde 1993). In Central America, other tree species that appear to be critical for conserving biodiversity in fragmented habitats include *Dipteryx panamensis,* which are the primary food source and nesting sites of the great green macaw (R. Bjork, pers. comm.), *Inga* spp. that are visited by a large number of migrant (especially Tennessee warblers, orioles) and resident nectarivores when flowering (Greenberg 1996), and plants in the Melastomataceae family (Luck and Daily 2003).

In some cases, isolated trees may help maintain some animal populations in fragmented landscapes by serving as stepping stones for both local and regional movement and as stopover points for shelter and resting during landscape-scale movements. For example, a study of birds visiting isolated fig trees in Veracruz, Mexico, found that the frequency of flight direction of birds arriving at the fig trees was highly correlated with the presence of live fences, other isolated trees, or other remnant vegetation, suggesting that bird movement patterns closely follow the arboreal elements in the landscape (Guevara and Laborde 1993). Trees in pastures in Australia similarly appear to serve as stepping stones for a variety of birds, such as Major Mitchell's cockatoo (*Cacatua leadbeateri;* Rowley and Chapman 1991), foliage-foraging birds, and some granivores and nectarivores (Fischer and Lindenmayer 2002b). Birds that undertake landscape-scale movements or migrations may use the isolated trees as stopover points for shelter and resting, as appears to be the case in Monteverde, Costa Rica, where the threatened three-wattled bellbird (*Procnias tricarunculata*) and the resplendent quetzal (*Pharomachrus moccino*) follow the fruiting patterns of Lauraceae and other trees as they migrate altitudinally from high, forested areas to lower, fragmented habitats (Harvey et al. 2000). Some frugivorous birds follow riparian corridors and isolated trees when they fly through the landscape, taking advantage of these elements for protection and food resources (Guevara et al. 1998; Slocum and Horvitz 2000). However, the matrix tolerance of each bird species also influences the degree to which they can use the agricultural habitat.

Fauna Associated with Windbreaks

Numerous detailed studies from temperate regions have shown that, depending on their floristic diversity, structural complexity, and management, windbreaks may help conserve a large number of plant and animal species, including a limited number of forest-dependent species, by providing food, cover from predators, refuge, and travel lanes (e.g., Arnold 1983; Osborne 1983; Fournier and Loreau 2001). Windbreaks tend to have the greatest conservation value if they contain a variety of native plant species and life forms, connect to intact forest or other natural vegetation, are wide (so that they contain some interior habitat), and are protected from grazing cattle (Arnold 1983; Capel 1988; Johnson and Beck 1988; Fritz and Merriam 1993, 1996; Burel 1996). In general, the greater the structural and floristic diversity, the more ecological niches are available for other plants and animals. When windbreaks connect forest fragments or other remnant vegetation, they may also serve as corridors for some animal species (Yahner 1983; Haas 1995). The modified microclimatic conditions in the windbreaks may be more favorable than those in the open pastures or fields and provide protection from weather extremes; however, these microclimatic conditions are likely to be spatially and temporarily variable throughout the length of the windbreak. Most of the species that benefit from the presence of windbreaks are edge species that are capable of using highly modified habitats; few forest interior species appear to take advantage of windbreak habitats (Burel 1996; Corbit et al. 1999).

In contrast to the abundant, detailed studies of the fauna associated with windbreaks in temperate areas, little is known about the importance of windbreaks for the conservation of tropical biodiversity. This reflects, in part, the shorter history of windbreaks and the still recent deforestation and conversion of forested areas to agricultural use. However, the emerging data suggest that tropical windbreaks may fulfill many of the same roles as their temperate counterparts.

There is some evidence that windbreaks may be important habitat for some tropical bird species. A 3-year study of birds in Monteverde, Costa Rica, found 64 bird species in planted windbreaks of three exotic species (*Casuarina equisetifolia, Cupressus lusitanica,* and *Croton niveus*) and one native species (*Montanoa guatemalensis*), compared with 74 bird species in natural windbreaks that were remnants of the original forest (Nielson and DeRosier 2000). Interestingly, the windbreaks consisting of natural vegetation appeared to serve as habitats and nesting sites for birds, whereas the planted windbreaks seemed to be only transient foraging sites and travel paths (D. Hamilton, unpublished data reported in Harvey et al. 2000), suggesting that natural windbreaks are more suitable habitats for birds than planted windbreaks. A similar study of planted windbreaks (consisting of *Eucalyptus camaldulensis, Leucaena leucocephala,* and *Tecoma stans*) in León, Nicaragua, reported a total of 35 bird species using the windbreaks (Alvarado et al. 2001). The most common species were widespread and open habitat species, but the windbreaks also harbored four species that are cited as threatened in the Convention on International Trade in Endangered Species Appendix II (Alvarado et al. 2001).

Although windbreaks may contribute to the conservation of some taxa, other species may not benefit from the presence of windbreaks or may even be affected negatively. For example, a study of fauna in planted windbreaks (consisting of one or two lines of *Eucalyptus* trees and grazed by cattle) and riparian areas in Queensland, Australia, found that whereas the windbreaks were useful for some bird species (except for rainforest species), they were insignificant habitat for mammals (Crome et al. 1994) and had much lower value as wildlife habitat than the riparian vegetation. A total of 37 bird species were found using the windbreaks, compared with 62 species in the riparian regrowth vegetation. Very few rainforest specialist bird species and no small mammals or arboreal mammals were caught in the windbreaks, perhaps reflecting their lower structural complexity and monospecific nature.

As linear elements that often connect forest patches or other remnant vegetation, windbreaks could serve as corridors or travel lanes for some animal species, especially if they are structurally and floristically similar to forest habitats and connected to patches of suitable habitat (Fritz and Merriam 1996; Bennett 1999), but whether animals use the windbreaks as linkages depends on their ability to disperse through the matrix and the structural characteristics of the windbreak. Studies in temperate regions show that a limited number of plant and animal species may be channeled through windbreaks, live fences, and other connecting networks (Yahner 1982a, 1982b; Bennett et al. 1994; Haas 1995). In the tropics, a few recent studies provide preliminary evidence that bird species may use windbreaks as corridors (DeRosier 1995). In tropical Australia, the establishment of a corridor (1.5 km by 100 m) to connect two forest patches has induced the rapid colonization and movement by a range of organisms (N. I. J. Tucker, pers. obs., 2002). Within a 3-year period, the corridor was colonized by 119 new plant species, 40 percent of which were not present in the surrounding agricultural matrix. The majority of new species were dispersed to the site by birds, although spectacled flying foxes (*Pteropus conspicillatus*) and other mammals were also implicated in dispersal. Avian communities in the restoration were almost identical to intact forests within 3 years, and a small mammal community comprising mainly forest species was also present (Tucker 2001). In addition, 18 morphospecies of wood-boring beetle (Coleoptera) colonized dead wood placed in the corridor before plant establishment (Grove and Tucker 2000), indicating that invertebrate colonization can also be quite rapid, although seasonal fluctuations in species diversity probably result from edge-related effects during the dry season.

Population Dynamics of Isolated Trees and Live Fences in Fragmented Landscapes

Despite the clear potential of agroforestry elements to maintain biodiversity in agricultural landscapes, there is concern that the diversity and density of trees in agricultural landscapes are slowly eroding through a combination of tree harvesting and natural death (Powell et al. 1989; Gijsbers et al. 1994; Harvey and Haber 1999). Many of the primary forest tree species in pastures that are relicts of the original forest do not regenerate in open habitats under current management systems and will not be replaced after they die or are harvested (Harvey and Haber 1999). Because many of the tree species in pastures occur at low densities (Guevara et al. 1998; Harvey and Haber 1999), the elimination or natural death of even a few trees can result in the local loss of that species from the landscape. The size distribution of primary forest trees in pastures and crop fields often reflects this lack of regeneration. Few individuals in the small size classes are found in pastures in Monteverde and Cañas, Costa

Rica (Harvey and Haber 1999; Morales and Kleinn 2001). The same trend toward lower tree densities and lower species richness on farmland also appears to be occurring in West African parklands (Gijsbers et al. 1994). In northeastern Queensland, Australia, *Ficus* trees are similarly being lost from pastures, and the lack of active replacement may affect both cattle production and conservation (N. I. J. Tucker, pers. obs., 2002). These trends toward lower tree densities on farmland could be changed if new management practices that favor tree establishment—such as reducing grazing, preventing fires, and fencing off areas to allow regeneration—were implemented.

In some tropical regions, the diversity of tree species used as live fences has also diminished in recent years. For example, in Costa Rica, where more than 100 species have been recorded in live fences, only a handful of species now dominates the landscape (Sauer 1979). In some areas, naturally regenerated live fences have been removed to make way for the construction of new roads or the expansion of agricultural land or for farm mechanization, reducing habitat available for wildlife.

What We Still Don't Know

Although the limited (but rapidly growing) literature suggests that a significant subset of the original flora and fauna may use live fences, isolated trees, and windbreaks as resources, habitats, corridors, or stepping stones, our understanding of the conservation value of these agroforestry elements is still in its infancy. Here we outline some of the key issues that warrant urgent attention if the conservation potential of these agroforestry elements is to be clearly understood and used.

First, more information is needed on the abundance, density, diversity, and spatial arrangements of windbreaks, live fences, and isolated trees and the consequences of these different arrangements for biodiversity conservation. In addition to documenting the distribution of agroforestry elements in the landscape, it is also critical to understand how these agroforestry elements complement remnant vegetation in the landscape and the degree to which animals or plants using them also depend on or use alternative habitats (especially remnant vegetation). Because almost all studies reported in this chapter (and in the literature) were conducted in landscapes where there was some remnant vegetation, it is not clear how well the biodiversity recorded in agroforestry elements reflects the habitat value of the agroforestry elements themselves or is a function of remaining remnant vegetation in the surrounding landscape. Further studies are needed to better elucidate the complementarity of agroforestry elements with remnant vegetation and how biodiversity in the agroforestry elements changes as the remaining forest cover increases or decreases in the landscape.

It will also be important to examine the scale at which live fences, windbreaks, and isolated trees are important for individual populations of plants and animals and whether they contribute to both local and regional biodiversity or only to local biodiversity, as Dunn (2000) found for isolated trees in pastures and ant communities. Information on how windbreaks, isolated trees, and live fences affect ecological processes such as animal dispersal, migration, seed dispersal, and pollen flow in fragmented landscapes is also sorely lacking (but see Nason et al. 1997; Thébaud and Strasberg 1997; Aldrich and Hamrick 1998; Chapter 12, this volume).

Detailed studies are also needed to determine how plants and animals use agroforestry elements and to what degree they depend on the agroforestry elements for food, shelter, or reproduction (relative to their dependence on other habitats in the landscape). The mere presence of animal or plant populations in live fences, windbreaks, and isolated trees does not indicate that these habitats are suitable for their persistence unless it is clear that they reproduce and survive in them. Even when animals or plants are reproducing in live fences, windbreaks, and isolated trees, it is not known whether survival rates are similar to those in the original forest or whether individuals suffer greater predation or competition. Additional information on the population biology of plant and animal species using or occurring in live fences, windbreaks, and isolated trees would allow us to know whether these habitats are population sinks or sources and whether populations are viable in the long term.

A few studies have indicated the potential importance of isolated trees, windbreaks, and live fences as conduits for animal movement, but there is still a need for more detailed studies of animal movement patterns and the factors that influence the use of agroforestry elements as corridors and stepping stones. In particular, it will be important to determine whether the presence of agroforestry elements increases gene flow in the fragmented landscape, colonization rates of unoccupied patches, and adaptive genetic variance for population fitness (Rosenberg et al. 1997). If these systems are indeed serving as corridors, it is also important to ensure that they are not facilitating the spread of exotic species or generalist species at the expense of forest interior species (Tucker 2000; see also Chapter 3, this volume).

To date, the available information shows a strong bias toward birds, bats, and other mammals, with few studies considering insects and belowground organisms. Yet because individual species and taxonomic groups respond differently to fragmented landscapes and to the agroforestry elements in them (depending on their behavior, dispersal capabilities, habitat needs, and ability to adapt to modified landscapes), it is important to study and compare a wide variety of organisms (Bennett 1999; Gascon et al. 1999) to determine which species or guilds will be able to take advantage of the live fences, isolated trees, or windbreaks and persist in fragmented landscapes, and, conversely, which organisms may be affected negatively.

Another gap in our limited understanding is how farmers design and manage live fences, windbreaks, and isolated trees and how farmer decisions influence their conservation value. Because live fences, windbreaks, and isolated trees are features of agricultural lands created and maintained by humans, any efforts to integrate these arboreal elements into conservation efforts must carefully understand their role in the farming system and the rural society that maintains them (Burel 1996; Schelhas 1996). In particular, it will be important to understand how farmers decide to retain, plant, or eliminate agroforestry elements, in what densities and arrangements they position them, which species they plant or retain, and how they manage them. Another key need is to identify the benefits and drawbacks of different agroforestry systems for farm productivity, including possible alterations in farm productivity and pest dynamics (Timm 1988), and the potential trade-offs or synergisms between retaining agroforestry elements in the landscape for conservation or agricultural purposes (see Chapter 19, this volume).

Finally, because tropical landscapes are dynamic entities, shaped by both socioeconomic and ecological processes, it is also important to understand how changes in the abundance, distribution, and diversity of agroforestry elements affect plant and animal populations. Of particular concern are the long-term consequences of the gradual loss of relict isolated trees in pastures and crop fields through natural death and harvesting.

Conclusions

The emerging data show that live fences, windbreaks, and isolated trees may contribute to biodiversity conservation and suggest that retaining or establishing trees in agricultural lands may be a critical component of conservation efforts in fragmented landscapes. The floristic diversity conserved in these agroforestry systems can be high, and a substantial number of animal species may exploit these habitats for feeding, movement, and in some cases reproduction, although the value of each agroforestry element depends on its structure, composition, management, and position in the landscape. Many species that benefit from agroforestry systems are generalist species, but some forest specialist species usually are also present. By forming networks of natural habitats, live fences, windbreaks, and isolated trees may also enhance landscape connectivity and contribute biodiversity conservation at different scales.

However, it should be emphasized that although these agroforestry elements are useful additions or complements to the conservation of natural habitats, they are not substitutes for the original vegetation. Live fences, windbreaks, and isolated trees are not complete ecological units and cannot provide the full array of habitats or services of the original habitat; consequently, the organisms in them are likely to depend, at least to some degree, on nearby remnant habitats. Efforts to conserve biodiversity in fragmented landscapes therefore should focus on developing landscape-scale strategies that integrate the retention and establishment of windbreaks, live fences, isolated trees, and other agroforestry elements with the conservation of forest fragments, the retention of riparian vegetation, the maintenance of connectivity in the agricultural landscape, and other conservation strategies (Vandemeer and Perfecto 1997; Harvey et al. 2000; Tucker 2000; Daily et al. 2001).

Acknowledgments

The authors thank Manuel Guariguata, Tamara Benjamin, Gary Luck, and Heraldo Vasconcelos for their suggestions on earlier drafts of this chapter. Partial funding for Celia Harvey was provided by the European Community Fifth Framework Programme, "Confirming the International Role of Community Research" (INCO, ICA4-CT-2001-10099). The authors are solely responsible for the material reported here; this publication does not represent the opinion of the European Community and the Community, is not responsible for any use of the data appearing herein.

References

Aldrich, P. R., and J. L. Hamrick. 1998. Reproductive dominance of pasture trees in a fragmented tropical forest mosaic. *Science* 281:103–105.

Alvarado, V., E. Anton, C. A. Harvey, and R. Martinez. 2001. Importancia ecológica de las cortinas rompevientos al este de la ciudad de León, Nicaragua. *Revista Agroforestería en las Américas* 8:18–24.

Arnold, G. W. 1983. The influence of ditch and hedgerow structure, length of hedgerows and area of woodland and garden on bird numbers of farmland. *Journal of Applied Ecology* 20:731–750.

Arnold, J. E. M., and P. A. Dewees. 1998. Trees in managed landscapes: factors in farmer decision making. Pages 277–294 in L. E. Buck, J. P. Lassoie, and E. C. M. Fernandes (eds.), *Agroforestry in sustainable agricultural systems.* Boca Raton, FL: CRC Press.

Baldwin, C. S. 1988. The influence of field windbreaks on vegetable and specialty crops. *Agriculture, Ecosystems and Environment* 22–23:191–203.

Barrance, A. J., L. Flores, E. Padilla, J. E. Gordon, and K. Schreckenberg. 2003. Tree and farming in the dry zone of southern Honduras I: campesino tree husbandry practices. *Agroforestry Systems* 59:97–106.

Baudry, J. 1988. Hedgerows and hedgerow networks as wildlife habitat in agricultural landscapes. Pages 111–123 in J. R. Park (eds.), *Environmental management in agriculture: European perspectives.* London: Belhaven.

Belsky, A. J. 1994. Influences of trees on savanna productivity: tests of shade, nutrients and tree-grass competition. *Ecology* 75:922–932.

Bennett, A. F. 1999. *Linkages in the landscape: the role of corridors and connectivity in wildlife conservation.* Gland, Switzerland: IUCN.

Bennett, A. F., K. Henein, and G. Merriam. 1994. Corridor use and the elements of corridor quality: chipmunks and fencerows in a farmland mosaic. *Biological Conservation* 68:155–165.

Bird, P. R., D. Bicknell, P. A. Bulman, S. J. A. Burke, J. F. Leys, J. N. Parker, F. J. van der

Sommen, and P. Voller. 1992. The role of shelter in Australia for protecting soils, plants and livestock. *Agroforestry Systems* 20:59–86.

Boffa, J. M. 2000. West African agroforestry parklands: keys to conservation and sustainable management. *Unasylva* 51:11–17.

Browder, J. O. 1996. Reading colonist landscapes: social interpretations of tropical forest patches in an Amazonian agricultural frontier. Pages 285–299 in J. Schelhas and R. Greenberg (eds.), *Forest patches in tropical landscapes*. Washington, DC: Island Press.

Budowski, G. 1987. Living fences in tropical America, a widespread agroforestry practice. Pages 169–178 in H. L. Gholz (ed.), *Agroforestry: realities, possibilities and potentials*. Dordrecht, the Netherlands: Martinus Nijhoff.

Budowski, G., and R. Russo. 1993. Live fence posts in Costa Rica: a compilation of the farmer's beliefs and technologies. *Journal of Sustainable Agriculture* 3:65–85.

Burel, F. 1996. Hedgerows and their role in agricultural landscapes. *Critical Reviews in Plant Sciences* 15:169–190.

Cajas-Giron, Y. S., and F. L. Sinclair. 2001. Characterization of multistrata silvopastoral systems on seasonally dry pastures in the Caribbean Region of Colombia. *Agroforestry Systems* 53:215–225.

Capel, S. W. 1988. Design of windbreaks for wildlife in the Great Plains of North America. *Agriculture, Ecosystems and Environment* 22–23:337–347.

Combe, J. 1981. Jaúl con pastos: práctica silvopastoril en el nivel submontano de Costa Rica. *Serie Técnica, Boletin Técnico* (CATIE), Turrialba, Costa Rica 14:71–75.

Corbit, M., P. L. Marks, and S. Gardescu. 1999. Hedgerows as habitat corridors for forest herbs in central New York, USA. *Journal of Ecology* 87:220–232.

Corrêa do Carmo, A. P., B. Finegan, and C. Harvey. 2001. Evaluación y diseño de un paisaje fragmentado para la conservación de biodiversidad. *Revista Forestal Centroamericana* 34:35–41.

Crome, F., J. Isaacs, and L. Moore. 1994. The utility to birds and mammals of remnant riparian vegetation and associated windbreaks in the tropical Queensland uplands. *Pacific Conservation Biology* 1:328–343.

Daily, G. C., P. R. Ehrlich, and G. A. Sanchez-Azofeifa. 2001. Countryside biogeography: use of human dominated habitats by the avifauna of southern Costa Rica. *Ecological Applications* 11:1–13.

DeRosier, D. 1995. *Agricultural windbreaks: conservation and management implications of corridor usage by avian species*. Master's thesis, School of the Environment, Duke University, Durham, NC.

Dix, M. E., and D. Leatherman. 1988. Insect management in windbreaks. *Agriculture, Ecosystems and Environment* 22–23:513–538.

Drone, S. I. 1988. Layout and design criteria for livestock windbreaks. *Agriculture, Ecosystems and Environment* 22–23:231–240.

Dunn, R. R. 2000. Isolated trees as foci of diversity in active and fallow fields. *Biological Conservation* 95:317–321.

Estrada, A., P. L. Cammarano, and R. Coates-Estrada. 2000. Bird species richness in vegetation fences and in strips of residual rain forest vegetation at Los Tuxtlas, Mexico. *Biodiversity and Conservation* 9:1399–1416.

Estrada, A., and R. Coates-Estrada. 1991. Howling monkeys (*Alouatta palliata*), dung beetles (Scarabaeidae) and seed dispersal: ecological interactions in the tropical rain forest of Los Tuxtlas, Veracruz, Mexico. *Journal of Tropical Ecology* 7:459–474.

Estrada, A., and R. Coates-Estrada. 2001. Bat species richness in live fences and in corridors of residual rain forest vegetation at Los Tuxtlas, Mexico. *Ecography* 24:94–102.

Estrada, A., R. Coates-Estrada, A. Anzures Dadda, and P. Cammarano. 1998. Dung and carrion beetles in tropical rain forest fragments and agricultural habitats at Los Tuxtlas, Mexico. *Journal of Tropical Ecology* 14:577–593.

Estrada, A., R. Coates-Estrada, and D. A. Merritt. 1994. Non flying mammals and landscape changes in the tropical rain forest region of Los Tuxtlas, Mexico. *Ecography* 17:229–241.

Estrada, A., R. Coates-Estrada, and D. A. Merritt. 1997. Anthropogenic landscape changes and avian diversity at Los Tuxtlas, Mexico. *Biodiversity and Conservation* 6:19–42.

Estrada, A., R. Coates-Estrada, D. Meritt Jr., S. Montiel, and D. Curiel. 1993. Patterns of frugivore species richness and abundance in forest islands and in agricultural habitats at Los Tuxtlas, Mexico. *Vegetatio* 107–108:245–257.

Estrada, A., A. Rivera, and R. Coates-Estrada. 2002. Predation of artificial nests in a fragmented landscape in the tropical region of Los Tuxtlas, Mexico. *Biological Conservation* 106(2):199–209.

FAO (Food and Agriculture Organization of the United Nations). 2000. Trees outside forests. *Unasylva* 51:1–68.

Finch, S. J. 1988. Field windbreaks: design criteria. *Agriculture, Ecosystems and Environment* 22–23:215–228.

Fischer, J., and D. B. Lindenmayer. 2002a. The conservation value of paddock trees for birds in a variegated landscape in southern New South Wales. 1. Species composition and site occupancy patterns. *Biodiversity and Conservation* 11:807–832.

Fischer, J., and D. B. Lindenmayer. 2002b. The conservation value of paddock trees for birds in a variegated landscape in southern New South Wales. 2. Paddock trees as stepping stones. *Biodiversity and Conservation* 11:833–849.

Forman, R. T. T., and J. Baudry. 1984. Hedgerows and hedgerow networks in landscape ecology. *Environmental Management* 8:495–510.

Fournier, E., and M. Loreau. 2001. Respective roles of recent hedges and forest patch remnants in the maintenance of ground-beetle (Coleoptera: Carabaidae) diversity in an agricultural landscape. *Landscape Ecology* 16:17–32.

Fritz, R., and G. Merriam. 1993. Fencerow habitats for plants moving between farmland forests. *Biological Conservation* 64:141–148.

Fritz, R., and G. Merriam. 1996. Fencerow and forest edge architecture in Eastern Ontario farmland. *Agriculture, Ecosystems and Environment* 59:159–170.

Fujisaka, S., C. Castilla, G. Escobar, V. Rodriguez, E. J. Veneklaas, R. Thomas, and M. Fisher. 1998. The effects of forest conversion on annual crops and pastures: estimates of carbon emission and plant species loss in a Brazilian Amazon ecology. *Agriculture, Ecosystems and Environment* 69:17–26.

Galindo-Gonzalez, J., S. Guevara, and V. J. Sosa. 2000. Bat and bird-generated seed rains at isolated trees in pastures in a tropical rainforest. *Conservation Biology* 14:1693–1703.

García Rodríguez, E., M. Jaime Najarro, B. Mejia Bendeck, L. Guillen, and C. A. Harvey. 2001. Caracterización de los árboles dispersos en las milpas del Municipio de Ilobasco, Cabañas, El Salvador. *Revista Agroforesteria en las Americas* 8:39–44.

Gascon, C., T. E. Lovejoy, R. O. Bierregaard Jr., J. R. Malcolm, P. C. Stouffer, H. L. Vasconcelos, W. F. Laurance, B. Zimmerman, M. Tocher, and S. Borges. 1999. Matrix habitat and species richness in tropical forest remnants. *Biological Conservation* 91:223–229.

Gijsbers, H. J. M., J. J. Kessler, and M. K. Knevel. 1994. Dynamics and natural regeneration of woody species in farmed parklands in the Sahel region (Province of Passore, Burkina Faso). *Forest Ecology and Management* 64:1–13.

Greenberg, R. 1996. Managed forest patches and the diversity of birds in southern Mexico. Pages 59–90 in J. Schelhas and R. Greenberg (eds.), *Forest patches in tropical landscapes.* Washington, DC: Island Press.

Greenberg, R., P. Bichier, and J. Sterling. 1997. Acacia, cattle and migratory birds in southeastern Mexico. *Biological Conservation* 80:235–247.

Grove, S. J., and N. I. J. Tucker. 2000. Importance of mature timber habitat in forest management and restoration: what can insects tell us?. *Ecological Management and Restoration* 1:68–69.

Guevara, S., and J. Laborde. 1993. Monitoring seed dispersal at isolated standing trees in tropical pastures: consequences for local species availability. *Vegetatio* 107–108:319–338.

Guevara, S., J. Laborde, D. Liesenfeld, and O. Barrera. 1997. Potreros y ganadería. Pages 43–58 in E. Gonzalez Soriano, R. Dirzo, and R. Vogt (eds.), *Historia natural de Los Tuxtlas.* Mexico City: UNAM, CONBIO.

Guevara, S., J. Laborde, and G. Sanchez. 1998. Are isolated remnant trees in pastures a fragmented canopy? *Selbyana* 19:34–43.

Guevara, S., J. Meave, P. Moreno-Casasola, and J. Laborde. 1992. Floristic composition and structure of vegetation under isolated trees in neotropical pastures. *Journal of Vegetation Science* 3:655–664.

Guevara, S., J. Meave, P. Moreno-Casasola, J. Laborde, and S. Castillo. 1994. Vegetación y flora de potreros en la Sierra de los Tuxtlas, Mexico. *Acta Botanica Mexicana* 28:1–27.

Haas, C. 1995. Dispersal and use of corridors by birds in wooded patches on an agricultural landscape. *Conservation Biology* 9:845–854.

Harvey, C. A. 2000a. The colonization of agricultural windbreaks by forest trees: effects of windbreak connectivity and remnant trees. *Ecological Applications* 10:1762–1773.

Harvey, C. A. 2000b. Windbreaks enhance seed dispersal into agricultural landscapes in Monteverde, Costa Rica. *Ecological Applications* 10:155–173.

Harvey, C. A., C. F. Guindon, W. A. Haber, D. Hamilton DeRosier, and K. G. Murray. 2000. The importance of forest patches, isolated trees and agricultural windbreaks for local and regional biodiversity: the case of Monteverde, Costa Rica. Pages 787–798 in *XXI IUFRO World Congress, 7–12 August 2000, Kuala Lumpur, Malaysia,* Subplenary sessions, Vol. 1. Kuala Lumpur, Malaysia: International Union of Forestry Research Organizations.

Harvey, C. A., and W. A. Haber. 1999. Remnant trees and the conservation of biodiversity in Costa Rican pastures. *Agroforestry Systems* 44:37–68.

Hellin, J., L. A. Welchez, and I. Cherrett. 1999. The Quezungual system: an indigenous agroforestry system from western Honduras. *Agroforestry Systems* 46:229–237.

Hietz-Seifert, U., P. Heitz, and S. Guevara. 1996. Epiphyte vegetation and diversity on remnant trees after forest clearance in southern Veracruz, Mexico. *Biological Conservation* 75:103–111.

Holl, K. D., M. E. Loik, E. H. V. Lin, and V. A. Samuels. 2000. Tropical montane forest restoration in Costa Rica: overcoming barriers to dispersal and establishment. *Restoration Ecology* 8:339–349.

Janzi, T., J. Schelhas, and J. P. Lassoie. 1999. Environmental values and forest patch conservation in a rural Costa Rican community. *Agriculture and Human Value* 16:29–39.

Jimenez, C. 2001. *Estudio preliminar del patrón de actividad crepuscular de los murciélagos, utilizando un traductor de frecuencia, en la región de Los Tuxtlas, Veracruz.* Master's thesis, Division de Ciencias Biológicas, Universidad Juárez Autónoma de Tabasco, Tabasco, Mexico.

Johnson, R. J., and M. M. Beck. 1988. Influences of shelterbelts on wildlife management and biology. *Agriculture, Ecosystems and Environment* 22–23:301–335.

Kleinn, C. 2000. On large-area inventory and assessment of trees outside forests. *Unasylva* 51:3–10.

Kowal, T. M. 2000. *Informe final de la consultoria en desarrollo agroforestal.* Proyecto Lempira Sur, GCP/HON/021/NET, October 2000 (not published).

Krishnamurthy, L., and M. Avila. 1999. *Agroforestería básica.* Mexico City: Programa de las Naciones Unidas para el Medio Ambiente (PUNMA), Serie Textos Básicos para la Formación Ambiental, No. 3.

Lagemann, J., and J. Heuveldop. 1983. Characterization and evaluation of agroforestry systems: the case of Acosta-Puriscal, Costa Rica. *Agroforestry Systems* 1:101–115.

Lewis, T. 1969. The diversity of the insect fauna in a hedgerow and neighboring fields. *Journal of Applied Ecology* 6:453–458.

Limpens, H. J. G. A., and K. Kapteyn. 1989. Bats, their behaviour and linear landscape elements. *Myotis* 29:63–71.

Luck, G. W., and G. C. Daily. 2003. Tropical countryside bird assemblages: richness, composition and foraging differ by landscape context. *Ecological Applications* 13(1):235–247.

Lynch, J. F. 1989a. Distribution of overwintering neoarctic migrants in the Yucatan Peninsula. I. General patterns of occurrence. *Condor* 91:515–544.

Lynch, J. F. 1989b. Distribution of overwintering neoarctic migrants in the Yucatan Peninsula II. Use of native and human-modified vegetation. Pages 287–298 in J. M. Hagan and D. W. Johnston (eds.), *Ecology and conservation of neotropical migrant landbirds.* Washington, DC: Smithsonian Institution Press.

Majer, J. D., and J. H. Delabie. 1999. Impact of tree isolation on arboreal and ground ant communities in cleared pastures in the Atlantic rain forest region of Bahia, Brazil. *Insectes Sociaux* 46:281–290.

Marmillod, A. 1989. Actitudes de los finqueros hacia los árboles. Pages 294–306 in J. W. Beer, H. W. Fassbender, and J. Heuveldop, *Avances en las investigaciones agroforestales.* Turrialba, Costa Rica: CATIE.

Molano, J. G., M. P. Quiceno, and C. Roa. 2002. El papel de las cercas vivas en un sistema de producción agropecuaria en el Pidemonte Llanero. In M. Sánchez and M. Rosales (eds.), *Agroforestería para la producción animal en América Latina II. Memorias de la Segunda Conferencia Electrónica de la FAO.* Rome: Estudio FAO de Producción y Sanidad Animal.

Morales, D., and C. Kleinn. 2001. El proyecto TROF: algunas experiencias preliminares en Centroamérica. In *Taller Latinoamericanos sobre Información de Arboles fuera de Bosque y Productos no Maderables del Bosque.* FAO, Caracas, Venezuela, 6–10 August 2001.

Murgueitio, E., and Z. Calle. 1999. Diversidad biológica en sistemas de ganaderia bovina en Colombia. Pages 53–87 in M. Sanchez and M. Rosales (eds.), *Agroforestería para la producción animal en América Latina.* Rome: Estudio FAO, Producción y Sanidad Animal 143.

Nason, J. D., P. R. Aldrich, and J. L. Hamrick. 1997. Dispersal and dynamics of genetic structure in fragmented tropical tree populations. Pages 304–320 in W. F. Laurance and

R. O. Bierregaard Jr. (eds.), *Tropical forest remnants: ecology, management, and conservation of fragmented communities*. Chicago: University of Chicago Press.

Nielson, K. B., and D. A. DeRosier. 2000. Agricultural windbreaks: conservation and management implications of corridor usage by avian species. Pages 448–450 in N. Nadkairni and N. T. Wheelwright (eds.), *The ecology and natural history of Monteverde*. New York: Oxford University Press.

Nieto, H., E. Somarriba, and M. Gómez. 2001. Contribución de *Acacia pennatula* (carbón) a la productividad agroforestal sostenible de la Reserva natural Miraflor-Moropotente, Estelí, Nicaragua. *Agroforestería en las Américas* 8:21–23.

Osborne, P. 1983. Bird numbers and habitat characteristics in farmland hedgerows. *Journal of Applied Ecology* 21:63–82.

Otero-Arnaiz, A., S. Castillo, J. Meave, and G. Ibarra-Manriquez. 1999. Isolated pasture trees and the vegetation under their canopies in the Chiapas Coastal Plain, Mexico. *Biotropica* 31:243–254.

Paap, P. F. 1993. *Farmers or foresters: the use of trees in silvopastoral systems of the Atlantic zone of Costa Rica*. The Atlantic Zone Programme CATIE-AUW-MAG no. 51. Turrialba, Costa Rica: CATIE.

Pezo, D., and M. Ibrahim. 1998. *Sistemas silvopastoriles*. Turrialba, Costa Rica: CATIE, Proyecto Agroforestal CATIE/GTZ.

Powell, G., J. H. Rappole, and J. A. Sader. 1989. Neotropical migrant landbird use of lowland Atlantic habitats in Costa Rica: a test of remote sensing for identification of habitat. Pages 287–298 in M. Hagan and D. W. Johnston (eds.), *Ecology and conservation of neotropical migrant landbirds*. Washington, DC: Smithsonian Institution Press.

Rappole, J. H. 1995. *The ecology of migrant birds*. Washington, DC: Smithsonian Institution Press.

Rosenberg, D. K., B. R. Noon, and E. C. Meslow. 1997. Biological corridors: form, function and efficacy. *BioScience* 47:677–687.

Rowley, I., and G. Chapman. 1991. The breeding biology, food, social organisation, demography and conservation of the Major Mitchell or pink cockatoo, *Cacatua leadbeateria,* on the margin of the Western Australian wheatbelt. *Australian Journal of Zoology* 39:211–261.

Saab, V., and D. Petit. 1994. Impact of pasture development on winter bird communities in Belize. *Condor* 94:66–71.

Sauer, J. D. 1979. Living fences in Costa Rican agriculture. *Turrialba* 29:225–261.

Schelhas, J. 1996. Land use choice and forest patches in Costa Rica. Pages 258–284 in J. Schelhas and R. Greenberg (eds.), *Forest patches in tropical landscapes*. Washington, DC: Island Press.

Slocum, M. G., and C. C. Horvitz. 2000. Seed arrival under different genera of trees in neotropical pasture. *Plant Ecology* 149:51–62.

Souza de Abreu, M. H., M. Ibrahim, C. A. Harvey, and F. Jimenez. 2000. Caracterización del componente arbóreo de los sistemas ganaderos de La Fortuna de San Carlos, Costa Rica. *Agroforestería en las Américas* 7:53–56.

Thébaud, C., and D. Strasberg. 1997. Plant dispersal in fragmented landscapes: a field study of woody colonization in rainforest remnants of the Macarene Archipelago. Pages 321–332 in W. F. Laurance and R. O. Bierregaard Jr. (eds.), *Tropical forest remnants: ecology, management, and conservation of fragmented communities*. Chicago: University of Chicago Press.

Timm, R. M. 1988. Vertebrate pest management in windbreak systems. *Agriculture, Ecosys-

tems and Environment 22–23:555–570.

Toh, I., M. Gillespie, and D. Lamb. 1999. The role of isolated trees in facilitating tree seedling recruitment at a degraded sub-tropical rainforest site. *Restoration Ecology* 7:288–297.

Tucker, N. I. J. 2000. Linkage restoration: interpreting fragmentation theory for the design of a rainforest linkage in the humid wet tropics of north-eastern Queensland. *Ecological Management and Restoration* 1:35–41.

Tucker, N. I. J. 2001. Wildlife colonisation on restored tropical lands: what can it do, how can we hasten it and what can we expect? Pages 279–295 in S. Elliott, J. Kerby, D. Blakesley, K. Hardwicke, K. Woods, and V. Anusarnsunthorn (eds.), *Forest restoration for wildlife conservation*. Chiang Mai, Thailand: Chiang Mai University.

Tucker, N. I. J., and T. M. Murphy. 1997. The effects of ecological rehabilitation on vegetation recruitment: some observations from the wet tropics of north Queensland. *Forest Ecology and Management* 99:133–152.

Vandermeer, J., and I. Perfecto. 1997. The agroecosystem: a need for the conservation biologist's lens. *Conservation Biology* 11:591–592.

Verboom, B., and H. Huitema. 1997. The importance of linear landscape elements for the pipistrelle *Pipistrellus pipistreullus* and the serotine bat *Eptesicus serotinus*. *Landscape Ecology* 12:117–125.

Westley, S. B. 1990. Living fences: a close-up look at an agroforestry technology. *Agroforestry Today* 2:11–13.

Wight, B. 1988. Farmstead windbreaks. *Agriculture, Ecosystems and Environment* 22–23:261–280.

Wilken, G. C. 1977. Integrating forest and small-scale farm systems in Middle America. *Agroecosystems* 3:291–302.

Williams-Linera, G., V. Sosa, and T. Plajas. 1995. The fate of epiphytic plants after fragmentation of a Mexican cloud forest. *Selbyana* 16:36–40.

Yahner, R. H. 1982a. Avian use of vertical strata and planting in farmstead shelterbelts. *Journal of Wildlife Management* 46:50–60.

Yahner, R. H. 1982b. Avian nest densities and nest site selection in farmstead shelterbelts. *Wilson Bulletin* 94:156–175.

Yahner, R. H. 1983. Small mammals in farmstead shelterbelts: habitat correlates of seasonal abundance and community structure. *Journal of Wildlife Management* 47:74–83.

Chapter 12

Agroforestry Systems: Important Components in Conserving the Genetic Viability of Native Tropical Tree Species?

David H. Boshier

In the tropics, human forest disturbance is omnipresent. The wide range of human uses of forests (e.g., timber, fuel, food, clearance for habitation, agriculture, grazing) vary in their impacts, depending on the type and intensity of use. However, increasing deforestation rates in recent decades have led to dramatic reductions in the area of forest (see Chapter 1, this volume) and its fragmentation into smaller patches of varying size and spatial isolation. For some tropical forest ecosystems, the remaining forests often are highly fragmented and below the size considered viable, such that the ideal of maintaining large, continuous reserves is impractical (Soulé 1987). The agricultural matrix in which many forest remnants now exist is in itself a complex mosaic of varying land use practices. They vary in their degree of tree cover from almost none (e.g., monocultures of crops such as sugarcane) to highly complex agroforests in which there is maintenance not only of a high degree of tree cover but also a variety of tree species (see Chapter 10, this volume). Therefore, in some cases conservation initiatives must consider approaches that depart from the traditional in situ conservation paradigm, involving protected wilderness areas, to ways in which managers can conserve the species of an already highly altered forest type by managing networks of small forest patches in such mosaics of land use types.

Deforestation and fragmentation may have obvious effects, such as the elimination of some species. However, there may also be less immediate effects on the longer-term viability of species through impacts on ecological and genetic processes. Managers must consider the reproductive and regenerative capacities of priority species and the perpetuation of management practices that allow natural or artificial regeneration to ensure that populations have a long-term future. The effects of fragmentation on remnant stands and trees, their gene pools, and consequent conservation value are the subject of debate (Saunders et al. 1991; Heywood and Stuart 1992; Young et al. 1996). At the pessimistic extreme are views characterizing remnant trees in agroecosystems as "living dead," of little conservation value because isolated from potential mating partners they may not produce offspring or offspring may fail to establish new generations (Janzen 1986). More optimistically, the possibility of extensive gene flow between isolated trees of many taxa, through pollen transport over long distances by animal or wind vectors, suggests that remnant forest patches and trees can be effective and important in conserving genetic diversity (Hamrick 1992).

This chapter examines the role that trees in agroforestry systems may play in conserving the genetic viability of native tropical tree species in protected areas, forest fragments, and the same agroforestry systems. Tree species found both in agroforestry systems and forest patches may contribute to both gene flow and the overall gene pool of those species. Species found only in agroforestry systems may still contribute to the conservation of forest tree species by providing habitat for pollinators and seed dispersers that facilitate gene flow in other tree species (Slocum and Horvitz 2000) or by creating an environment that favors seedling regeneration.

The chapter explores what is known about the level and nature of intraspecific genetic variation actively conserved in trees in agroforestry systems, the distance at which forest fragments become genetically isolated, the particular types of agroforestry systems that favor gene flow and the tree species most likely to profit from it, and the consequences of gene flow between managed and remnant natural populations. This discussion leads to consideration of tree planting and natural regeneration in agroforestry systems, biological corridor design that combines target species conservation and sustainable use compatibility, better targeting of resources to more critically threatened species, and research and education needs.

At the outset we need to consider what we want to conserve. Effective conservation entails clear definition of objectives, which may range from preservation of actual diversity to conservation of evolutionary potential (Eriksson et al. 1993). However, genetic variation and processes are dynamic and respond to changing conditions. A pragmatic objective is one that maintains options for future generations while satisfying present needs (WCED 1987), such that sufficient genetic variation is conserved for tree populations and species to continue to adapt in the future. Achieving both short- and long-term goals entails an understanding of the basic processes of tree reproductive biology (sexual systems, incompatibility mechanisms, flowering patterns, and pollination processes) and how they combine to produce observed patterns of gene flow and genetic variation. Identifying the potential for agroforestry systems to facilitate, critically alter, or endanger these processes is key. Reducing the possibility or impact of inbreeding and maintaining diversity in naturally outcrossing tree species is important, and maintenance of breeding system flexibility is a priority for species that naturally combine outcrossing and inbreeding.

Gene Flow and Mating Patterns in Tropical Tree Populations

Understanding of gene flow and mating patterns in tropical forest trees has progressed over the last 50 years from theories to direct estimates based on field studies and molecular markers (Young et al. 2000). Trees in tropical forests, with low to medium species densities, were once thought likely to be self-pollinated because large interplant distances and asynchronous flowering would reduce the chance of successful cross-pollination (Corner 1954; Federov 1966). Subsequent studies based primarily on hand pollination to determine self- and cross-compatibility and observations of pollinator behavior indicated strong barriers to selfing and led to the conclusion that tropical trees are predominantly outcrossed (Janzen 1971; Bawa 1974; Zapata and Arroyo 1978; Bawa et al. 1985). In fact, the extent and pattern of gene flow through pollen depend on a number of factors:

- *Sexual system:* Tropical trees have a diverse array of sexual systems (Bawa and Beach 1981; Loveless and Hamrick 1984; Bawa et al. 1985; Loveless 1992). In dioecious species male and female flowers occur on different trees, so these species are unable to self-pollinate, being obligate outcrossers. In hermaphroditic species (trees with both male and female function), individuals may have monoecious (single-sex) or hermaphrodite (both sexes) flowers, such that self-pollination may occur. In hermaphroditic species, however, self-pollination and self-fertilization may be reduced or prevented through a variety of mechanisms (e.g., differential maturation of female and male phases on the same tree).
- *Mating system:* Mating may be predominantly outcrossing in some hermaphroditic and monoecious species where incompatibility mechanisms (physiological or genetic barriers) prevent selfing (self-fertilization) despite self-pollination. However, the lack of such an incompatibility mechanism does not mean that a species will be obligately selfed. Some self-compatible tropical tree species show mixed mating (both selfing and outcrossing), such as *Cavanillesia platanifolia* (Murawski and Hamrick 1992a), *Ceiba pentandra* (Murawski and Hamrick 1992b; Gribel et al. 1999), and some *Shorea* spp. (Murawski et al. 1994).
- *Mechanism of pollen dispersal:* Depending on the type of vector (e.g., wind, bees, hummingbirds, or bats), pollen may be dispersed over different distances, while pollinator behavior such as traplining (preferential movement along corridors or between precociously flowering trees), as seen in some

bee, bat, and butterfly species (Frankie and Baker 1974; Gilbert 1975; Ackerman et al. 1982), has a significant effect on dispersal patterns.

- *Spatial distribution and density of trees:* The spatial distribution and density of mature trees (Murawski et al. 1990; Murawski and Hamrick 1991, 1992b) and flowering synchrony (Boshier et al. 1995b) may also influence pollen dispersal and hence the extent of outcrossing. Hamrick (1992) typified mating in neotropical tree populations as showing two contrasting trends: individual trees receive pollen from relatively few pollen donors, but the genetic composition of the pollen varies greatly from tree to tree; and although a high proportion of fertilization is affected by nearest neighbors, a significant proportion of pollen movement occurs over relatively long distances. Consequently, the effective breeding area of an individual tree of a common tropical tree species is large (25–50 ha; Hamrick and Murawski 1990).

Gene flow in trees also occurs through seed dispersal. Ashton (1969) argued that tropical trees would be genetically structured (i.e., neighboring trees would be more closely related—with more alleles in common—than more distant trees) as a consequence of limited seed dispersal from mother trees, a high degree of selfing and near neighbor pollinations, and selection for adaptation to the local environment. Recent studies show that some local genetic structure (over tens of meters) in tree populations is indeed typical (Boshier et al. 1995a; Hamrick et al. 1993), particularly for taxa with wind-dispersed seed, where most seed falls out within a short distance of the mother tree (Augspurger 1984). Where such local genetic structure occurs, the relative extent of pollen and seed dispersal influences the extent of related mating (inbreeding). If pollen and seed dispersal are similar, much mating will be between related individuals, whereas if pollen dispersal is much greater than that of seed, the amount of mating between closely related trees will decrease and outcrossing will predominate.

In natural populations of *Eucalyptus* spp. such family groups appear to lead consistently to a degree of inbreeding, with outcrossing rates averaging about 0.75 (Eldridge et al. 1993). Outcrossing rates (t_m) theoretically range from 0 (complete selfing) to 1.0 (outcrossed to a random sample of the population's pollen pool). Values significantly lower than 1.0 indicate a degree of inbreeding, which may result from selfing and mating between related individuals. In contrast to natural populations, in plantations the use of collected seed, which is normally mixed from a number of mother trees, breaks up such family structure, and mating there shows a corresponding increase in outcrossing (e.g., *E. regnans:* 0.74 for a natural stand, 0.91 for a plantation; Moran, Bell, et al. 1989a). However, molecular marker studies have found that pollen dispersal generally is much more extensive than any local genetic structure, so the majority of tropical tree species avoid such related mating (Stacy et al. 1996; Nason et al. 1997; Boshier 2000) and show high levels of outcrossing (Nason and Hamrick 1997; Lepsch-Cunha et al. 2001). However, if naturally out-crossing species are forced by human disturbance (e.g., from increased physical isolation) to inbreed (increased related mating and selfing), there may be associated risks of reduced fertility, growth, and environmental tolerance and greater susceptibility to pests and diseases (Sim 1984; Griffin 1990). Maintenance of genetic diversity is vital for the long-term viability and adaptability of populations of many tree species.

Intraspecific Genetic Variation Conserved in Agroforestry Systems

From both conservation and use viewpoints we need to know about the extent of genetic variation (allelic richness) within a species and the distribution of genetic variation (allelic evenness, i.e., whether populations have the same alleles or different ones). Tree taxa generally show high levels of genetic diversity (allelic richness) in comparison with nonwoody plants, with most alleles (typically 70–80 percent) common across most populations (Loveless 1992; Hamrick 1992; Hamrick et al. 1992; Moran 1992). However, levels of genetic diversity vary by mating system, with higher levels (high allelic richness) maintained in the predominantly outcrossing species because of high levels of mating between unrelated individuals. In contrast, inbred species show lower levels (low allelic richness) in inbreeding populations but greater interpopulation variation because of the more limited gene flow that occurs with inbreeding. Gene flow may also be limited by geographic separation (i.e., gaps in a species' natural distribution), such that species with disjunct distributions often show high genetic differentiation between the disjunct areas (e.g., *Acacia mangium;* Moran, Muona, et al. 1989). Widespread species with continuous distributions may be characterized by a hierarchy of population structure, such that whereas there is little differentiation between nearby populations, geographically distant populations diverge genetically. Thus a larger proportion of the total genetic variation within the species often is between physically distant regions, sometimes corresponding to geographic regions (e.g., Pacific and Caribbean watershed divide for *Cedrela odorata;* Gillies et al. 1997) rather than between populations within regions.

Although genetic differentiation between populations is low, it is often of major significance for adaptation or production. This is evident from provenance trials of many tropical tree species, where trees from different seed sources often show differential performance on a common trial site (Zobel and Talbert 1984; Eldridge et al. 1993). Significant interactions between genotype and environment generally occur only with large environmental site differences (e.g., dry and wet zones, alkaline and acidic soils; Boshier and Billingham 2000). Therefore, conservation of different populations is also important to tropical tree genetic resources. However, protected area design and management are determined mostly by political, social, and economic constraints, with reserves often located on slopes, on sites of lower fertility, and in stands of lesser economic value, which in turn biases their composition and limits their value for genetic resource conservation (Ledig 1988). Selective land clearance for agriculture has also decimated populations of many tree species on flat, fertile soils. Consequently, where there is genetic adaptation of populations to specific soil types, remnant trees and their offspring in agricultural fields, pastures, and agroforests may be the sole representatives and opportunity for conserving particular gene pools of some species. Therefore, the question is how adequate such tree populations might be for genetic conservation. The answer depends on the stage at which forest fragments and trees become genetically isolated, the extent of genetic adaptation to particular sites, and whether farmers' management practices, especially the conservation of remnant trees and their regeneration, maintain population gene pools and are therefore compatible with conservation objectives. Although few studies specifically address this scenario, there is plenty of relevant research that allows us to extrapolate some principles.

At What Distance Do Forest Fragments Become Genetically Isolated?

To attempt to answer this question we must look at a number of studies that have been conducted in recent years and show a range of results. The first study concerns *Swietenia humilis,* a mahogany species listed in Appendix II of the Convention on International Trade in Endangered Species (CITES), growing in secondary dry forest patches and as remnant trees in pastures that replaced the original dry forest on the Honduran Pacific coastal plains. Forest remnants were 1 to 4.5 km apart, varying in size from 10 to 150 ha (containing 8 to 44 *S. humilis* trees), while a continuous forest area was also studied as a control. *S. humilis* is self-incompatible, such that geographically isolated trees should be "living dead" unless there is pollen exchange with other trees. Molecular markers showed that at this degree of isolation, fragmentation did not impose a genetic barrier between remnants but increased levels of long-distance pollen flow into the smaller fragments, resulting in a network of pollen exchange over a 16-km^2 area (White and Boshier 2000). In both the continuous forest and fragments there was a predominance of near neighbor mating (within 300 m of the maternal tree). In the forest fragments, 53–62 percent of the pollen donors were from the same fragment, indicating that 38–47 percent of the pollen was imported by pollinators from other fragments. With such extensive pollen exchange, there was no evidence of increased inbreeding even in the smallest fragments. One tree, separated by 1.2 km of pasture from the nearest *S. humilis* trees, in accord with the species' self-incompatibility, showed 100 percent external pollen sources, with more than 70 percent from trees in a forest more than 4.5 km away. In addition, seed production was much higher and more reliable in pasture and other disturbed environments than in the closed forest (Boshier et al. 2003).

A study of *Dinizia excelsa,* a canopy-emergent tree, also showed an extensive network of pollen flow in Amazonian pastures and forest fragments with increased seed production, even in the absence of native pollinators (Dick 2001). African honeybees were the predominant floral visitors in fragmented habitats and replaced native insects in isolated pasture trees. Molecular markers showed that genetic diversity was maintained across habitats, with gene flow over as much as 3.2 km of pasture, although there was a slight increase in selfing in the pasture trees (t_m=0.85) as compared with trees in forest fragments and those in continuous forest (t_m=0.95).

Similar results were found in a study of pollen flow into continuous forest and five island populations (Lake Gatun, Panama Canal) of *Spondias mombin,* a self-incompatible, insect-pollinated tree (Nason and Hamrick 1997). The control forest showed pollen immigration rates of 45 percent from more than 100 m distance within the forest, whereas in the island populations 60–100 percent of the effective pollination was from at least 80-1,000 m away. However, the more isolated islands (1-km isolation) showed lower seed set and germination rates than the control forest, apparently because a lack of effective cross-pollination led to higher rates of self-fertilization. The inbred seed either abort or fail to germinate, and viable seed are predominantly those produced by long-distance pollen dispersal. Although this suggests that isolation reduces the species' ability to regenerate, it is likely that the same degree of fragmentation in a terrestrial land use mosaic would have less severe consequences because pollinator movement between fragments would probably be better than between true islands separated by water.

A negative effect on seed production as a result of reduced cross-pollination of isolated trees was also visible in the Southeast Asian timber tree *Shorea siamensis* (Table 12.1; Ghazoul et al. 1998). High-intensity logging resulted in much lower fruit set, although the number of flowers pollinated was similar. This resulted from the lower frequency of intertree movements by pollinators in the more open environment increasing the self-pollination frequency of this self-incompatible species.

In *Enterolobium cyclocarpum,* a self-incompatible, dominant tree of seasonally dry forests and associated pastures in Central America pollinated by bees and hawkmoths, there was no difference in the outcrossing rate between trees in continuous forest (t_m=1.00) and those in pasture (t_m=0.99; Rocha and Aguilar 2001a). There was extensive pollen flow into fragments separated by 250–500 m, while isolated pasture trees experienced more pollen donors than trees located in tree clumps (Apsit et al. 2001; Rocha and Aguilar 2001a). This contrasts with earlier predictions that spatially isolated trees are more likely to

Table 12.1. Fruiting in *Shorea siamensis* under disturbance resulting from different levels of timber extraction.

	Heavy	*Moderate*	*Undisturbed*
Tree density (trees/ha)	22.0	86.0	205.0
Flowering trees/ha			
1996	9.0	62.0	96.0
1997	5.0	59.0	76.0
Percentage fruit set			
1996	0.7	2.2	2.5
1997	1.5	5.5	5.5
Percentage stigmas with >5 pollen grains	62.0	59.0	79.0

Source: Ghazoul et al. (1998).

deviate from random mating and receive pollen from fewer donors (Murawski and Hamrick 1991). However, the isolated trees did show more year-to-year variation in their pollen donors than trees located in less disturbed habitats. Trees grouped in clumps showed less temporal variation, often because of within-clump mating (Hamrick 2002). Similar year-to-year variation was also found in *Hymenaea courbaril* (bat-pollinated) and *Spondias purpurea* (small insect-pollinated), although the magnitude varied between trees.

Cecropia obtusifolia, a dioecious, wind-pollinated pioneer tree, showed extensive pollen flow from natural forest (27 percent and 10 percent from 6 and 14 km, respectively) into a pristine forest reserve in Veracruz, Mexico, although there was apparently none from forest fallows at the same distances (Kaufman et al. 1998). The precise reason for the lack of pollen flow from this human-disturbed system, despite the high density of the species there, is uncertain. Trees are notably shorter and tree densities much higher in fallows than in tree fall gaps in natural forest, and it may be that the pollen is simply intercepted by the surrounding vegetation in the fallows. Apparently the dynamics of gene flow in such wind-pollinated species under disturbance may differ from those of animal-pollinated species.

Studies of how outcrossing rates vary with tree or flowering tree density in natural forest may also be informative because the range of tree densities in agroforestry systems varies. Differences between and annual variation in outcrossing rates for individual trees of several neotropical tree species have been reported to be consistent with changes in local flowering densities and the spatial patterns of flowering individuals (Murawski and Hamrick 1991). Species occurring at low densities appeared to combine significant levels of biparental mating (each maternal tree mates primarily with one other tree) with long-distance gene flow, whereas higher-density species showed more random mating, generally over shorter distances. In three neotropical tree species (*Calophyllum longifolium, Spondias mombin,* and *Turpinia occidentalis*) occurring naturally at low densities, mating patterns were strongly affected by the spatial distribution of reproductive trees, although they still showed high levels of outcrossing. Where trees were clumped, the majority of matings were with near neighbors, whereas with evenly spaced trees a large proportion of matings was over several hundred meters and well beyond the nearest reproductive neighbors (Stacy et al. 1996). The degree of flowering synchrony between neighbors may also increase the tendency toward inbreeding, such that in an outcrossing population of *Cordia alliodora* (self-incompatible) some trees, surrounded by a few trees of similar genotype with which flowering was highly synchronous, showed related mating, whereas other asynchronous flowerers in the same group were outcrossed (Boshier et al. 1995b). In the self-compatible species *Cavanillesia platanifolia,* outcrossing rates were lower where flowering levels were lower in different years (t_m=0.57 with 74 percent trees flowering, 0.35 with 49 percent, and 0.21 with 32 percent; Murawski et al. 1990; Murawski and Hamrick 1992b). In years of greater flowering, more floral rewards are available, such that there is a greater tendency for pollinators to move between trees, resulting in cross-pollination. However, when few trees are flowering there is a greater tendency for self-pollination.

The evidence to date clearly supports the idea that trees in agroforestry systems can be important in facilitating pollen flow between forest fragments. It is apparent that for some tree species under fragmentation, pollination occurs over much greater distances than are often considered and more in accord with distances previously identified by entomologists (Janzen 1971; Roubik and Aluja 1983), with the potential to maintain genetic variation. This contrasts with traditional views of the genetic effects of fragmenting populations, where increases in spatial isolation and population size reduction have been considered to reduce gene flow between fragments, leading to losses in genetic diversity (Saunders et al. 1991).

The potential to move between patches depends on the behavioral response of pollinators to the resultant mosaic of land use types. Some bat species move preferentially down forest tracks and pathways (Estrada et al. 1993). Many tree species are pollinated by bees, particularly social bees. It is likely that some cases of enhanced pollen flow in degraded tropical ecosystems result from domestic or African honeybees replacing native bees as principal pollinators (Dick 2001). However, bees typically live in habitats where nesting and floral resources are patchily distributed, such that all but perhaps the smallest bees normally move between resource patches isolated in an unrewarding matrix (Cane 2001). Bees of medium body size regularly fly 1–2 km from nest site to forage sites (Cane 2001), with some moving more than 4 km across agroecosystems between forest patches (Frankie et al. 1976; Raw 1989).

Changes in pollinator assemblages in fragmented landscapes may strongly affect patterns of gene flow and reproduction in remnant tree populations, such that considerations of pollinator management (e.g., provision of alternative food sources) may be as important as that of trees. Concerns that declines in pollinator populations in agroecosystems may eventually limit tree reproduction require monitoring of numbers and evidence of pollinator limitation (Allen-Wardell et al. 1998). Pollen flow dynamics in the many tree species that have a range of nonspecialist pollinators are probably far less susceptible to habitat disturbance than those with more specialist or limited-range pollinators (e.g., small beetles on *Virola* sp.). However, even *Ficus* spp., with species-specific wasp pollinators, were shown to form extensive metapopulations in fragmented landscapes (Nason et al. 1998).

However, there is obviously a distance beyond which genetic isolation will occur, with associated problems for population viability and adaptation (Young et al. 1996). Although determination is experimentally problematic, thresholds will vary between species depending on pollinator characteristics and availability, the specificity of the tree-pollinator relationship, and the presence and strength of any self-incompatibility mechanism. Whether greater physical isolation of trees results in increases in selfing appears to be controlled mainly by whether the species of interest has a self-incompatibility mechanism. Self-compatible species that normally show some level of outcrossing (Murawski and Hamrick 1992b) or are only weakly self-incompatible are likely to show higher levels of inbreeding at much shorter distances of separation (lower thresholds) than strongly self-incompatible species. The latter are more likely to show evidence of a threshold through reduced seed production.

Possible Consequences of Gene Flow between Managed and Remnant Natural Tree Populations

Although the studies reviewed earlier provide evidence that trees in agroforestry systems can be important mediators of pollen flow across fragmented agroecosystem landscapes, it is equally important to consider the possible consequences of this gene flow. What is the impact on the level and quality (genetic diversity and viability) of seed produced? A study of mating patterns and regeneration of the tree species *Symphonia globulifera* in fragmented and continuous forest in Costa Rica (Aldrich and Hamrick 1998) shows that the impacts may be more complex than might at first be apparent. This self-compatible species showed a predictable increase in selfing, as with *Cavanillesia platanifolia*, among remnant trees growing at low densities in pasture (t_m=0.74) as compared with trees in both continuous and forest fragments (t_m=0.9). The forest fragments were superficially healthy, showing much higher seedling densities than in the control forest (Table 12.2), with trees in the surrounding pasture playing an important role in pollination. However,

Table 12.2. Sources of *Symphonia globulifera* seedling production in fragmented forest.

	Continuous	*Fragmented*
Seedling densities	27.5 per ha	152.3 per ha
Origins of reproduction (%)		
Same plot or fragment	31.5	4.5
Other plot or fragment	15.0	15.0
External to assessed forest or fragment	53.5	12.5
Pasture	0.0	68.0

Source: Aldrich and Hamrick (1998).

52.5 percent of seedlings in the forest patches were fathered by two pasture "super adults." Such reproductive dominance by a few trees reduces effective population size, that is, the number of trees effectively contributing to reproduction, leading to losses of genetic diversity in following generations.

With reduced or no crown competition, many trees in agroforestry systems show greater crown size and exposure, with the potential for the formation of many more flower initials than in closed canopy forest. Higher seed production in pasture trees was also shown for *S. humilis* (in each year some 72 percent of pasture trees showed moderate to heavy seed production, compared with 12 percent of closed forest trees), but not for *Bombacopsis quinata* (Boshier et al. 2003). Trees of *Dinizia excelsa* in pasture and forest fragments produced more than three times as many pods per tree as trees from adjacent continuous forest populations, although there was no difference in fecundity between those in pasture and those in forest fragments (Dick 2001). In *Samanea saman,* the total number of seeds per fruit and the number of sound seeds were similar regardless of location in the landscape (i.e., isolated pasture trees or continuous forest; Cascante et al. 2002). In contrast, *Enterolobium cyclocarpum* flowers from trees in continuous forests were more likely to have pollen deposited on their stigmas than flowers from trees in pastures (52 and 32 percent, respectively), with trees from continuous forests almost six times more likely to set fruits and produce more seeds per fruit than trees in pastures (Rocha and Aguilar 2001a).

As well as potentially dominating the pollen pool by high flower production, such trees may also be a more attractive food source to pollinators, increasing the proportion of pollinations that originate from these trees. Thus, although trees in agroforestry systems may facilitate pollen flow between forest fragments, they may also reduce the effective population sizes of forest trees by dominating regeneration (where fragments are within seed dispersal range) and the pollen pool by producing more flowers or attracting a higher percentage of pollinators (where fragments are within pollen dispersal range).

In two studies (*Enterolobium cyclocarpum*, Rocha and Aguilar 2001a; *Cedrela odorata*, Navarro 2002), although the function of pasture trees in pollen movement between fragments has been recognized, seeds from pasture trees have shown less vigorous growth than those from trees of the same species found at higher densities, with the suggestion that the seeds are not appropriate for use in plantations. In *Enterolobium cyclocarpum*, the tendency for some fathers to be overrepresented in the seed crop (Rocha and Aguilar 2001b) and the greater yearly variation in pollen of pasture trees (Hamrick 2002) suggest a need for rigorous procedures when making seed collections for plantations, with broad sampling of each tree, a high number of trees, and avoidance of seed from poor-flowering years.

The genetic origin of trees in agroforestry systems also raises potential issues for genetic conservation and of gene flow from agroforestry systems to natural populations. Levels of genetic diversity are influenced by the means of tree establishment. Trees originating from natural regeneration may show some level of related mating caused by interactions between a number of factors rather than any particular factor (e.g., spatial and temporal genetic structure associated with incompatibility mechanisms, variation in flowering, and stand composition and density). Provided that seed production levels are not adversely affected, any abnormal, increased levels of inbreeding may be unimportant from an evolutionary viewpoint, with selfed individuals selected against at various stages of regeneration (seed production, seedling establishment, and growth). There is good evidence that genetic diversity is maintained in these on-farm populations (e.g., Chamberlain et al. 1996). However, increased levels of inbreeding may be critical in terms of the levels of diversity that are sampled for planting, ex situ conservation, or tree-breeding programs. Where trees are planted, the levels of genetic diversity maintained also depend on species and the seed collection process. Apart from the exceptions for self-compatible species already outlined, genetic diversity is likely to be maintained where normal seed collection protocols are followed (Schmidt 2001). However, in species that produce large quantities of seed per tree there is a tendency to make collections from a limited part of the crown and from a small number of trees, which leads to limited sampling of the gene pool (Boshier et al. 1995a). Consequently, through use of a reduced gene pool or future domination of the pollen pool (as in *S. globulifera*), some tree planting in agroforestry systems may be less beneficial, from a genetic conservation viewpoint, than might be expected.

Such considerations are not specific to any particular agroforestry system. Instead, the species characteristics, tree density, and origin (natural regeneration or planted) in any system are the most important factors influencing gene pools. However, the living fenceline is one agroforestry system in which the method of establishment can greatly influence the size of the breeding population and gene pool of native tree species. The use of living fencelines is common practice in many tropical countries (see Chapter 11, this volume; Kass et al. 1993). A wide range of species (e.g., *Bombacopsis quinata, Erythrina berteroana, Gliricidia sepium,* and *Spondias mombin*) and management practices are used, depending on local preferences. However, a constant factor is that the species used are vegetatively propagated from large stake cuttings. This feature has led to their characterization as lacking in genetic diversity and as sources of selfed seed, with possible adverse effects on the growth of subsequent trees. In some land use mosaics they may form a very high proportion of the tree component and therefore will dominate the pollen pool. Understanding how they influence tree gene pools (e.g., genetic variation, pollen flow, outcrossing rate) therefore is important in understanding the potential role of agroforestry for conservation.

An evaluation of genetic diversity in two living fencerows of *B. quinata* in Costa Rica found a smaller genetic base in comparison with material from a seed orchard (Table 12.3; Sandiford 1998). The first fencerow showed only 8 genotypes among the 42 trees sampled, with 57 percent of the trees represented by one genotype (clone). A second fencerow was more variable, with 20 different genotypes among the 42 trees sampled. However, both fencerows showed a high outcrossing rate (t_m=1.023, SE=0.060), comparable with those found for *B. quinata* in natural and fragmented populations (Sandiford et al. 2003) and with no evidence of selfing or inbreeding between related individuals. However, differences in allele frequencies between the pollen and ovule pools were evidence of a degree of nonrandom mating resulting from a combination of characteristics common to other fencerow species and particulars of the reproductive biology of *B. quinata* (maternal differences in fertility, nature of the self-incompatibility mechanism, and selection against homozygotes at seed maturation phase; Sandiford 1998). These characteristics are accentuated by the behavior of the pollinators (long-tongue bat, *Glossophaga soricina*) in terms of their preferential (nonrandom) visits to certain trees and their long-distance flights, which facilitate extensive pollen flow from outside the seed population.

Table 12.3. Number of maternal genotypes of *Bombacopsis quinata* for two living fencerows in Costa Rica and a seed orchard in Honduras.

Site	Number of Trees	Number of Genotypes	Most Common Genotype
Fencerow 1	42	8	57%
Fencerow 2	42	20	12%
Seed orchard	12	11	

Source: Sandiford (1998).

To what extent are such results typical of other species used in fencerows? By the nature of their establishment method, fencerow populations are not random populations, with the small and biased number of genotypes meaning there is little likelihood that they are in genetic (Hardy-Weinberg) equilibrium, with consequent heterozygote excesses or deficiencies. Seed collections obviously include mainly those trees with abundant fruiting, whereas those that fail to fruit are not represented or are represented only in the pollen pool. Under such conditions, only a relatively small number of genotypes may act as mothers, whereas a larger majority will act as fathers, unbalancing the allelic frequencies for pollen and ovules, as does the entry of pollen from outside the fence populations, unlike the mating patterns typical of natural forest. A low degree of flowering synchrony between trees also increases the probability of nonrandom mating. Although a species incompatibility mechanism may prevent selfing, in fencerows there may be many individuals of the same clone such that there are high levels of self-pollination with possible failures of fruit production (see the *Shorea siamensis* and *Spondias mombin* studies mentioned earlier) if there is insufficient cross-pollination.

Promotion of the planting and use (e.g., for faster growth) in agroforestry systems of exotic species or provenances at the expense of natural regeneration of local populations or species may have deleterious effects for genetic conservation. The replacement of native species (e.g., *Leucaena salvadorensis* by *Leucaena leucocephala;* Hughes 1998) may reduce population sizes or even eliminate particular populations of threatened species. Hybridization of introduced species with native species is particularly prevalent in certain genera (e.g., *Leucaena,* Hughes 1998; *Prosopis,* Carney et al. 2000) and has obvious implications for conservation of native gene pools (see also Chapter 15, this volume). Evidence to date of outbreeding depression (reduced growth or fertility from the breakup of co-adapted allelic complexes or dilution of adapted alleles; Ledig 1992) from crossing between different populations of the same tree species is inconclusive because of a lack of studies. Controlled crossing between populations of *B. quinata* saw outbreeding depression (reduced seed set) only when populations as genetically distinct as those from Honduras and Colombia were crossed. In contrast, there was no outbreeding depression in *S. humilis* when populations from a 500-km distance were crossed (Billingham 1999). For species of *Syzygium* and *Shorea* in southwestern Sri Lanka, there was a significantly lower fruit set in crosses with the most distant pollen donor (approximately 12 km; Stacy 1998). The author suggested the apparent outbreeding depression at a fairly small scale was more likely to result from spatial heterogeneity in the selective environment than from isolation by distance, with the geographic heterogeneity of the study area possibly of a finer scale than that of the majority of tropical forested landscapes.

Which Agroforestry Systems Favor Gene Flow, and What Types of Tree Species Are Favored by Gene Flow?

As the previous examples show, the assessment of the benefits of agroforestry systems for genetic conservation of tree species entails integration of genetic, ecological, and management information to reflect the complexities of the systems. Such complexity suggests that it may be very difficult to predict when and for what pairs of forest fragments connectivity will be a critical issue and to assess the connectivity of different systems. Conservation strategies to address the issues posed by fragmentation have generally been based on island biogeography theory, leading to the idea that fragments linked by a corridor of similar suitable habitat are likely to have greater conservation value than the same fragments if isolated (Diamond 1975; Wilson and Willis 1975). More recent developments recognize that forests do not exist as islands in a sea of completely hostile, biodiversity-poor environments but as a mosaic of modified land uses and habitats that vary in their ability to fulfill the original ecological functions and consequently vary in their value as corridors (see also Chapters 1 and 3, this volume).

Connectivity has two distinct but related components under broad headings of gene flow and migration related to home range. Some habitats are suitable for certain species to live in, others may not be but may not inhibit movement, some may allow movement only seasonally, while others may be totally inhospitable. Connectivity may be sought to reduce the susceptibility of small populations to a variety of impacts, including genetic stochasticity (e.g., genetic drift, inbreeding depression, and predator and competitor population fluctuations).

The integration of landscape models that use spatially explicit information on habitat type mosaics with metapopulation models that describe a set of connected populations within a landscape (see Chapters 2 and 3, this volume) offers a means to examine the influence of landscapes and habitats on genetic processes and structure of populations. Although metapopulation theory provides a conceptual model for understanding population dynamics in fragmented environments, there is currently limited evidence of its practical value in conservation management. There is limited knowledge of the scale of many animal species' movements, their habitat needs, disturbance tolerance, or the other impacts of fragmentation and hence limited information on the relative effectiveness of different systems to act as migration sources or sinks or as connectors between populations. Connecting genetic and demographic models at landscape scales entails adopting scales of study that are more relevant than those over which migration is currently measured and that are sensitive to recent changes in gene flow. Direct parentage analysis methods have generally been applied over relatively small spatial scales (less than 100 ha), whereas the studies reviewed here suggest that under fragmentation, pollen flow distances may increase by factors greater than 10 (kilometers rather than hundreds of meters). In addition, although pollen and seed movement may influence genetic structure differentially, from the perspective of demographic processes (i.e., colonization) in metapopulation and landscape models, seed dispersal data may be as important as pollen dispersal, requiring the use of a range of markers for direct comparisons between relative gene flow levels resulting from pollen and seed dispersal (Sork et al. 1998).

Against this background land managers are asked to select, design, and manage landscape links that will be effective in conserving biodiversity. The studies summarized here suggest reasons for optimism about the connectivity value of trees in agroforestry systems. They clearly support a broad vision of corridor design that embraces a range of land use mosaics rather than just continuous corridors of intact forest. The emphasis therefore should be on connectivity (for genes, species, and ecological processes) of landscape mosaics through maintenance and improvement of land use patterns that promote connectivity and conservation of biodiversity more generally. Corridor design, management, and monitoring should thus involve assessment of different land use types in terms of how well, individually and in combination, they meet the biological criteria of connectivity, amongst others, and how the balance of land use types may need modification to maintain or improve connectivity. This can be done, at least as a first approximation, in a simple way (Laurance et al. 1997; see also Chapter 3, this volume) by qualitatively scoring or ranking agroforestry practices and other land uses for their likely contribution to connectivity (capacity to allow movement and gene flow for species, or wildlife habitat provision), as well as other conservation and environmental amelioration values (e.g., protection from fires and exotic species, softened edge effects; see Chapter 2, this volume).

Any assessments are inevitably site specific given the variable connectivity, species, and sustainable land use aims of different areas and their differing degrees of resilience to disturbance. Such assessments can be summarized as matrices, specific to each area, in which land uses relevant to that area are ranked for each service of potential interest. They can help to identify priority agroforestry practices that show high connectivity and sustainable use compatibility and those that do not. In an area of high forest cover, agroforestry systems may be assessed principally for gene flow, whereas in much more highly deforested areas a fuller complement of benefits may be sought from particular systems, with their specific location in the corridor zone also being important. Thus, in the highly deforested dry forest zone of western Honduras the traditional Quezungual fallow system (Kass et al. 1993), in which farmers manage naturally regenerated shrubs, fruit trees, and timber trees among their crops, is likely to provide a variety of genetic conservation benefits for a range of native tree species (see also Chapter 8, this volume). Other complex systems, such as traditional shaded coffee or jungle rubber, may rate highly for all the possible genetic conservation benefits (see Chapters 9 and 10, this volume). In contrast, simpler agroforestry systems such as pasture trees and living fencerows offer fewer genetic conservation benefits and are unlikely to prove effective mediators of pollen flow for species without a self-incompatibility mechanism (see Chapter 11, this volume).

In most cases, however, assessments of the genetic conservation benefits of agroforestry systems are less likely to be system specific than species specific, taking account of the farming systems context of an area, the density of trees, and their origin (natural regeneration or planted). For example, maintaining native timber trees over large areas of coffee is likely to have beneficial genetic effects for gene flow, population numbers, and conservation of particular populations. In contrast, the same system in only a small area may lead to a reduced genetic base in seed production through related or biparental mating. Thus, the area or management unit should be measured in numbers of participating households or numbers of land units in which agroforestry land uses beneficial to target species conservation are practiced (Boshier et al. 2004). Given the speed with which land management practices may change in response to market prices, this measure in itself may require monitoring.

Identifying the factors that leave some species genetically susceptible to human disturbance requires extensive reproductive and regeneration ecology and genetic data. The lack of information, resource limitations, and the need for more immediate action in many situations necessitate pragmatic best-guess approaches to identify which species will be favored by gene flow between agroforestry areas and which will not. The ability to extrapolate from results from model species to make more general recommendations for species management groups (combining ecological guild, spatial distribution, and reproductive biology) depends on the existence of basic biological information (e.g., incompatibility and pollination mechanisms, dispersal, and seedling regeneration) that enables species to be classified (Jennings et al. 2001).

Consideration of available information suggests that the following species types are unlikely to show genetic conservation benefits from agroforestry systems: outcrossing species that are self-compatible, slow-growing species that reproduce only when they are large (extreme of monocarpic species, i.e., those that flower only once in their life), species with poor regeneration under human disturbance, species with highly specific pollinators or seed dispersers susceptible to disturbance, rare species with low population densities, and species with highly clumped distributions. Inevitably such generalizations will be qualified by the range of factors that have been shown to influence patterns of genetic variation in trees.

Conclusions

Evidence suggests that for many species, populations, and individuals of tropical trees, gene flow may be high across agroforestry landscapes with little
apparent forest cover. The view of forest fragmentation as producing genetic
isolation may be more a human perception than a true reflection of actual
gene flow. It is therefore important to recognize the complementary role that
maintenance of trees on farms is already playing to in situ conservation. Trees
in a whole range of agroforestry practices may play an important but varied
role in the long-term genetic viability of many native tree species, facilitating
gene flow between existing reserves, conserving particular genotypes not
found in reserves, maintaining minimum viable populations, and acting as
intermediaries and alternative host habitat for pollinators and seed dispersers
(Harvey and Haber 1999). Underestimating the capacity of many species to
persist in large numbers in these agroecosystems under current practices could
lead to the misdirection of limited conservation resources toward species not
under threat (Boshier et al. 2004). Agroforestry tree populations may represent a considerable conservation resource, which if taken into consideration
may show species to be thriving that are currently assumed to be threatened
by habitat loss (Vandermeer and Perfecto 1997).

However, although they undoubtedly contribute to reproduction in remnant forests, the benefits and effects are more complex than at first might be
predicted and vary from species to species. Uneven representation and over-
representation in pollen pools and mating may lead to nonrandom mating,
with reductions in genetic diversity in subsequent generations. Evidence of the
quantity and quality of seed produced is variable and currently insufficient to
draw more general conclusions, although of the range of agroforestry practices
only living fencelines and very low-density trees in pastures are likely to cause
problems.

However, we should not overestimate the extent to which agroforestry systems will benefit the genetic conservation of forest tree species. In addition to
some of the complications raised in the studies reviewed here, it is evident that
many of the tree species found in agroforestry areas already exist in adequate
numbers in existing reserves. Similarly, some of the species threatened by low
population numbers are not of the type that will easily persist in such systems.
The greatest potential role of agroforestry will be in highly deforested areas
where reserves are very small or nonexistent and where the trees maintained in
agroforestry systems represent an important part of a particular population's or
species' gene pool. In such circumstances, the fact that many tree species that
live in such disturbed vegetation can be conserved through existing agroforestry practices can free resources for the conservation of more critically
threatened species requiring more conventional, resource-intensive approaches.

A multidisciplinary vision is needed to establish the general potential for integrating conservation and development and, more specifically, which species are or could be sustainably conserved in such systems, from both biological and human management perspectives. Efforts to maintain genetic diversity and adaptive capacity are irrelevant if current management drastically alters population persistence.

We still don't know at what distance forest fragments become genetically isolated. Fragmentation thresholds for gene flow must be determined and the possible selection pressures exerted by farmers elucidated. The complementary benefits of different agroforestry practices for genetic conservation must be further evaluated, recognized, and promoted. There is a need to raise awareness among development professionals of the value of natural regeneration as both a conservation and a socioeconomic resource. Pushing of a limited range of species, often exotics, by development agencies may reduce the potential genetic benefits of such systems, besides creating potential problems of invasiveness (see Chapter 15, this volume). However, there is also a need for conservation planners, more accustomed to in situ methods, to consider the possibility that tree populations found outside protected areas have a role in biodiversity conservation (Boshier et al. 2004). This in turn necessitates the direct involvement of development organizations in biodiversity conservation and an effective interaction between them and traditional conservation organizations to ensure both conservation and development benefits.

Acknowledgments

I would like to thank Götz Schroth and two anonymous reviewers for their incisive comments on drafts of the manuscript. This chapter is an output from research projects partly funded by the United Kingdom Department for International Development (DFID) for the benefit of developing countries (R6168, R6516 Forestry Research Programme). The views expressed are not necessarily those of DFID.

References

Ackerman, J. D., M. R. Mesler, K. L. Lu, and A. M. Montalvo. 1982. Food foraging behaviour of male euglossini (Hymenoptera: Apidae): vagabonds or trapliners? *Biotropica* 14:241–248.

Aldrich, P. R., and J. L. Hamrick. 1998. Reproductive dominance of pasture trees in a fragmented tropical forest mosaic. *Science* 281:103–105.

Allen-Wardell, G., P. Bernhardt, R. Bitner, A. Burquez, S. Buchmann, J. Cane, et al. 1998. The potential consequences of pollinator declines on the conservation of biodiversity and stability of food crop yields. *Conservation Biology* 12:8–17.

Apsit, V. J., J. L. Hamrick, and J. D. Nason. 2001. Breeding population size of a fragmented population of a Costa Rican dry forest tree species. *The Journal of Heredity* 92: 415–420.

Ashton, P. S. 1969. Speciation among tropical forest trees: some deductions in the light of recent evidence. *Biological Journal of the Linnean Society of London* 1:155–196.

Augspurger, C. K. 1984. Seedling survival of tropical tree species: interactions of dispersal distance, light gaps, and pathogens. *Ecology* 65:1705–1712.

Bawa, K. S. 1974. Breeding systems of tree species of a lowland tropical community. *Evolution* 28:85–92.

Bawa, K. S., and J. H. Beach. 1981. Evolution of sexual systems in flowering plants. *Annals of the Missouri Botanical Gardens* 68:254–274.

Bawa, K. S., D. R. Perry, and J. H. Beach. 1985. Reproductive biology of tropical lowland rain forest trees. I. Sexual systems and incompatibility mechanisms. *American Journal of Botany* 72:331–345.

Billingham, M. R. 1999. *Genetic Structure, localised adaptation and optimal outcrossing distance in two neotropical tree species.* D.Phil. thesis, University of Oxford, UK.

Boshier, D. H. 2000. Mating systems. Pages 63–79 in A. Young, D. H. Boshier, and T. J. Boyle (eds.), *Forest conservation genetics: principles and practice.* Melbourne: CSIRO; Wallingford, UK: CAB International.

Boshier, D. H., and M. R. Billingham. 2000. Genetic variation and adaptation in tree populations. Pages 267–289 in M. J. Hutchings, E. A. John, and A. J. A. Stewart (eds.), *Ecological consequences of habitat heterogeneity.* Oxford, UK: Blackwell Science.

Boshier, D. H., M. R. Chase, and K. S. Bawa. 1995a. Population genetics of *Cordia alliodora* (Boraginaceae), a neotropical tree. 2. Mating system. *American Journal of Botany* 82:476–483.

Boshier, D. H., M. R. Chase, and K. S. Bawa. 1995b. Population genetics of *Cordia alliodora* (Boraginaceae), a neotropical tree. 3. Gene flow, neighborhood, and population substructure. *American Journal of Botany* 82:484–490.

Boshier, D. H., J. E. Gordon, and A. J. Barrance. 2004. Prospects for *circa situm* tree conservation in Mesoamerican dry forest agro-ecosystems. Pages 210–226 in G. W. Frankie, A. Mata, and S. B. Vinson (eds.), *Biodiversity conservation in Costa Rica, learning the lessons in the seasonal dry forest.* Berkeley: University of California Press.

Cane, J. H. 2001. Habitat fragmentation and native bees: a premature verdict? *Conservation Ecology* 5:3. Online: http://www.consecol.org/vol5/iss1/art3.

Carney, S. E., D. E. Wolf, and L. H. Rieseberg. 2000 Hybridisation and forest conservation. Pages 167–182 in A. Young, D. H. Boshier, and T. J. Boyle (eds.), *Forest conservation genetics: principles and practice.* Melbourne: CSIRO; Wallingford, UK: CAB International.

Cascante, A., M. Quesada, J. Lobo, and E. A. Fuchs. 2002. Effects of dry tropical forest fragmentation on the reproductive success and genetic structure of the tree *Samanea saman. Conservation Biology* 16:137–147.

Chamberlain, J. R., C. E. Hughes, and N. W. Galwey. 1996. Patterns of variation in the *Leucaena shannonii* alliance (Leguminosae: Mimosoideae). *Silvae Genetica* 45:1–7.

Corner, E. J. H. 1954. The evolution of tropical forest. Pages 34–46 in J. S. Huxley, A. C. Hardy, and E. B. Ford (eds.), *Evolution as a process.* London: Allen and Unwin.

Diamond, J. M. 1975. The island dilemma: lessons of modern biogeographic studies for the design of natural reserves. *Biological Conservation* 7:129–146.

Dick, C. W. 2001. Genetic rescue of remnant tropical trees by an alien pollinator. *Proceedings of the Royal Society of London* Series B 268:2391–2396.

Eldridge, K., J. Davidson, C. Harwood, and G. van Wyk. 1993. *Eucalypt domestication and breeding.* Oxford, UK: Clarendon Press.

Eriksson, G., G. Namkoong, and J. H. Roberds. 1993. Dynamic gene conservation for uncertain futures. *Forest Ecology and Management* 62:15–37.

Estrada, A., R. Coates-Estrada, D. Meritt Jr., S. Montiel, and D. Curiel. 1993. Patterns of frugivore species richness and abundance in forest islands and in agricultural habitats at Los Tuxtlas, Mexico. *Vegetatio* 107–108:245–257.

Federov, A. A. 1966. The structure of tropical rainforest and speciation in the humid tropics. *Journal of Ecology* 60:147–170.

Frankie, G. W., and H. G. Baker. 1974. The importance of pollinator behaviour in the reproductive biology of tropical trees. *Anales del Instituto de Biología Universidad Nacional Autónoma de México Serie Botánica* 45:1–10.

Frankie, G. W., P. A. Opler, and K. S. Bawa. 1976. Foraging behaviour of solitary bees: implications for outcrossing of a neotropical forest tree species. *Journal of Ecology* 64:1049–1057.

Ghazoul, J., K. A. Liston, and T. J. B. Boyle. 1998. Disturbance induced density-dependent seed set in *Shorea siamensis* (Dipterocarpaceae), a tropical forest tree. *Journal of Ecology* 86:462–473.

Gilbert, L. E. 1975. Ecological consequences of a co-evolved mutualism between butterflies and plants. Pages 210–240 in L. E. Gilbert and P. H. Raven (eds.), *Coevolution of animals and plants.* Austin: University of Texas Press.

Gillies, A. C. M., J. P. Cornelius, A. C. Newton, C. Navarro, M. Hernandez, and J. Wilson. 1997. Genetic variation in Costa Rican populations of the tropical timber species *Cedrela odorata* L., assessed using RAPDs. *Molecular Ecology* 6:1133–1145.

Gribel, R., P. E. Gibbs, and A. L. Queiroz. 1999. Flowering phenology and pollination biology of *Ceiba pentandra* (Bombaceae) in Central Amazonia. *Journal of Tropical Ecology* 15:247–263.

Griffin, A. R. 1990. Effects of inbreeding on growth of forest trees and implications for management of seed supplies for plantation programmes. Pages 355–374 in K. S. Bawa and M. Hadley (eds.), *Reproductive ecology of tropical forest plants.* Carnforth, UK: Parthenon.

Hamrick, J. L. 1992. Distribution of genetic diversity in tropical tree populations: implications for the conservation of genetic resources. Pages 74–82 in *Breeding tropical trees,* Proceedings of the IUFRO S2.02-08 Conference, 9–18 October 1992, Cali, Colombia.

Hamrick, J. L. 2002. Temporal variation in the pollen pools of three neotropical dry-forest tree species. In *Population genetics of neotropical plant species.* Association of Tropical Biology annual meeting, Panama City, July 29, 2002. Online: http://www.stri.org/atb2002/Sym_Population_genetics.htm#hamrick.

Hamrick, J. L., M. J. W. Godt, and S. L. Sherman-Broyles. 1992. Factors influencing levels of genetic diversity on woody plant species. *New Forests* 6:95–124.

Hamrick, J. L., and D. A. Murawski. 1990. The breeding structure of tropical tree populations. *Plant Species Biology* 5:157–165.

Hamrick, J. L., D. A. Murawski, and J. D. Nason. 1993. The influence of seed dispersal mechanisms on the genetic structure of tropical tree populations. *Vegetatio* 107–108:281–297.

Harvey, C. A., and W. A. Haber. 1999. Remnant trees and the conservation of biodiversity in Costa Rican pastures. *Agroforestry Systems* 44:37–68.

Heywood, V. H., and S. N. Stuart. 1992. Species extinctions in tropical forests. Pages

91–117 in T. C. Whitmore and J. A. Sayer (eds.), *Tropical deforestation and species extinction*. London: Chapman & Hall.

Hughes, C. E. 1998. *Leucaena: a genetic resources handbook*. Tropical Forestry Paper 37. Oxford, UK: Oxford Forestry Institute.

Janzen, D. H. 1971. Euglossine bees as long-distance pollinators of tropical plants. *Science* 171:203–205.

Janzen, D. H. 1986. Blurry catastrophes. *Oikos* 47:1–2.

Jennings, S. B., N. D. Brown, D. H. Boshier, T. C. Whitmore, and J. C. A. Lopes. 2001. Ecology provides a pragmatic solution to the maintenance of genetic diversity in sustainably managed tropical rain forests. *Forest Ecology and Management* 154:1–10.

Kass, D. C. L., C. Foletti, L. T. Szott, R. Landaverde, and R. Nolasco. 1993. Traditional fallow systems of the Americas. *Agroforestry Systems* 23:207–218.

Kaufman, S. R., P. E. Smouse, and E. R. Alvarez-Buylla. 1998. Pollen-mediated gene flow and differential male reproductive success in a tropical pioneer tree, *Cecropia obtusifolia* Bertol. (Moraceae): a paternity analysis. *Heredity* 81:164–173.

Laurance, W. F., R. O. Bierregaard Jr., C. Gascon, R. K. Didham, A. P. Smith, A. J. Lynam, et al. 1997. Tropical forest fragmentation: synthesis of a diverse and dynamic discipline. Pages 502–514 in W. F. Laurance and R. O. Bierregaard Jr. (eds.), *Tropical forest remnants: ecology, management, and conservation of fragmented communities*. Chicago: University of Chicago.

Ledig, F. T. 1988. The conservation of genetic diversity in forest trees. *BioScience* 38:471–479.

Ledig, F. T. 1992. Human impacts on genetic diversity in forest ecosystems. *Oikos* 63:87–108.

Lepsch-Cunha, N., C. Gascon, and P. Kageyama. 2001. The genetics of rare tropical trees: implications for conservation of a demographically heterogeneous group. Pages 79–95 in R. O. Bierregaard Jr., C. Gascon, T. E. Lovejoy, and R. Mesquita (eds.), *Lessons from Amazonia: the ecology and conservation of a fragmented forest*. New Haven, CT: Yale University Press.

Loveless, D. 1992. Isozyme variation in tropical trees: patterns of genetic organization. *New Forests* 6:67–94.

Loveless, M. D., and J. L. Hamrick. 1984. Ecological determinants of genetic structure in plant populations. *Annual Review of Ecology and Systematics* 15:65–95.

Moran, G. F. 1992. Patterns of genetic diversity in Australian tree species. *New Forests* 6:49–66.

Moran, G. F., J. C. Bell, and A. R. Griffin. 1989. Reduction in levels of inbreeding in a seed orchard of *Eucalyptus regnans* F. Muell. compared with natural populations. *Silvae Genetica* 38:32–36.

Moran, G. F., O. Muona, and J. C. Bell. 1989. *Acacia mangium:* a tropical forest tree of the coastal lowlands with low genetic diversity. *Evolution* 43:231–235.

Murawski, D. A., B. Dayanandan, and K. S. Bawa. 1994. Outcrossing rates of two endemic *Shorea* species from Sri Lankan tropical rain forests. *Biotropica* 26:23–29.

Murawski, D. A., and J. L. Hamrick. 1991. The effect of the density of flowering individuals on the mating systems of nine tropical tree species. *Heredity* 67:167–174.

Murawski, D. A., and J. L. Hamrick. 1992a. Mating system and phenology of *Ceiba pentandra* (Bombacaceae) in central Panama. *Journal of Heredity* 83:401–404.

Murawski, D. A., and J. L. Hamrick. 1992b. The mating system of *Cavanillesia platanifolia* under extremes of flowering-tree density: a test of predictions. *Biotropica* 24:99–101.

Murawski, D. A., J. L. Hamrick, S. P. Hubbell, and R. B. Foster. 1990. Mating systems of two bombacaceous trees of a neotropical moist forest. *Oecologia* 82:501–506.

Nason, J. D., P. R. Aldrich, and J. L. Hamrick. 1997. Dispersal and the dynamics of genetic structure in fragmented tropical tree populations. Pages 304–320 in W. F. Laurence and R. O. Bierregaard (eds.), *Tropical forest remnants: ecology, management, and conservation of fragmented communities*. Chicago: University of Chicago Press.

Nason, J. D., and J. L. Hamrick. 1997. Reproductive and genetic consequences of forest fragmentation: two case studies of neotropical canopy trees. *Journal of Heredity* 88:264–276.

Nason, J. D., E. A. Herre, and J. L. Hamrick. 1998. The breeding structure of a tropical keystone plant resource. *Nature* 391:685–687.

Navarro, C. 2002. *Genetic resources of* Cedrela odorata *L. and their efficient use in Mesoamerica*. Doctoral dissertation, Faculty of Agriculture and Forestry, Department of Applied Biology, University of Helsinki, Finland.

Raw, A. 1989. The dispersal of euglossine bees between isolated patches of eastern Brazilian wet forest (Hymenoptera, Apidae). *Revista Brasileira de Entomologia* 33:103–107.

Rocha, O. J., and G. Aguilar. 2001a. Reproductive biology of the dry forest tree *Enterolobium cyclocarpum* (guanacaste) in Costa Rica: a comparison between trees left in pastures and trees in continuous forest. *American Journal of Botany* 88:1607–1614.

Rocha, O. J., and G. Aguilar. 2001b. Variation in the breeding behavior of the dry forest tree *Enterolobium cyclocarpum* (guanacaste) in Costa Rica. *American Journal of Botany* 88:1600–1606.

Roubik, D. W., and M. Aluja. 1983. Flight ranges of *Melipona* and *Trigona* in tropical forest. *Journal of the Kansas Entomological Society* 56:217–222.

Sandiford, M. 1998. *A study of the reproductive biology of* Bombacopsis quinata *(Jacq.) Dugand*. D.Phil. thesis, University of Oxford, UK.

Sandiford, M., M. R. Billingham, and D. H. Boshier. 2003. Sistemas de apareamiento de *Bombacopsis quinata* en bosque, huerto semillero y cerca viva. Pages 39–47 in J. Cordero and D. H. Boshier (eds.), *Bombacopsis quinata*. Tropical Forestry Paper no. 39. Oxford, UK: Oxford Forestry Institute.

Saunders, D. A., R. J. Hobbs, and C. R. Margules. 1991. Biological consequences of ecosystem fragmentation: a review. *Biological Conservation* 5:18–32.

Schmidt, L. 2001. *Guide to handling of tropical and subtropical forest seed*. Humlebaek, Denmark: DANIDA Forest Seed Centre.

Sim, B. L. 1984. The genetic base of *Acacia mangium* Willd. in Sabah. Pages 597–603 in R. D. Barnes and G. L. Gibson (eds.), *Provenance and genetic improvement strategies in tropical forest trees*. Harare, Zimbabwe: Commonwealth Forestry Institute, Oxford and Forest Research Centre.

Slocum, M. G., and C. C. Horvitz. 2000. Seed arrival under different genera of trees in a neotropical pasture. *Plant Ecology* 149:51–62.

Sork, V. L., D. Campbell, R. Dyer, J. Fernandez, J. Nason, R. Petit, P. Smouse, and E. Steinberg. 1998. *Proceedings from a workshop on gene flow in fragmented, managed, and continuous populations*. Research Paper no. 3. Santa Barbara, CA: National Center for Ecological Analysis and Synthesis.

Soulé, M. E. 1987. *Viable populations for conservation*. Cambridge, UK: Cambridge University Press.

Stacy, E. A. 1998. Cross-compatibility in tropical trees: associations with outcrossing distance, inbreeding, and seed dispersal. *American Journal of Botany* 85(6 supplement):62.

Stacy, E. A., J. L. Hamrick, J. D. Nason, S. P. Hubbell, R. B. Foster, and R. Condit. 1996. Pollen dispersal in low-density populations of three neotropical tree species. *American Naturalist* 148:275–298.

Vandermeer, J., and I. Perfecto. 1997. The agroecosystem: a need for the conservation biologist's lens. *Conservation Biology* 11:591–592.

WCED (World Commission on Environment and Development). 1987. *Our common future.* Oxford, UK: Oxford University Press.

White, G. M., and D. H. Boshier. 2000. Fragmentation in Central American dry forests: genetic impacts on *Swietenia humilis.* Pages 293–311 in A. G. Young and G. Clarke (eds.), *Genetics, demography and the viability of fragmented populations.* Cambridge, UK: Cambridge University Press.

Wilson, E. O., and E. O. Willis. 1975. Applied biogeography. Pages 523–534 in M. L. Cody and J. M. Diamond (eds.), *Ecology and evolution of communities.* Cambridge, MA: Belknap Press.

Young, A., D. H. Boshier, and T. J. Boyle. (eds.) 2000. *Forest conservation genetics: principles and practice.* Melbourne: CSIRO; Wallingford, UK: CAB International.

Young, A., T. Boyle, and T. Brown. 1996. The population genetic consequences of habitat fragmentation for plants. *Trends in Ecology and Evolution* 10:413–418.

Zapata, T. R., and M. T. K. Arroyo. 1978. Plant reproductive ecology of a secondary deciduous tropical forest in Venezuela. *Biotropica* 10:221–230.

Zobel, B. J., and J. T. Talbert. 1984. *Applied forest tree improvement.* New York: Wiley.

PART IV

Biodiversity as Burden and Natural Capital: Interactions between Agroforestry Areas, Natural Ecosystems, and Rural Communities in Tropical Land Use Mosaics

The conservation of tropical biodiversity is highly valued by global society, especially by many people living in wealthy countries in temperate regions. This support is crucial for influencing international political agendas and generating funds for conservation programs in the tropics. However, of more immediate importance for the conservation of tropical biodiversity is how it is valued by the rural populations and farmers who live in direct contact with native wildlife and plant species. On one hand, biodiversity is part of their natural capital and contributes to the natural resource base from which they derive products (e.g., timber, fuelwood, fruits, game animals) and services (e.g., clean water, pollinators, and biocontrol agents for their agricultural crops). For these reasons, biodiversity should also be valued by tropical farmers, and there should be substantial opportunities for alliances between farmers and conservationists.

On the other hand, wild plants, animals, and microorganisms may impose costs to tropical farmers and threaten their crops, their property, or even their lives. Although these conflicts are most obvious for large predators such as tigers and jaguars and megafauna species such as elephants, it is also true, more subtly but no less importantly, for crop pests, diseases, and weeds that may spread from native ecosystems into agricultural areas, causing crop loss. Consequently, tropical farmers, especially those who live in direct contact with natural ecosystems, such as in buffer zones around parks or near forest frontiers, carry the burdens and costs of maintaining high biodiversity. The extent to which these farmers derive benefits and incur costs from biodiversity strongly affects their attitudes toward conservation programs and specifically toward efforts to conserve and promote agroforestry practices that favor the presence and movement of native species in the agricultural matrix between forest reserves.

In tropical land use mosaics, rural populations and their farming systems not only are influenced by the presence of natural habitat and wild biodiversity but also may influence these natural ecosystems in various ways that may differ from the interactions between simpler agricultural or pasture systems and natural habitat. Although natural ecosystems bordering on agroforestry areas may be less exposed to fire, wind, and microclimatic extremes than those in a matrix dominated by pasture and annual cropping systems, other threats may originate from diversified and tree-dominated land use systems, including invasions of native habitat by introduced tree species, cross-infections of diseases between native and cultivated trees, and increased hunting pressure in agroforestry areas, which may act as a sink for wildlife populations in adjacent natural ecosystems. These considerations are crucial in evaluating the usefulness of different agroforestry practices in integrated conservation programs.

The purpose of this section is to discuss such interactions between natural ecosystems, agroforestry areas, and rural populations in tropical land use mosaics, with emphasis on the role of wild biodiversity as both natural capital and burden to tropical farmers, and the risks associated with diversified, tree-dominated land use systems in buffer zones and mosaic landscapes for native biodiversity.

Chapter 13 explores the attitudes and interactions of farmers and rural communities living near parks in three contrasting settings (the Peruvian Amazon, the highlands of Kalimantan in Indonesia, and Uganda in East Africa) with the wildlife conserved in these parks. It contrasts the perceptions of farmers, who profit from wildlife through hunting but also suffer from crop damage by raiding animals, with those of park managers, who perceive only benefits from conserving biodiversity, and stresses the need for open communication and information exchange between stakeholders as a precondition for effective buffer zone management. It also shows that hunting is a significant factor influencing both the attitudes of local people toward and their impact on wildlife.

Whether hunting in agroforestry land use mosaics and forest fallows can be sustainable and can reduce the hunting pressure on natural forest is discussed in Chapter 14. With examples from west-central Africa and Southeast Asia, the authors show that although hunting produces substantial amounts of food in some areas and is an important natural resource for local populations, offtake levels in both agroforestry areas and tropical forests are very often unsustainable and could lead to forests (and agroforests) full of trees but empty of wildlife. Alternatives to unsustainable hunting, which may help to conserve wildlife and its natural capital function for future generations, are also discussed.

When agroforestry is used as a tool for increasing biodiversity, it is essential that introduced agroforestry species do not interact negatively with native biota and ecosystems or invade the surrounding landscape. Chapter 15 addresses the largely neglected problem of invasiveness of exotic agroforestry trees, reviewing experiences with invasive tree species and genera from different climates and use groups and providing recommendations for avoiding future invasions. The use of native species in agroforestry plantings, especially near conservation areas, and the systematic collection of data on the invasiveness of potential agroforestry species are critical to these efforts.

Finally, Chapter 16 addresses the role of landscape diversity on crop disease dynamics, a topic that has been largely neglected in previous research but could be of tremendous importance for the success of landscape-scale conservation strategies using agroforestry in tropical land use mosaics. Although agroforestry and diversified land use mosaics could reduce the spread of disease propagules, increase biocontrol options, and contribute to the conservation and in situ evolution of resistance genes, there is also the risk of cross-infection between agricultural crops and native vegetation, with detrimental effects on either side. In view of the disease risks associated with the use of exotic crop and tree species, the value (from both an agronomic and a conservation perspective) of using native species is stressed.

Chapter 13

Wildlife Conservation in Agroforestry Buffer Zones: Opportunities and Conflict

Lisa Naughton-Treves and Nick Salafsky

Conservationists now recognize the need to work beyond protected areas if they are to sustain viable populations of wildlife and large-scale ecological processes. Given that many tropical forest parks and reserves are surrounded by some form of agriculture, finding more space for wildlife is difficult. Too often, conservationists draw ambitious maps extending wildlife corridors and buffer zones far beyond protected area boundaries without considering the practical and political feasibility of promoting wildlife in preexisting land use systems. Unless such maps are drawn with participation and input from resident populations, conservationists risk creating paper buffer zones. By ignoring local attitudes toward wildlife, protected area managers may miss opportunities to build alliances or, worse, antagonize people and turn them against conservation.

In this chapter we survey the opportunities and conflicts associated with wildlife conservation in agroforestry buffer zones. Drawing on several cases from across the tropics, we reveal how the local social and physical context shapes the viability of wildlife management and the value wildlife has for forest farmers and protected area managers. We also show that the relative costs and benefits of wildlife in agroforestry buffer zones depend on individual perspectives. For example, a mountain gorilla foraging on crops outside a park may be a precious endangered animal to the conservationist but a menacing pest to the local farmer. Sensitivity to local context and local perspectives is essential to finding workable arrangements for winning space for wildlife beyond protected area boundaries. Managers must think creatively and build alliances with local communities while they ameliorate conflict. They must also recognize that their authority beyond protected area boundaries may be uncertain, making the need to incorporate local residents as planners and decision makers all the more important.

In the tropics, protected forest areas and agriculture meet and mix in innumerable ways. A towering stand of old growth may abut a tea plantation, or a grove of fruit trees may gradually blur into secondary forest. As in this book's other chapters, we adopt a broad definition of agroforestry and consider an agroforestry buffer zone as any land use system combining trees with agricultural crops that lies adjacent to a park or reserve. The nature of the park-agriculture interface shapes wildlife survival and local tolerance of wildlife (Newmark et al. 1994; Hoare and du Toit 1999). For example, wildlife may not venture far outside a protected area surrounded by densely settled agriculture, but when it does, it causes intense conflict (Woodroffe and Ginsburg 1998). Conversely, in a sparsely inhabited extractive reserve, wildlife may range throughout a forest-agriculture mosaic and be managed as game by local farmers (although here too there is the potential for wildlife to cause conflict; Naughton et al. 1999). Equally important to understanding local attitudes are the access rules to wildlife in the agroforestry system. Farmers probably will respond differently to an animal they are free to hunt than to a strictly protected one. Taken together, these physical and social parameters shape local attitudes toward wildlife and its status as a pest or valued resource. Therefore, as we analyze the costs and benefits of wildlife for different stakeholders, we explain how the local physical and sociopolitical context shapes these appraisals. In this way, we hope to provide fresh insights on the viability of wildlife conservation beyond protected area boundaries in different situations.

Case Study Sites on a Continuum from Forest-Agriculture Mosaics to Hard Edges

To identify key factors shaping the viability of wildlife management in agroforestry buffer zones, we describe human-wildlife interactions around three national parks, each from a different tropical region, each surrounded by different types of agroforestry land uses (Table 13.1). The most remote of the three sites is Bahuaja-Sonene National Park (BSNP), located in the southeast Peruvian Amazon (Figure 13.1). BSNP is a 1-million-ha uninhabited park bordered by the 252,000-ha Tambopata National Reserve (TNR).[1] Adjacent to these protected areas is a sparsely inhabited (less than 1 person per square kilometer) 272,582-ha buffer zone (Figure 13.1) whose residents plant rice, cassava, and maize in shifting cultivation fields. The second site is Gunung Palung National Park (GPNP), in West Kalimantan, Indonesia (Figure 13.2). GPNP is a 90,000-ha uninhabited park surrounded by production forests and agricultural lands of varying population densities where local residents grow lowland and upland rice, other grains, and fruits. The third site is the 76,000-ha Kibale National Park, located in western Uganda. Kibale is an uninhabited park surrounded by densely inhabited land (90–242 people per square kilometer) used for permanent agriculture (Figure 13.3). Kibale is an example of

Table 13.1. Social and physical attributes of three study sites.

	Bahuaja-Sonene National Park, Peru	Gunung Palung National Park, Indonesia	Kibale National Park, Uganda
Year established	1996[a]	Early 1900s	1933[b]
Forest type	Lowland rainforest to cloud forest	Lowland rainforest to cloud forest	Premontane humid forest
Size of park (ha)	1,000,000	90,000	76,000
Elevation range (m above sea level)	220–2,700	0–1,000	1,110–1,590
Rainfall (mm yr^{-1})	1,200–4,000	4,000	1,100–1,600
Legal uses of national park	Hunting by indigenous people using traditional technology, ecotourism	Research, ecotourism	Water collection, ecotourism, medicinal plant harvest, wild coffee harvest
Number of residents in park	0	0	0
Threats to park	Hunting, mining, logging, oil exploration[c]	Small-scale logging, conversion to agricultural land, non-timber forest product harvesting	Hunting, fire, agricultural encroachment, charcoal manufacture, pit-sawing
Legal status of buffer zone	Ambiguous[d]	Production forest, agricultural lands	None[e]
Land use in buffer zone	Cultivating rice, maize, cassava, and plantains in 0.5- to 1-ha shifting cultivation plots	Forest gardens	Cultivating maize, sweet potatoes, plantains, and about 20 other crops in short fallow fields, eucalypts, and fruit trees in small plots
Population density in buffer zone (individuals km^{-2})	< 1	0[f]	92–242
Average landholding size in buffer zone (ha)	40	1.2	1.4

[a]Originally 325,000 ha when established, expanded to 1,000,000 ha in year 2000.

[b]Originally established as a forest reserve for logging, reclassified as a national park in 1993.

[c]Exxon-Mobile occupied an exploratory concession for oil and gas in 1996–2000. Commercially viable reserves were not found, and the land under concession was added to the national park.

[d]Buffer zones are not included in the categories of protected areas defined by the Peruvian Institute of Natural Resources (INRENA). According to government officials, INRENA can state an opinion over land use in the buffer zone and has the "final voice" (C. Landeo, director Bahuaja-Sonene National Park [BSNP], pers. comm., March 2002). Currently, 271,582 ha are designated as a buffer zone, adjoining the Tambopata National Reserve (252,000 ha), an uninhabited area that in turn neighbors BSNP.

[e]The Ugandan Wildlife Authority is undertaking a planning exercise to establish a multiple-use buffer zone just inside the park boundary (size has yet to be determined; A. Mugisha, pers. comm.., 2001).

[f]The forest gardens are largely on hills above the village. People generally do not live in the forest gardens themselves but in cleared village areas below.

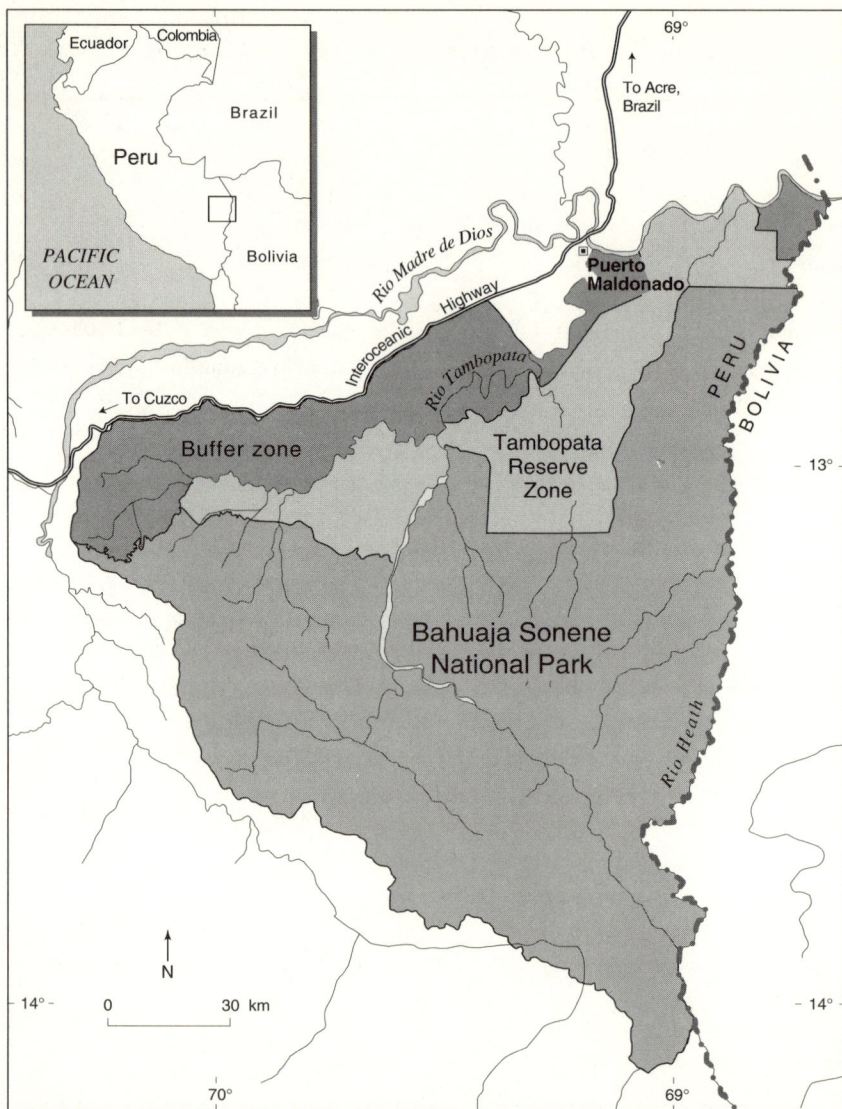

Figure 13.1. Bahuaja-Sonene National Park (BSNP), in the southeast Peruvian Amazon.

a hard edge, where forest conditions and land use types change abruptly at the park boundary. In addition to marked variation in human population density and land use intensity, these three sites vary by access rules to wildlife and forest resources and cultural values of wildlife. In Table 13.1 we list key social and ecological attributes for each site. Detailed site descriptions are available in other publications (Salafsky 1993, 1994; Foster 1994; Lawrence et al. 1995; Chapman and Chapman 1997; Struhsaker 1997; Naughton-Treves 1998,

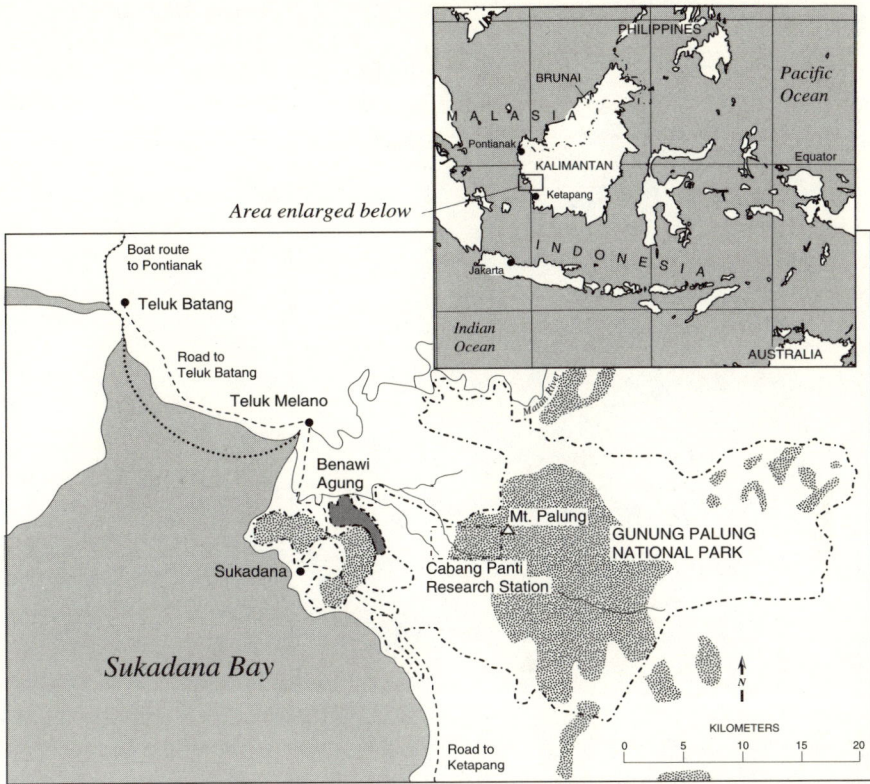

Figure 13.2. Gunung Palung National Park (GPNP), West Kalimantan, Indonesia.

1999; Chicchón 2000). Our information is drawn from semistructured interviews, participatory rural appraisal exercises (PRAs), and systematic measure-
ment of wildlife presence on farms.[2]

Costs and Benefits of Wildlife beyond Protected Area Boundaries

Residents of agroforestry buffer zones hold varied attitudes toward wildlife.
Their attitudes are shaped by their culture, economic situation, political
context, education level, and geographic location. Some people value
wildlife primarily in utilitarian terms and hunt animals to obtain food or
money for school fees. For other residents, particularly indigenous groups,
wildlife holds cultural and symbolic value. Among colonists and indigenous groups alike are people who value the beauty of wildlife and have
moral reasons for letting them survive. Our observation from extensive
interviews at these three sites is that most buffer zone residents support
wildlife conservation in principle, but they do not necessarily want wild

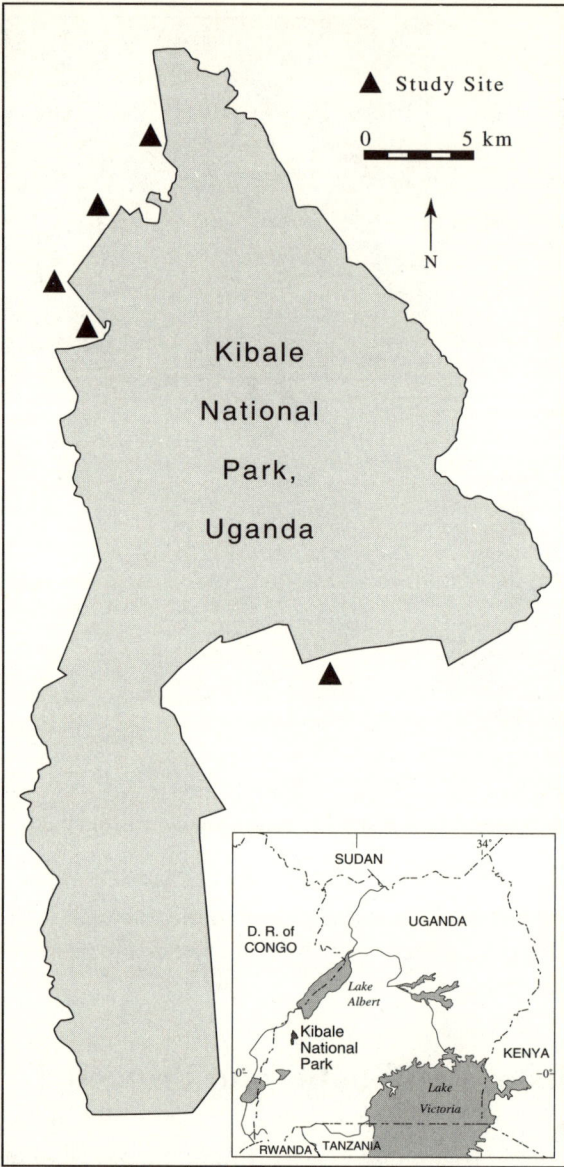

Figure 13.3. Kibale National Park, western Uganda.

animals on their land. Or they may desire the presence of certain species but dislike others.

To understand the complex attitudes among residents of buffer zones, we focus on the relative costs and benefits of wildlife presence in agroforestry landholdings outside parks in the three case studies. In each of these sites, wild species cross park boundaries to forage or hunt in surrounding agroforestry zones. This boundary crossing produces both costs and benefits for different stakeholders. In Table 13.2 we list costs and benefits of maintaining wildlife

Table 13.2. Costs and benefits of maintaining wildlife in buffer zones.

Stakeholder Group	Babuaja-Sonene National Park		Gunung Palung National Park		Kibale National Park	
	Benefits	Costs	Benefits	Costs	Benefits	Costs
Agroforestry holders	Game meat (legal) Environmental services (e.g., seed dispersal of Brazil nut trees by agoutis)	Crop loss to agoutis, pacas, peccaries Livestock loss Damage to Brazil nuts by macaws	Environmental services (e.g., pollination by flying foxes) Hunting (practiced by non-Muslim citizens) Water source for villages	Crop damage Poultry losses	Game meat (illicit) Tourism revenue School support via environmental education and revenue sharing Environmental services (e.g., tree seed dispersal by primates to regenerating fallows)	Crop damage Labor and health costs of guarding Human lives threatened or lost Fear of eviction
Park managers and conservationists	Increased habitat area Increased populations Corridor maintenance	Hunting	Increased habitat area Corridor maintenance Increased populations Spatial buffer zones	Hunting Local animosity toward park	Increased habitat area Wildlife provisioning Corridor maintenance	Local conflict, animosity toward park Risk of disease transmission (human-wildlife or livestock-wildlife) Hunting snares maim or kill wildlife Rise of weedy species

in buffer zones for two principal stakeholders: forest farmers (i.e., people residing in agroforestry buffer zones or cultivating crops there) and protected area managers. In the discussion that follows, the contrasting perspectives of these two stakeholder groups emerge.

Bahuaja-Sonene National Park: Hunting Opportunities for Neighboring Farmers

Among the three sites, BSNP presents the best conditions for sustainable hunting in the surrounding buffer. BSNP is located in the Department of Madre de Dios, one of the most remote and biodiverse regions of Peru (Figure 13.1; Foster 1994; Ascorra et al. 1999). This lowland, forested region was isolated from external markets until the rubber boom of the late 1800s. The arrival of 6,000 rubber workers gave rise to *ribereño* society (Amazonian residents of mixed ancestry), while enslavement and epidemics decimated indigenous populations (e.g., the Ese'eja peoples; Varcarcel 1993; Chicchón et al. 1997). After the collapse of the rubber industry in the early 1900s, the local population was stable until the mid-1960s, when a road was constructed into Madre de Dios. Andean peasants were drawn to the region by gold, land availability, and economic incentives for ranching and farming (Chicchón et al. 1997; Alvarez and Naughton-Treves 2003). Tambopata's population grew fivefold in 25 years, reaching 76,610 in 1997, with roughly half the population residing in the capital city of Puerto Maldonado (GESUREMAD 1998). Despite rapid population growth, Madre de Dios continues to have the lowest population density for Peru (0.9 inhabitant per square kilometer), and the largest tracts of undisturbed forest (GESUREMAD 1998).

BSNP is vast and uninhabited (Table 13.1). Large mammals that are endangered or rare elsewhere in the Amazon are abundant in the park, including white-lipped peccaries (*Tayassu pecari*), giant otters (*Pteronura brasiliensis*), tapirs (*Tapirus terrestris*), and large-bodied primates (Foster 1994). Intact primary forest blankets most of the park and most of the adjacent TNR. Forest predominates in the buffer zone, but it includes logged forest, regenerating shifting cultivation fallows, and forest under extractive use. In the buffer zone, about 3,000 farmers clear 0.5-ha shifting cultivation plots to plant rice, maize, and cassava. Roughly 10 percent of the buffer zone residents are indigenous Ese'eja people who live on communally owned land. This case study focuses on *ribereños,* the dominant social group in the buffer zone. *Ribereño* landholdings average about 40 ha in size, of which 21 ha is under mature forest, 7 ha fallow land, 5 ha annual crops, and 5 ha pasture (Alvarez 2001). Agriculture is the most common economic activity among local residents, but like most Amazonian residents, they pursue other economic activities (e.g., fishing, mining, Brazil nut [*Bertholletia excelsa*] collecting, and logging) in response to resource availability and boom and bust economic cycles. Roughly 20 percent of residents in the buffer zone keep cattle (average 15 head per household), primarily as an investment. Most people (more than 90 percent) raise small livestock for home consumption (average 20 animals per household, including pigs, poultry, and guinea pigs). Precise income figures for local residents are unavailable, but the minimum wage in 2000 was 5 soles (US$1.40 in 2000) per day.

Roughly half of buffer zone residents hunt (Chicchón 1996). No single law pulls together hunting rules in Peru; instead, regulations are tied to forestry and other laws and often change from year to year (Ascorra 1996). Local residents and park guards have confused understandings of what animals can be killed where, particularly in light of recent changes in protected area boundaries at BSNP. But in general, Peruvian law designates wildlife as national patrimony and places responsibility for its protection with the national government. All Amazonian species are now protected from hunting except for 15 game species. The game species include red brocket deer (*Mazama americana*), peccaries (*Tayassu pecari* and *Tayassu tayacu*), tapirs, pacas (*Agouti paca*), agoutis (*Dasyprocta variegata*), capybaras (*Hydrochaeris hydrochaeris*), armadillos (*Dasypus novemcinctus*), turtles, and some game birds (Varese 1995). For these 15 game species, hunting is allowed for subsistence and local sale in small communities, including those in reserves and buffer zones. Hunting is also legal whenever wildlife threatens crops or livestock (Ascorra 1996). Local residents call this form of hunting *cacería sanitaria* ("sanitary hunting"), particularly in reference to predator removal (Ascorra 1996). The majority of hunters use shotguns to kill prey. The trapping of animals is rare.

Local Agriculturalists' Perspective

Local residents around BSNP commonly call wildlife species as *presa* ("game") or *plaga* ("pest"; Ascorra 1996). Their classification depends on the marketability of the species and its potential to cause significant crop or livestock losses (Table 13.3). The local abundance of a species and the farmer's ability and interest in hunting also shape individual attitudes towards wildlife. *Ribereños* and indigenous people are more likely to hunt than are colonists, who lack the requisite detailed knowledge of the forest (Redford and Robinson 1987; Naughton-Treves 2002). Even colonists living in game-rich areas hunt less frequently than *ribereños* or indigenous people (Loja et al. 2000). Hunters at Tambopata generally are more positive than nonhunters toward having wildlife on their farms, particularly high-value or large species such as pacas, brocket deer, tapirs, and white-lipped peccaries. Those who live adjacent to the park are fortunate to hunt large and high-value game on their farms, easily offsetting the crop losses caused by these and other wild animals. But farmers residing farther from the park boundary (more than 700 m), are

Table 13.3. Wildlife pests and valued animals according to agroforestry farmers and park managers at Bahuaja-Sonene National Park (BSNP), Gunung Palung National Park (GPNP), and Kibale National Park.

Stakeholders	Valued Animals			Pests		
	BSNP	GPNP	Kibale	BSNP	GPNP	Kibale
Agroforestry farmers	Pacas	Flying foxes	Bushpigs[a]	Jaguars	Macaques	Elephants
	Peccaries	Deer	Cane rats	Pumas	Pigs	Baboons
	Tapirs	Pigs (valued by non-Muslims)	Game birds	Tayras	Deer	Leopards
	Howler monkeys			Ocelots	Weasels	Chimpanzees
				Macaws	Cats	Bushpigs[a]
				Agoutis		
Park managers	Jaguars	Orangutans	Chimpanzees	None	None	Elephants[b]
	Giant otter	Proboscis monkeys	Monkeys			
	Tapirs	Gibbons	Endangered bird species			
	Harpy eagles	Hornbills	Elephants[b]			
	Primates	Rare birds				
	Macaws					

[a] Attitudes toward bushpigs varied between communities: people residing near park headquarters complained bitterly about bushpigs because there hunting rules were strictly enforced and they felt powerless to defend their farms. In more remote areas, people explained that if bushpigs caused too much damage, they simply "ate them" (L. Naughton-Treves, unpublished data, 1995).

[b] Managers recognize the importance of Kibale's elephants given that elephant populations are endangered or extirpated throughout much of Uganda. Elephants draw tourists, also. But where elephant densities are high in the park, natural forest regeneration is suppressed.

more likely to encounter only smaller, highly adaptable species such as armadillos (*Dasypus* sp.), collared peccaries (*T. tajacu*), tayras (*Eira barbara*), and agoutis (*Dasyprocta* spp.). Agoutis are responsible for the greatest amount of damage to maize and cassava in the buffer zone, and people complain that this 4-kg rodent is "not even worth a bullet" (Naughton-Treves 2002).

Just as hunting benefits are not equally shared in the local community, the costs of raiding are unevenly distributed. On average, residents in the buffer zone lose negligible amounts of crops (maize, cassava, rice, and plantain). A study of wildlife damage from 1998 to 2000 found that wildlife damage averages less than 3 percent by area or roughly $13 per planting season (for details, see Naughton-Treves 2002), but average values mask the skewed distribution of loss. Most farmers lost little to wildlife, whereas a few lost significant amounts (up to 47 percent). Generally, losses to prime game species, such as pacas, tapirs, and peccaries, are better tolerated than losses to small game (e.g., agoutis) or nongame species (e.g., tayras, a member of the weasel family that thrives in agroforestry mosaics; Bisbal 1993). Residents who hunt in shifting cultivation fields or fallows in the buffer zone (i.e., garden hunters) capture on average 9 kg or about $14 of game meat per hunter during a planting season (about 5 months) (Naughton-Treves 2002). (Game meat values are expressed here in U.S. dollars to allow comparison to crop damage values; in reality some hunters share the meat with their families, others sell it to neighbors, and others illicitly sell it to intermediaries for eventual sale in Puerto Maldonado.) The average gains are roughly equivalent to the average crop losses to wildlife. But systematic monitoring of 24 garden hunters in the broader region revealed that only three earned more income from meat than they lost in crop damage. These hunters lived close to the reserve boundary and were able to shoot tapirs and white-lipped peccaries in remote areas, including heavily forested areas in the buffer zone and the reserve. For farmers further from the reserve boundary, it is more difficult to balance crop losses with hunting gains given the scarcity of big game. Moreover, many hunters do not enjoy hunting in the brushy, hot fallows and fields, where visibility is poor. They prefer to hunt in unsettled forests, where they can fell larger animals. Several hunters describe hunting in agroforests as an activity for the old and weak.

The one type of wildlife universally considered a pest in Tambopata is predators. Neither hunters nor farmers tolerate predators such as jaguars (*Panthera onca*), pumas (*Felis concolor*), ocelots (*Leopardus pardalis*), and jaguarundis (*Felis yagouaroundi*) on their land. In a random sample of 60 farmers in the buffer zone, the majority (75 percent) reported losing poultry or pigs to wild predators. The wild animals most frequently blamed were ocelots (identified by 31 percent of farmers) and hawks (28 percent), followed by jaguars (5 percent). Also mentioned were tayras, jaguarundis, pumas, and bush dogs (*Speothos venaticus*). Farmers residing close to the reserve reported losing more domestic animals to a greater variety of predators than did those farther from the reserve. The value of the livestock reported lost per farmer for those complaining of jaguar attacks averaged $118 per year (range $6–$294, SD = $121, n = 9). The value of livestock lost to other carnivores averaged $49 per year (range $6–$194, SD = $54.3, n = 24). Among those suffering losses, jaguar attacks were reported every 2.6 years, compared with every 1.1 years for other predators. (Note: These reported losses were not verified and offer only an estimate of losses.) Hunters reported losing fewer domestic animals to predators than did nonhunters (Naughton-Treves 2002).

In addition to direct benefits from hunting, wildlife potentially provides buffer zone residents indirect benefits. The economic concerns of many of Tambopata's buffer zone residents are tied to Brazil nut production. Roughly 20 percent of local residents collect Brazil nuts for commercial sale (Chicchón 1996). Brazil nut trees need healthy populations of euglossine bees if they are to be pollinated and yield abundant fruit. These bees in turn depend on an intact forest. Therefore, the yield of Brazil nut trees is tied to the health of local bee populations, which in turn need an intact forest. On the other hand, people often shoot macaws and parrots that arrive to feed on ripening Brazil nuts. Biologists report that contrary to local belief, parrots and macaws cause negligible damage to Brazil nut yields, and they are now leading a campaign against macaw hunting, explaining that shooting macaws costs more in ammunition than it saves in fruits (Ortiz 1995). When the fruits mature and fall to the ground, agoutis open the fruits and bury the nuts one by one, usually within 100 m of the tree (Ortiz 1995). Very few seeds germinate because most are later consumed by agoutis or other animals who discover the agoutis' caches. Thus, Brazil nut trees need agoutis for seed dispersal, but an overabundance of agoutis probably would inhibit their regeneration. Instead of intentionally conserving or managing agoutis on their concessions, Brazil nut harvesters hunt them with shotguns or machetes whenever the opportunity arises. In this way, agoutis contribute to protein needs of Brazil nut harvesters. Brazil nuts are an example of the interdependence of insects, trees, wildlife, and local people in managed tropical forests.

Finally, wildlife is a potential source of income as an ecotourism spectacle. Ecotourism is booming in the Tambopata region. However, tourists generally bypass inhabited regions, and their dollars flow to a minority of people who run lodges in the buffer zone or reserve. Hunting, agriculture, and ecotourism cannot all be managed in the same place. One indigenous community in the buffer zone, the Esse'eja, negotiated an agreement with a tour operator that guaranteed them 60 percent of profits and half the decision-making authority, and in turn they have agreed not to hunt or farm in 4,000 ha of their land (Stronza 1998). It is more difficult for the individual landholder to do this. Given that most of the people manage their land independently in 40-ha parcels, zoning must come from higher representative institutions (e.g., a farmers' union).

Park Managers' Perspective

The institution officially responsible for managing BSNP and TNR is Peru's Institute for Natural Resources (INRENA). Its authority over the buffer zone is less certain. According to Peruvian law, buffer zones are not protected areas, but INRENA staff claim to have the final voice in land use planning in the buffer zone (C. Landeo, pers. comm., 2002). In practice, INRENA has limited control over human activities in all three areas, given the large size of the areas and given INRENA's meager budget (e.g., $35,000 for BSNP in 2000; A. Bruner, pers. comm., 2002). The area has so far been protected mainly because it is so remote. Given these conditions, INRENA's primary management goal is to stabilize land use along colonization fronts. To this end, it is working with local stakeholders (e.g., environmental nongovernment organizations [NGOs], agriculturalists' unions, indigenous federations, and tourism companies) to establish zones in the TNR where land use of varying intensity is allowed (Ascorra et al. in preparation). Managers view the national reserve neighboring the park as a buffer region where economic development and biodiversity conservation can be combined. They also see the reserve and the buffer zone beyond as a source of additional habitat for wildlife. But managers fear that hunters are already eliminating large and slowly reproducing species throughout the buffer region (Loja et al. 1999). Large mammal populations outside the park, even in sparsely inhabited areas, already show signs of over-exploitation. In essence, the buffer zone is acting as a sink for some vulnerable species (e.g., large-bodied primates, jaguars, tapirs, white-lipped peccaries; Novarro et al. 2000). In the long run, this dynamic is acceptable to conservationists, provided that source areas are sufficiently large and well protected in the park. Hill and Padwe (2000) estimate a ratio of 7:1 for source area to sink area for hunting in the sink to be sustainable.

An NGO called Conservación Internacional–Peru has been working with two communities in the buffer zone to promote more sustainable hunting practices. Urban demand for game meat and uncertain property rights make sustainability elusive, but some promising examples exist of communities voluntarily setting aside no-hunting areas and monitoring game populations in their area. Paradoxically, although a minority of avid hunters threaten the survival of game populations, they are the ones most interested in programs to conserve wildlife (Naughton-Treves 2002). Most farmers hunt only occasionally, or not at all, and for them wildlife is not a significant resource. These people support wildlife conservation programs in general, although they would rather not have wildlife visiting their fields.

Gunung Palung National Park: A Spectrum of Land Uses

GPNP, in West Kalimantan, Indonesia, contains a range of tropical habitats from mangrove swamps to lowland forest to cloud forest at 1,000 m above sea level (Figure 13.2). The park is home to a wide range of flora and fauna including dipterocarp trees, orangutans (*Pongo pygmaeus*), gibbons (*Hylobates agilis*), sun bears (*Helarcto malayanus*), and several hornbill species (*Buceros rhinoceros, Anorrhinus* spp.). Many of the animal species in the park use a range of different habitats throughout the year. Over the past few decades, the forested lands around the park have been logged or converted into agricultural lands, leaving the park increasingly isolated.

The villages on the western edge of the park along the coast have been inhabited for several centuries, largely by Islamic families, most of whom are of Malay ancestry, and by some families of Chinese ancestry who migrated to the area in the early twentieth century. In the past few decades, a number of transmigrants from Bali and Java have also settled in the area. Villages on the interior borders of the park are inhabited largely by Dayak families.

Residents of the villages on the western edge of the park have developed a complex land use system that includes farms, homegardens, forest gardens, and extractive areas from which they harvest rattan, specialty timbers, and other forest products in a gradient that leads away from the village toward the park (Salafsky 1994; Lawrence et al. 1995). Key crops that are grown in the forest gardens include durian, rubber, coffee, and other crops for market sale (for the structure and biodiversity of such complex agroforests see also Chapter 10, this volume). In the farm areas, the main crops include paddy rice, maize, and vegetables. Officially, the border between the park and the villages on the western edge is the bottom of several small hills that are in the park; there is no officially recognized buffer zone. Over the past few decades, however, local villagers have been steadily expanding their forest gardens up the hill, taking over land that is technically part of the park. This case thus presents an example in which the agroforestry buffer zone is encroaching on the forest lands of the park.

Forest Garden Owners' Perspective

In a study of forest garden owners' perspectives on wildlife, the most critical factor determining attitudes toward wildlife was religion. In villages on the western edge of the park, although some families hunted larger deer (*Cervus unicolor* and *Muntiacus* spp.) and the Balinese and Chinese families hunted pigs (*Sus barbatus*), the majority of the village residents did not eat much bushmeat at least in part because of Islamic religious prohibitions against eating wild animals other than deer. As a result, one could commonly encounter gibbons, hornbills, and other large diurnal animals in the forest gardens located quite close to Islamic villages. In Dayak villages on the northern edge of the park, by contrast, people hunted all kinds of animals (Lawrence et al. 1995). As a result, there are far fewer animals in the forest gardens near Dayak villages.

Based on a survey of community members in the Islamic villages (Salafsky 1993), animals that were common in the forest gardens included maroon langurs (*Presbytis rubicunda*), long-tailed macaques (*Macaca fasicularis*), gibbons, various squirrels (*Callosciurus prevostii* and *Ratufa affinis*), civets (*Paradoxurus* spp.), flying foxes (*Pterpus vampyrus*), and sambar deer (*Cervus unicolor*). Residents also said that orangutans and sun bears were seasonally present in the forest gardens but did not venture into the farm areas. Pigs, deer, and macaques were also frequently found in farm areas, although they generally only foraged there and then returned to the adjacent forest.

Local residents clearly distinguished between benign species and crop raiders (Table 13.3). Specific animals that were most problematic included langurs, long- and pig-tailed macaques, squirrels, rats and mice, pigs, and sambar deer, which caused minor to severe damage to planted crops. Residents claimed that weasels (*Mustela* spp.) and leopard cats (*Felis bengalensis*) killed their chickens and ducks. Residents also believed that the flying foxes ate the durian tree flowers, not recognizing the pollination role that they play. On average, residents reported that they lost 9.0 percent (*SD* = 10.4 percent of their gross income (the range was 0–50 percent) to damage caused by all animals (including rats, mice, and flocks of small birds). This figure does not include the time, labor, and equipment that residents invested in guarding crops before harvesting them. For example, residents slept in the fields for a week or two before harvest to protect their rice crop from being eaten. A few residents (mostly men who had spent a great deal of time working in the forest) expressed enjoyment or interest in observing and discussing what animals did. However, most residents had ambivalent or even negative attitudes concerning the presence of animals.

Park Managers' Perspective

GPNP is home to a wide variety of life, including endemic proboscis monkeys (*Nasalis larvatus*) and the largest remaining orangutan population in Kalimantan. It also contains one of the last remaining corridors of natural habitat from the ocean to the cloud forest in West Kalimantan. Unfortunately, as described earlier, the park is fast becoming an island of forest surrounded by agricultural lands. Furthermore, forest within the park is being steadily cleared at its edges by timber harvesting and conversion of land to agricultural uses.

As in much of Indonesia, the national park authority has traditionally had only a weak presence on the ground near the park. There are only a handful of park guards, and given their meager salaries, it is easy for the guards to become engaged in illegal resource extraction activities including timber and nontimber forest product harvesting and fishing. Agroforestry buffer zones are also vulnerable to illicit harvesting. In fact, buffer zones were created on what used to be park lands as a means to deal with ongoing resource extraction. Although these zones are less diverse than primary forest, they at least maintain forest cover and therefore are preferable to monocultures. In other words, these buffer zones provide spatial buffering that increases the distance between the villages and the forest. The buffer zones also provide habitat for some wildlife species and can help create corridors to permit migration of key animal species.

Kibale National Park: Conflict and Compromise along a Hard Edge

Located in western Uganda, Kibale is a 76,600-ha remnant of midaltitude forest that is much celebrated for its exceptional diversity and density of primates, including chimpanzees (*Pan troglodytes*), eight monkey species, and three prosimians (Struhsaker 1997). Also present at Kibale are species notorious for raiding crops, such as olive baboons (*Papio anubis*), red-tail monkeys (*Cercopithecus ascanius*), bushpigs (*Potamochoerus porcus*), and elephants (*Loxodonta africana*), a species reduced by more than 90 percent in Uganda over the past 30 years (Hill 1998; Amooti 1999). Hunting in Kibale is illegal, although snares are regularly encountered in the park. Outside the park, citizens may hunt only "vermin" (baboons, vervet monkeys [*Cercopithecus aethiops*], and bushpigs) and only with permission from the Ugandan Wildlife Authority (see Naughton-Treves 1999 for a description of colonial underpinnings of Uganda's hunting rules and recent efforts at reform).

Roughly 54 percent of land within 1 km of Kibale's boundary is used in smallholder agriculture (Mugisha 1994). Agriculturalists in the area belong to two predominant ethnic groups: the Batoro, whose presence in the area dates to the 1890s, and the Bakiga, who began settling in the area in the 1950s (Turyahikayo-Rugyema 1974; Naughton-Treves 1999). The Batoro chiefs at the time allocated land to immigrants on the outskirts of their settlements, hoping to buffer Batoro farmers from crop damage by wildlife. Today, both groups plant a mixture of more than 30 species of subsistence and cash crops, including bananas, maize, beans, yams, coffee, and fruit trees. Farm sizes are small (1.4 ha on average) and are managed individually. In this diverse farming system, various animals forage on crops, resulting in much frustration and resentment against the park among local cultivators. In a 1992–1994 study, the most crop damage was observed within 200 m of the park boundary, and losses averaged between 4 and 7 percent by area per season. However, as in BSNP in Peru, the distribution of loss was highly uneven. More than half of all farmers within 500 m of the boundary lost no crops to wildlife, but 7 percent lost more than half their crops (Naughton-Treves 1997, 1998). Certain villages were particularly vulnerable to elephant damage and suffered the highest losses.

Forest Farmers' Perspective

Farmers appreciate the drinking water, fuelwood, and medicinal plants they gather from Kibale (Naughton-Treves 1998). They also acknowledge receiving indirect environmental services from the forest, and they believe the forest is important for maintaining clean air and abundant rainfall. But most respondents did not recognize the role of wildlife in maintaining forest ecosystem function. "Why can't the government move the animals to some other park?" some respondents asked. Another typical outburst was, "These animals leave us poor and hungry. Why should we starve so that baboons may eat?" (see Table 13.3). Their perception of risk reflects extreme damage events, not average losses. Perceptions oriented toward extreme events may also explain why farmers complain less about small animals, such as redtail monkeys, than about large animals, such as elephants. Animals such as redtail monkeys visit many farms and may cause greater aggregate damage but do not destroy an entire field in a single raid; their damage is self-limiting (Naughton-Treves 1997). Elephants, meanwhile, affect fewer farmers but can cause catastrophic damage and pose a physical threat to farmers. Among the various crop-raiding species, only elephants are capable of causing damage so severe as to cause people to abandon their farms around Kibale.

Farmers residing in the buffer area adjoining Kibale vary in their capacity to cope with crop loss to wildlife. By far the most common defensive strategy is guarding (60 percent). Half of the farmers leave land fallow at the forest edge, where the risk of loss is high. Farmers are reluctant to admit killing wildlife, as it is illegal, but snares or poison were encountered on 15 percent of farms along the boundary (Naughton-Treves 1997). Some farmers are able to mitigate risk by creating buffers within their farms. For example, affluent owners of large farms occasionally use pasture or plant coffee or tea to separate their food crops from the forest. But the owner of a small farm has little leeway for arranging crops of different palatability to wildlife and may wind up planting maize directly on the boundary. Similarly, more affluent farmers may employ others (often others' children) to guard their fields, whereas poorer farmers must either face the risk of crop loss with no guard posted or sacrifice other opportunities such as schooling to leave a child guarding crops. An added cost of guarding is increased exposure to malaria, given that most raiding occurs at dawn or dusk, when *Anopheles* mosquitoes are active. Ultimately, these passive defense measures were considered costly by most farmers and only partially effective (Naughton-Treves 1998).

Although farmers around Kibale do not work communally to defend their crops, an individual's vulnerability is influenced by his or her neighbors' activities. For example, a farmer living in a village where several others hunt in their fields probably will suffer lower levels of damage by bushpigs, even if he or she does not hunt (Naughton-Treves 1998). Ultimately, a farmer's best defense against losing crops to wildlife is to have a neighbor's crops between his or her farm and the forest, so that the damage is incurred on the neighbor's land rather than his or her own. Some large landholders (more than 8 ha) take advantage of this defense by leasing plots to other farmers immediately along the forest boundary. Researchers elsewhere in Africa have also noted that a densely settled band of farms forms the best barrier to wildlife incursions deep into agricultural lands (Bell 1984; Hawkes 1991; Hill 1997).

Beyond spatial patterns of risk, tolerance to wildlife at Kibale is shaped by the political context. Many buffer zone residents bitterly call crop-raiding animals "the government's livestock" and believe the government must help guard, cull animals, or build a fence. Farmers repeatedly draw the analogy of the government being a bad neighbor, allowing its "livestock" to damage other people's crops. They point out that under customary rules, farmers must reimburse their neighbors for any damage caused by their livestock. Seldom do they mention that many farmers often graze their cattle and goats illegally in the park. Obviously, the traditional local social contracts regarding grazing rights and restitutions for animal crop damage are not operating between farmers and Kibale National Park authorities. Some people's complaints against wildlife are magnified by their general resentment of the park. When Kibale was upgraded to a national park from a reserve in 1993, thousands of people were forcibly evicted from the Kibale game corridor and resettled elsewhere. Residents remaining in the area are apprehensive that any park intervention on their land could result in more evictions.

Park Managers' Perspective

From a manager's perspective, crop losses of 4–7 percent in a narrow band of farms appear to be a trivial price for maintaining endangered wildlife and forest habitat. In fact, the zone of heaviest crop loss (about 200 m beyond the forest boundary) could be considered 3,000 ha of extra wildlife habitat at Kibale (Mugisha 1994). But living in this extra habitat are approximately 4,000 frustrated farmers who protest vehemently against the use of their land as "a park for grazing wild animals." This resentment is an obstacle to alliances between conservationists and local residents. However, most managers recognize that there is no alternative but to reach out to local communities. Wild animals inevitably cross park boundaries, and when they do, they are vulnerable to snares and poison. Up to 20 percent of one community of chimpanzees in Kibale have lost a foot or hand to snares they picked up when foraging in crops outside the boundary (R. Wrangham, pers. comm., 1996). More broadly, Uganda is moving toward decentralized management of resources, motivated as much by the donors' emphasis on community participation as by budget shortfalls in national agencies. All these factors together make it more important to raise public support for wildlife.

To date, managers have experimented with planting nonpalatable crops at the park boundary, such as soybeans, sunflowers, tobacco, tea, and Mauritius thorn (*Caesalpinia decapetala*). To be successful as a buffer, a cultivar must be profitable, unpalatable to wildlife, and planted over a large enough area to reduce the attractiveness of crops beyond. Most farmers around Kibale own small landholdings (1.4 ha) and do not cooperate with neighbors with regard to crop selection, planting, or maintenance. This limits the success of buffers. For example, when a single farmer planted tea on the forest boundary and maize 100 m beyond, baboons simply traveled through his neighbor's fallow land to reach the maize. Given the small landholding on the edge of Kibale, a buffer is a viable option only if neighbors collaborate in their planting. Better results have been achieved in communal efforts to plant Mauritius thorn barriers along the boundary. Tea is a popular buffer crop in highland Africa because it is not consumed by any wildlife species. But a tea buffer must be planted continuously and extensively and pruned frequently. Such a planting regime is beyond the scope of an individual farmer and would take collective or corporate ownership. Where tea buffers were planted around Kenya parks, the land cleared came from the national park, a significant sacrifice given the small and isolated nature of most highland parks. The barrier that many Kenyan park managers have resorted to is an electric fence. Many of Kibale's neighbors have demanded a fence, particularly those who live in sites vulnerable to elephant raids. But fences are costly ($1,000–$2,000 per kilometer in moist forest environments; Hoare 1995) and are anathema to conservation biologists striving to connect ecosystems and reduce the isolation of wildlife populations.

To raise local tolerance to wildlife, Kibale managers have also launched tourism revenue-sharing programs. Kibale is visited by roughly 1,000 tourists a year, each of whom pays US$10 for a guided forest walk that offers an opportunity to see chimpanzee (Archabald and Naughton-Treves 2001). To date, 5 out of 27 parishes neighboring Kibale have participated in revenue-sharing projects. Together, they received $3,000 of tourism revenue in a 3-year period to support schools and clinics. Although the sum is modest, park managers enjoyed better relationships with residents from recipient communities. It is uncertain whether these communities receiving community revenue hunt less than those who do not, although at a neighboring park, recipient communities assisted in the capture of mountain gorilla (*Gorilla gorilla beringei*) poachers (Archabald and Naughton-Treves 2001). Interestingly, many participants ranked receiving revenue sharing as a greater advantage of being a park neighbor than gaining access to nontimber forest products (Archabald and Naughton-Treves 2001). In addition to raising significant revenue, selecting the appropriate community in the buffer zone to enjoy tourism revenue sharing is a serious challenge, echoing general dilemmas in integrated conservation and development project (ICDP) design (Agrawal 1997).

Managing Agroforestry Buffers for Wildlife Conservation: Conditions of Success

Managers aiming to promote wildlife survival in agroforestry landscapes beyond protected area boundaries must carefully consider the social and physical parameters of their site and tailor their approach to local context. Where wildlife populations are fairly abundant and human population densities in surrounding buffers are low (e.g., BSNP), managers can encourage the maintenance of wildlife habitat by promoting sustainable hunting among local residents (see also Chapter 14, this volume). Although this hunting may enhance the local value of wildlife for hunters, it is critical that access rules to game (e.g., who gets to hunt which animals and where) be clearly communicated and enforced or else wildlife probably will be exhausted as an open access resource. Managers must also realize that hunting opportunities are not evenly distributed in a buffer zone, and people who live close to a park and have ecological knowledge will best be able to exploit wildlife. On the other hand, many farmers have no interest in hunting (e.g., recent colonists who lack knowledge or experience in tropical forests). Rarely is hunting a collective, organized activity for rural communities in the way that harvesting other nontimber forest products (e.g., rattan, Brazil nuts, or firewood) may be. Furthermore, the ecological viability of hunting in agroforestry buffer zones depends on the size and growth rate of source game populations in the park and the hunting intensity in the surrounding buffer zone (Novarro et al. 2000). Forest interior species and large-bodied species probably will be depleted in agroforestry zones unless there are stringent cultural or legal prohibitions on hunting (see Chapter 14, this volume). Some adaptable species with high reproductive rates may thrive in agroecosystems. These tend to be smaller, cosmopolitan species such as agoutis, bushpigs, baboons, cane rats (*Thryonomys* spp.), and macaques.

In areas such as Gunung Palung, where parks and forests are becoming increasingly isolated islands, agroforestry practices can be used in buffer zones to provide spatial buffering for the protected area and can, at least in theory, provide corridors to connect forest areas. However, the utility of these lands depends in large part on the degree to which farmers can be protected from crop-raiding animals and wildlife from local hunting pressure.

In high-conflict, high-risk situations, such as Kibale, there is less room to maneuver given the endangered status of some crop raiders (e.g., chimpanzees) and conditions of land scarcity and poverty at the protected area edge. At such sites, expensive and management-intensive interventions such as revenue-sharing schemes, buffer crop planting, land buyouts, and fences may be appropriate (Naughton-Treves 1997). Compensation and insurance schemes have a dismal track record in most tropical countries because of corruption, administrative inefficiency, and other problems (Booth et al. 1992), but they deserve consideration, particularly in sites where there are highly endangered species and secure conservation funding. Compensation is a standard practice for conserving wildlife outside protected area boundaries in the United States and Europe, and it has been used to build political support for conserving wildlife that threaten livestock and crops, such as timber wolves and bears (Naughton-Treves et al. 2003).

Kibale is also a difficult site because of the presence of elephants. Elephants and other megafauna (animals weighing more than 1,100 kg, e.g., hippos [*Hippopotamus amphibus*], buffalo [*Syncerus caffer*]) present special problems in buffer zones because they can cause catastrophic crop damage and threaten lives. Large carnivores present similar challenges. Conservationists should not expect people to accept these animals on their farms unless they are compensated for losses. Even in celebrated examples of community-based wildlife management, farmers turn to barriers. For example, in the CAMPFIRE (Communal Areas Management Programme for Indigenous Resources) program, 20 percent of safari hunting revenue is spent on electric fences encircling farms (World Wildlife Fund 1998). From a conservation standpoint, it is better to enclose agriculture than wildlife, but rarely is this politically or financially feasible.

Conclusions

Human-wildlife interactions show wide variation both between and within the three sites described in this chapter. Therefore, there are no one-size-fits-all recommendations for using agroforestry to promote wildlife conservation. However, there are a few basic concepts that managers may want to keep in mind in determining whether this approach will be useful under local conditions.

Distance from Natural Forest Matters

In all three sites, the population farming closest to the protected area boundary pays the greatest costs and receives the greatest benefits of wildlife. In the case of BSNP in Peru, many protected area neighbors enjoy hunting large game on their farms, easily offsetting crop losses to wildlife. In Kibale, where hunting is illegal, large species threaten the lives and livelihoods of those residing on the boundary. In Indonesia, the forest gardens and fields nearest the park suffer the most damage from crop-raiding animals. Furthermore, in all three cases, with increasing distance from the park, large and interior forest-dwelling wildlife species are replaced by adaptable, fast-reproducing species such as rodents, cervids, wild pigs, and (in the paleotropics) semiterrestrial primates. This observation is confirmed in the broader literature on hunting in forest gardens (Koch 1968; Nietschmann 1973; Peterson 1981; Denevan et al. 1984; Irvine 1987; Bahuchet and Garine 1990; Dove 1993; Fairhead and Leach 1996). Even in sparsely inhabited, heavily forested buffer zones (e.g., BSNP), hunting pressure can quickly reduce endangered or rare species (see Chapter 14, this volume). As a result, the value of the agroforestry systems for wildlife may diminish as one moves away from areas protected from hunting. Other factors that are important in determining the value of agroforestry for wildlife include the composition and structure of the agroforestry areas and the degree to which they are connected to remnant forest patches.

Hunting Is a Critical Variable

Across all three sites, hunting is a critical variable that influences both local attitudes toward wildlife and the effect local people have on wildlife. In BSNP, indigenous groups and *ribereño* communities are more likely to value wildlife than are recent colonists, who rarely hunt and have poor knowledge of the forest. But although local hunters may value game, they are overexploiting many species. Wildlife is an open access resource around BSNP, and until there are publicly accepted rules governing access to wildlife, wildlife will be vulnerable in the buffer zone and accessible areas of the reserve. In Indonesia, religion is a major factor determining who hunts and who does not. Although the people who hunt may value wildlife more, greater numbers of animals can be found around the villages where only limited hunting takes place. At Kibale, strict prohibitions on hunting amplify local feelings of vulnerability to crop loss and lead to resentment of the "government's livestock." The small size and individual nature of farms around Kibale also limit people's options to buffer themselves from damage. Although collective losses to wildlife are low for buffer zone residents, the risk of catastrophic individual loss to elephant raiding and resentment over park evictions of corridor residents have led to generally negative feelings. The experience at all three sites indicates that negotiating hunting rules and refuges is important if game species are to survive in agroforestry systems.

Collective Action to Counter Crop Raiding May Be Needed

Another challenge facing those promoting wildlife in agroforestry buffer zones is the potential for collective action by buffer zone residents. Given the natural tendency of wild animals to cross property boundaries, it is important that there be consensus and enforcement of hunting rules, or wildlife will be vulnerable to overexploitation by individual hunters. Similarly, individual farmers alone cannot protect themselves effectively from raiding by wildlife, particularly by animals such as elephants. Managers should protect and promote collective land management practices to build effective barriers or guarding regimes. In forest farming communities, individuals have variable capacity to take advantage of the presence of these species (or cope with pests). Affluent farmers with larger landholdings can better mitigate their risks from wildlife, although they are often the worst complainers and may have greater political influence (Naughton-Treves 1999).

Local People's Environmental Agendas May or May Not Include Wildlife Conservation

Agroforestry farmers in all three sites were concerned with environmental protection. They recognized the vital environmental services provided by natural forests and agroforestry, including soil and water protection, firewood and construction materials, and medicinal plants. But farmers were less likely to acknowledge the role of wildlife in maintaining ecosystem function via pollination, seed dispersal, and predation. Recent colonists in rainforests are likely to be unaware of these indirect functional roles. Long-term residents may understand complex plant-animal interactions in rainforests, but if they are so poor as to be unable to meet their subsistence needs, it is unlikely that they will be greatly concerned about a long-term decline in tree species or the other consequences of removing wildlife from forests.

Culture and politics are equally important in understanding local tolerance and the viability of wildlife conservation in buffer zones. Local communities are heterogenous, and their members probably will have different values for wildlife depending on the degree of benefit they get from hunting and the crop and other losses they suffer. Other stakeholders (e.g., loggers, ecotourists, and rural populations beyond the agroforestry buffer zone) will hold still different values and may use their political influence or directly intervene to shape the composition and abundance of wildlife in agroforestry areas.

Overall, it is clear that protected area managers and local people may have very different perspectives about the value of wildlife in the buffer zone areas (Table 13.3). Protected area managers and conservationists aiming to promote wildlife survival in agroforestry buffer zones need to relinquish appealing but impractical notions of smallholder agriculturalists welcoming all wildlife on their farms. Neighboring farmers may support the general ideal of wildlife conservation, but they probably will respond to wildlife on their farms according to their individual economic needs and cultural values. Those who hunt may be most tolerant of wildlife, but they probably will also have the greatest direct impact on wildlife populations. As a result, managers need to work closely with local communities to develop solutions that work for both wildlife and humans. An important first step toward collaborative wildlife management in agroforestry landscapes is the exchange of information. Park managers need to be educated about the urgent economic needs and particular environmental agendas of farmers at their site. Agroforestry farmers may be more likely to accept wildlife if there is collaboration between them and park managers.

Endnotes

1. In 2000, Peruvian authorities expanded Bahuaja-Sonene National Park to 1 million ha and reduced the previous 1.5-million-ha transient zone called Tambopata-Candamo Reserve Zone (TCRZ) to a 252,000-ha Tambopata National Reserve. The human-wildlife interactions reported here were observed from 1997 to 1999 in the TCRZ in the area that was later designated as the buffer zone (Figure 13.1).
2. BSNP data collected from 1997 to 2000: 65 interviews, 60 farms monitored, 12 PRAs. GPNP data collected from 1989 to 1992: 88 households interviewed (15 percent), 42 forest gardens mapped. KNP data collected from 1992 to 1997: 245 interviews, 103 farms monitored, more than 20 PRAs.

References

Agrawal, A. 1997. *Community in conservation: beyond enchantment and disenchantment.* Gainesville, FL: Conservation and Development Forum.

Alvarez, N. 2001. *Deforestation in the southeastern Peruvian Amazon: linking remote sensing analysis to local views of landscape change.* Madison: Geography Department, University of Wisconsin.

Alvarez, N., and L. Naughton-Treves. 2003. Linking national agrarian policy to deforestation in the Peruvian Amazon: a case study of Tambopata, 1986–1997. *Ambio* 32:269–274.

Amooti, N. 1999. *Elephants reduced to 6%.* Kampala, Uganda: The New Vision.

Archabald, K., and L. Naughton-Treves. 2001. Tourism revenue-sharing around national parks in western Uganda: early efforts to identify and reward local communities. *Environmental Conservation* 28:135–149.

Ascorra, C. 1996. *Evaluacion de fauna en sistemas agroforestales para el manejo sostendio.* Lima: Conservación Internacional–Peru.

Ascorra, C., R. Barreda, A. Chicchon, A. M. Chonati, L. Davalos, L. Espinel, A. Gironda, G. Llosa, E. Mendoza, C. Mitchell, C. Mora, M. Mora, P. Padilla, R. Piland, C. Ponce, J. Ramirez, and M. Varese. 1999. *Zona reservada de Tambopata-Candamo.* Lima: Conservación Internacional–Peru.

Bahuchet, S., and I. Garine. 1990. The art of trapping in the rainforest. Pages 25–49 in C. M. Hladik, S. Bahuchet, and I. Garine (eds.), *Food and nutrition in the African rainforest.* Paris: UNESCO.

Bell, R. H. V. 1984. The man-animal interface: an assessment of crop damage and wildlife control. Pages 387–416 in R. H. V. Bell and E. McShane-Caluzi (eds.), *Conservation and wildlife management in Africa.* Lilongwe, Malawi: U.S. Peace Corps Office of Training and Program Support.

Bisbal, F. 1993. Impacto human sobre los carnivoros de Venezuela. *Studies on Neotropical Fauna and Environment* 28:145–156.

Booth, V. R., N. N. Kipuriand, and J. M. Zonneveld. 1992. *Environmental impact of the proposed fencing programme in Kenya.* Nairobi: Elephant and Community Wildlife Programme, Kenya Wildlife Service.

Chapman, C. A., and L. J. Chapman. 1997. Forest regeneration in logged and unlogged forests of Kibale National Park, Uganda. *Biotropica* 29:396–412.

Chicchón, A. 1996. *Subsistence system improvement and conservation in buffer areas of the Bahuaja-Sonene National Park, Madre de Dios and Puno, Peru.* Lima: Conservación Internacional–Peru.

Chicchón, A. 2000. Conservation theory meets practice. *Conservation Biology* 14:138–139.

Chicchón, A., M. Glave, and M. Varese. 1997. La lenta colonizacion del Inambari y el Tambopata: uso del espacio en la selva sur del Peru. Pages 551–587 in E. G. Olarte, B. Revesz, and M. Tapia (eds.), *Peru. El problema agrario en debate.* Lima: SEPIA VI.

Denevan, W. H., J. M. Treacy, J. B. Alcorn, C. Padoch, C. Denslow, J. Flores, and S. Paitan. 1984. Indigenous agroforestry in the Peruvian Amazon: Bora management of swidden fallows. *Interciencia* 9:346–357.

Dove, M. R. 1993. The responses of Dayak and bearded pig to mast-fruiting in Kalimantan: an analysis of nature-culture analogies. Pages 113–123 in C. M. Hladik, A. Hladik, O. F. Linares, H. Pagezy, A. Semple, M. Hadley (eds.), *Tropical forests, people and food: biocultural interactions and application to development.* Paris: Parthenon.

Fairhead, J., and M. Leach. 1996. *Misreading the African landscape.* Cambridge, UK: Cambridge University Press.

Foster, R. 1994. *The Tambopata-Candamo Reserved Zone of southeastern Peru.* Washington, DC: Conservation International.

GESUREMAD (Gerencia Sub-Regional de Madre-de-Dios). 1998. *Diagnóstico del Departamento de Madre de Dios.* Puerto Maldonado, Peru: Gerencia Sub-Regional de Madre de Dios.

Hawkes, R. K. 1991. *Crop and livestock losses to wild animals in the Bulilimamangwe Natural Resources Management Project Area.* Harare: Center for Applied Social Sciences, University of Zimbabwe.

Hill, C. 1998. Conflicting attitudes towards elephants around the Budongo Forest Reserve, Uganda. *Environmental Conservation* 25:244–250.

Hill, C. M. 1997. Crop-raiding by wild animals: the farmers' perspective in an agricultural community in western Uganda. *International Journal of Pest Management* 43:77–84.

Hill, K., and J. Padwe. 2000. The sustainability of Ache hunting in the Mbaracayu Reserve. Pages 79–105 in J. Robinson and E. Bennett (eds.), *Hunting for sustainability in tropical forests.* New York: Columbia University Press.

Hoare, R. 1995. Options for the control of elephants in conflict with people. *Pachyderm* 19:54–63.

Hoare, R. E., and J. T. du Toit. 1999. Coexistence between people and elephants in African savannas. *Conservation Biology* 13:633–639.

Irvine, D. 1987. *Resource management by the Runa Indians of the Ecuadorian Amazon.* Stanford, CA: Anthropology, Stanford University.

Koch, H. 1968. *Magie et chasse dans la foret camerounaise.* Paris: Berger-Lerrault.

Lawrence, D. C., M. Leighton, and D. R. Peart. 1995. Availability and extraction of forest products in managed and primary forest around a Dayak village near Gunung Palung, Indonesia. *Conservation Biology* 9:76–88.

Loja, J., A. Gironda, and L. Guerra. 2000. *Biologia y uso de la fauna silvestre en Tambopata: un caso de estudio.* Lima: Conservación Internacional–Peru.

Loja, J., A. Pinedo, and F. Torres. 1999. *Manejo sostenible de fauna silvestre en la comunidad nativa infierno.* Puerto Maldonado, Peru: Conservación Internacional–Peru.

Mugisha, S. 1994. *Land cover/use around Kibale National Park.* Kampala, Uganda: MUIENR, RS/GIS Lab.

Naughton, L., R. Rose, and A. Treves 1999. *The social dimensions of human-elephant conflict in Africa.* Gland, Switzerland: IUCN. Online: http://iucn.org/themes/ssc/sgs/afesg/hectf/index.html.

Naughton-Treves, L. 1997. Farming the forest edge: vulnerable places and people around Kibale National Park, Uganda. *Geographical Review* 87:27–47.

Naughton-Treves, L. 1998. Predicting patterns of crop damage by wildlife around Kibale National Park, Uganda. *Conservation Biology* 12:156–168.

Naughton-Treves, L. 1999. Whose animals? A history of property rights to wildlife in Toro, western Uganda. *Land Degradation and Development* 10:311–328.

Naughton-Treves, L. 2002. Wild animals in the garden. *Annals of the Association of American Geographers* 92:488–506.

Naughton-Treves, L., R. Grossberg, and A. Treves. 2003. Paying for tolerance? The impact of depredation and compensation payments on rural citizens' attitudes toward wolves. *Conservation Biology* 17:1500–1512.

Newmark, W. D., D. N. Manyanza, D. Gamassa, M. Sariko, and I. Hashan. 1994. The conflict between wildlife and local people living adjacent to protected areas in Tanzania: human density as a predictor. *Conservation Biology* 8:249–255.

Nietschmann, B. 1973. *Between land and water: the subsistence ecology of the Miskito Indians, eastern Nicaragua.* New York: Seminar Press.

Novarro, A., K. Redford, and R. Bodmer. 2000. Effect of hunting in source-sink systems in the neotropics. *Conservation Biology* 14:713–721.

Ortiz, E. G. 1995. Survival in a nutshell. *Americas* 47:6–17.

Peterson, J. T. 1981. Game, farming and interethnic relations in northeastern Luzon, Philippines. *Human Ecology* 9:1–22.

Redford, K., and J. G. Robinson. 1987. The game of choice: patterns of indian and colonist hunting in the neotropics. *American Anthropologist* 89:650–667.

Salafsky, N. 1993. Mammalian use of a buffer zone agroforestry system bordering Gunung Palung National Park, West Kalimantan, Indonesia. *Conservation Biology* 7:928–933.

Salafsky, N. 1994. Forest gardens in the Gunung Palung Region of West Kalimantan, Indonesia: defining a locally-developed, market-oriented agroforestry system. *Agroforestry Systems* 28:237–268.

Stronza, A. 1998. Cutting edge. The impact of ecotourism on the Peruvian rainforest. *The Times Higher Education Supplement* 1359:34.

Struhsaker, T. T. 1997. *Ecology of an African rain forest.* Gainesville: University Press of Florida.

Turyahikayo-Rugyema, B. 1974. *The history of the Bakiga in southwestern Uganda and northern Rwanda, 1500–1930.* Ann Arbor: University of Michigan.

UWA (Ugandan Wildlife Authority). 2000. *Community-protected areas institution policy.* Kampala: Uganda Wildlife Authority Policy Paper.

Varcarcel, M. 1993. *Madre de Dios, un espacio en formación.* Lima: Centro de Investigaciones Antropológicas de la Amazonía Peruana.

Varese, M. 1995. *El mercado de productos de fauna silvestre en la ciudad de Puerto Maldonado, Madre de Dios.* Puerto Maldonado: Conservación International–Peru.

Woodroffe, R., and J. Ginsburg. 1998. Edge effects and the extinction of populations inside protected areas. *Science* 280:2126–2128.

World Wildlife Fund. 1998. *Wildlife electric fencing projects in communal areas of Zimbabwe: current efficacy and future role.* Harare, Zimbabwe: Price Waterhouse Coopers, for World Wildlife Fund.

Chapter 14

Hunting in Agroforestry Systems and Landscapes: Conservation Implications in West-Central Africa and Southeast Asia

David S. Wilkie and Robert J. Lee

Growing human demand for land, goods, and services is increasing incentives to convert highly diverse natural landscapes into agroecosystems where a larger proportion of the nutrients and energy is captured by the few plants and animals of direct use to people. Although this process increases the production of goods for humans, it progressively fragments the landscape, creating islands of natural habitat embedded in a matrix of human-modified plant communities that range from the simple to the complex. If these remnant islands of natural habitat are not of sufficient size and composition to meet the habitat needs of certain wild species, then the land cover characteristics and land use practices within the matrix between islands become particularly important as a source of food and shelter or as a threat to long-term wildlife conservation.

In addition to the fragmentation and degradation of the world's forests, the growing demand for protein is driving many large-bodied forest mammals to local extinction, particularly in areas where wild animals are hunted for food and hunting is unregulated. Loss of these large mammals is likely to have profound and permanent effects on forest composition and function because the ecological services they provide (grazing, browsing, trampling, seed dispersal, and excavation) are disproportionately large relative to their total numbers. Loss of these megafauna risks the same cascade of extinctions of smaller animals that occurred in the late Pleistocene, when early hominids hunted out the large mammals of North and South America (Martin 1973).

At present, although nearly 30 percent of the world's terrestrial landscape is made up of forests (FAO 2000), a mere 6–10 percent is under formal protection (James et al. 1999). Given the wide-ranging behavior of large-bodied mammals such as elephants, rhinos, buffalos, and giant forest hogs, even large protected areas are seldom large enough to provide all the habitat needs of these area-demanding species. Therefore, it is critical from a biodiversity conservation perspective to examine whether agroecosystems and, more specifically, agroforestry systems promote or hinder long-term persistence of wildlife populations.

Other chapters in this volume tackle the issue of whether agroforestry systems provide habitat and movement corridors suitable for wildlife conservation. This chapter examines agroforestry systems by looking at one land use practice: hunting. We discuss hunting in agroforestry systems and landscapes through a lens directed at West and Central Africa and Southeast Asia. These regions are very different ecologically, culturally, and economically but show common patterns associated with hunting. We consider a number of factors relevant to hunting such as land cover, animal densities in different habitats, reasons for hunting, and human population density and economies. With these factors and patterns, we illustrate what sort of hunting is sustainable under what conditions and practices. Finally, we examine the trade-offs that planners and conservationists must consider with regard to hunting and agroforestry land uses and potential approaches to reconcile human demand for resources and conservation of wild animals and their habitats.

Brief Overview of Hunting

People hunt wild animals in both forested and savanna regions in the tropics to provide meat and medicine, to control agricultural crop pests, to reduce perceived threats to human safety, and to collect trophies. In Africa and Southeast Asia, the meat of wild animals in rural areas and in many urban areas typically is less expensive and more available than is the meat of domesticated animals (Caldecott 1987; Caspary 1999; Barnett 2000; Bennett and Robinson 2000). Not surprisingly, wild animals often are the primary source of animal protein in the diets of low-income rural and urban households (Chin 1981; Redford 1993; Chardonnet 1995; Alvard 2000). In Africa, hunting also provides higher-than-average annual incomes to hunters and to many traders (Dethier 1995; Ngnegueu and Fotso 1998; Auzel and Wilkie 2000; Barnett 2000). In some traditional Asian cultures, wild meat and animal parts are an economic and social currency (Ellen 1975; Dryer 1985; Bennett et al. 1997).

In certain regions of the tropics where hunting pressure is high, defaunation is a more immediate threat to wildlife populations than is habitat loss and disturbance (Robinson et al. 1999; Wilkie and Carpenter 1999; Bodmer and Lozano 2001). When hunting pressure exceeds the reproductive output of a given population, that population will decline, potentially to local extinction (Winterhalder and Lu 1997). Because tropical forests are an order of magnitude less productive than tropical savannas in terms of wildlife biomass, forest wildlife populations are more prone to overexploitation (Robinson and Bennett 2000).

Hunting in Central and West Africa

In regions of the Democratic Republic of Congo and the Republic of Congo where rural human populations are sparse (Table 14.1), agroforestry landscapes composed of agricultural fields and plantations dotted in more extensive forested fallow lands make up a small percentage of the landscape (Table 14.2), and hunting pressure is not high, large-bodied animals are abundant and constitute the bulk of the wildlife biomass harvested (Wilkie and Carpenter 1999). However, when consumer demand is high and wild animals are progressively overhunted, we see a decline in the average adult body size of wild species harvested and sold in markets (Steel 1994; Fa 1999; Barnett 2000). In a decade-long study of wildlife hunting for food in equatorial Guinea, John Fa and his colleagues showed that as large game were progressively depleted by overhunting, the proportion of rodents and other small game being harvested increased until they constituted 37 percent of the wild meat biomass sold in markets (Fa et al. 2000). In Ghana a combination of forest loss and overhunting has resulted in small game (less than 5 kg average adult body weight) now constituting 90 percent of all individuals and 54 percent of total biomass harvested by hunters in forested regions of the nation (Ntiamoa-Baidu 1998). Similarly, in the Côte d'Ivoire small game make up 68 percent of the total biomass of wildlife hunted for food (Caspary 1999). In Nigeria rodents make up 50 to 60 percent of wild animals sold for meat in urban markets (Martin 1983; Anadu et al. 1988).

Hunting in Southeast Asia

With the exception of Borneo and Papua New Guinea, which still retain large blocks of intact forest, most of Southeast Asia is made up of small patches of forest surrounded by agroforestry landscapes composed of agricultural fields, 1- to 10-year fallow areas, and older secondary forests. This is not surprising because human population density in Southeast Asia averages 100 people per square kilometer, and population growth is almost 2 percent.

Five Southeast Asian countries (Indonesia, Malaysia, Cambodia, Laos, and Vietnam) are considered hotspots for conversion of forest frontiers into agricultural land (FAO 2000). Based on present rates of population growth and deforestation, by 2010 forests are expected to cover 45 percent of Southeast Asia, compared with 47 percent at present (Table 14.3), although 7 percent of the land classified as forest by the Food and Agriculture Organization of the United Nations (FAO) is actually postagricultural secondary vegetation.

The ambiguity in FAO's forest classification is reflected by the inhabitants of Indonesia's outer islands, northern Thailand, Cambodia, and Laos, who sometimes make little distinction between forests and gardens. In particular, shifting cultivators see forests and gardens as part of a dynamic agricultural

Table 14.1. Demographic and economic indices.

Country or Area	Land Area Total (thousands of hectares), 1998	Land Area Total (thousands), 1999	Population Density (n/km²), 1999	Population Annual Change (%), 1995–2000	Rural (%), 1999	Per Capita Gross National Product (US$), 1997	Annual GDP Change (%), 1997
Benin	11,063	5,937	53.3	2.7	58.5	381	5.6
Côte d'Ivoire	31,800	14,526	45.7	1.8	54.1	727	6.0
Ghana	22,754	19,678	86.5	2.7	62.2	384	4.2
Guinea	24,572	7,360	30.0	0.8	68.0	552	4.8
Guinea-Bissau	3,612	1,187	42.2	2.2	76.7	232	5.0
Liberia	11,137	2,930	30.4	8.6	52.7	—	—
Nigeria	91,077	108,945	119.6	2.4	56.9	239	3.9
Sierra Leone	7,162	4,717	65.9	3.0	64.1	150	−20.2
Togo	5,439	4,512	83.0	2.7	67.3	337	4.7
Total West Africa	208,616	169,792	61.8	3.0	62.3	375.3	1.8
Cameroon	46,540	14,693	31.6	2.7	51.9	587	5.1
Central African Republic	62,297	3,550	5.7	1.9	59.2	341	5.1
Congo	34,150	2,864	8.4	2.8	38.3	633	−1.9
Democratic Republic of the Congo	226,705	50,335	22.2	2.6	70.0	114	−5.7
Equatorial Guinea	2,805	442	15.8	2.5	52.9	892	76.1
Gabon	25,767	1,197	4.6	2.6	45.9	3,985	4.1
Total Central Africa	398,264	73,081	14.7	2.5	53.0	513.41	13.8
Cambodia	17,652	10,945	62.0	2.3	77.2	303	1.0
Indonesia	181,157	209,255	115.5	1.4	60.8	1,096	4.9
Lao People's Democratic Republic	23,080	5,297	23.0	2.6	77.1	414	6.5
Malaysia	32,855	21,830	66.4	2.0	43.5	4,469	7.8
Thailand	51,089	60,856	119.1	0.9	78.8	2,821	−0.4
Vietnam	32,550	78,705	241.8	1.6	80.3	299	8.8
Total Southeast Asia	338,383	386,888	104.6	1.8	69.6	1,567.0	4.8

Source: FAO (2000).

Table 14.2. U.S. Geological Survey land cover characteristics of a sample of nations.

Country	Forest	Forest Fallow	Crops and Pasture	Wetlands and Water	Grasslands	Other
West Africa						
Côte d'Ivoire	14.91%	11.64%	8.36%	2.77%	62.30%	0.03%
Ghana	6.56%	17.84%	4.25%	7.21%	64.00%	0.14%
Central Africa						
Congo	50.94%	3.49%	1.98%	3.40%	40.14%	0.06%
Democratic Republic of the Congo	64.35%	7.63%	2.66%	3.74%	21.27%	0.34%
Southeast Asia						
Cambodia	39.34%	11.93%	41.63%	3.38%	1.54%	2.18%
Lao People's Democratic Republic	80.68%	8.37%	6.70%	0.94%	2.52%	0.79%

system that continually rotates through the processes of ecological succession to meet the needs of local agricultural production (Boserup 1966). In a study that extended over 40 years, Fox (1999) used aerial photographs to track changes in the landscape in northeastern Cambodia. The results showed that the percentage of forest cover remained constant as a result of long fallow periods between crop rotations. Although trees apparently remained on the landscape during fallow-crop rotations, it is unclear whether these agroforestry areas support anything like the full range and abundance of wildlife found in undisturbed forests (Padoch and Peter 1993; see also Chapter 8, this volume). Like the agroforestry areas of West Africa, these anthropogenic landscapes are unlikely to support large mammals that compete with humans for space but may be breeding grounds for small, rapidly reproducing human commensals such as rodents.

In Southeast Asia wild animals are hunted for food and medicinal use. In the past decades, trade in wildlife has paralleled the region's economic growth, and wildlife traders in Southeast Asian countries have strengthened an already extensive export network that reaches Eastern Asia (Luxmoore and Groombridge 1989; Groombridge and Luxmoore 1991; Wenjun et al. 1996; Van Dijk et al. 2000). In Lao People's Democratic Republic, yearly wildlife sales at major markets in the capital city included carcasses of 10,000 mammals, 6,000–7,000 birds, and 3,000–4,000 reptiles totaling more than 33,000 kg (Duckworth 1999). Although most meat is eaten, animal parts such as horn, bones, skin, and dried internal organs are shipped to Eastern Asian countries for the traditional Chinese medicine markets (Fujita and Tuttle 1985; Duck-

Table 14.3. Forest cover and forest loss statistics.

Country or Area	Total Land Area (thousands of hectares)	Total Forest (thousands of hectares), 1990	Total Forest (thousands of hectares), 2000	Forest Cover (%), 2000	Forest Cover Change, 1990–2000 Annual Change (thousands of hectares)	Annual Change Rate (%)	Estimated Forest (thousands of hectares), 2010
Benin	11,063	3,349	2,650	24%	−70	−2.3	2,100
Côte d'Ivoire	31,800	9,766	7,117	22%	−265	−3.1	5,194
Ghana	22,754	7,535	6,335	28%	−120	−1.7	5,337
Guinea	24,572	7,276	6,929	28%	−35	−0.5	6,590
Guinea-Bissau	3,612	2,403	2,187	61%	−22	−0.9	1,998
Liberia	11,137	4,241	3,481	31%	−76	−2.0	2,844
Nigeria	91,077	17,501	13,517	15%	−398	−2.6	10,387
Sierra Leone	7,162	1,416	1,055	14.7%	−36	−2.9	749
Togo	5,439	719	510	9%	−21	−3.4	361
Total West Africa	*201,454*	*52,790*	*42,726*	*29%*	*−672*	*−1.54*	*34,811*
Cameroon	46,540	26,076	23,858	51%	−222	−0.9	21,796
Central African Republic	62,297	23,207	22,907	37%	−30	−0.1	22,679
Congo	34,150	22,235	22,060	65%	−17	−0.1	21,840
Democratic Republic of the Congo	226,705	140,531	135,207	60%	−532	−0.4	129,895
Equatorial Guinea	2,805	1,858	1,752	62%	−11	−0.6	1,650
Gabon	25,767	21,927	21,826	85%	−10	0.0	21,804
Total Central Africa	*398,264*	*235,834*	*227,610*	*60%*	*−822*	*−0.42*	*219,664*
Cambodia	17,652	9,896	9,335	53%	−56	−0.6	8,790
Indonesia	181,157	118,110	104,986	58%	−1,312	−1.2	93,047
Lao People's Democratic Republic	23,080	13,088	12,561	54%	−53	−0.4	12,068
Malaysia	32,855	21,661	19,292	59%	−237	−1.2	17,098
Thailand	51,089	15,886	14,762	29%	−112	−0.7	13,761
Vietnam	32,550	9,303	9,819	30%	52	0.5	10,321
Total Southeast Asia	*338,383*	*187,944*	*170,755*	*47%*	*−1,718*	*−0.60*	*155,085*

Source: FAO (2000).

worth 1999). Between 1992 and 1998, as much as 578,607 kg of hard-shelled turtle shells and 120,438 kg of soft-shelled turtle shells were shipped from Southeast Asian countries to Taiwan alone (Chen et al. 2000). The use of wildlife for medicines in particular is important in Asia because of the volume and range of species exploited.

In the context of Southeast Asia we cannot look simply at whether protein or calorie needs are being met through hunting when assessing whether agroforestry areas can provide enough wildlife for consumptive use. We must ask whether hunted species are hunted for food or medicine. If hunting includes the latter, conservation plans cannot be confined simply to producing a higher volume of meat.

Hunter Preference for Large-Bodied Animals

As long as humans have hunted, they have preferred large-bodied animals. Some argue that early hominids hunted to extinction almost all the megafauna of North and South America in the late Pleistocene (Martin 1973). Loss of the gomphotheres (prehistoric elephants) from the rainforests of South America more than 10,000 years ago may be a reason for the present patchy and limited distribution of plants with large seeds (Janzen and Martin 1982). Moreover, slaughter of these and other megaherbivores and the loss of their pivotal role in biodiversity maintenance may be why at the end of the Pleistocene it was not just the megafauna themselves that became extinct but also a wide variety of smaller mammals and birds. In Java, less than half the mammal fauna survived the mid-Pleistocene and the arrival of humans (McNeely 1978).

Large mammals yield a high volume of meat, trophies, and other byproducts such as horns, hides, and bones. Therefore, they typically generate the most value per unit effort invested in hunting (Redford 1993; Wilkie and Godoy 1996; Freese 1998). Most studies of hunting show that large-bodied mammals have decreased in numbers and face extirpation in areas of high hunting pressure. Lee (2000) showed that wild meat demand was depleting large mammal populations in the forests of North Sulawesi, Indonesia, and hunters were moving farther south to find the preferred large-bodied species. Hunters switch to hunting small game only when large game densities have declined significantly (Barnett 2000; Fa et al. 2000).

However, as predicted by optimal foraging theory (Stephens and Krebs 1986), although a large-bodied and valuable animal may become scarce, a hunter will attempt to capture it whenever he encounters it. Similarly, even the least common antelope will be killed if it steps into and springs a leg snare. As a result, hunting is likely to result in the local extinction of large game as long as small game densities continue to make hunting wildlife for food economically worthwhile (Wilkie and Godoy 2001), and large animal abundance tends to decrease with increasing proximity to human settlements (Dethier 1995; Ngnegueu and Fotso 1998; Noss 1998; Fimbel et al. 2000) and the age of settlements (Naughton-Treves 1997), as does hunter success (Blake 1994; Auzel 1996). Although hunters prefer large animals that are typically more abundant far from settlements, that is not to say that they do not hunt close to their settlements and that they do not hunt small animals; in fact, they do both (Lee 2000). The question we need to resolve is whether hunting close to settlements in disturbed habitat and agroforestry areas promotes or hinders wildlife conservation.

Conservation Value of Hunting in Agroforestry Systems

For hunting in agroforestry systems to have a conservation payoff, it must accomplish one or all of the following: provide a sufficient and sustainable source of animal protein to eliminate the need to hunt in natural habitat, compete with hunters' labor such that they are unable to hunt elsewhere, and not exert so much pressure as to prevent the safe passage of animals that depend on agroforestry lands as movement corridors between patches of undisturbed forest. We focus on the first criterion because if this is not met, the second is moot, and the third is addressed elsewhere in this book (see Chapter 13, this volume).

Do Agroforestry Systems Produce Sufficient Wildlife to Meet Protein Demand?

The forested regions of West and Central Africa (see Table 14.1) offer an interesting comparative perspective for reviewing the role that hunting in agroforestry landscapes plays in conservation of forest wildlife. To narrow the scope of the discussion and to best reflect the sparse available information, this section focuses on Ghana, the Côte d'Ivoire, the Republic of Congo, and the Democratic Republic of Congo.

According to the U.S. Geological Survey global land cover characteristics database (Table 14.2), forests in Ghana and the Côte d'Ivoire cover 24 percent and 27 percent of the landscape, respectively (FAO 2000 estimates: 28 percent and 22 percent). Yet most of these forests are a mosaic of crops and forest fallow (73 percent and 44 percent, respectively; Table 14.2). In contrast, forests in the Republic of Congo and the Democratic Republic of Congo still cover 54 percent and 72 percent of the landscape, respectively, and only 6 percent and 11 percent of that is disturbed (Table 14.2). Consequently, whereas the landscapes of Ghana and the Côte d'Ivoire are dominated by seas of disturbed agroforestry areas with small islands of intact forest, the opposite is true for the two Congo countries where agroforestry areas are limited to narrow bands along major rivers and roads. This difference in relative abundance of agro-forestry and fallow-forest lands between nations in West and Central Africa allows us to compare and assess heuristically and empirically the conservation implications of hunting in these human-dominated forested landscapes.

In a review of available literature on the sustainability of hunting of wildlife for food in intact tropical forests, Robinson and Bennett (2000) note that "the carrying capacity for people depending exclusively on game meat will not greatly exceed one person per km^2, even under the most productive cir-cumstances" (p. 24). We also know that tropical savannas are an order of mag-nitude more productive for herbivores than are forests (Robinson and Bennett 2000). Agroforestry landscapes are regularly and extensively disturbed; as a result, they may be more productive for herbivores than are intact forests. That said, even if herbivore productivity was four times that of intact forests, human carrying capacity might still only be four to eight people per square kilometer.

Using published data on human population densities, human protein needs, and wildlife productivity, we should be able to assess whether agro-forestry landscapes are likely to be able to sustainably supply sufficient wildlife to meet the human demand for protein.[1] To estimate potential maximum wildlife productivity in agroforestry landscapes, we need to determine what species are likely to persist in such systems and at what densities. For this chap-ter we assume that humans will not tolerate and therefore will extirpate from agroforestry areas large-bodied species such as elephants (*Loxodonta africana*), buffalos (*Syncerus caffer*), mandrills (*Papio sphinx*), and baboons (*Papio* spp.) that can destroy farmers' crops.

We know from the literature that the density of large-bodied animals increases with distance from hunter settlements (Wilkie 1989; Ngnegueu and Fotso 1998; Fimbel et al. 2000). If we assume that hunters have largely extir-pated large game from forest fallows and farm bush that typically surround vil-lages, then these areas will support primarily rodents and the smallest ante-lope. Therefore, expected annual production of wildlife, based on available density and productivity estimates, would be less than 7 kg per hectare per year (15 kg per hectare per year if one includes all forest species in Table 14.4). We assume that short-lived species with high reproductive rates can tolerate a harvest intensity of 60 percent of annual production (Robinson and Redford 1994). Thus human carrying capacity in Central and West African forest areas that support only rodents and other small-bodied wildlife harvested for food would be four people per square kilometer. This figure rises to six people per square kilometer if all species in Table 14.4 are included and sustainable off-take is reduced to 40 percent of annual production to account for the slower growth and reproductive rates of larger-bodied species. The situation is likely to be similar in South America, where human carrying capacity using the

Table 14.4. Production of commonly exploited species in Central and West Africa.

Species		Weight (kg)	Density (# ha⁻¹)	Production (# ha⁻¹ yr⁻¹)	Biomass (kg ha⁻¹)	Production (kg ha⁻¹ yr⁻¹)	r	Longevity (yr)
Cricetomys emini	Gambian giant rat	1.95	1.340	0.812	2.61	1.58	2.01	7.80
Thryonomys swinderianus	Cane rat	7.00	1.00	0.60	7.00	4.20	2.00	
Atherurus africanus	Brush-tailed porcupine	2.88	0.550	0.271	1.58	0.78	1.82	22.90
Cercocebus galeritus	Crested mangabey	7.90	0.020	0.002	0.16	0.02	1.19	20.00
Cercocebus albigena	Grey-cheeked mangabey	7.70	0.069	0.008	0.53	0.06	1.19	20.00
Colobus pennanti ellioti	Red colobus	8.20	0.267	0.024	2.19	0.20	1.15	30.00
Cercopithecus ascanius	Black-cheeked guenon	3.60	0.189	0.014	0.68	0.05	1.12	30.80
Cercopithecus mona	Mona monkey	3.80	0.231	0.017	0.88	0.06	1.12	30.80
Cercopithecus mitis	Blue monkey	6.00	0.242	0.016	1.45	0.10	1.11	30.80
Potomochoerus porcus	Red river hog	54.00	0.154	0.122	8.32	6.59	2.32	
Cephalophus monticola	Blue duiker	4.70	0.242	0.091	1.14	0.43	1.63	7.00
Cephalophus nigrifrons	Black-fronted duiker	13.90	0.017	0.006	0.24	0.08	1.54	8.00
Cephalophus leucogaster	White-bellied duiker	16.70	0.044	0.014	0.73	0.24	1.54	8.00
Cephalophus callipygus	Peter's duiker	17.70	0.063	0.020	1.12	0.36	1.54	8.00
Cephalophus dorsalis	Bay duiker	22.00	0.032	0.004	0.70	0.09	1.22	8.00
Cephalophus sylvicultor	Yellow-backed duiker	68.00	0.016	0.005	1.09	0.35	1.54	10.30
Rodents and *C. monticola*					12.33	6.99		
All species					30.42	15.19		

r = the maximum finite rate of increase of the population.

Table 14.5. Production of commonly exploited species in Central and South America.

Species		Weight (kg)	Density (# ha^{-1})	Production (# ha^{-1} yr^{-1})	Biomass (kg ha^{-1})	Production (kg ha^{-1} yr^{-1})	r	Longevity (yr)
Cebus apella	Brown capuchin monkey	2.60	0.400	0.036	1.04	0.09	1.15	0.14
Alouatta spp.	Howler monkey	6.00	0.300	0.030	1.80	0.20	1.19	0.17
Ateles spp.	Spider monkey	7.00	0.250	0.011	1.75	0.08	1.07	0.07
Dasypus novemcinctus	Nine-banded armadillo	3.50	0.220	0.131	0.77	0.46	1.99	0.69
Hydrochaeris hydrochaeris	Capybara	45.00	0.016	0.010	0.72	0.43	1.99	0.69
Dasprocta spp.	Agouti	4.00	0.052	0.063	0.21	0.25	3.00	1.10
Agouti paca	Paca	8.00	0.035	0.020	0.28	0.16	1.95	0.67
Tapirus terrestris	Tapir	160.00	0.005	0.001	0.80	0.11	1.22	0.20
Tayassu pecari	White-lipped peccary	35.00	0.030	0.024	1.05	0.83	2.32	0.84
Tayassu tajacu	Collared peccary	25.00	0.056	0.084	1.40	2.09	3.49	1.25
Mazama americana	Red brocket deer	30.00	0.026	0.008	0.78	0.23	1.49	0.40
Rodents and pigs					3.81	3.76		
All species					10.60	4.93		

r = the maximum finite rate of increase of the population.

wildlife production figures in Table 14.5 is two people per square kilometer under both the small mammal and all species scenarios.

Now that we have estimated human carrying capacity in disturbed forest areas we can determine, at least for Congo, the Democratic Republic of Congo, Ghana, and the Côte d'Ivoire, whether human population density is likely to exceed the capacity of wildlife to sustainably meet their protein needs.

Using land cover data from the U.S. Geological Survey global land cover characteristics database and population data from the Center for International Earth Science Information Network, Colombia University, we were able to show that population density in all forest areas of the Côte d'Ivoire and Ghana exceeds estimated carrying capacity for people who depend exclusively on wildlife for protein. In contrast, in Congo forests only in the extreme south and in the southwest on the border with Gabon does human population density exceed five people per square kilometer and thus human carrying capacity. It is important to note that in estimating human carrying capacity we assume that all animals harvested are consumed directly by the hunter's family and are not traded extralocally. If wild animals are hunted for extralocal markets, then estimates of carrying capacity based on in situ consumption alone will underestimate hunting pressure and overestimate carrying capacity. In the Democratic Republic of Congo, the human population in most of the central basin forests is below carrying capacity for local consumption of wildlife as a primary source of protein.

These data suggest that wild animals in agroforestry land use mosaics in the Côte d'Ivoire and Ghana are highly unlikely to meet a significant portion of residents' daily protein needs. In contrast, human population density throughout much of the forests of Congo and the Democratic Republic of Congo is low enough to suggest that hunting for local consumption could meet residents' dietary protein needs. Again, we would like to reiterate that these estimates are based on the assumption that wild animals are consumed locally and are not traded to meet urban demand and other extralocal markets.

Does Hunting in Agroecosystems Limit Hunting in Intact Forests?

Areas in Congo, the Democratic Republic of Congo, Ghana, and the Côte d'Ivoire under intact and disturbed forest were estimated using the U.S. Geological Survey global land cover characteristics database. Because forest fallow (crop, forest, and grassland mosaic) makes up only 3.5 percent and 7.6 percent of total land cover in Congo and the Democratic Republic of Congo, respectively, it is highly unlikely that hunting in agroecosystems alone would be sufficient to meet the protein needs of forest residents.

Wilkie (1989) showed that in the Ituri forest of northeastern Democratic Republic of Congo in the early 1980s, 59 percent of all wild animals were killed in forest fallows. However, these areas generated only 39 percent of total wildlife biomass, and hunting in intact forests provided most of the protein consumed by subsistence foragers. Fallow forests were limited to within 3 km of settlements, constituted only 26 percent of land cover, and occurred in small patches surrounded by a sea of old-growth forest. Although fallow forests were repeatedly and successfully hunted, they were probably being continuously replenished with wildlife from the more abundant old-growth forests (Novaro et al. 2000). Thus, in areas where fallow forests are sparse relative to old-growth forest it is unlikely that wildlife production in these areas will be sufficient to meet household demand for protein, and surrounding old-growth forests will continue to be hunted, reducing their capacity to serve as a source pool of wildlife for intensively exploited forest fallows that lie closer to settlements.

In Ghana, the area of fallow forest is more than twice the area of intact forests. In the Côte d'Ivoire, the area of disturbed forest and intact forest is roughly equivalent (Table 14.2). Not surprisingly, rodents and other small game constitute approximately 90 percent of wildlife carcasses sold in markets for food in Ghana (Ntiamoa-Baidu 1998) and 68 percent in the Côte d'Ivoire (Caspary 1999). Although fallow forests clearly provide large numbers of wild animals, this does not prevent hunters from entering parks and reserves to hunt, and poaching wildlife for food is still considered the primary threat to wildlife conservation in Ghana and the Côte d'Ivoire.

In Southeast Asia intact forests are rare. Although closed forests as defined by FAO make up a large percentage of Asian forests, reality contradicts what is seen on paper. Hunting trails cross commonly through "intact" forests. Throughout Southeast Asia, forests and agricultural areas are intermixed in a mosaic of small landscape patches. If we look at protected forests in Indonesia, for example, we find that most are smaller than 50,000 ha and tend to be oblong, with more edge than interior surface area. As a result, many families find it convenient to use forests as an extension of their gardens and make little distinction between the two. In Indonesia, agricultural fields are interspersed throughout forests, attracting commensal animals such as pigs, monkeys, deer, rodents, and bats. Rather than using rifles, a labor-intensive hunting technique, farmers use passive techniques that can be incorporated into their daily farming routines. Snares and traps are placed near gardens. Therefore, hunting achieves three tasks at once: cultivating crops, obtaining wild meat, and preventing crop predation. Linares (1976) called this garden hunting, a practice common in the forest-garden mosaic of Sulawesi, Indonesia (Lee 2000).

Given the mosaic nature of most of the forests of Southeast Asia, if we consider all small forests (less than 10,000 ha) as parts of agroforestry landscapes, then clearly people are using agroforestry landscapes to hunt wildlife. The question remains: if people are hunting in these agroforestry landscapes, are wildlife populations withstanding the hunting pressures? In general, the answer is "no." Examples from Malaysia, Indonesia, Thailand, and the Philippines indicate that wildlife populations are being reduced by hunting. A number of studies in Sarawak show a reduction in animal densities and geographic range associated with intensive hunting (Medway 1977; Rabinowitz 1995) and density (Bennett 1992; Bennett and Dahaban 1995; Bennett and Robinson 2000). Studies from Sulawesi (Alvard 2000; Lee 2000; O'Brien and Kinnaird 2000) show significant reduction of large mammals from hunting. O'Brien and Kinnaird (pers. comm., 2001) show an inverse relationship between density of mammals and distance from park borders at Bukit Barisan National Park in Sumatra. Most striking, results from small mammal trapping at the Crater Mountain Wildlife Management Area in Papua New Guinea showed a clear difference in small mammal diversity and abundance between intact forest and agroforestry areas. Wright and Mack (pers. comm., 2001) caught 115 individuals from 10 species, including a bandicoot (*Peroryctes raffrayana*) and cuscus (*Phalanger carmelitae*), in forest as compared with five individuals and three species in adjacent agroforests using comparable trapping intensity. Lynam (pers. comm., 2001) shows that browser and grazer density and, in turn, large carnivore density, are significantly lower in hunted areas. Reports from Laos indicate that wildlife hunted for food consists mainly of small animals including small birds such as myna (*Acridotheres* spp.), flowerpecker (*Dicaeum* spp.), and thrushes (*Turdus* spp.; Duckworth 1999). All these findings indicate directly and indirectly that large-bodied animals have been lost in most forest areas and that hunters have been forced, through overhunting, to target small-bodied animals.

Alternatives to Wild Meat Consumption

Wild meat consumption is not unique to developing countries. It has been a common part of human existence for millennia. However, commercial wildlife hunting for meat is almost never sustainable (Freese 1997, 1998; Wilkie and Carpenter 1999). In fact, of all the animal species known to have become extinct worldwide since 1660 and whose causes of extinction were known, 27 percent went extinct as a direct consequence of hunting (Smith et al. 1995), including the Steller's sea cow (*Hydrodamalis gigas*), the great auk (*Pinguinus impennis*), the dodo (*Raphus cuculatus*), and the heath hen (*Tympanuchus cupido cupido*).

Almost everywhere that wild animals are hunted for meat, consumption patterns follow the same trajectory as they did in the United States: wildlife are viewed as a gift of nature and are hunted as food until animal populations are so scarce that wild meat prices finally exceed the costs of raising livestock. In the United States, this transition began after the Civil War in the 1860s, when the western grasslands were converted to cattle ranches and the

transcontinental railroad provided ready access to eastern and western markets. Once people switch to eating the meat of domestic animals, wildlife, for
the majority of consumers, shifts from a necessity to a luxury item, eaten only
on special occasions. That said, in the United States today, wild animals
remain, somewhat remarkably, an important source of meat for poor families
in the forests of New England, Appalachia, and the boreal North. If history is
left to repeat itself, wild animals in most parts of the world will be hunted to
extinction, at which time all but the poorest families will switch to eating the
meat of domestic animals. The questions we have to ask, therefore, are
whether history is likely to be repeated in developing countries where wildlife
still exist in high numbers and what policy levers are available to change the
historical trajectory of the wild meat trade.

The Growing Demand for Meat

At present, the world is experiencing a livestock revolution, particularly in
developing countries (Delgado et al. 1999), with livestock production growing faster than human population (Table 14.6). From the beginning of the
1970s to the mid-1990s, meat consumption in developing countries increased
by 70 million metric tons, a volume more than twice the increase in developed
nations (Delgado et al. 1999). However, in West Africa, Central Africa, and
Southeast Asia production of cattle, sheep, and goats in all nations other than
the Central African Republic does not come close to meeting the protein
needs of the human population (about 106 kg of undressed animal per person
per year). Moreover, as economies modernize and incomes rise, demand for
meat typically increases faster than population growth. From 1982 to 1994,
the demand for meat in developing countries grew by 5.4 percent per year.

Central and West Africa reflect this trend. Many people in the Congo Basin
eat as much meat as do Europeans and North Americans (about 70 kg per person per year), and 60–80 percent of the meat that rural families eat comes from
wildlife (Wilkie and Carpenter 1999). Wildlife consumption may have been
sustainable at the end of the nineteenth century, when the population of Africa
was 100 million. However, by 1990 the human population had increased to
800 million and is expected to double again by 2025. In the Congo Basin
today, more than 30 million people eat more than 1 million metric tons of
wildlife each year, the equivalent of more than 4 million cattle or 200 million
blue duikers (*Cephalophus monticola;* Wilkie and Carpenter 1999).

Wilkie and Godoy (2001) showed that, at least in neotropical forests, consumption of wildlife as food resembles a Kuznets curve (inverted *U*) with
household income. That is, consumption rises from a low level as household
incomes grow but begins to decline with income when families have reached
a certain level of wealth. In economic terms, wildlife protein is a necessity for
poor families and an inferior good for wealthy families. If we assume that the

Table 14.6. Meat production annual average, 1986–1996.

Country or Area	Beef			Sheep and Goats			Total
	Thousands of Metric Tons	Kilograms per Capita	Percentage Change, 1986–1996	Thousands of Metric Tons	Kilograms per Capita	Percentage Change, 1986–1996	Kilograms per Capita
Benin	20	34	50	6	10	20	44
Côte d'Ivoire	40	28	42	10	7	15	35
Ghana	21	11	7	12	6	24	17
Guinea	15	20	81	4	5	77	25
Guinea-Bissau	4	26	46	2	13	39	39
Liberia	1	3	−25	1	3	0	6
Nigeria	290	27	11	240	22	68	49
Sierra Leone	6	13	28	1	2	12	15
Togo	7	16	46	7	16	39	32
Total West Africa	404	20	32	283	9	33	29
Cameroon	88	60	43	31	21	48	81
Central African Republic	54	152	58	8	23	81	175
Congo	2	7	−10	1	3	17	10
Democratic Republic of the Congo	16	3	−37	24	5	65	8
Equatorial Guinea	0	0	7	0	0	13	0
Gabon	1	8	61	1	8	26	16
Total Central Africa	161	38	20	65	10	42	48
Cambodia	41	37	84	0	0	0	37
Indonesia	350	17	78	101	5	30	22
Lao People's Democratic Republic	13	24	156	0	0	99	24
Malaysia	18	8	59	1	0	0	8
Thailand	213	35	34	1	0	2	35
Vietnam	83	11	32	5	1	45	12
Total Southeast Asia	718	22	74	108	1	29	23

Source: World Resources Institute (2002).

same is true in Africa and Southeast Asia, then economic development is likely to result in increased demand for wildlife until household incomes rise sufficiently for demand to switch to substitutes. In the case of wildlife use for medicinals, substitutes may not exist, and the demand curve may not resemble an inverted *U*.

Given present human population and economic growth rates in Africa and Asia, demand for meat is likely to increase by at least 3 percent per year and to double in 20 years. Because most wildlife populations are not growing as fast as is demand, hunting wildlife for food and medicines will become increasingly untenable unless a significant portion of demand for meat is supplied by domestic livestock and demand for wildlife medicines met with modern pharmaceuticals.

Unsustainable hunting for meat and medicines will mean the loss of a valuable source of food and income for the large number of families involved in wildlife trade. Finding ways to conserve threatened and endangered wildlife species without compromising the health and welfare of poor rural and urban families is a challenge. Shifting demand to regionally produced alternatives to wildlife protein and modern pharmaceuticals and revitalizing the traditional agricultural economies of recent entrants into the wildlife trade are two key steps in curbing the commercial wildlife trade without jeopardizing the health and security of West-Central Africans and Southeast Asians.

That said, before donors, governments, and conservation organizations try to increase small and large livestock production in regions of the world where wildlife constitutes a significant component of household diets, we need to understand the role of livestock raising in rural and urban households, the relative costs of livestock production and hunting, the influence of incentives to increase livestock production in rural and periurban tropical forested areas on forest clearing rates, and whether promotion of livestock production would merely substitute deforestation as the primary threat to wildlife conservation. We now turn to a brief discussion of these points.

Raising Domesticated Livestock and Not Wildlife

One approach to reducing the unsustainable use of wildlife as food is to promote consumer access to substitute sources of protein. A number of projects have been started to domesticate and raise selected wildlife species (e.g., cane rats, duikers, forest antelope, bush pigs), under the assumption that families in the region like the taste of wildlife so much that only if captive-bred wildlife is raised and offered for sale will the need to hunt wild animals decline (Rahm 1962; Tewe and Ajaji 1982; Codjia and Heymans 1990; Zongo et al. 1990). Unfortunately, the logic behind captive breeding of wildlife species is flawed for several reasons and therefore is unlikely to significantly reduce the demand for wildlife or decrease the hunting of wild animals for food.

First, there is little evidence that families in the region would continue to eat meat from wildlife if other sources of protein were available and cheaper. In fact, outside of urban areas people appear to eat wild meat because it is almost always the cheapest source of meat in markets (Wilkie and Carpenter 1999; Barnett 2000). Furthermore, preliminary evidence from Bolivia and Honduras (Wilkie and Godoy 2001) shows that consumers are very price sensitive and that as the price of wildlife substitutes drop, consumption of wildlife meat declines even more rapidly.

Second, captive breeding of wildlife makes little sense for low-productivity species such as large antelope, primates, and most reptiles. Even production rates of cane rats (*Thryonomys swinderianus*), with a gestation of 5 months and 6–13 months to reach an adult size of 4–5 kg (Houben 1999), are far lower than for domestic pigs and chickens (D. Messinger, pers. comm., 2001). Raising mollusks and reptiles as a staple food is unlikely to be cost effective because they are slow to reach slaughter size and inefficient at transforming food into meat. For example, a green iguana (*Iguana iguana*) consumes as much food as a chicken but takes 3 years instead of 4 months to reach a slaughter weight of 3 kg (Werner 1991). Similarly, Smythe (1991) calculated that captive raising of pacas (*Agouti paca*), though feasible, was economically irrational because the meat would have to be sold for more than $20 per kilogram to cover costs. Feer (1993) argues that meat productivity decreases from pigs to zebu cattle to cane rat to duikers. Consequently, increasing the supply of meat through husbandry of truly domesticated livestock such as pigs, goats, chickens, and ducks, which have been selectively bred for more than 5,000 years to convert feed into meat most efficiently, makes much more sense than attempting to raise wildlife in captivity, which is merely the first step in the long process of domestication.

Small livestock production (NRC 1991; Branckaert 1995; Hardouin 1995) such as rabbit raising has been adopted by households in Cameroon in areas where wild animals are already scarce (HPI 1996). Raising small domesticated animals such as rabbits and chickens is attractive in that methods of husbandry and veterinary care are well known. Small animal raising has been shown to be viable in periurban areas that are close to sources of demand and where proximal wildlife species populations have already been depleted (Lamarque 1995). However, pig or rabbit rearing as an alternative to wildlife hunting is likely to be successful only when the labor and capital costs of production are less than the costs of wildlife hunting and marketing (i.e., when game becomes too scarce to be worth searching for and transportation costs are not prohibitive). Of course, if domestic meat production becomes economically viable only after wildlife have become so scarce as to be unprofitable to hunt, the strategy clearly is ineffective as a conservation measure.

Promoting small livestock raising in periurban areas will disrupt the flow of economic benefits from urban consumers to poor rural producers of wildlife and may, perversely, encourage intensification of wildlife hunting to maximize profits before prices drop as domestic substitutes enter the market in increasing quantities.

The situation is even more complex in Southeast Asia, where wildlife is hunted for medicines as well as food. Raising domestic livestock may help reduce the latter but clearly will do little to reduce demand for wildlife as traditional Chinese medicine. Only education and enforcement are likely to convince or pressure consumers to change their preferences. However, there are hopes that the widespread availability of inexpensive Western medicines such as Viagra may help reduce demand for traditional wildlife-based aphrodisiacs in China.

Addressing Differences in Production Costs

In many areas of the world regulatory mechanisms to control the number of hunters and the quantity of wildlife harvested are weak or nonexistent, so hunters have free access to wildlife. Under these circumstances, production costs of wildlife are simply the costs to acquire hunting technology and the opportunity costs of labor associated with traveling to hunting areas, searching for wildlife, and transporting killed animals back to settlements or markets. In contrast, the cost of livestock husbandry includes the capital to acquire breeding stock, the labor invested in protecting animals from predators or thieves, the labor and capital used to provide food for livestock (e.g., grazing lands, forage crops), and the risk of losing animals to disease, predators, and theft. Although no empirical data exist that explicitly assess the relative costs of producing wildlife meat and the meat of livestock in the same community, it is likely that wildlife meat production in areas were access is unregulated or poorly regulated is much less costly than is livestock rearing. In this context the price of wildlife meat is always likely to be less than the price of livestock meat, and consumers will continue to opt for the cheaper alternative.

Given this, policymakers have at least three options: use law enforcement to reduce the supply of wildlife meat in markets, thus raising the price relative to that of alternatives; promote livestock production in areas where wild species have already been severely depleted and the costs of hunting exceed or equal those of livestock rearing; and subsidize the production or sale of livestock in areas of high biodiversity conservation value where wildlife meat hunting is still the most economically rational mode of meat production. If the first alternative were undertaken in periurban areas that are close to centers of demand for meat, transportation costs would be kept below those for wildlife meat harvested in more distant, intact forests. Moreover, because periurban areas typically have already been deforested for fuelwood or crop production, promoting pasture development and livestock production in these areas would not increase forest degradation or clearing rates.

Ranching for Food, Not Insurance

In many rural areas of the world where wild animals are a significant source of household protein, livestock are raised not primarily as a source of food but as savings and insurance to smooth consumption after a shock (Fafchamps and Gavian 1998; Godoy et al. 1998). This suggests that providing livestock producers with formal capital markets and other alternative ways to save and insure against shocks may be a prerequisite to increasing livestock raising for food. It should be possible to assess the utility of this proposal by looking at how successful microcredit schemes such as Grameen banking have changed the use of livestock in communities that traditionally used them for savings or insurance.

Avoiding Forest Depletion

Some conservationists are likely to cringe at the suggestion of promoting livestock production to reduce pressure on wildlife that is hunted for food because they are concerned that this will increase incentives to convert forest to pasture and result in a surge in deforestation rates. Certainly, pasture lands throughout the United States, Western Europe, and the Brazilian Amazon are good examples of this. We know that herbivore productivity in tropical forests is less than 10 percent that of tropical savannas (Robinson and Bennett 2000). Consequently, raising livestock on grasslands is likely to generate 10 times the biomass of meat that could be harvested in the same area of forest. Loss of a small percentage of total forest cover may be a reasonable price to pay to ensure the conservation of wildlife populations in the remaining forest. Moreover, in most nations with tropical forests there are also abundant savannas, many of which are underused for livestock production. This is particularly the case in Central Africa but is also true in much of South America. That said, any attempts to increase extensive or intensive livestock production to reduce wildlife meat consumption should assess the direct and indirect impact on forest lands.

Conclusions

We must be very clear in what is happening throughout the forests of the world. Through conversion of forest into agricultural or agroforestry lands, wildlife populations are increasingly being confined to small patches of forest, mostly in protected areas. And hunting in these refuges also creates a forest filled with trees but empty of wildlife.

Empirical studies that look exclusively at hunting or wildlife production in fallow forests are few, so it is exceedingly difficult to assess how much wildlife might be generated for consumption in these areas and how much people could rely on fallow forest wildlife as a source of dietary protein. What is clear, however, is that there is an inverse relationship between human population density and the percentage of dietary protein that can be supplied sustainably by wildlife. Rough estimates presented in this chapter suggest that fallow forests even under the best conditions are unlikely to be able to provide the primary source of meat for people in areas where human population density exceeds four people per square kilometer. Moreover, because savannas are an order of magnitude more productive in terms of herbivore biomass it makes much more sense to promote intensive livestock production in grasslands than extensive production of wildlife in forests if the goal is to produce meat for human consumption, although livestock production also has ecological impacts (e.g., vegetation clearing, riparian degradation, and water pollution).

In the end, the question to consider is not only one of the economic costs and benefits of hunting but also one of cultural change. Clearly, wildlife consumption—whether for economic or cultural reasons—cannot continue at its present per capita level. The argument that eating wild animals is a cultural tradition and right is moot. At current levels of consumption, many wildlife species will soon face local extinction, and people will no longer be able to practice what was once thought of as an integral part of their culture. In light of rapid wildlife declines, conservation organizations and governments must find ways to help people to modify their consumption preferences so that wildlife can be harvested at sustainable levels. By doing so, wildlife consumers will ultimately ensure that both the wildlife and their traditions persist.

Endnote

1. The World Health Organization has modified its estimates of human daily protein needs several times since 1936. In this chapter we assume that the average human needs approximately 1 g of protein per kilogram body weight per day. From U.S. Department of Agriculture food composition tables we estimate that the protein content of game meat is approximately 20 percent and that wastage during the dressing of a carcass does not exceed 40 percent of the initial carcass weight. We did not have anthropometry data from a representative population in Central or West Africa. However, using a sample of 1,329 Tsimane' individuals of all ages from the lowland forests of Bolivia we estimated average body weight across all age classes to be 36 kg. Given this, the average person with a body weight of 40 kg needs 106 kg of undressed wildlife biomass per year to completely meet protein needs.

References

Alvard, M. S. 2000. The impact of traditional subsistence hunting and trapping on prey populations: data from Wana horticulturalists of upland Central Sulawesi, Indonesia. Pages 214–230 in J. G. Robinson and E. L. Bennett (eds.), *Hunting for sustainability in tropical forests*. New York: Columbia University Press.

Anadu, P. A., P. O. Elamah, and J. F. Oates. 1988. The bushmeat trade in southwestern Nigeria: a case study. *Human Ecology* 16:199–208.

Auzel, P. 1996. *Evaluation de l'impact de la chasse sur la faune des forêts d'Afrique Centrale, Nord Congo. Mise au point de méthodes basées sur l'analyse des pratiques et les résultats des chasseurs locaux.* Bomassa, Republic of Congo: Wildlife Conservation Society/GEF Congo.

Auzel, P., and D. S. Wilkie. 2000. Wildlife use in northern Congo: hunting in a commercial logging concession. Pages 413–426 in J. G. Robinson and E. L. Bennett (eds.), *Hunting for sustainability in tropical forests.* New York: Columbia University Press.

Barnett, R. 2000. *Food for thought: the utilization of wild meat in Eastern and Southern Africa.* Nairobi, Kenya: TRAFFIC/WWF/IUCN.

Bennett, E. L. 1992. *A wildlife survey of Sarawak.* Kuala Lumpur, Malaysia: Wildlife Conservation Society and World Wildlife Fund.

Bennett, E. L., and Z. Dahaban. 1995. Wildlife responses to disturbances in Sarawak and their implications for forest management. Pages 66–85 in R. B. Primack and T. E. Lovejoy (eds.), *Ecology, conservation and management of Southeast Asian rainforests.* New Haven, CT: Yale University Press.

Bennett, E. L., A. J. Nyaoi, and J. Sompud. 1997. Hornbills and culture in northern Borneo: can they continue to co-exist? *Biological Conservation* 82:41–46.

Bennett, E. L., and J. G. Robinson. 2000. *Hunting of wildlife in tropical forests: implications for biodiversity and forest peoples.* Washington, DC: The World Bank.

Blake, S. 1994. *A reconnaissance survey of the Kabo logging concession south of the Nouabale-Ndoki National Park northern Congo.* New York: Wildlife Conservation Society.

Bodmer, R. E., and E. P. Lozano. 2001. Rural development and sustainable wildlife use in Peru. *Conservation Biology* 15:1163–1170.

Boserup, E. 1966. *The conditions of agricultural growth: the economics of agrarian change under population pressure.* Chicago: Aldine.

Branckaert, R. D. 1995. Minilivestock: sustainable animal resource for food security. *Biodiversity and Conservation* 4:336–338.

Caldecott, J. 1987. *Hunting and wildlife management in Sarawak.* Washington, DC: World Wildlife Fund.

Caspary, H.-U. 1999. *Utilisation de la faune sauvage en Côte d'Ivoire et en Afrique de l'Ouest: potentiels et contraintes pour le développement.* Abidjan, Côte d'Ivoire: The World Bank.

Chardonnet, P. 1995. *Faune sauvage Africaine: la ressource oubliée.* Luxembourg: International Game Foundation, CIRAD-EMVT.

Chen, T., H. Lin, and H. Chang. 2000. Current status and utilization of chelonians in Taiwan. Pages 145–151 in P. P. Van Dijk, B. L. Stuart, and A. G. J. Rhodin (eds.), *Asian turtle trade: proceedings of a workshop on conservation and trade of freshwater turtles and tortoises in Asia.* Lunenberg, MA: Chelonian Research Foundation.

Chin, L. 1981. *Agriculture and subsistence in a lowland rainforest Kenyah community.* Ph.D. thesis, Yale University, New Haven, CT.

Codjia, J. T. C., and J. C. Heymans. 1990. Experimental breeding of giant rats (*Cricetomys gambianus, C. emini*). *Nature et Faune* 6:62–66.

Delgado, C. L., M. W. Rosegrant, H. Steinfeld, S. Dhui, and C. Courbois. 1999. *Livestock to 2020: the next food revolution.* Washington, DC: International Food Policy Research Institute.

Dethier, M. 1995. *Etude chasse villageoise, forêt de Ngotto.* Yaounde, Cameroon: ECOFAC.

Dryer, P. 1985. The contribution of non-domesticated animals to the diet of Etolo, southern highlands province, Papua New Guinea. *Ecology of Food and Nutrition* 17:101–115.

Duckworth, W. 1999. *Wildlife in Lao DPR: a status report.* Lao DPR: International Union for the Conservation of Nature and Natural Resources.

Ellen, R. 1975. Non-domesticated resources in Nuaulu ecological relations. *Social Science Information* 14:129–150.

Fa, J. E. 1999. Hunted animals in Bioko Island, West Africa: sustainability and future. Pages 168–198 in J. G. Robinson and E. L. Bennett (eds.), *Hunting for sustainability in tropical forests.* New York: Columbia University Press.

Fa, J. E., J. E. G. Yuste, and R. Castelo. 2000. Bushmeat markets on Bioko Island as a measure of hunting pressure. *Conservation Biology* 14:1602–1613.

Fafchamps, M., and S. Gavian. 1998. *The spatial integration of livestock markets in Niger.* Nairobi, Kenya: International Livestock Research Institute.

FAO (Food and Agriculture Organization of the United Nations). 2000. *Global forest resources assessment: 2000.* FAO Forestry Paper 140. Rome: United Nations Food and Agriculture Organization.

Feer, F. 1993. The potential for sustainable hunting and rearing of game in tropical forests. Pages 691–708 in C. M. Hladik, A. Hladik, O. F. Linares, H. Pagezy, A. Semple, and M. Hadley (eds.), *Tropical forests, people and food: biocultural interactions and applications to development.* Paris: UNESCO.

Fimbel, C. C., B. Curran, and L. Usongo. 2000. Enhancing the sustainability of duiker hunting through community participation and controlled access in the Lobeke region of southeastern Cameroon. Pages 356–374 in J. G. Robinson and E. L. Bennett (eds.), *Hunting for sustainability in tropical forests.* New York: Columbia University Press.

Fox, J. 1999. *Mapping a conservation landscape: land use, land cover and resource tenure in northeastern Cambodia.* Honolulu: East-West Center.

Freese, C. H. 1997. *Harvesting wild species: implications for biodiversity conservation.* Baltimore, MD: Johns Hopkins University Press.

Freese, C. H. 1998. *Wild species as commodities: managing markets and ecosystems for sustainability.* Washington, DC: WWF/Island Press.

Fujita, M., and M. D. Tuttle. 1985. Flying foxes (Chiroptera: Pteropodidae): threatened animals of key ecological and economic importance. *Conservation Biology* 5:455–463.

Godoy, R. A., M. Jacobson, and D. S. Wilkie. 1998. Strategies of rain-forest dwellers against misfortunes: the Tsimane' Indians of Bolivia. *Ethnology* 37:55–69.

Groombridge, B., and R. J. Luxmoore. 1991. *Pythons in Southeast Asia: a review of distribution, status, and trade in three selected species.* Lausanne, Switzerland: CITES.

Hardouin, J. 1995. Minilivestock: from gathering to controlled production. *Biodiversity and Conservation* 4:220–232.

Houben, P. 1999. Elevage d'aulacodes au Gabon. *Canopée* 15:7–8.

HPI (Heifer Project International). 1996. *Boyo Rural Integrated Farmer's Alliance, Cameroon: project summary.* Little Rock, AR: Heifer Project International.

James, A. N., K. J. Gaston, and A. Balmford. 1999. Balancing the earth's accounts. *Nature* 401:323–324.

Janzen, D. H., and P. S. Martin. 1982. Neotropical anachronisms: the fruits the gomphotheres ate. *Science* 215:19–27.

Lamarque, F. A. 1995. The French co-operation's strategy in the field of African wildlife. Pages 267–270 in J. A. Bissonette and P. R. Krausman (eds.), *Integrating people and wildlife for a sustainable future.* Bethesda, MD: The Wildlife Society.

Lee, R. J. 2000. Impact of subsistence hunting in North Sulawesi, Indonesia, and conservation options. Pages 455–472 in J. G. Robinson and E. L. Bennett (eds.), *Hunting for sustainability in tropical forests.* New York: Columbia University Press.

Linares, O. F. 1976. "Garden hunting" in the American tropics. *Human Ecology* 4:331–349.

Luxmoore, R. J., and B. Groombridge. 1989. *Asian monitor lizards: a review of distribution, status, and exploitation and trade in four selected species.* Lausanne, Switzerland: CITES.

Martin, G. H. G. 1983. Bushmeat in Nigeria as a natural resource with environmental implications. *Environmental Conservation* 10:125–134.

Martin, P. 197? The discovery of America. *Science* 179:969–974.

McNeely, J. A. 1978. Dynamics of extinction in Southeast Asia. *BIOTROP Special Publication* 8:137–158.

Medway, L. 1977. *Mammals of Borneo.* Kuala Lumpur, Malaysia: MBRAS.

Naughton-Treves, L. 1997. Farming the forest edge: vulnerable places and people around Kibale National Park, Uganda. *Geographical Review* 87:27–46.

Ngnegueu, P. R., and R. C. Fotso. 1998. *Chasse villageoise et conséquences pour la conservation de la biodiversité dans la Réserve de Biosphère du Dja.* Yaounde, Cameroon: ECOFAC.

Noss, A. J. 1998. Cable snares and bushmeat markets in a Central African forest. *Environmental Conservation* 25:228–233.

Novaro, A. J., K. H. Redford, and R. E. Bodmer. 2000. Effect of hunting in source-sink systems in the neotropics. *Conservation Biology* 14:713–721.

NRC (National Research Council). 1991. *Microlivestock: little-known small animals with a promising economic future.* Washington, DC: National Academy Press.

Ntiamoa-Baidu, Y. 1998. *Sustainable harvesting, production and use of bushmeat.* Vol. 6. Accra, Ghana: Wildlife Department, Ministry of Lands and Forestry.

O'Brien, T. G., and M. F. Kinnaird. 2000. Differential vulnerability of large birds and mammals to hunting in North Sulawesi, Indonesia, and the outlook for the future. Pages 199–213 in J. G. Robinson and E. L. Bennett (eds.), *Hunting for sustainability in tropical forests.* New York: Columbia University Press.

Padoch, C., and C. Peter. 1993. Managed gardens in West Kalimantan, Indonesia. Pages 167–176 in C. S. Potter, J. I. Cohen, and D. Janczewski (eds.), *Perspectives on biodiversity: case studies of genetic resource conservation and development.* Washington, DC: Association for the Advancement of Science Press.

Rabinowitz, A. 1995. Helping a species go extinct: the Sumatran rhino in Borneo. *Conservation Biology* 9:482–488.

Rahm, U. 1962. L'elevage et la reproduction en captivité de l'*Atherurus africanus* (Rongeurs, Hystricidae). *Mammalia* 26:1–9.

Redford, K. H. 1993. Hunting in neotropical forests: a subsidy from nature. Pages 227–246 in C. M. Hladik, A. Hladik, O. F. Linares, H. Pagezy, A. Semple, and M. Hadley (eds.), *Tropical forests, people and food: biocultural interactions and applications to development.* Paris: UNESCO.

Robinson, J. G., and E. L. Bennett. 2000. Carrying capacity limits to sustainable hunting in tropical forests. Pages 13–30 in J. G. Robinson and E. L. Bennett (eds.), *Hunting for sustainability in tropical forests.* New York: Columbia University Press.

Robinson, J. G., and K. H. Redford. 1994. Measuring the sustainability of hunting in tropical forests. *Oryx* 28:249–256.

Robinson, J. G., K. H. Redford, and E. L. Bennett. 1999. Wildlife harvest in logged tropical forest. *Science* 284:595–596.

Smith, F. D., G. C. Daily, and P. Erhlich. 1995. Human population dynamics and biodiversity loss. Pages 125–141 in T. M. Swans (ed.), *The economics of biodiversity: the forces driving global change*. Cambridge, UK: Cambridge University Press.

Smythe, N. 1991. Steps toward domesticating the paca (Agouti: *Cuniculus paca*) and prospects for the future. Pages 202–216 in J. G. Robinson and K. H. Redford (eds.), *Neotropical wildlife use and conservation*. Chicago: University of Chicago Press.

Steel, E. A. 1994. *Study of the value and volume of bushmeat commerce in Gabon*. Libreville, Gabon: World Wildlife Fund.

Stephens, D. W., and J. R. Krebs. 1986. *Foraging theory*. Princeton, NJ: Princeton University Press.

Tewe, G. O., and S. S. Ajaji. 1982. Performance and nutritional utilization by the African giant rat (*Cricetomys gambianus*, W.) on household waste of local foodstuffs. *African Journal of Ecology* 20:37–41.

Van Dijk, P. P., B. L. Stuart, and A. G. J. Rhodin. 2000. *Asian turtle trade: proceedings of a workshop on conservation and trade of freshwater turtles and tortoises in Asia*. Lunenberg, MA: Chelonian Research Foundation.

Wenjun, L., T. K. Fuller, and W. Sung. 1996. A survey of wildlife trade in Guanxi and Guandong, China. *TRAFFIC Bulletin* 16:9–16.

Werner, D. I. 1991. The rational use of green iguanas. Pages 181–201 in J. G. Robinson and K. H. Redford (eds.), *Neotropical wildlife use and conservation*. Chicago: University of Chicago Press.

Wilkie, D. S. 1989. Impact of roadside agriculture on subsistence hunting in the Ituri forest of northeastern Zaire. *American Journal of Physical Anthropology* 78:485–494.

Wilkie, D. S., and J. F. Carpenter. 1999. Bushmeat hunting in the Congo Basin: an assessment of impacts and options for mitigation. *Biodiversity and Conservation* 8:927–955.

Wilkie, D. S., and R. A. Godoy. 1996. Trade, indigenous rain forest economies and biological diversity: model predictions and directions for research. Pages 83–102 in M. Ruiz Perez and J. E. M. Arnold (eds.), *Current issues in non-timber forest products research*. Bogor, Indonesia: Center for International Forestry Research.

Wilkie, D. S., and R. A. Godoy. 2001. Income and price elasticities of bushmeat demand in lowland Amerindian societies. *Conservation Biology* 15:1–9.

Winterhalder, B., and F. Lu. 1997. A forager-resource population ecology model and implications for indigenous conservation. *Conservation Biology* 11:1354–1364.

Zongo, D., M. Coulibaly, O. H. Diambra, and E. Adjiri. 1990. Document on the breeding of the giant African snail *Achatina achatina*. *Nature et Faune* 6:62–66.

Chapter 15

Invasive Agroforestry Trees: Problems and Solutions

David M. Richardson, Pierre Binggeli, and Götz Schroth

Thousands of plant species have been and continue to be transported by humans to areas far from their natural habitats. Some are moved accidentally, but more important are the many species that are intentionally introduced and cultivated to serve human needs (Ewel et al. 1999). In many parts of the world, some alien plant species (a small sample of all nonnative species in a given region) cause problems as invaders, spreading from sites of introduction and cultivation to invade natural or seminatural ecosystems, where they sometimes cause widespread damage. Biological invasions are now viewed as one of the main threats to global diversity (Sala et al. 2000; McNeely et al. 2001). Invasive plant species, and woody plants in particular, have major impacts on ecosystem structure and functioning (Versfeld and van Wilgen 1986).

The terminology of alien plant invasions used in this chapter follows Richardson, Pysek, et al. (2000; see also Rejmánek et al. 2004 and Figure 15.1):

- *Alien plants:* Plant taxa whose occurrence in a given area results from their introduction (intentionally or accidentally) by human activity (synonyms: "exotic plants," "nonnative plants," "nonindigenous plants").
- *Casual plants:* Alien plants that may flourish in an area but do not persist for more than one life cycle without further introductions (includes taxa labeled in the literature as "waifs," "occasional escapes," and "persisting after cultivation").
- *Naturalized plants:* Alien plants that reproduce and sustain populations over more than one life cycle without direct intervention by humans (or despite human intervention); they often recruit offspring freely, but often just near adult plants, and do not necessarily invade natural, seminatural, or human-made ecosystems.
- *Weeds:* Plants (not necessarily alien) that are undesirable from a human point of view. These are usually taxa with detectable economic or environmental

371

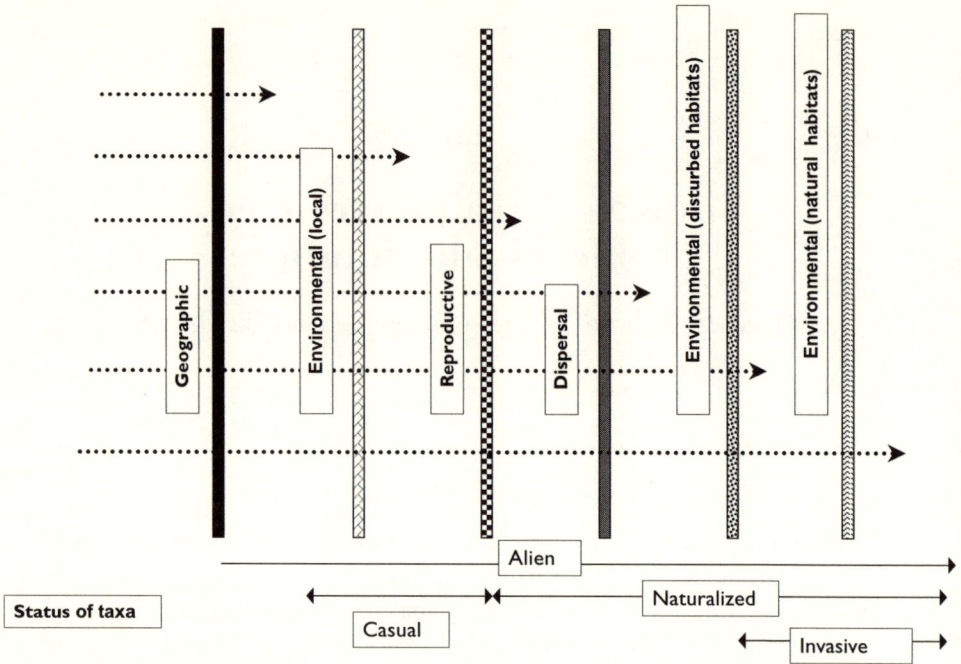

Figure 15.1. A conceptualization of the naturalization-invasion process, showing successive barriers that a species has to overcome to become naturalized or invasive (after Richardson, Pysek, et al. 2000). Each barrier provides options for management of invasive agroforestry trees.

effects (synonyms: "pests," "harmful species," "problem plants"). Environmental weeds are alien plant taxa that invade natural vegetation, usually adversely affecting native biodiversity or ecosystem functioning.

• *Invasive plants:* Alien plants that recruit reproductive offspring, often in very large numbers, at considerable distances from parent plants and thus have the potential to spread rapidly.

• *Transformer species:* A subset of invasive plants that change the character, condition, form, or nature of ecosystems over a substantial area relative to the extent of that ecosystem.

Many agroforestry systems, particularly those that rely on tree planting in or near treeless landscapes, rely heavily on alien plant taxa. As is the case in all endeavors based largely on nonnative species, problems arise when these organisms spread from sites of introduction and cultivation to invade areas where their presence is, for various reasons, deemed inappropriate. In some areas, problems caused by the spread of agroforestry trees from sites set aside for this land use pose a serious threat to biodiversity that may reduce or negate any biodiversity benefit of the agroforestry enterprise. The actual or potential impacts caused by such invasions must be weighed carefully against the actual or potential benefits deriving from the use of these alien species.

Very little has been published about invasions as a direct result of agroforestry. Indeed, the very concept of invasion is meaningless or absurd to some agroforesters, especially those working in dry and severely degraded environments. In these situations, recruitment of any woody plant is considered a bonus. A review of the agroforestry literature, discussions with agroforesters, and the authors' experience in many parts of the world reveal that some commonly used trees and shrubs in agroforestry are harmful invaders in some localities, under certain situations. Others that have not yet been reported as invasive are likely, because of their growth and reproductive characteristics, to invade when they are introduced to suitable environments or after the time lag that is usually observed between when a species is introduced and cultivated or disseminated and when it starts to invade. For more than two decades, some of the agroforestry literature has warned about the threats of alien species spreading uncontrollably (BOSTID 1980, 1983). However, Hughes (1994) and Hughes and Styles (1989) are exceptional among recent contributions in that they also provide practical guidelines how to avoid invasions from agroforestry trees.

Problems with invasions in agroforestry are not nearly as well reported in the international literature as is the case for invasions associated with commercial forestry (see reviews in Richardson 1998a, 1998b, 1998c), although several recent publications on agroforestry list low invasiveness as a selection criterion for agroforestry trees (Young 1997; Elevitch and Wilkinson 2001), warn against uncritical distribution of seed lots (Huxley 1999), and mention reported invasiveness in species descriptions (Salim et al. 1998). However, detailed accounts of invasiveness for a range of tree species, including agroforestry trees, have only recently become available (Binggeli et al. 1998; CABI 2000).

The fact that invasions have received little attention in the agroforestry literature may be because invasions are simply not perceived or reported to the same extent in agroforestry as they are in forestry. It may also be because the ecosystems that are affected by agroforestry often are degraded sites where concerns regarding impacts of invading plants are not high priorities. However, where agroforestry is practiced in buffer zones around protected areas (Chapters 17 and 18, this volume), invasiveness of agroforestry trees could negate any advantages gained from the provision of alternative timber and fuel sources, buffering, and interconnecting forest fragments (see Chapters 2 and 3, this volume).

Despite the lack of a complete global picture of the dimensions of plant invasions associated with agroforestry, it seems prudent to make a preliminary assessment of the situation by reviewing what information exists and drawing insights from advances in the understanding of plant invasions in general.

Advances in Plant Invasion Ecology Applied to Agroforestry

We begin by discussing recent advances in the understanding of the ecology of alien plant invasions in general. Biological invasions are notoriously idiosyncratic, leading some authors to suggest that attempts to predict the outcomes of introduction are futile. Nonetheless, there has been much progress over the past few decades toward developing a toolbox for understanding (Rejmánek et al. 2004) and managing (Wittenberg and Cock 2001) plant invasions. Rejmánek et al. (2004) discuss a number of robust generalizations that have emerged from recent research on plant invasions (Appendix 15.1). Some of the points in Appendix 15.1 are interrelated, some relate to complicated processes that are beyond the scope of this chapter, and some relate primarily or specifically to nonwoody plants. However, several points have important practical implications for actual or potential invasions of agroforestry trees and shrubs and ways of managing such invasions.

Invasion Success Is Positively Correlated with Propagule Pressure and Time Since Introduction

One robust generalization is that problems with invasions increase as the size of the propagule pool and the time since introduction increase. For agroforestry this means that we should look at the oldest and largest plantings when seeking generalizations. Importantly, the lack of invasions from recent (perhaps experimental) plantings does not necessarily mean that these species will not invade at these sites in the future, or when planted on a larger scale. Many mistakes are made in assessing the invasiveness of species after even a few decades of cultivation in new habitats. The passage of time changes many parameters such as the likelihood of favorable chance events, the likelihood of encountering mutualistic symbionts, genetic adaptation, and natural population growth. Very large numbers of propagules can result in successful invasions, even if the environment is suboptimal for establishment of the species. This is the result of a mass effect whereby potential establishment sites are swamped, allowing some propagules to capitalize on rare facilitating events.

Some Species Are Inherently Better Invaders Than Others

Recent syntheses of plant invasion ecology and regional and global compendia provide us with global lists of the most invasive taxa. Many interesting and useful generalizations can be drawn from such lists. For example, there is clear evidence that if a species is highly invasive in one part of the world, there is a high risk of it replicating its invasiveness in similar environments elsewhere. Also, some plant taxa are much more likely to invade when introduced to new habitats than others. Such empirical evidence is very useful for management. For example, it can be used to compile "black lists" of species known to be invasive, which either should not be used in new agroforestry operations or warrant special management attention. The use of "white lists" (safe species) is much more problematic because time lags, introduction history, and many other factors make it difficult to know when any species can confidently be considered safe (Rejmánek et al. 2004).

Genetic Change Caused by Introduction and Cultivation History Can Favor Invasiveness

Changes in the genetic makeup of introduced species can have a marked effect on their ability to invade. This may be as a result of the evolution of landraces that are more suited to local conditions than original introductions, increased genetic diversity as a result of the introduction of new genotypes, spontaneous hybridization in situ (e.g., *Prosopis* taxa in South Africa, Poynton 1990; *Leucaena* taxa, Hughes 1998), or human-mediated breeding programs aimed at genetic improvement. Spontaneous interspecific hybridization is important for the evolution of invasiveness in plants (Ellstrand and Schierenbeck 2000). Hybridization can change the rules for an alien organism and may enhance its ability to become established and invasive through the greater vitality of the hybrids compared with the parent species.

Prolific Seed Production Spells Trouble

Most invasive agroforestry species regenerate from seeds. Various aspects of seed biology are important determinants of invasiveness (Rejmánek et al. 2004). Heavy seed production in the absence of natural enemies is a crucial factor in many plant invasions. Very large seed numbers can swamp regeneration microsites, thus reducing the potential effect of biotic (and even abiotic) resistance (Rejmánek et al. 2004). Heavy seed production also affects dispersal in several ways. More seeds usually result in more offspring farther from parent plants. Importantly, very large seed numbers greatly increase the probability of seeds traveling great distances (many orders of magnitude farther than the mode for all seeds) and establishing satellite populations. Such isolated populations are disproportionately important in initiating invasions (Higgins and Richardson 1999) and greatly complicate the task of containment. Biological control using seed-attacking insects has great potential to reduce seed production of desirable (but invasive) agroforestry species without affecting other features of the plant. For example, much progress has been made in South Africa with biocontrol of *Acacia* spp. (Dennill et al. 1999) and *Prosopis* spp. (Impson et al. 1999; see also Richardson 1998a).

The rapid onset of seed production, a key criterion for the mass production of planting material, also confers invasiveness in many situations. Early maturity in plants usually is associated with other traits that confer colonizing ability (Richardson et al. 1990; Rejmánek and Richardson 1996). It is difficult to select for one trait without getting the "whole package" (Richardson 1998c).

Mutualisms Are Critically Important, and Reshuffling the World's Biota Is Making Ecosystems More Open to Invasion by More Species

Many invasions depend on mutualistic interactions between the introduced plant species and other organisms in the new habitat. Among the most important of these are animal-mediated pollination and seed dispersal and interactions between plant roots and mycorrhizal fungi and nitrogen-fixing bacteria (Richardson, Allsop, et al. 2000). Generalist vertebrate seed dispersers such as cattle, goats, and sheep often are a component of agroforestry systems and provide a reliable mechanism for seed movement in the new habitat. Added to this is the fact that propagules of many agroforestry trees are widely disseminated by humans. These factors contribute to enhanced long-distance dispersal and the establishment of new foci for invasion. Potential barriers to establishment (and invasion beyond planting sites) are overcome for many agroforestry trees and shrubs when appropriate mycorrhizal symbionts and bacteria are introduced. Such inoculations enable the alien agroforestry species to grow productively in the new habitat and also radically enhance the suitability of surrounding areas for establishment and invasion by the alien species (Richardson, Allsop, et al. 2000).

Potential Impacts of Invaders Often Are Related to the Functions and Services That Make These Species Desirable Subjects for Agroforestry

Alien species used in agroforestry are selected for the new functions and services that they bring to the system, functions and services that cannot be provided (as well) by native species. It is often exactly these functions and services (e.g., rapid biomass accumulation, nitrogen fixation) that cause harmful impacts when these species invade beyond sites intended for agroforestry. Such species have been called ecosystem engineers (Crooks 2002).

Experimentation with Many Species Worldwide Ensures Better Species-Site Matching Than in the Past

Improved communication between agroforesters in many parts of the world has resulted in the rapid and widespread dissemination of news of highly successful agroforestry species (e.g., the many species of "miracle trees"). Such information, based on the natural experiment of the planting of hundreds of species across the world, is providing empirical evidence on species-site matching. Rather than needing to experiment with a large number of potential species, agroforesters are able to select from a small number of species with a very high chance of success in their area. Species selection following this process, in many cases, is also selecting for invasiveness.

Many Facets of Global Change Alter Ecosystems and Trigger, Facilitate, or Sustain Invasions

The many facets of global change (e.g., global warming, elevated CO_2 concentrations, and altered nutrient-cycling and disturbance regimes) are widely expected to greatly exacerbate the nature and magnitude of problems with biological invasions in the next few decades (Mooney and Hobbs 2000; see also Chapter 20, this volume).

Trees Typically Used in Agroforestry: Current Levels of Invasiveness, Perceptions, and Approaches

Detailed monographs of a number of key agroforestry species, several known as highly invasive, have been published recently (e.g., *Acacia karroo,* Barnes et al. 1996; *Acacia seyal,* Hall and McAllan 1993; *Balanites aegyptiaca,* Hall and Walker 1991; *Calliandra calothyrsus,* Chamberlain 2001; *Cordia alliodora,* Greaves and McCarter 1990; *Gliricidia sepium,* Stewart et al. 1996; *Leucaena* spp., Hughes 1998; *Parkia biglobosa,* Hall et al. 1997; *Prosopis* spp., Pasiecznik et al. 2001; *Vitellaria paradoxa,* Hall et al. 1996). Where appropriate (i.e., *A. karroo, C. calothyrsus, Leucaena* spp., *Prosopis* spp.), these monographs include sections on the introduction, invasive potential, and impacts of the species in the introduced range.

Here we review the experience with invasiveness for some taxa of trees and shrubs that are commonly used in agroforestry. We have divided the taxa into groups corresponding to their typical functions in agroforestry land uses. This is because when aiming to substitute noninvasive taxa for invasive (or potentially invasive) taxa, one needs to find functionally equivalent rather than taxonomically related species.

Fast-Growing, Nitrogen-Fixing Legume Trees

This group includes a large number of pioneer species that are commonly used in agroforestry because of their ability to grow on nitrogen-deficient sites where they improve fertility with their litter and prunings, to produce large amounts of fuelwood, and to provide protein-rich foliage and pods. The many uses of such trees make them typical multipurpose trees that are used in reforestation of degraded land, improved fallows, contour hedgerows, and silvopastoral systems. Many of these species cast a light shade and resprout readily even after severe crown pruning, which makes them ideal shade trees for tree crops such as coffee and cocoa.

The best-known example of this group is *Leucaena leucocephala,* a small tree from Central America that has been introduced to most parts of the tropics, starting more than 400 years ago when the Spanish brought it to the Philippines (Binggeli et al. 1998). It has been used as a shade tree for coffee, cocoa, and tea, for windbreaks and firebreaks, and for many other purposes. Since 1983 the species has been attacked by the psyllid insect *Heteropsylla cubana,* which causes defoliation and tree death especially in dry regions. This has inspired the search for alternatives, including other *Leucaena* species. Other Latin American legume trees used extensively outside their home range include *Gliricidia sepium* and *Calliandra calothyrsus. G. sepium* often is used for live fences, as cocoa shade, and for many of the same purposes as *L. leucocephala. C. calothyrsus* was brought from Guatemala to Java in 1936 as a potential shade tree in coffee but was then mainly used by villagers for fuelwood plantations, as fodder plant and bee pasture, and as a planted fallow tree for the regeneration of agricultural soils (NRC 1983a). More recently, *C. calothyrsus* has also been used in short-rotation fallows and as a fodder tree in East Africa. In Cameroon it is used in rotational tree fallow systems where *L. leucocephala* is no longer used because of its invasiveness (Kanmegne and Degrange 2002).

Leucaena leucocephala is highly invasive and has been recorded as a weed in West, East, and South Africa, India, Southeast Asia, Australia, and several Pacific, Indian Ocean, and Caribbean islands (Binggeli et al. 1998; Hughes and Jones 1998; Meyer 2000; Henderson 2001; Randall 2002). It invades open and disturbed habitats such as roadsides and abandoned fields and pastures, where it may form dense, monotypic stands, currently covering about 5 million ha worldwide (Binggeli et al. 1998). The species is very common in the roadside vegetation in the forest zone of the Côte d'Ivoire, where it was used as shade tree in coffee (R. Peltier, pers. comm., 2001). It replaces open forest in Hawaii and threatens endemic species on several oceanic islands (Binggeli et al. 1998). It is not known to invade undisturbed closed forests (Hughes and Jones 1998) but may invade disturbed dry forest (Binggeli et al. 1998). In a fallow improvement trial with several introduced legume tree species on a degraded soil in the Côte d'Ivoire, *L. leucocephala* had the densest understory of all species under consideration, consisting almost entirely of its own regeneration (Schroth et al. 1996). The abundant regeneration of this tree species was recognized as a problem when it was used as coffee and tea shade in Indonesia and inspired research into sterile hybrids (Dijkman 1950; Hughes and Jones 1998).

Other widely used species from this group have caused fewer problems. For *Leucaena diversifolia,* the second most widely planted species in this genus, this may simply be a consequence of the shorter time of extensive plantings and its more limited use. The two species share the same weedy traits (precocious flowering and fruiting, abundant seed production, self-fertility, hard seed coat, and ability to resprout after fire or cutting; Hughes 1998; CABI 2000), suggesting that new introductions generally should be avoided. In contrast, *C. calothyrsus* has been used in agroforestry for more than half a century, including for planting around state forestland in Java (NRC 1983a), which would be an excellent starting point for invading these forests once they have been logged. Yet it has rarely been reported as invasive (http://www.hear.org/pier/cacal.htm), possibly because of its low seed production (CABI 2000). However, careful observation of the species in areas where it has been introduced more recently is needed. Similarly, despite its widespread use in agroforestry throughout the tropics, *Gliricidia sepium* has also not caused widespread problems as an invader, except in Jamaica (Holm et al. 1979). It is also known to be naturalized on Koolan Island off the tropical Western Australian coast (Keighery et al. 1995). It is listed as potentially weedy in the Pacific Islands, where it was recently introduced (Thaman et al. 2000). Sources collated by Randall (2002) list *G. sepium* as "weed," "quarantine weed," "naturalized," "garden escape," "environmental weed," and "cultivation escape," suggesting its potential to cause bigger problems in the future. Much work has been done on genetic improvement of *G. sepium* (e.g., Chamberlain and Pottinger 1995), and new genotypes may well prove to be more invasive.

Another group of fast-growing, nitrogen-fixing agroforestry trees with somewhat different properties is the Australasian acacias, especially *Acacia auriculiformis* and *A. mangium*. Because of their ability to grow and produce large amounts of fuelwood even on severely degraded soils and to colonize grassland, *A. auriculiformis* and to a lesser extent *A. mangium* have been introduced to many tropical countries. The use of these species in agroforestry is recent and not yet common but is likely to increase in the future because of their exceptional characteristics. Both species are listed in Randall's (2002) *Global Compendium of Weeds* and have the potential to become more invasive in the future (CABI 2000; http://www.hear.org/pier/acman.htm; http://www.hear.org/pier/acaur.htm). *A. mangium* seeds and regenerates prolifically, although disturbance of the understory, as by fire, usually is necessary for large-scale natural regeneration (NRC 1983b). Several closely related *Acacia* species are highly invasive in southern Africa, namely, *A. cyclops, A. longifolia, A. melanoxylon, A. mearnsii, A. pycnantha,* and *A. saligna* (Richardson et al. 1997; Henderson 2001). Randall (2002) lists 288 *Acacia* species, about a fourth of this large genus, with the status of casual alien, "quarantine weed," or worse.

Trees for Dry Zones

In semiarid regions, cutting and lopping of trees for timber, fuelwood, and fodder in combination with browsing, recurrent fires, and a harsh climate have caused widespread degradation of vegetation and soils. Increasing tree cover in such areas has been the objective of many development projects. Because tree growth is generally slow and mortality high under these conditions, selecting site-adapted species is particularly important. Promising species such as *Acacia nilotica, Azadirachta indica* (neem), and *Prosopis* spp., have been introduced into many countries outside their native home range (Salim et al. 1998). Because of their outstanding growth characteristics, they have often proved superior competitors to the native vegetation and become invasive.

Acacia nilotica, native to India, Pakistan, and much of Africa, was introduced to western Queensland, Australia, as a shade and fodder species in the 1920s and initially caused no problems as an invader. It first became naturalized along creeks and boredrains and became widespread and invasive after a series of years with above-average rainfall in the 1970s. It now forms stands of several thousand individuals per hectare in pastures in coastal areas (Carter 1994; Tiver et al. 2001; van Klinken and Campbell 2001).

Azadirachta indica is native to Southeast Asia and India but has been widely planted outside its range. It is known to be invasive in Ghana (http://www.green.ox.ac.uk/cnrd/jo.htm#darwin), spreads from rural plantings into undisturbed bush in northern Australia (A. A. Mitchell, pers. comm., 2002), and is also listed as a potential environmental weed in Australia (Cshurhes and Edwards 1998). Various sources list *A. indica* as "weed," "naturalized," "garden escape," or "environmental weed" (Randall 2002).

The neotropical *Prosopis juliflora* and closely related species have been widely used in the dry tropics to halt desertification because they tolerate low rainfall, great heat, and poor and saline soils and because of their ability to stabilize sand dunes (e.g., in the Sudan and Pakistan) and enrich the soil through their nitrogen fixation. Other products include good-quality fuelwood and charcoal, pods as fodder and food, and seed gum. Because of the low palatability of its foliage, *P. juliflora* is suitable for use as a live fence (CABI 2000). It has been widely introduced throughout the dry tropics and has spread over large tracts of Africa, Asia, and South America (Hulme n.d.; Jadhav et al. 1993; Sharma and Dakshini 1998; Tiwari and Rahmani 1999). Its invasive potential has long been known but has until recently gained little attention or, in very degraded areas, even been seen as a bonus. Referring to Africa, Baumer

(1990, p. 170) stated that "one would be only too happy in certain very degraded, not to say denuded, regions to find an invasive plant with as many qualities as *Prosopis,*"a view that reflects particularly well the ambivalence of the use of aggressive colonizers in areas badly in need of tree cover (see also Coppen 1995; El Fadl 1997). These authors also argued that the spread of *P. juliflora* must be checked in areas that are not degraded and suggested that this could be achieved through good management but did not provide concrete guidelines.

Although the deleterious impacts of root competition and allelopathy of *P. juliflora* in agricultural areas have been noted earlier (Baumer 1990; Coppen 1995), the environmental and human impacts of invasions by this species have only recently gained more attention (e.g., review by Pasiecznik et al. 2001). In 1999, an Ethiopian workshop on agricultural weeds concluded, after much debate, that the species was on balance detrimental to the environment and should be eradicated (Anonymous 1999). In neighboring Kenya, after *P. juliflora* introduction in the early 1980s and subsequent rapid spread, it has been reported that toxicity and even death have occurred in livestock after pod ingestion. The tree locust (*Anacridium melanorhodon arabafrum*), a well-known African pest, feeds on *P. juliflora,* and it is feared that this insect, hitherto not a problem in the Lake Turkana region, might become established in this part of Kenya (Anonymous 1997). In northeastern Brazil, *P. juliflora* has spread from managed agricultural systems into arid shrublands (caatinga) rich in endemic species, where its impact on biodiversity is viewed as highly detrimental (Hulme n.d.). In Gujarat, India, *P. juliflora* is viewed as negatively affecting pastures, cattle health, and water resources and is viewed as a threat to number of bird and mammal species (Jhala 1993). However, it does play a role in erosion control and provides people with a source of income from charcoal, pods, and honey. It is therefore thought that the tree should be contained rather than eradicated (Tiwari and Rahmani 1999). A detailed study by Gold (1999) of an Indian rural village community confronted with deforestation, where an aid program had been established in 1993 to plant *P. juliflora* on the hilly wastelands, revealed that the species had become the only source of fuelwood. However, local people also identified a number of significant drawbacks: it colonizes agricultural land and is hard to remove, its thorns cause dangerous infections and play havoc with bicycle tires, the leaves are unappealing to goats, and no grass or crops grow in its shade. Consequently, local people considered *P. juliflora* to have fewer uses than native trees.

The genus *Prosopis* has an interesting history in South Africa. Several species have been widely used as amenity trees, mainly for livestock fodder and shade, in the arid parts of the country. After the spontaneous hybridization of several taxa (notably *P. glandulosa* var. *torreyana* and *P. velutina*), *Prosopis* spp. rapidly spread over huge areas, making large tracts of rangeland unproductive

(Harding and Bate 1991). There is an ongoing debate on whether *Prosopis* is a friend or foe in South Africa, but opinions are converging on the latter view. The national "Working for Water Programme" (http://www-dwaf.pwv.gov.za/wfw) has spearheaded an innovative management program for invasive *Prosopis* spp., involving mechanical and chemical control, the management of livestock, and biological control using seed-attacking insects (Richardson 1998a). *Prosopis* species are also among the "weeds of national significance" in Australia and cause serious impacts over some 800,000 ha, mainly in northern Australia. A strategic plan for managing these invasions aims to remove the current stands and to prevent impacts by coordinating and maintaining management at a national level, containing all core infestations and subjecting them to sustained management aimed at eventually eradicating them, removing all isolated and scattered stands, and preventing further spread (Agriculture & Resource Management Council of Australia and New Zealand 2001).

Nonlegume Service Trees

The Latin American pioneer tree *Cecropia peltata,* which forms a complex with *Cecropia pachystachya* and *C. concolor,* is a particularly interesting case because it shows that an alien agroforestry tree may initially appear harmless but may become invasive once human activities or climate change have created more favorable conditions for its spread. It was introduced to the Côte d'Ivoire in 1910 as a shade tree in coffee plantations and spread very slowly in the first six decades. Subsequent large-scale destruction of the forest cover (see Chapter 6, this volume) created the open, disturbed conditions necessary for its rapid spread in competition with native pioneer species (Binggeli et al. 1998). The species has also become invasive in other places in Africa, Southeast Asia (Binggeli et al. 1998), French Polynesia (Meyer 2000), and Hawaii (http://www.hear.org/pier3/cepel.htm). A related species, *Cecropia obtusifolia,* is highly invasive in the Cook Islands, where it is also widely used as a shade species in coffee plantations (Meyer 2000).

Fast-Growing Timber Trees

This group includes a number of tree species whose main characteristic is their fast growth. Such trees are in high demand in deforested regions for the production of fuelwood and poles, and as windbreaks; use as shade trees is less common. The genus *Eucalyptus* includes a number of fast-growing tree species that are commonly used not only in plantation forestry but also in agroforestry in both dry and humid tropical regions. *Eucalyptus camaldulensis* and *E. tereticornis* are planted as windbreaks in dry regions (CABI 2000), and *E. globulus* and several other *Eucalyptus* spp. are commonly planted around cultivated fields and in farm woodlots in the deforested Ethiopian highlands (Demel Teketay 2000) and are also used in many other tropical countries (CABI 2000). In humid tropical Costa Rica, *E. deglupta* is increasingly used as coffee shade because of its fast growth, homogeneous light shade, and low pruning needs compared with the traditional legume shade trees (Tavares et al. 1999). *Eucalyptus* spp. also were promoted for fuelwood plantations in buffer zones around protected forests in Uganda and Burundi in the 1980s, although the respective projects later shifted toward the use of local tree species (van Orsdol 1987).

Eucalypts are naturalized or (usually marginally) invasive in many parts of the world. Randall (2002) lists no fewer than 67 *Eucalyptus* species that have been listed as "weed," "sleeper weed," "quarantine weed," "noxious weed," "naturalized," "native weed," "garden escape," "environmental weed," "cultivation escape," or "casual alien" in various parts of the world. *Eucalyptus* species given four or more status categories in Randall's (2002) compendium, and therefore possibly the most widely naturalized, are *E. botryoides, E. camaldulensis, E. cladocalyx, E. conferruminata, E. globulus, E. grandis, E. lehmannii, E. leucoxylon, E. maculata, E. paniculata, E. polyanthemos, E. saligna,* and *E. sideroxylon.* In South Africa, *E. camaldulensis, E. cladocalyx, E. diversicolor, E. grandis, E. lehmannii, E. paniculata,* and *E. sideroxylon* are considered invasive (Henderson 2001), although only *E. camaldulensis, E. grandis,* and *E. lehmannii* are unquestionably "invasive" as defined earlier (Forsyth et al. 2004). Despite their appearance in many weed lists, eucalypts have fared poorly as invaders when compared with other tree genera that have been planted to a similar extent in many parts of the world. *Pinus,* the obvious comparison (although not widely used in agroforestry), has been orders of magnitude more successful. For example, Richardson and Higgins (1998) list 19 *Pinus* species that are clearly invasive (*sensu* Richardson, Pysek, et al. 2000) in the Southern Hemisphere. Rejmánek et al. (2004), in reviewing invasiveness in *Eucalyptus* (or the lack thereof), concluded that propagule pressure (large seed pools) explains much more of the variance in observed invasiveness in eucalypt taxa than any known combination of ecological factors. This suggests that the risk of invasiveness is much smaller when eucalypts are used in windbreaks or other agroforestry practices involving small tree numbers per hectare than when they are used in plantation forestry. The use of eucalypts as shade trees in coffee over large areas represents an intermediate case. Further work is needed to quantify this relationship and the invasion ecology of *Eucalyptus* in general.

Casuarina equisetifolia, an actinorhizal tree species native to Australasia, has been widely planted in Africa, South and Central America, and the Caribbean. Its nitrogen-fixing ability allows it to grow on very poor sandy soils, and it is often used for sand dune stabilization and in shelterbelts, especially in coastal areas. It readily invades disturbed vegetation in many parts of the world. It has spread in Hawaii and has invaded the Everglades National

Park in Florida, where its roots also interfere with the nesting of sea turtles on foreshore dunes (Geary 1983; Binggeli et al. 1998). This species, and *C. cunninghamiana,* are also invasive over large parts of the eastern half of South Africa, especially along the eastern seaboard (Henderson 2001).

High-Value Timber Trees

Several high-value timber species, which are used in agroforestry in their native ranges, have become invasive in parts of the tropics to which they had been introduced for forestry plantations. The planting of high-value timber species on farms is likely to increase in the near future when the supply of such timbers from natural forests ceases and could in fact be a way of simultaneously increasing farm incomes and reducing pressure on remaining forests. In future agroforestry programs, experience with invasive plantation species must be taken into account to avoid introducing potentially invasive species into sensitive areas, such as buffer zones. For example, *Swietenia macrophylla* (mahogany, Meliaceae) is planted by farmers in association with fruit trees both in its native Amazonia and in Indonesia, where it is alien (Michon and de Foresta 1995); invasion by this species of native forests especially after disturbance (Commonwealth Agricultural Bureau International 2000) has been observed in Sri Lanka, Asia, and the Pacific Islands. *Cordia alliodora,* a common local shade tree in coffee and cocoa plantations in Latin America (Beer et al. 1998), has been widely planted in Africa for timber production (Richardson 1998a). It was planted in Vanuatu to provide timber; however, regular cyclones caused havoc with the trees and subsequent heavy cattle grazing enhanced their regeneration and spread, thus reducing the grazing potential of the land (Tolfts 1997; Meyer 2000).

Fruit Trees

Fruit trees are important components of agroforestry systems such as homegardens and mixed tree crop plantations and often provide an important basis for farmers' subsistence and income. As with other agricultural crops, the use of alien species is common, and the invasion of native ecosystems has been reported for a few species. The possibilities of exchanging an alien for a native species or an invasive alien for a noninvasive alien are much more limited for fruit trees than for shade or soil-improving trees because of the restrictions on species selection imposed by consumption preferences and local markets for fruits.

The most notorious case of an alien fruit tree becoming a weed is guava (*Psidium guajava*). This small tree, native to the American tropics, has been introduced throughout the tropics since colonial times. It is a pasture weed in

Central America and a serious weed in arable plantation and pastureland on the Fiji Islands. The seeds are spread by cattle, making it difficult to eradicate from pastures. The impact of introduced guava on the native vegetation has been best documented on the Galapagos Islands, where it invades disturbed forest and outcompetes endemic plant species (BOSTID 1983; Binggeli et al. 1998). This species is also highly invasive in natural vegetation in eastern South Africa (Richardson et al. 1997; Henderson 2001). A related species, *Psidium cattleianum,* is one of the worst invaders in many islands in the Indian Ocean (notably on La Réunion; Macdonald et al. 1991) and the Pacific (Meyer 2000).

Another example is *Passiflora mollissima* (banana passion fruit or curuba), a woody vine up to 20 m in length originating from the Andes that has been cultivated for its fruits and as an ornamental in several tropical countries and has become invasive in Hawaii, New Zealand, and South and East Africa. In Hawaii it has spread into native forest, scrub vegetation, pastures, forestry plantations, and lava flows, suppressing tree regeneration and killing trees through shading, thereby reducing species richness. The prolific fruit production increases populations of feral pigs. The species is also naturalized in South African forests and grows at forest edges and in clearings in East Africa (Binggeli et al. 1998). In Paraguay, *Citrus* species (*C. aurantium, C. sinensis,* and intermediate types) readily spread into undisturbed forest and become a characteristic feature of their understory, a process that is favored by grove abandonment (Gade 1976).

Invasive Agroforestry Trees: Scenarios and Management Options

Effective management of invasive agroforestry trees demands action at global, regional, national, and local scales. The recent *Global Strategy on Invasive Alien Species* (McNeely et al. 2001) provides a good starting point for global attention to the problem. A global plan should build management capacity; build research capacity; promote information sharing; develop economic policies and tools; strengthen national, regional, and international legal and institutional frameworks; institute a system of environmental risk analysis; build public awareness; prepare national strategies and plans; build alien species into global change initiatives; and promote international cooperation.

Various options are available for reducing actual or potential impacts of invasive agroforestry trees in different situations. The conceptualization of the naturalization and invasion process (Figure 15.1) provides a template for planning interventions to strengthen different barriers to prevent widespread invasion and impacts. For example, excluding known invasive alien species (or, ideally, all alien species) from agroforestry projects strengthens the geographic barrier; combinations of burning, grazing, and weeding can strengthen the local environmental barrier and limit regeneration of invasive species; and management of livestock and other seed dispersal agents can strengthen reproductive and dispersal barriers, as could genetic engineering to produce sterile trees, although the usefulness of this approach for agroforestry has not yet been explored. Ledgard and Langer (1999) provide an excellent example of practical ways to reduce invasive spread from forestry plantations in New Zealand. Similar guidelines could be produced for different agroforestry land uses. In this section we propose some guidelines, based on the information reviewed in this chapter, for reducing or mitigating problems associated with invasions of agroforestry trees.

Wherever Possible, Use Local Tree Species in Agroforestry

An obvious way to avoid risks of invasiveness is to use only native species. Avoiding nonnative species is most feasible in the case of service trees (e.g., species for shade or soil conservation), where several alternative species may be available in a region, though perhaps not with identical properties. Trees grown for commercial products with established markets are much more difficult to substitute, but the introduction of invasive species to new regions or their promotion in sensitive areas such as agroforestry buffer zones should nevertheless be avoided. Where agroforestry is intended also to serve conservation purposes, as in buffer zones or biodiversity-friendly coffee or cocoa production systems, native tree species also have the advantage of creating habitat for native fauna. In mainly deforested landscapes, the promotion of native species will contribute to the in situ conservation of local tree germplasm (see Chapter 12, this volume). Furthermore, native species provide goods and services with which the population is familiar (e.g., medicinal uses).

The feasibility of this approach depends on the availability of species in the native flora that are suitable for the intended uses, the ease of obtaining seeds or vegetative propagation material, knowledge of site needs, growth rates, conditions for germination, inoculation requirements, and establishment and management methods. Working with little-known species is clearly a risk and may delay the generation of demonstrable results. These difficulties decrease as more information on local fruit, timber, and service trees becomes available from national and international research centers (Salim et al. 1998; CABI 2000). In many cases, however, a local species that is as effective for a given purpose as the best alien species may simply not be available or may remain to be identified. In such cases, using the second- or third-best (local) choice may mean sacrificing farmers' interests to avoid the risk of invasiveness. However, it should also be mentioned that the focus on very fast-growing "miracle trees" in agroforestry has often led to problems of competition with crops; such problems may be less severe when local tree species with intermediate growth rates are used. This conflict of interest in searching for the best-performing species while trying to avoid introduced species can be particularly severe in strongly degraded, arid environments, where species with outstanding growth characteristics such as *Prosopis* spp. have often been introduced and become invasive. Where agroforestry plantings are intended to reduce pressure on overused native ecosystems, especially in arid regions, a fast-growing but non-invasive alien tree species could be more beneficial for biodiversity than a native species that produces timber, fuelwood, and other tree products at lower rates. Objective estimates of the risks of invasiveness for potential species are clearly needed for this approach.

Subject Species Selection to Formal Cost-Benefit Analysis and Environmental Impact Assessment

Potential costs associated with the spread of agroforestry species beyond sites set aside for planting and the associated damage to natural ecosystems (including a wide range of ecosystem services) must be assessed before an agroforestry project is initiated. The polluter-pays principle is being advocated as an element of the solution to the problem of invasive forestry tree species in South Africa (Richardson 1998a), and it should be explored for agroforestry. But who is the polluter who pays? The small farmer who accepted the introduced tree? The research institution, perhaps a foreign university, whose 3-year project ended long before the species became invasive? The seed bank that responded to the request of a local research station and sent *Leucaena* seeds? The principle is clearly easier to apply with a forestry company establishing commercial plantations than with typically decentralized, small-scale, and noncommercial agroforestry projects.

Continued use of known invasive species in areas where they are already planted and new plantings of such species should be done with due cognizance of potential harmful effects of invasions. In sensitive areas such as buffer zones of protected areas, conservation payments or access to preferential markets for biodiversity-friendly products may be the best way to persuade farmers to do without economically attractive but potentially invasive alien species.

Establish Criteria for Rational Risk Assessment for Agroforestry Worldwide

Apart from the agroforestry species widely known to be invasive, there are many other species that are likely to become invasive in the future. Research is needed to establish criteria for the objective assessment of the potential risk of any alien species becoming invasive at a given locality. Global black lists have limited value because they are generally too restrictive; in fact, there is reason to tag even some well-known invasive species as safe for use in certain environments or under certain conditions. For example, some species grow well but do not produce fruits in certain climates, such as *Gliricidia sepium* in climates without a well-defined dry season. Some species may show poor growth outside agricultural areas because of low adaptation to infertile soil or may need fire for regeneration. As with other predictions of invasiveness, it should be noted that conditions (e.g., nutrient conditions, climate, and fire frequency) at a site change. For this reason, assessments must be time- and site-specific.

Provide Easy Access to Up-to-Date, Objective Information on Invasiveness of Agroforestry Species

Many species that are used in agroforestry are widely known to be invasive in all or most sites where they are used. Such information already resides in widely available databases such as International Centre for Research in Agroforestry's Agroforestree Database (Salim et al. 1998), CABI's *Forestry Compendium* (CABI 2000), and R. P. Randall's (2002) *Global Compendium of Weeds*. Further efforts are needed to ensure that such databases are kept up to date and that assessments are based on objective criteria.

Include Explicit Considerations of Invasiveness in Standard Assessments for Species-Site Matching

Detailed assessments of the suitability of particular species for use in agroforestry (e.g., Durr 2001 for *Samanea saman*), as a matter of course, should include objective assessments of the potential for the species to become invasive.

Incorporate Considerations Relating to Invasion in Standard Management Protocols

For example, pruning trees to prevent them from producing fruits is done in many agroforestry situations and could possibly be a practical prevention strategy in some cases. However, relying on this strategy is risky because trees may remain unpruned in some areas or years, and the strategy would fail when an area is abandoned temporarily or permanently. Where tree felling for wood collection is legally restricted (e.g., parks), restrictions could be relaxed to allow felling of invasive alien species (J. Healey, pers. comm., 2001).

Conclusions

Our knowledge of plant invasions that are the direct result of agroforestry is fragmentary. We know that some tree and shrub species that are (or have until recently been) widely used in agroforestry are among the most widespread and damaging of plant invaders. We also know that our knowledge base is changing rapidly. Some species that were deemed safe based on available information even a decade ago are now known to be invasive. The natural experiment involving the plantings of thousands of species in many types of environment is ongoing. Guidelines for reducing the problems, without excluding every potential agroforestry species, must be reviewed at regular intervals. A thorough global survey of problems and the perspectives of interested and affected parties is urgently needed. Such a global perspective is essential for rational planning in countries or regions.

Effective management of plant invasions at a site entails the integration of approaches for dealing with species that have already spread over large areas and for assessing other alien species already present in an area but perhaps not showing signs of invasion. Also critically important is the need to screen new introductions to identify species that have a high risk of invading if introduced.

Biotechnology has been proposed as a way of reducing problems of invasiveness by controlling flowering and thereby reducing or eliminating seed production in forestry plantations (Meilan et al. 2001; Strauss et al. 2001). The use of this approach for agroforestry trees warrants careful consideration, taking into account problems of both feasibility (small-scale, often noncommercial agroforestry with many different tree species as opposed to the large-scale use of a few species in commercial plantation forestry) and desirability (e.g., dependency of farmers on external seed suppliers, poorly understood environmental risks of genetic engineering).

Because the invasion-related problems faced by agroforestry are not unique, it seems logical to strive for coordinated efforts in the various plant-related enterprises that rely heavily on alien species. Efforts in commercial forestry have already been mentioned (Richardson 1998a, 1998b, 1998c; Rouget et al. 2002). Attempts are also being made to understand the various pathways affecting invasions as a result of aquaculture (Naylor et al. 2001) and horticulture (Reichard and White 2001), both of which also rely heavily on alien species. Agroforestry is lagging behind, and a clear strategy for dealing with the increasing problem of invasive species is urgently needed.

Acknowledgments

We thank the following people for useful discussions and for comments on various drafts of the chapter: Chris Buddenhagen (Charles Darwin Research Station, Galapagos Islands), Colin Hughes (Department of Plant Sciences, University of Oxford), Jeffrey A. McNeely (IUCN, The World Conservation Union, Switzerland), Andrew Mitchell (Australian Quarantine and Inspection Service), and Sue Milton (University of Stellenbosch, South Africa).

Appendix 15.1

A summary of generalizations regarding alien plant invasions, based on information reviewed in Rejmánek et al. (2004). Not all these points apply equally to all types of plants or in all situations.

1. Which taxa invade?
   1.1. Stochastic approach
      1.1.1. The probability of invasion success increases with initial population size and the number of introduction attempts (propagule pressure).
      1.1.2. Long residence time improves the chances of invasion.
      1.1.3. Many or most plant invasions are preceded by a lag phase that may last many decades.
   1.2. Empirical, taxon-specific approach
      1.2.1. If a species is invasive anywhere in the world, there is a good chance that it will invade similar habitats in other parts of the world.
      1.2.2. Among invasive plants, some families (Amaranthaceae, Brassicaceae, Fabaceae, Hydrocharitaceae, Papaveraceae, Pinaceae, Poaceae, and Polygonaceae) are significantly overrepresented.
   1.3. The role of biological characters
      1.3.1. Fitness homeostasis (maintenance of fitness over a range of environmental conditions) is an important determinant of invasiveness.
      1.3.2. Genetic change can facilitate invasions, but many species have sufficient phenotypic plasticity to exploit new environments.
      1.3.3. Small genome size (SGS) has value as an indicator of invasiveness in closely related taxa (SGS seems to be a result of selection for short minimum generation time, and because it is also associated with small seed size, high leaf area ratio, and high relative growth rate of seedlings in congeners, it may be an ultimate determinant or at least an indicator of invasiveness).
      1.3.4. Several characters linked to reproduction and dispersal are key indicators of invasiveness (e.g., simple or flexible breeding systems, small seed mass, short juvenile period, short intervals between large seed crops, long flowering and fruiting periods).
      1.3.5. Seed dispersal by vertebrates is implicated in many plant invasions.
      1.3.6. Low relative growth rate of seedlings and low specific leaf area (the ratio of leaf area to leaf dry mass) are good indicators of low plant invasiveness in many environments.
      1.3.7. Large native range is an indicator of potential invasiveness.
      1.3.8. Vegetative reproduction is responsible for many plant invasions.
      1.3.9. Alien taxa are more likely to invade a given area if native members of the same genera (and family) are absent, partly because many herbivores and pathogens cannot switch to phylogenetically distant taxa.
      1.3.10. The ability to use generalist mutualists (seed dispersers, pollinators, mycorrhizal fungi, nitrogen-fixing bacteria) greatly improves an alien taxon's chances of becoming invasive.
      1.3.11. Efficient competitors for limiting resources are likely to be the best invaders in natural and seminatural ecosystems.

1.3.12. Characters favoring passive dispersal by humans (e.g., small, soil-stored seeds) greatly improve an alien plant taxon's chance of becoming invasive.

1.4. Environmental compatibility

1.4.1. Climate matching is a useful first step in screening alien species for invasiveness.

1.4.2. Resource enrichment and release, often just intermittent (e.g., exceptionally wet years, canopy opening through logging, fire), initiate many invasions.

1.4.3. Propagule pressure (see 1.1.1) can override biotic or abiotic resistance of a community to invasion.

1.4.4. Determinants of invasibility (macro-scale climate factors, microclimatic factors, soils, and various community or ecosystem properties) interact in complicated ways; therefore, evaluation of invasibility must always be context-specific.

1.5. Relationship between species richness and invasibility

1.5.1. At the landscape scale, invasibility seems to be positively correlated with native plant species richness; at smaller (neighborhood) scales, the correlation seems to be negative.

2. How fast?

2.1. Spread is determined primarily by reproduction and dispersal, but various extrinsic factors interact with these factors to mediate spread rates.

2.2. Spread rates based on local dispersal mechanisms (e.g., wind, birds, or mammals) greatly underestimate spread potential.

2.3. Rare, long-distance dispersal (often via mechanisms that cannot be predicted from an assessment of the ecology of a species) is hugely important for explaining population growth and spread over medium and long time scales.

3. Impact

3.1. Predicting the impact of invasive alien plants is much more difficult than predicting invasiveness.

3.2. Alien species that add a new function (e.g., nitrogen fixation) to an invaded ecosystem are much more likely to have big impacts than those that merely alter existing resource use levels.

4. Control, contain, or eradicate?

4.1. Early detection and initiation of management can make the difference between being able to use feasible offensive strategies (eradication) and the need to retreat to a more expensive defensive strategy (e.g., mitigation, containment).

References

Agriculture & Resource Management Council of Australia and New Zealand, Australian & New Zealand Environment & Conservation Council and Forestry Ministers. 2001. *Weeds of national significance. Mesquite (*Prosopis *species) strategic plan.* Launceston, Australia: National Weeds Strategy Executive Committee.

Anonymous. 1997. Prosopis: a desert resource or a menace? *National Museums of Kenya Horizons* 1:14.

Anonymous. 1999. EWSS 6th conference discussed invasive weeds. *AgriTopia* 14:15–16.

Barnes, R. D., D. L. Filer, and S. L. Milton. 1996. *Acacia karroo:* monograph and annotated bibliography. Tropical Forestry Paper 32. Oxford, UK: Oxford Forestry Institute.

Baumer, M. 1990. *The potential role of agroforestry in combating desertification and environmental degradation with special reference to Africa.* Wageningen, the Netherlands: CTA.

Beer, J., R. Muschler, D. Kass, and E. Somarriba. 1998. Shade management in coffee and cacao plantations. *Agroforestry Systems* 38:139–164.

Binggeli, P., J. B. Hall, and J. R. Healey. 1998. *An overview of invasive woody plants in the tropics.* School of Agricultural and Forest Sciences Publication No. 13. Bangor: University of Wales. Online: http://www.bangor.ac.uk/~afs101/iwpt/web1-99.pdf.

Board on Science and Technology for International Development (BOSTID). 1980. *Firewood crops.* Washington, DC: National Academy of Sciences.

BOSTID (Board on Science and Technology for International Development). 1983. *Firewood crops,* Vol. 2. Washington, DC: National Academy of Sciences.

CABI. 2000. *Forestry compendium global module,* CD-ROM. Wallingford, UK: CAB International.

Carter, J. O. 1994. *Acacia nilotica:* a tree legume out of control. Pages 338–351 in R. C. Gutteridge and H. M. Shelton (eds.), *Forage tree legumes in tropical agriculture.* Wallingford, UK: CAB International. Online: http://www.fao.org/ag/AGP/AGPC/doc/Publicat/Gutt-shel/x5556e0v.htm.

Chamberlain, J. R. (ed.). 2001. Calliandra calothyrsus: *an agroforestry tree for the humid tropics.* Tropical Forestry Paper 40. Oxford, UK: Oxford Forestry Institute.

Chamberlain, J. R., and A. J. Pottinger. 1995. Genetic improvement of *Calliandra calothyrsus.* Pages 250–257 in *Nitrogen fixing trees for acid soils, workshop proceedings.* Morrilton, AR: Winrock International and Nitrogen Fixing Tree Association.

Coppen, J. J. W. 1995. *Gums, resins and latexes of plant origin.* Rome: Food and Agriculture Organization of the United Nations

Crooks, J. A. 2002. Characterizing ecosystem-level consequences of biological invasions: the role of ecosystem engineers. *Oikos* 97:153–166.

Cshurhes, S., and R. Edwards. 1998. *Potential environmental weeds in Australia. Candidate species for preventative control.* Canberra: Environment Australia, Canberra.

Demel Teketay. 2000. Facts and experiences on eucalypts in Ethiopia and elsewhere: ground for making wise and informed decision. *Walia* 20:25–46.

Dennill, G. B., D. Donnelly, K. Stewart, and F. A. C. Impson. 1999. Insect agents used for the biological control of Australian *Acacia* species and *Paraserianthes lophantha* (Willd.) Nielsen (Fabaceae) in South Africa. *African Entomology Memoir* 1:45–54.

Dijkman, M. J. 1950. *Leucaena:* a promising soil erosion control plant. *Economic Botany* 4:337–349.

Durr, P. A. 2001. The biology, ecology and agroforestry potential of the raintree, *Samanea saman* (Jacq.) Merr. *Agroforestry Systems* 51:223–237.

Elevitch, C. R., and K. Wilkinson. 2001. *The overstory book*. Holualoa, HI: Permanent Agriculture Resources.

El Fadl, M. A. 1997. *Management of* Prosopis juliflora *for use in agroforestry systems in the Sudan*. Ph.D. thesis, Department of Forest Ecology, University of Helsinki, Finland.

Ellstrand, N. C., and K. A. Schierenbeck. 2000. Hybridization as a stimulus for the evolution of invasiveness in plants? *Proceedings of the National Academy of Sciences* 97:7043–7050.

Ewel, J. J., D. O'Dowd, J. Bergelson, C. C. Daehler, C. M. D'Antonio, L. D. Gomez, D. R. Gordon, R. J. Hobbs, A. Holt, K. R. Hopper, C. E. Hughes, M. LaHart, R. B. Leakey, W. G. Lee, L. L. Loope, D. H. Lorence, S. M. Louda, A. E. Lugo, P. B. McEvoy, D. M. Richardson, and P. M. Vitousek. 1999. Deliberate introductions of species: research needs. *BioScience* 49:619–630.

Forsyth, G. G., D. M. Richardson, P. J. Brown, and B. W. van Wilgen. In press. A rapid assessment of the invasive status of *Eucalyptus* species in two South African provinces. *South African Journal of Science* (in press).

Gade, D. W. 1976. Naturalization of plant aliens: the volunteer orange in Paraguay. *Journal of Biogeography* 3:269–279.

Geary, T. F. 1983. Casuarinas in Florida (USA) and some Caribbean Islands. Pages 107–109 in S. J. Midgley, J. W. Turnbull, and R. D. Johnston (eds.), *Casuarina: ecology, management and utilization*. Melbourne: Commonwealth Scientific and Industrial Research Organization.

Gold, A. G. 1999. From wild pigs to foreign trees: oral histories of environmental change in Rajasthan. Pages 20–58 in S. T. Madsen (ed.), *State, society and the environment in South Asia*. Richmond, VA: Curzon.

Greaves, A., and P. S. McCarter. 1990. Cordia alliodora: *a promising tree for tropical agroforestry*. Tropical Forestry Paper 22. Oxford, UK: Oxford Forestry Institute.

Hall, J. B., D. P. Aebischer, H. F. Tomlinson, E. Osei-Amaning, and J. R. Hindle. 1996. Vitellaria paradoxa: *a monograph*. Bangor: University of Wales.

Hall, J. B., and A. McAllan. 1993. Acacia seyal: *a monograph*. Bangor: University of Wales.

Hall, J. B., H. F. Tomlinson, P. I. Oni, M. Buchy, and D. P. Aebischer. 1997. Parkia biglobosa: *a monograph*. Bangor: University of Wales.

Hall, J. B., and D. H. Walker. 1991. Balanites aegyptiaca: *a monograph*. Bangor: University of Wales.

Harding, G. B., and G. C. Bate. 1991. The occurrence of invasive *Prosopis* species in the north-western Cape, South Africa. *South African Journal of Science* 87:188–192.

Henderson, L. 2001. *Alien weeds and invasive plants. A complete guide to declared weeds and invaders in South Africa*. Plant Protection Research Institute Handbook 12. Pretoria, South Africa: Plant Protection Research Institute, Agricultural Research Council.

Higgins, S. I., and D. M. Richardson. 1999. Predicting plant migration rates in a changing world: the role of long-distance dispersal. *American Naturalist* 153:464–475.

Holm, L., J. V. Pancho, J. P. Herberger, and D. L. Plucknett. 1979. *A geographical atlas of world weeds*. New York: Wiley.

Hughes, C. E. 1994. Risks of species introductions in tropical forestry. *Commonwealth Forestry Review* 73:243–252.

Hughes, C. E. 1998. *Leucaena. A genetic resources handbook*. Tropical Forestry Paper 37. Oxford, UK: Oxford Forestry Institute.

Hughes, C. E., and R. J. Jones. 1998. Environmental hazards of *Leucaena*. Pages 61–70 in H. M. Shelton, R. C. Gutteridge, B. F. Mullen, and R. A. Bray (eds.), Leucaena:

adaptation, quality and farming systems. Canberra: Australian Centre for International Agricultural Research.

Hughes, C. E., and B. T. Styles. 1989. The benefits and risks of woody legume introductions. *Monographs in Systematic Botany, Missouri Botanic Garden* 29:505–531.

Hulme, P. E. (n.d.). Prosopis *invasion: implications for the biodiversity of Caatinga in Northeast Brazil.* Online: http://www.dur.ac.uk/Ecology/phrpro.htm.

Huxley, P. 1999. *Tropical agroforestry.* Oxford, UK: Blackwell Science.

Impson, F. A. C., V. C. Moran, and J. H. Hoffmann. 1999. A review of the effectiveness of seed-feeding bruchid beetles in the biological control of mesquite, *Prosopis* species (Fabaceae), in South Africa. *African Entomology Memoir* 1:81–88.

Jadhav, R. N., M. M. Kimothi, and A. K. Kandya. 1993. Grassland mapping monitoring of Banni, Kachchh (Gujarat) using remotely-sensed data. *International Journal of Remote Sensing* 14:3093–3103.

Jhala, Y. V. 1993. Predation on blackbuck by wolves in Velavadar National Park, Gujarat, India. *Conservation Biology* 7:874–881.

Kanmegne, J., and A. Degrande. 2002. From alley cropping to rotational fallow: farmers' involvement in the development of fallow management techniques in the humid forest zone of Cameroon. *Agroforestry Systems* 54:115–120.

Keighery, G. J., N. Gibson, K. F. Keneally, and A. A. Mitchell. 1995. Biological inventory of Koolan Island, Western Australia 1. Flora and vegetation. *Records of the Western Australian Museum* 17:237–248.

Ledgard, N. J., and E. R. Langer. 1999. *Wilding prevention. Guidelines for minimizing the risk of unwanted wilding spread from new plantings of introduced conifers.* Christchurch: New Zealand Forest Research Institute Limited.

Macdonald, I. A. W., C. Thébaud, W. A. Strahm, and D. Strasberg. 1991. Effects of alien plant invasions on native vegetation remnants on La Réunion (Mascarene Islands, Indian Ocean). *Environmental Conservation* 18:51–61.

McNeely, J. A., H. A. Mooney, L. E. Neville, P. Schei, and J. K. Waage (eds.). 2001. *A global strategy on invasive alien species.* Gland, Switzerland: International Union for the Conservation of Nature and Natural Resources (IUCN).

Meilan, R., A. M. Brunner, J. S. Skinner, and S. H. Strauss. 2001. Modification of flowering in transgenic trees. Pages 247–256 in N. Morohoshi and A. Komamine (eds.), *Molecular breeding of woody plants.* Amsterdam: Elsevier Science.

Meyer, J.-Y. 2000. Preliminary review of the invasive plants in the Pacific Islands (SPREP Member Countries). Pages 85–114 in G. Sherley (ed.), *Invasive species in the Pacific: a technical review and draft regional strategy.* Samoa: South Pacific Regional Environmental Programme.

Michon, G., and H. de Foresta. 1995. The Indonesian agro-forest model. Pages 90–106 in P. Halladay and D. A. Gilmour (eds.), *Conserving biodiversity outside protected areas.* Gland, Switzerland: IUCN.

Mooney, H. A., and R. J. Hobbs (eds.). 2000. *Invasive species in a changing world.* Washington, DC: Island Press.

Naylor, R. L., S. L. Williams, and D. R. Strong. 2001. Aquaculture: a gateway for exotic species. *Science* 294:1655–1656.

NRC (National Research Council). 1983a. Calliandra: *a versatile small tree for the humid tropics.* Washington, DC: National Academy Press.

NRC (National Research Council). 1983b. Mangium *and other fast-growing acacias for the humid tropics.* Washington, DC: National Academy Press.

Pasiecznik, N. M., P. Felker, P. J. C. Harris, L. N. Harsh, G. Cruz, J. C. Tewari, K. Cadoret, and L. J. Maldonado. 2001. *The* Prosopis juliflora–Prosopis pallida *complex: a monograph.* Coventry, UK: HDRA Consultants.

Poynton, R. J. 1990. The genus *Prosopis* in southern Africa. *South African Forestry Journal* 152:62–66.

Randall, R. P. 2002. *A global compendium of weeds.* Meredith, Victoria, Australia: R. G. & F. J. Richardson.

Reichard, S. H., and P. White. 2001. Horticulture as a pathway of invasive plant introduction in the United States. *BioScience* 51:103–113.

Rejmánek, M., and D. M. Richardson. 1996. What attributes make some plant species more invasive? *Ecology* 77:1655–1661.

Rejmánek, M., D. M. Richardson, S. I. Higgins, M. J. Pitcairn, and E. Grotkopp. 2004. Ecology of invasive plants: state of the art. In H. A. Mooney, J. A. McNeely, L. Neville, P. J. Schei, and J. Waage (eds.), *Invasive alien species: searching for solutions.* Washington, DC: Island Press.

Richardson, D. M. 1998a. Commercial forestry and agroforestry as sources of invasive alien trees and shrubs. Pages 237–257 in O. T. Sandlund, P. J. Schei, and A. Viken (eds.), *Invasive species and biodiversity management.* Dordrecht, the Netherlands: Kluwer.

Richardson, D. M. 1998b. Forestry trees as invasive aliens. *Conservation Biology* 12:18–26.

Richardson, D. M. 1998c. Invasive alien trees: the price of forestry. In *Invaders from planet Earth. World Conservation* 4/97–1/98:14–15. Gland, Switzerland: IUCN.

Richardson, D. M., N. Allsopp, C. M. D'Antonio, S. J. Milton, and M. Rejmánek. 2000. Plant invasions: the role of mutualisms. *Biological Reviews* 75:65–93.

Richardson, D. M., R. M. Cowling, and D. C. Le Maitre. 1990. Assessing the risk of invasive success in *Pinus* and *Banksia* in South African mountain fynbos. *Journal of Vegetation Science* 1:629–642.

Richardson, D. M., and S. I. Higgins. 1998. Pines as invaders in the Southern Hemisphere. Pages 450–473 in D. M. Richardson (ed.), *Ecology and biogeography of* Pinus. Cambridge, UK: Cambridge University Press.

Richardson, D. M., I. A. W. Macdonald, J. H. Hoffmann, and L. Henderson. 1997. Alien plant invasions. Pages 535–570 in R. M. Cowling, D. M. Richardson, and S. M. Pierce (eds.), *Vegetation of Southern Africa.* Cambridge, UK: Cambridge University Press.

Richardson, D. M., P. Pysek, M. Rejmánek, M. G. Barbour, D. F. Panetta, and C. J. West. 2000. Naturalization and invasion of alien plants: concepts and definitions. *Diversity and Distributions* 6:93–107.

Rouget, M., D. M. Richardson, J. A. Nel, and B. W. van Wilgen. 2002. Commercially-important trees as invasive aliens—towards spatially explicit risk assessment at a national scale. *Biological Invasions* 4:397–412.

Sala, O. E., S. F. Chapin III, J. J. Armesto, E. Berlow, J. Bloomfield, R. Dirzo, E. Huber-Sanwald, L. F. Huenneke, R. B. Jackson, A. Kinzig, R. Leemans, D. M. Lodge, H. A. Mooney, M. Oesterheld, N. LeRoy Poff, M. T. Sykes, B. H. Walker, M. Walker, and D. H. Wall. 2000. Global biodiversity scenarios for the year 2100. *BioScience* 287:1770–1774.

Salim, A. S., A. J. Simons, A. Waruhiu, C. Orwa, and C. Anyango. 1998. *Agroforestree database,* CD-ROM. Nairobi, Kenya: International Centre for Research in Agroforestry.

Schroth, G., D. Kolbe, P. Balle, and W. Zech. 1996. Root system characteristics with agroforestry relevance of nine leguminous tree species and a spontaneous fallow in a semideciduous rainforest area of West Africa. *Forest Ecology and Management* 84:199–208.

Sharma, R., and K. M. M. Dakshini. 1998. Integration of plant and soil characteristics and the ecological success of two *Prosopis* species. *Plant Ecology* 139:63–69.

Stewart, J. L., G. E. Allison, and A. J. Simons (eds.). 1996. Gliricidia sepium: *genetic timberlake resources for farmers.* Tropical Forestry Paper 33. Oxford, UK: Oxford Forestry Institute.

Strauss, S. H., P. Coventry, M. M. Campbell, S. N. Pryor, and J. Burley. 2001. Certification of genetically modified forest plantations. *International Forestry Review* 3:85–102.

Tavares, F. C., J. Beer, F. Jiménez, G. Schroth, and C. Fonseca. 1999. Experiencia de agricultores de Costa Rica con la introducción de árboles maderables en plantaciones de café. *Agroforestería en las Américas* 6:17–20.

Thaman, R. R., C. R. Elevitch, and K. M. Wilkinson. 2000. *Multipurpose trees for agroforestry in the Pacific Islands.* Agroforestry Guides for Pacific Islands no. 2. Holualoa, HI: Permanent Agriculture Resources.

Tiver, F., M. Nicholas, D. Kriticos, and J. R. Brown. 2001. Low density of prickly acacia under sheep grazing in Queensland. *Journal of Range Management* 54:382–389.

Tiwari, J. K., and A. R. Rahmani. 1999. An army of mad trees: profile of a tree that is changing the face of Kutch. *Down to Earth* 7:32–34. Online: http://www.oneworld.org/cse/html/dte/dte990415/dte_cross1.htm.

Tolfts, A. 1997. *Cordia alliodora:* the best laid plans. . . . *Aliens* 6:12–13.

Van Klinken, R. D., and S. D. Campbell. 2001. The biology of Australian weeds. 37. *Prosopis* L. species. *Plant Protection Quarterly* 16:2–20.

van Orsdol, K. G. 1987. *Buffer zone agroforestry in tropical forest regions.* Washington, DC: Forestry Support Program, U.S. Agency for International Development.

Versfeld, D. B., and B. W. van Wilgen. 1986. Impact of woody aliens on ecosystem properties. Pages 239–246 in I. A. W. Macdonald, F. J. Kruger, and A. A. Ferrar, *The ecology and management of biological invasions in Southern Africa.* Cape Town, South Africa: Oxford University Press.

Wittenberg, R., and M. J. W. Cock. 2001. *Invasive alien species: a toolkit of best prevention and management practices.* Wallingford, UK: CAB International on behalf of the Global Invasive Species Programme.

Young, A. 1997. *Agroforestry for soil management.* Wallingford, UK: CAB International.

Chapter 16

Diseases in Tropical Agroforestry Landscapes: The Role of Biodiversity

Ulrike Krauss

Diverse land use systems can contribute significantly to biodiversity conservation in tropical landscapes (see Part III, this volume). Furthermore, functional diversity can help to stabilize ecosystems against disturbance, including pest and disease outbreaks (Schmidt 1978). Although diseases are common in natural plant communities (Allen et al. 1999), epidemics in undisturbed habitats usually subside without leaving behind devastation. In contrast, anthropogenic disturbance and associated diseases can overwhelm even highly diverse ecosystems.

Agroforestry practices are particularly well suited to create functionally diverse landscapes consisting of a range of land use systems with a diverse and conspicuous tree component. Landscape diversity and especially the presence of trees can influence disease dynamics in several ways. Landscape mosaics can influence the spread of disease propagules by wind, water, and animal vectors. The spatial patterns of host and nonhost vegetation may influence the buildup of inoculum but also of biological control agents. Diversified landscapes may also harbor wild populations of plants that provide the genetic resources for resistance breeding programs for cultivated species. Depending on the relative importance of the effects of landscape diversity and increased tree cover, crops and trees in agroforestry landscapes may either be more or less exposed to disease risks than in less diversified landscapes. Consequently, arguments related to crop disease pressures and consequent risks of crop failure may either support or oppose the promotion of diversified land use systems for conservation purposes.

To judge the potential of agroforestry landscapes for the conservation of biodiversity, it is important to know their impact on the production of agricultural and forestry commodities and, especially in the context of smallholder livelihoods, on income security. In the tropics, the latter depends strongly on the risk of disease outbreaks, as can be seen, for example, with the catastrophic socioeconomic and concomitant environmental effects of the witches' broom (*Crinipellis perniciosa*) epidemic on Bahian cocoa (Trevizan 1996).

The aim of this chapter is to review the available information regarding the effect of landscape mosaics, particularly the presence of agroforestry elements, on disease dynamics in cultivated and wild plants. The mutual association of agroforestry elements, crops, and wild plants offers opportunities to exploit the positive effects of landscape diversity on plant health such as increased system stability and reduced losses but can also entail disease risks for both cultivated and wild plants. Recommendations on how to avoid these risks will also be presented.

The Symbiosis-Disease Continuum

If plant-infecting microorganisms reduce the productivity, fertility, or longevity of a plant, they are called pathogens; the symptoms provoked by them are the disease. An infection that is beneficial not only to the microbe but also to the plant is called symbiosis. However, the distinction between disease and symbiosis is not always as clear-cut as one may expect. Some microbes can either be detrimental or beneficial to the infected plant, depending on the circumstances. Examples are mycorrhizal fungi. A surprisingly high proportion of early infections by mycorrhizae kills the infected plant, especially in small-seeded species that depend on mycorrhiza formation for successful germination. Surviving plants benefit through increased uptake of nutrients and water and protection of plant roots from soilborne diseases (Harley 1959).

In other cases, infection by microbes may be beneficial for the plant but reduce its utilitarian value. This is the case with some endophytes, that is, fungi that grow asymptomatically in plants. Endophytes often are closely related to pathogens and may have evolved from them via extension of a latency (i.e., symptom-free) period and reduction of virulence (Saikkonen et al. 1998). Infection of temperate grasses by certain endophytes is known to contribute to longevity and vigor of the fungus and also of the plant through protection against herbivory (Carroll 1988). Although advantageous to the plant, endophyte infection may be either desirable or undesirable for its user. For example, toxicosis in livestock, hoof gangrene (Saikkonen et al. 1998), and grass sickness in horses (Lacey 1975) have been attributed to toxins produced by pasture endophytes. Another example are heart rots of trees, which destroy the inner core of their stems and render them unmarketable but may actually benefit the tree and possibly the ecosystem. Resources of the inner, nonfunctional stem tissue are recycled while the flexible and living outer cylinder exhibits increased resistance to storm damage. The hollow stem cylinder left by *Fomitopsis pinicola* provides a good habitat for ectomycorrhizal fungi and nitrogen-fixing bacteria (Rayner 1995).

On the other hand, some endophytes increase the economic value of the plant. Taxol, the world's first billion-dollar anticancer drug, is produced in the bark of the Pacific yew (*Taxus brevifolia*) in its natural habitat, Montana, in the presence of the endophytic fungus *Taxomyces andreanae* and also by *Pestalotiopsis microspora,* an endophyte isolated from Nepalese yew (*Taxus wallachiana*) from the Himalayan foothills. Other fungi isolated from yews planted outside their home ranges failed to produce taxol at levels that could be exploited (Strobel and Long 1998). Thus, the habitat seems to play a role in the expression of the symbiosis.

Plant Disease Dynamics in Agroforestry Landscapes

The population dynamics of plant-infecting microbes, and thus the spread of diseases, are influenced by two main factors: by the way the organism is dispersed and by landscape patterns. Whereas the former is largely intrinsic to the pathogen, the latter can be influenced by agroforestry techniques. The extent to which this is possible is outlined in this section.

Principally, canopy stratification and climatic conditions govern the movement of infectious propagules in the landscape. Host density and accessibility determine the outbreak of secondary infections. Microbial immigration and emigration are the main driving forces of pathogen population dynamics, and thus the spread of disease, on the landscape scale. Commonly, wind- and water-dispersed organisms are distinguished, but the transition is gradual and the two media often play a synergistic role in disease epidemiology. For example, rain plays a pivotal role in spore deposition even for principally wind-dispersed organisms as it removes particles, including spores, from the air (Fitt et al. 1989).

In still air, splash dispersal is of localized importance only. If a susceptible host is surrounded by nonhost species, diversity in an agroforestry system can slow disease progress. This may partly explain reports of reduced disease in crops located in some high-interface configurations such as alley cropping and contour hedgerows. For example, in an alley cropping experiment with *Sesbania sesban* in the Rwanda highlands, the progress of maize rust (*Puccinia sorghii*) at the tree-crop interface was reduced by 24 percent (Yamoah and Burleigh 1990). Stem canker (*Puccinia cordiae*) of *Cordia alliodora* is important in monoculture plantations but not in silvopastoral or other agroforestry systems (Greaves and McCarter 1990). Mixed cropping and multiline varieties (Browning and Frey 1969), with high within-field diversity, can help to minimize the risk of losses by providing a mixture of resistant and susceptible cultivars to any one pathogen race present in the system.

In other cases, however, beneficial effects of diversity may be masked by microclimatic alterations such as increased humidity and prolonged leaf wetness, leading to higher disease incidence at the interface (Schroth et al. 1995, 2000). In forest clearings 30–50 m in diameter in the Congo, humidity remained at an average of 91 percent (Geiger 1961), a microclimate highly conducive to disease development. Therefore, in mosaic landscapes, such as smallholder fields surrounded by forest, disease progress could be more rapid than in a landscape with less tree cover once the pathogen has gained access to a patch of a susceptible crop via immigration, although pathogen spread from one field to the next may be reduced by the tree component.

In the presence of wind, waterborne inoculum can travel great distances. *Xanthomonas campestris* in cotton is splash-dispersed to a distance rarely exceeding 1 m horizontally from its source in still air but as far as 10 m with windspeeds of only 2.5 m s^{-1} (Fitt et al. 1989). For disease progress in citrus, wind speeds greater than 8 m s^{-1} and inoculum concentrations of at least 10^5 propagules mL^{-1} are needed. The magnitude of these values means that *X. campestris* can be controlled by agroforestry structures such as windbreaks, which reduce windspeeds and thereby reduce the distances traveled by inoculum (Gottwald and Timmer 1995).

Wind dispersal of pathogens is more important at the landscape scale than splash dispersal because of its further reach. Gusty winds are most effective in spore detachment. The speed of gusts in canopies can be several times higher than the average ambient windspeed (Aylor 1990). Windbreaks have very limited effectiveness in preventing spore detachment because of localized gusts, although the average windspeed is lowered. In fact, the denser the canopy (i.e., the more efficient the windbreak at the macro scale), the more turbulent eddies are, which can mediate spore detachment. After detachment, a substantial proportion of spores, especially those that originate in the lower canopy, are deposited close to the source. If a susceptible host is encountered, this can lead to intense, clustered infection foci. Low host densities can mitigate this effect. Thus, intercropping can prevent the formation of these intense foci. However, the host plant density in a plantation is of little importance to wind-driven spread throughout the whole system because wind dispersal is on the order of kilometers rather than meters. Associating cocoa clones differing in their susceptibility to the wind-dispersed witches' broom fungus did not reduce the incidence of the disease (Evans 1998), and combining rubber (*Hevea brasiliensis*) clones susceptible to different isolates of the wind-dispersed South American leaf blight fungus (*Microcyclus ulei*) was equally ineffective in controlling its spread in a plantation (Junqueira et al. 1989). This indicates that associations of these species with other non–host tree or crop species would be equally ineffective in limiting the spread of such highly mobile propagules, at least on the plantation scale.

Up to 50 percent of detached spores can escape from the canopy and can be carried large distances by horizontal winds. Ascospores of *Mycosphaerella fijiensis,* causing black Sigatoka of banana (*Musa* spp.), have been monitored over distances exceeding 50 km, and no significant dilution of inoculum was observed in a 4-km radius from a single inoculum source (Calvo and Romero 1998). Spores of coffee rust (*Hemileia vastatrix*) were detected at heights of 100 m and at horizontal distances of 150 km from the nearest diseased coffee (Meredith 1973). This high mobility of wind-dispersed spores renders the manipulation of wind dispersal of plant pathogens not amenable to agroforestry planting designs.

Typically water-dispersed microorganisms can become windborne in the form of aerosols when the water droplets in which they are suspended evaporate (Fitt et al. 1989). Aerosol transport is more important in pathogen spread on the landscape scale than rain splash because fine mists remain suspended in the air for longer periods of time and therefore are subjected to more cumulative wind. Aerosols are particularly important in the upward movement of infectious propagules. Rising fog is believed to be responsible for the effective uphill spread of white pine blister rust (*Cronartium ribicola*) in mountainous terrain in the United States (Martin 1944). In contrast to truly windborne inoculum, aerosols can be intercepted effectively by windbreaks. Given a fog density of 0.8 g m^{-3} and a windspeed of 4 m s^{-1}, the fog interception at a forest edge is approximately 0.5 mm h^{-1}, six times more than would precipitate over grassland (Geiger 1961). Thus, hedgerows and vegetation belts on the windward side of a field can intercept much aerosol-borne inoculum and, if resistant species are chosen, protect susceptible species on the leeward side.

Pathogens often are transported by other organisms such as humans and wildlife. Such vectoring is more difficult to predict and control on a landscape scale than purely physical dispersal. Wildlife and humans can carry pathogens over large distances and circumvent both natural and human-made barriers such as mountain chains, vegetation buffer strips, and windbreaks. Connecting forest corridors help protect wildlife (see Chapter 3, this volume), and vertebrate diversity often is considered a positive indicator of ecosystem health. However, domestic and wild animals not only can damage crops directly but also can vector diseases in apparently intermittent patterns that defy the rules of wind and water dispersal. Feathers and fur trap fungal spores very effectively, and soil adheres to feet. For example, feral pigs disseminated *Phytophthora cinnamomi* in Hawaii (Wallace 1978), and birds have been implicated in the spread of blister blight (*Exobasidion vexans*) of tea, chestnut blight (*Cryphonectria parasitica*), citrus dieback (*Deuterophoma tracheiphila;* Warner and French 1970), and coconut bud rot (Johnston 1912). If the productive agroforestry component experiences increased losses because of pests and diseases transmitted by wildlife, this could discourage the conservation of forest corridors unless other benefits outweigh the risks.

Whereas pathogen spread by wildlife may be an unavoidable consequence of biodiversity-friendly landscapes, pathogen dispersal by humans often can be prevented by planning and education. Pereira (1996) reported spread of

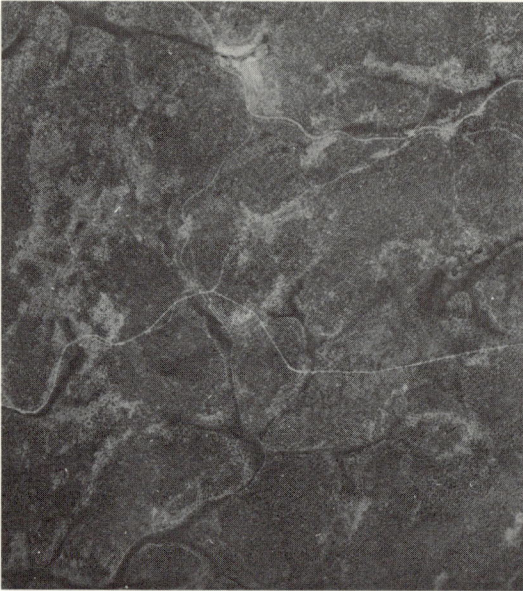

Figure 16.1. Aerial photograph of an Australian jarrah forest dominated by *Eucalyptus marginata* and *Eucalyptus calophylla* in the upper stratum and *Banksia grandis* in the understory. Medium gray patches of granular texture are healthy stands. Pale gray areas of blotchy texture are affected by dieback, caused by *Phytophthora cinnamomi,* and follow roads (white lines) and drainage canals. Dark gray, densely matted areas are swamps and riparian flats with field-resistant *Eucalyptus megacarpa* and *Eucalyptus patens* (reproduced from Podger 1972, with permission of the American Phytopathological Society).

outbreaks of the typically wind-dispersed witches' broom fungus in cocoa downstream along riverbanks in Bahia at a rate of 2 km per month. He attributed the atypical water-associated transport to farmers' disposing of diseased material in rivers. This spread could be prevented by burrowing or burning the brooms, which entails additional farmer labor. Podger (1972) presented compelling aerial photographs of jarrah (*Eucalyptus* spp.) dieback, caused by *P. cinnamomi,* following logging roads and drainage canals in Australia (Figure 16.1). This soilborne pathogen was most effectively vectored on heavy-duty vehicles. Therefore, in areas with known foci of an aggressive disease, road planning and community development should consider epidemiological aspects. This obviously entails a multidisciplinary approach to land management.

Cross-Infection between Agroforestry Plants and Native Flora

The native flora in and around agroforestry areas can host numerous pathogens that are usually overlooked for various reasons. First, diseases of wild plants receive little attention except when the plants are subject to extractive use. Second, infections by many viruses, but also some fungi and bacteria, are difficult to detect because they have long latent periods in wild hosts, and symptoms high in tree canopies are not easily observed even if present. If pathogens of the spontanecus flora threaten crops and agroforestry trees, this would discourage the conservation of natural vegetation in agroforestry land-scapes. Similarly, if infective inoculum can build up on cultivated plants and spread to wild plants, this would threaten the native flora and counteract con-servation efforts. In this section, I analyze the mutual risks the cultivated and spontaneous flora pose to each other and provide recommendations as to how these risks can be minimized.

A large proportion of plant pathogens are highly host specific, but some possess a wide host range. Viruses and some soilborne pathogens and nema-todes are particularly nonspecific. Such broad host range parasites have the ability to cross-infect between spontaneous and cultivated flora, with poten-tially pernicious effects on either. Cross-infection is more common between closely related species, and such associations carry a higher risk, but the selec-tion of unrelated species for agroforestry associations is no guarantee against disease losses. Examples of both scenarios have been compiled by Schroth et al. (2000).

However, the most serious economic losses and ecosystem destruction are almost invariably associated with anthropogenic disturbance such as deforesta-tion or the introduction of exotic plants or pathogens. Clearing forest alters habitats and thereby influences disease dynamics. For example, cucumber mosaic virus often infects bananas in new plantings where aphid vectors have been disturbed by destruction of a preferred wild host (Ploetz et al. 1994). If leaving belts of natural vegetation can alleviate the negative effect of deforesta-tion on disease dynamics, this would encourage the preservation of at least part of the forest.

Exotic Species and Cross-Infection

Whereas major crops and their pathogens have been moved between conti-nents for thousands of years, germplasm movement of agroforestry trees is more recent but remains a high-risk decision (see Chapter 15, this volume). The most damaging movement of exotic germplasm results when the host plant is co-introduced with its pathogen from the common center of origin to an area lacking coevolved biocontrol agents. This happened with the Califor-nian *Cupressus macrocarpa* and *Pinus radiata,* which were first promoted in East Africa but are no longer planted because of canker (*Rhynchosphaeria cupressi*) and needle blight (*Dothistroma pini*), respectively (Boa 2000).

But even with a carefully conducted introduction with adequate quaran-tine measures, the same historic pattern is often observed: the exotic species performs well for years in its new habitat, but at some time a disease erupts and escalates into a second-order epidemic, the most expansive kind, which is typical for exotic species (Zadoks and van der Bosch 1994). Within a few years after the introduction of *Eucalyptus grandis* and *Eucalyptus saligna* for a refor-estation project in Suriname, these trees succumbed to a regional pathogen,

Endothia havanensis, with devastating mortality rates of 90–100 percent (Boerboom and Maas 1970). *Cordia alliodora* is a popular agroforestry tree worldwide and faces few serious problems outside its native range in tropical America. However, on the island of Pentecost in Vanuatu, infection by the fungus *Phellinus noxius* is particularly severe. This has been attributed to dense stands of an alternative host, *Myristica fatua,* in the local forest (Greaves and McCarter 1990). Several exotic tree and fruit crops that are popular in agroforestry worldwide acquired pernicious diseases from symptomless or unidentified forest hosts in their new habitat. Examples include South American leafspot (*Mycena citricolor*) on coffee (Sequeira 1958) and Moko disease (*Ralstonia solanacearum* race 2) of banana, plantain, and bluggoe (*Musa* spp.) in the Americas (Ploetz et al. 1994); Sumatra disease (*Pseudomonas syzygii*) of cloves (*Syzygium aromaticum*) in Sumatra and Java (Lomer et al. 1992); ring spot virus of passion fruit (Fauquet and Thouvenel 1980); and cocoa swollen-shoot virus in West Africa, where outbreaks in cocoa plantations were most severe near the boundary with forest reserves (Posnette et al. 1950). Thus, the introduction of exotic germplasm harbors a substantial risk acquiring a previously unknown disease from the local forest, with a devastating economic effect on the exotic species.

On the other hand, in some cases the introduction of exotic germplasm is successful for many years. For example, 74,000 rubber seeds collected from a single area in Amazonia in 1876 gave rise to all the rubber in Southeast Asia (Dean 1987), where the high-risk speculation of cultivating this crop in vast monocultures continues to pay off. In the 1930s, "improved" high-yielding varieties were brought back to the Americas, where they failed because of the South American leaf blight fungus (*Microcyclus ulei*), a pathogen absent from Asia and not considered in the breeding program (Imle 1978).

The Spontaneous Flora at the Receiving End

It is a tenet of conservation that agroforestry practices do not threaten the native flora. However, introduced pathogens can seriously affect the indigenous flora in ecosystems disturbed by human activity. The most spectacular example is the fungus *Phytophthora cinnamomi,* a pantropical, human-vectored pathogen of Southeast Asian origin. It is presumed to have entered a Mexican village on mango seedlings, and within 12 years, thousands of trees in a 300-ha mixed oak forest were dead, and nance (*Brysonima crassifolia*), a woody understory species, had all but disappeared (Tainter et al. 2000). When this pathogen was accidentally introduced to Australia, it converted an evergreen forest into floristically impoverished and open vegetation of greatly reduced productivity. Jarrah, which once provided 70 percent of raw materials for the forest industry and protected watersheds, died at a rate of 12,000 ha per year (Podger 1972). Ten years later, more than 200,000 ha had disappeared, and jarrah was restricted to the overstory of ridges and divides (Shea et al. 1983). Sclerophyllous woodland and shrubland and moorland heath assemblages were obliterated in favor of a few resistant species such as *Melaleuca parviflora* and *Nuytsia floribunda*. No other anthropogenic disturbance, including logging and fire, has had a comparably dramatic effect on speciation in Australia (Podger 1972).

These examples indicate that international germplasm movement, even under the theoretical assumption of effective quarantine, is a high-risk speculation in the long term. Initially, exotic species often can be produced with better results than in their native range or than native species because there are no coevolved pathogens. However, once a disease outbreak occurs, it is often catastrophic, with potentially serious repercussions for the cultivated and the native, spontaneous flora. Therefore, the use of exotic agroforestry species should be discouraged in favor of local species, especially in buffer zones and other ecologically sensitive areas. When exotic germplasm is unavoidable or introductions are already ongoing, a responsible research and development project should not rely solely on government institutions to implement existing laws but should procure its own containment facilities where seedlings can be observed for longer periods of time before transplanting or go through quarantine facilities in a third (temperate) country. The repeated introduction of different strains of the same species increases pathogen diversity and thus renders control more difficult, as has been shown for *Sphaeropsis sapinea* populations co-introduced repeatedly on *Pinus radiata* into South Africa (Burgess et al. 2001). Furthermore, education efforts are needed to caution rural developers with good intentions but limited ecological knowledge about the risks associated with introduced planting material (see also Chapter 15, this volume).

In Situ Conservation

Wild plants and their pathogens coevolve with the constant selection pressure of disease resistance, whereas cultivated plants often are selected with little regard to their disease susceptibility. Therefore, pathogens on wild plants tend to be particularly aggressive, and once they succeed in jumping onto a cultivated species with fewer defense mechanisms, pathogen populations can explode and lead to devastating epidemics.

Plants and their pathogens have a particularly high genetic diversity in their center of origin because the plants constantly adapt to new, aggressive pathogen races. Therefore, the center of origin of a crop provides a source of dynamically evolving resistance genes. This phenomenon is best investigated for subtropical cereals. In the Middle East, the genetic center of barley, oats, and wheat, several rusts and mildews, which are normally highly host-specific, infected not only the cereal host but also numerous distantly related grasses and members of the Hyacinthaceae and Rhamnaceae in cropland and grassland adjacent to shrubland (Browning 1974). Because new pathogen races can also arise on the alternative host, the cereal population is particularly challenged to evolve new resistance mechanisms. As a result, this area is a superb source of a variety of resistance genes in natural multiline populations. Browning (1974) suggests that live gene pools should be established and protected for all major crops in their center of origin. In situ conservation entails conservation of the pathogen against which resistance is being sought (Allen et al. 1999). No comparable examples for agroforestry trees are documented, but it seems logical to assume that a huge untapped potential for in situ conservation of resistance genes also exists in forests and other natural ecosystems, and these provide additional arguments for conservation areas (O'Neill et al. 2001). The present narrow genetic base of selected agroforestry species is cause for concern (Boa 1998).

Extensification and Abandonment of Agricultural Land

Both the extensification of production and the abandonment of a plantation can lead to the uncontrolled buildup of pathogen inoculum. This probability is higher in tropical perennial crops, which are a continuous source of inoculum, and in highly diversified, low-risk production systems because the grower is less obliged to dedicate effort to managing the crop (and diseases) if, for instance, prices decline. An example is the abandonment of more than 50 percent of the area under cocoa in Peru in response to frosty pod (Servicio Nacional de Sanidad Agraria 2000). The causal fungus, *Crinipellis roreri,* can produce up to 7 billion wind-dispersed spores over a period of 9 months on a single pod remaining suspended in the canopy because of lack of phytosanitation and can readily infect nearby trees (Evans et al. 1977). Witches' brooms in abandoned cocoa and cupuaçu (*Theobroma grandiflorum*) plantations are another example. Where the abandonment of part of the managed land is desired for conservation purposes (see Chapter 7, this volume) but the same crops are produced elsewhere in the region, it may be necessary to eliminate certain plant species from the system before abandonment to prevent them from serving as pathogen sources. This strategy can be successful only if no wild hosts of the pathogen exist in the adjacent forest, such as wild Sterculariaceae in the case of *C. roreri.*

Biological Control

Biological control is the practice or process whereby the undesirable effects of an organism are reduced through the activity of another organism that is not the host plant, the pathogen, or humans (Deacon 1983). Natural biocontrol prevents destructive epidemics in undisturbed environments. Currently, plant pathologists exploit the principles of natural biocontrol to protect crops and trees. If promising biocontrol agents are to be preserved for possible future use, it will be important to conserve their habitats.

Exotic plants can develop into serious weeds that aggressively invade crops or replace the indigenous flora when introduced into new habitats that lack their natural enemies. Classic biocontrol has successfully been used to curtail weed populations below the threshold at which they threaten the native flora, as in the biocontrol of water hyacinth (*Eichhornia crassipes*) by host-specific weevils. This aquatic weed of South American origin is now a cosmopolitan throughout tropical wetlands (Wittenberg and Cock 2001). In classic biocontrol, coevolved pests or pathogens are obtained in the center of origin of the exotic weed and released into the new habitat to restore the plant-parasite balance (Evans 1999). This strategy also entails in situ conservation of both pathogen and host.

Many epiphytic and endophytic microorganisms contribute to natural biocontrol of plant diseases. Such natural biocontrol is particularly common in systems with perennial plants and high microbial diversity (Allen et al. 1999). Populations of antagonistic microbes are augmented on cultivated plants to shift the balance in favor of the saprophyte (inundative biocontrol). According to Thurston (1998), fragmented landscapes with a large interface between planted and spontaneous flora facilitate the immigration of biocontrol agents, such as native epiphytes and endophytes, and reduce losses caused by disease. In temperate regions, elm trees infected with the endophyte *Phomopsis oblonga* are avoided by the elm bark beetle (*Scolytus* spp.), which transmits the deadly Dutch elm disease (*Ophiostoma ulmi;* Webber 1981). Similarly, the endophyte *Lophodermium congenum* excludes the pathogen *Lophodermium seditiosum* from needles of Scots pines (Carroll 1988). Increasing evidence suggests that endophytes may also protect tropical woody perennials against disease (Arnold 1999). Endophyte transmission in woody perennials usually is horizontal via water-dispersed spores (Carroll 1988), and infection depends strongly on the surrounding flora (Rodrigues et al. 1995). The result can be an extraordinarily high degree of endophyte diversity in tropical woody perennials (Arnold et al. 2000). Similarly, nonpathogenic epiphytes on orange leaves significantly increased when the trees were grown downwind from plant species other than citrus. Epiphyte populations were highest on the windward side of the orchard and declined with distance from the other plant species (Lindow and Andersen 1996). Thus, preserving natural forest around agroforestry plots can be instrumental in biological disease control.

Whereas earlier biocontrol approaches searched for "superstrains" from a large initial collection of isolates, recent strategies make increasing use of mixtures of antagonists that can cover a wider range of environmental conditions and diverse pathogen populations. Mixed inocula were consistently superior in controlling banana crown rot than the best single strain (Krauss et al. 2001) and led to the highest yield increase in cocoa afflicted simultaneously by frosty pod, witches' broom, and black pod (*Phytophthora* sp.; Krauss and Soberanis 2001). According to Boa (2000), agroforesters should aim for a high degree of plant structural complexity (habitat diversity) and functional complexity (trophic links and nutrient cycling) to achieve the more complex and stable diversity needed to reduce the risk from pests and diseases rather than merely to reduce the risk of crop failure.

Plant biodiversity may be pivotal in maintaining a high functional diversity of potentially useful plant-associated microorganisms. Given the increasing interest in coevolved pathogens, endophytes, and antagonists, conservation of the biodiversity of host plants in their natural habitat and research into plant-microbe interactions are gaining importance, with likely applications in agriculture and biotechnology.

Conclusions

Agroforestry offers a range of tools to reduce the risk of disease losses, achieve sustainable production, and aid conservation. These tools must be applied wisely. Indiscriminate use of agroforestry techniques, such as the introduction of susceptible exotic tree and crop germplasm, can be highly counterproductive. In contrast, creating a heterogeneous agroforestry landscape consisting of a multitude of plant species with a strong presence of woody perennials can help to limit disease losses by slowing the spread of water- and aerosol-dispersed pathogens and preventing the formation of intense infection foci of wind-dispersed pathogens, especially if arranged in high interface configurations. Compatible plant species must be selected carefully for this purpose so that pathogen propagules produced on one species do not encounter alternative hosts in their vicinity. Natural forest surrounding agroforestry plots can also lead to disease reduction by hosting naturally occurring biocontrol agents and by luring virus vectors away from the crop. However, the adjacent forest can also create a microclimate favorable for disease development and be a source of vertebrate pests and pathogen vectors.

Abandonment of agricultural areas, for economic or conservation reasons, can augment the inoculum of wind-dispersed pathogens and pose a threat to surrounding farms. The dissemination of wind-dispersed propagules is not readily curbed by agroforestry techniques such as windbreaks, and direct intervention to eradicate certain susceptible crop species from abandoned areas may be needed. However, agroforestry can help to spread risks of disease losses in surrounding agricultural areas by producing a variety of crops that do not share the same pathogens. Obviously, the optimum design of land use mosaics with abandoned and productive areas takes a multidisciplinary team effort.

The most destructive disease epidemics are the result of anthropogenic disturbance of a natural ecosystem. We have to learn from mistakes committed during the globalization of crops for centuries, such as the worldwide spread of many cereal pathogens, to avoid repeating these errors with trees that we started domesticating only recently. When designing agroforestry landscapes, planners should use native species because devastating epidemics of both the planted and the wild flora can be associated with movement of exotic germplasm. Therefore, germplasm movement has implications for the sustainability of production and conservation of native flora. If exotic species are necessary or already present in the system from past introductions, utmost care should be practiced not to co-introduce an exotic pathogen with its host (e.g., in follow-up importations).

There is still a need for basic ecological research into habitat-plant-microbe interactions, particularly on the landscape scale. We need to understand what determines whether a microorganism acts as a pathogen or a symbiont. Our knowledge of endophytes, especially those of tropical perennials, is still in its infancy. Emerging evidence suggests that these microorganisms have potential as biocontrol agents and as pharmaceuticals. Future research should investigate how populations of beneficial microbes can be manipulated for increased productivity and reduced risk of crop failure in agroforestry landscapes. Similarly, in situ conservation offers an underexplored source of resistance genes for plant breeding and of coevolved pathogens for the biocontrol of invasive alien weeds. With a holistic approach to the entire ecosystem, in situ conservation can add economic value to protected areas.

Decision makers in rural development should take a holistic and multidisciplinary approach to landscape planning. Disciplines apparently unrelated to plant pathology, such as road construction and wildlife management, can have a profound impact on disease dynamics in agroforestry landscapes and therefore the sustainability and acceptability of recommendations and conservation measures.

Acknowledgments

This chapter was prepared during an alternative crop and diversification project funded by the United States Department of Agriculture through CABI Bioscience and the Tropical Agricultural Research and Higher Education Center (CATIE), Costa Rica. The constructive comments of an anonymous reviewer, Götz Schroth, and Celia Harvey were invaluable. I also wish to thank many colleagues for useful discussion and support, especially Betsy Arnold, Roy Bateman, Eric Boa, Harry Evans, Julie Flood, André George, Allen Herre, Jean Lodge, Serge Savary, Frank Tainter, and Martijn ten Hoopen.

References

Allen, D. J., J. M. Lenné, and J. M. Waller. 1999. Pathogen biodiversity: its nature, characterization and consequences. Pages 123–153 in D. Wood and J. M. Lenné (eds.), *Agrobiodiversity. Characterization, utilization and management.* Wallingford, UK: CAB International.

Arnold, A. E. 1999. *Sustainable cocoa: the fungal community component.* Online: http://sun1.oardc.ohio-state.edu/cocoa.sustain.htm.

Arnold, A. E., Z. Maynard, G. S. Gilbert, P. D. Coley, and T. A. Kursar. 2000. Are tropical fungal endophytes hyperdiverse? *Ecology Letters* 3:267–274.

Aylor, D. E. 1990. The role of intermittent wind in the dispersal of fungal pathogens. *Annual Review of Phytopathology* 28:73–92.

Boa, E. R. 1998. Diseases of agroforestry trees: do they really matter? *Agroforestry Forum* 9:19–22.

Boa, E. R. 2000. *Tree health and agroforestry.* Final Technical Report. London: DIFD Crop Protection Programme, R7499.

Boerboom, J. H. A., and P. W. T. Maas. 1970. Canker of *Eucalyptus grandis* and *E. saligna* in Surinam caused by *Endothia havanensis. Turrialba* 20:94–99.

Browning, J. A. 1974. Reliance of knowledge about natural ecosystems to development of pest management programs for agro-ecosystems. *Proceedings of the American Phytopathological Society* 1:191–199.

Browning, J. A., and K. J. Frey. 1969. Multiline cultivars as a means of disease control. *Annual Review of Phytopathology* 7:355–382.

Burgess, T., B. D. Wingfield, and M. J. Wingfield. 2001. Comparison of genotypic diversity in native and introduced populations of *Sphaeropsis sapinea* isolated from *Pinus radiata. Mycological Research* 105:1331–1339.

Calvo, C., and R. Romero. 1998. Evaluación del gradiente de dispersión de la enfermedad de la Sigatoka negra del banano (*Musa* AAA). *Corbana* 23:51–56.

Carroll, G. 1988. Fungal endophytes in stems and leaves: from latent pathogen to mutualistic symbiont. *Ecology* 69:2–9.

Deacon, J. W. 1983. *Microbial control of plant pests and diseases.* Wokingham, UK: Van Nostrand Reinhold.

Dean, W. 1987. *Brazil and the struggle for rubber.* Cambridge, UK: Cambridge University Press.

Evans, H. C. 1998. Disease and sustainability in the cocoa agroecosystem. In *Proceedings of the First International Workshop on Sustainable Cocoa Growing, Panama City, 30 March to 2 April 1998.* Smithsonian Tropical Research Institute, Smithsonian Migratory Bird Center. Online: http://www.si.edu/smbc/evans.htm.

Evans, H. C. 1999. Classical biological control. Pages 29–37 in U. Krauss and P. Hebbar (eds.), *Research methodology in biocontrol of plant diseases with special reference to fungal diseases of cocoa. Workshop manual.* CATIE, Turrialba, Costa Rica.

Evans, H. C., D. F. Edwards, and M. Rodríguez. 1977. Research on cocoa diseases in Ecuador: past and present. *PANS* 23:68–80.

Fauquet, C., and J. C. Thouvenel. 1980. *Viral diseases of crop plants in Ivory Coast.* Documentation Techniques 46. Paris: ORSTOM.

Fitt, B. D. L., H. A. McCartney, and P. J. Walklate. 1989. The role of rain dispersal of pathogen inoculum. *Annual Review of Phytopathology* 27:241–270.

Geiger, R. 1961. *Das Klima der bodennahen Luftschicht.* Braunschweig, Germany: Friedrich Vieweg und Sohn.

Gottwald, T. R., and L. W. Timmer. 1995. The efficacy of windbreaks in reducing the spread of citrus canker caused by *Xanthomonas campestris* pv. *citri. Tropical Agriculture* 72:194–201.

Greaves, A., and P. S. McCarter. 1990. Cordia alliodora. *A promising tree for tropical agroforestry.* Tropical Forestry Paper 22. Oxford, UK: Oxford Forestry Institute.

Harley, J. L. 1959. *The biology of mycorrhizae.* London: Plant Science Monographs, Leonard Hill Ltd.

Imle, E. P. 1978. *Hevea* rubber: past and future. *Economic Botany* 32:264–277.

Johnston, J. R. 1912. The history and cause of coconut bud-rot. *USDA Bureau of Plant Industry Bulletin* 228:1–175.

Junqueira, N. T. V., L. Gasparotto, R. Lieberei, M. C. S. Normando, and M. I. P. M. Lima. 1989. Physiological specialization of *Microcyclus ulei* on different *Hevea* species: identification of pathotype groups. *Fitopatologia Brasileira* 14:147.

Krauss, U., P. Matthews, R. Bidwell, M. Hocart, and F. Anthony. 2001. Strain discrimination by fungal antagonists of *Colletotrichum musae:* implications for biocontrol of crown rot of banana. *Mycological Research* 105:67–76.

Krauss, U., and W. Soberanis. 2001. Biocontrol of cocoa pod diseases with mycoparasite mixtures. *Biological Control: Theory and Application* 22:149–158.

Lacey, J. 1975. Airborne spores in pastures. *Transaction of the British Mycological Society* 64:265–281.

Lindow, S. E., and G. L. Andersen. 1996. Influence of immigration on epiphytic bacterial populations on navel orange leaves. *Applied and Environmental Microbiology* 62:2978–2987.

Lomer, C. J., S. J. Eden-Green, E. R. Boa, and Supriadi. 1992. Evidence for a forest origin of Sumatra disease of cloves. *Tropical Science* 32:95–98.

Martin, J. F. 1944. *Ribes* eradication efficiently controls white pine blister rust. *Journal of Forestry* 42:255–260.

Meredith, D. S. 1973. Significance of spore release and dispersal mechanisms in plant disease epidemiology. *Annual Review of Phytopathology* 11:313–342.

O'Neill, G. A., I. Dawson, C. Sotelo-Montes, L. Guarino, M. Guariguata, D. Current, and J. C. Weber. 2001. Strategies for genetic conservation of trees in the Peruvian Amazon. *Biodiversity and Conservation* 10:837–850.

Pereira, J. L. 1996. Renewed advance of witches' broom disease of cocoa: 100 years later. Pages 87–91 in *Proceedings of the 12th International Cocoa Research Conference,* November, 1996, Salvador, Bahía, Brazil.

Ploetz, R. C., G. A. Zentmyer, W. T. Nishijima, K. G. Rohrbach, and H. D. Ohr. 1994. *Compendium of tropical fruit diseases.* St. Paul, MN: APS Press.

Podger, F. D. 1972. *Phytophthora cinnamomi,* a cause of lethal disease in indigenous plant communities in Western Australia. *Phytopathology* 62:972–981.

Posnette, A. F., N. F. Robertson, and J. M. Todd. 1950. Virus diseases of cocoa in West Africa. V. Alternative host plants. *Annals of Applied Biology* 37:229–240.

Rayner, A. D. M. 1995. Fungi, a vital component of ecosystem function in woodland. Pages 231–254 in D. Allsopp, R. R. Colwell, and D. L. Hawksworth (eds.), *Microbial diversity and ecosystem function.* Wallingford, UK: CAB International.

Rodrigues, K. F., O. Petrini, and A. Leuchtermann. 1995. Variability among isolates of

Xylaria cubensis as determined by isozyme analysis and somatic incompatibility tests. *Mycologia* 87:592–596.

Saikkonen, K., S. H. Faeth, M. Helander, and T. J. Sullivan. 1998. Fungal endophytes: a continuum of interactions with host plants. *Annual Review of Phytopathology* 29:319–343.

Schmidt, R. A. 1978. Diseases in forest ecosystems: the importance of functional diversity. Pages 287–315 in J. G. Horsfall and E. B. Cowling (eds.), *Plant diseases: an advanced treatise,* Vol. 2: *How disease develops in populations.* New York: Academic Press.

Schroth, G., P. Balle, and R. Peltier. 1995. Alley cropping groundnut with *Gliricidia sepium* in Côte d'Ivoire: effects on yields, microclimate and crop diseases. *Agroforestry Systems* 29:147–163.

Schroth, G., U. Krauss, L. Gasparotto, J. A. Duarte Aguilar, and K. Vohland. 2000. Pests and diseases in agroforestry systems in the humid tropics. *Agroforestry Systems* 50:199–241.

Sequeira, L. 1958. The host range of *Mycena citricolor. Turrialba* 8:136–147.

Servicio Nacional de Sanidad Agraria. 2000. *The cocoa's moniliasis control program in Peru.* Proceedings of the Annual Meeting of the American Cocoa Research Institute, Miami, Florida, 7–8 February 2000.

Shea, S. R., B. L. Shearer, J. T. Tippett, and P. M. Deegan. 1983. Distribution, reproduction, and movement of *Phytophthora cinnamomi* on sites highly conducive to jarrah dieback in south Western Australia. *Plant Disease* 67:970–973.

Stobel, G. A., and D. M. Long. 1998. Endophytic microbes embody pharmaceutical potential. *ASM News* 64:263–268.

Tainter, F. H., J. G. O'Brien, A. Hernández, F. Orozco, and O. Rebolledo. 2000. *Phytophthora cinnamomi* as cause of oak mortality in the state of Colima, Mexico. *Plant Disease* 84:394–398.

Thurston, J. L. 1998. *Tropical plant diseases.* 2nd edition. St. Paul, MN: APS Press.

Trevizan, S. D. P. 1996. Mudanças no sul da Bahía associadas à vassoura-de-bruxa do cacau. Pages 1109–1116 in *Proceedings of the 12th International Cocoa Research Conference, November, 1996,* Salvador, Bahía, Brazil.

Wallace, H. R. 1978. Dispersal in time and space: soil pathogens. Pages 181–202 in J. G. Horsfall and E. B. Cowling (eds.), *Plant diseases: an advanced treatise,* Vol. 2: *How disease develops in populations.* New York: Academic Press.

Warner, G. M., and D. W. French. 1970. Dissemination of fungi by migratory birds: survival and recovery of fungi from birds. *Canadian Journal of Botany* 48:907–910.

Webber, J. 1981. A natural biological control of Dutch elm disease. *Nature* 292:449–451.

Wittenberg, R., and M. J. W. Cock (eds.). 2001. *Invasive alien species: a toolkit of best prevention and management practices.* Wallingford, UK: CAB International.

Yamoah, C. F., and J. R. Burleigh. 1990. Alley cropping *Sesbania sesban* (L) Merill with food crops in the highland region of Rwanda. *Agroforestry Systems* 10:169–181.

Zadoks, J. C., and F. van den Bosch. 1994. On the spread of plant disease: a theory of foci. *Annual Review of Phytopathology* 32:503–521.

PART V

Matrix Management in Practice: Agroforestry Tools in Landscape Conservation

The preceding parts of this book have reviewed the scientific basis, from both an economic and a biological perspective, for using agroforestry as a conservation tool, to complement the protection of natural ecosystems in the tropics. This part presents some practical examples of the use of agroforestry in conservation strategies in different tropical regions. It ends with a discussion of the potential applications of agroforestry in mitigating climate change effects on tropical biodiversity.

Chapter 17 describes the use of linear agroforestry plantings to buffer forest edges and the establishment of agroforestry patches as stepping stones between forest fragments in the highly deforested and fragmented Atlantic rainforest zone of São Paulo State, Brazil. The implementation of such agroforestry-for-conservation projects with farmers who have little tradition in forest resource management is emphasized.

Chapter 18 focuses on farm forestry plantings in the Atherton Tablelands of north Queensland, Australia, another humid tropical region that has suffered extensive forest clearing mainly for pasture. It discusses past experiences with farm forestry plantations and evaluates possibilities of increasing the biodiversity value of plantations with commercial timber species, thereby creating synergies between private production objectives and conservation benefits in a largely pasture-dominated landscape. The chapter also provides extensive recommendations for tree species selection for such plantings.

Increasing agricultural productivity while improving the habitat value of pasture areas, especially for migratory birds, is the focus of Chapter 19. This chapter reviews different options for increasing tree cover and thus the habitat value of tropical pasture landscapes and then discusses a silvopastoral practice that has spontaneously developed over large areas of Central America, with a focus on Nicaragua. This practice is based on a native tree species, *Acacia pennatula*. This species finds favorable germination and growth conditions in pastures, where local farmers retain and manage it because of the fodder value of its pods, and increases the food basis for a diversified bird fauna that feeds on arthropods living on the trees.

Finally, Chapter 20 reviews the potential role of agroforestry in mitigating the effects of climate change on tropical biodiversity as a component of climate change–integrated conservation strategies. As global temperature and moisture conditions change, plant and animal populations in remnant forest fragments or protected areas will become increasingly vulnerable to extinction unless they can adjust their ranges to include areas that meet their physiological needs. Agroforestry land uses in the matrix surrounding parks and natural habitats could increase the permeability of the matrix, enabling some species to adjust their home ranges and move into new habitats. In this manner, agroforestry practices could be a valuable complement to protected areas of sufficient size and covering a range of site conditions in long-term conservation strategies under conditions of climate change.

Chapter 17

Agroforestry Buffer Zones and Stepping Stones: Tools for the Conservation of Fragmented Landscapes in the Brazilian Atlantic Forest

Laury Cullen Jr., Jefferson Ferreira Lima, and Tiago Pavan Beltrame

The Brazilian Atlantic Forest (Mata Atlântica) is one of the most endangered ecosystems on the planet, at risk of wholesale destruction. When Europeans first arrived in Brazil in the sixteenth century, the highly diverse Atlantic forest covered 1 million km² of the eastern and southern coast, representing 12 percent of the Brazilian territory. These forests have been fragmented and reduced to about 7 percent of their original area (SOS Mata Atlântica and INPE 1993). The Mata Atlântica harbors great biological diversity, containing nearly 7 percent of the world's species, many of which are endemic and threatened with extinction.

The Atlantic forest domain can be subdivided into two major regions based on vegetation types and geographic features (Eiten 1974; Fonseca 1985). The first type, tropical evergreen mesophytic broadleaf forest, originally covered most of the Brazilian east slope extending to the coast. This type is found at low to medium elevations with mean annual precipitation around 2,000 mm and mean annual temperatures of 16°–19°C (Hueck 1972). The second major type, tropical semideciduous mesophytic broadleaf forest (Eiten 1974), extends to the western range of the coastal hills, stretching to the Plateau region. This vegetation type originally covered large areas of the states of Minas Gerais, Rio de Janeiro, São Paulo, and Paraná. Plateau forests (Mata de Planalto) occur in areas of lower annual rainfall (1,000–1,500 mm) with a pronounced dry season of 5–6 months, corresponding to the winter season, when average monthly rainfall is around 50 mm (Passos 1992). Despite lower precipitation, tall forests are still present, containing both evergreen and semideciduous species (Eiten 1974; Alonso 1977).

Currently, most of the remaining forest cover in the Mata Atlântica is found on hillsides along the coast. Very little forest remains in the Plateau region because agricultural and industrial expansion have resulted in the loss of more than 98 percent of the forest cover (Figure 17.1). Only about 280,000 ha remains of the most fragmented and threatened ecosystem of the Atlantic forest domain (SOS Mata Atlântica and INPE 1993; Dean 1995). Although many of these forest remnants are small (Table 17.1), they nevertheless support a very diverse flora and fauna (Quintela 1990), including one of the most endangered primates of the world, the black lion tamarin (*Leontopithecus chrysopygus*).

Nearly all of the Plateau forests that still exist are found in the Pontal do Paranapanema region, located in the western part of the State of São Paulo (Figure 17.2). This region alone comprises 84 percent of the remaining Plateau

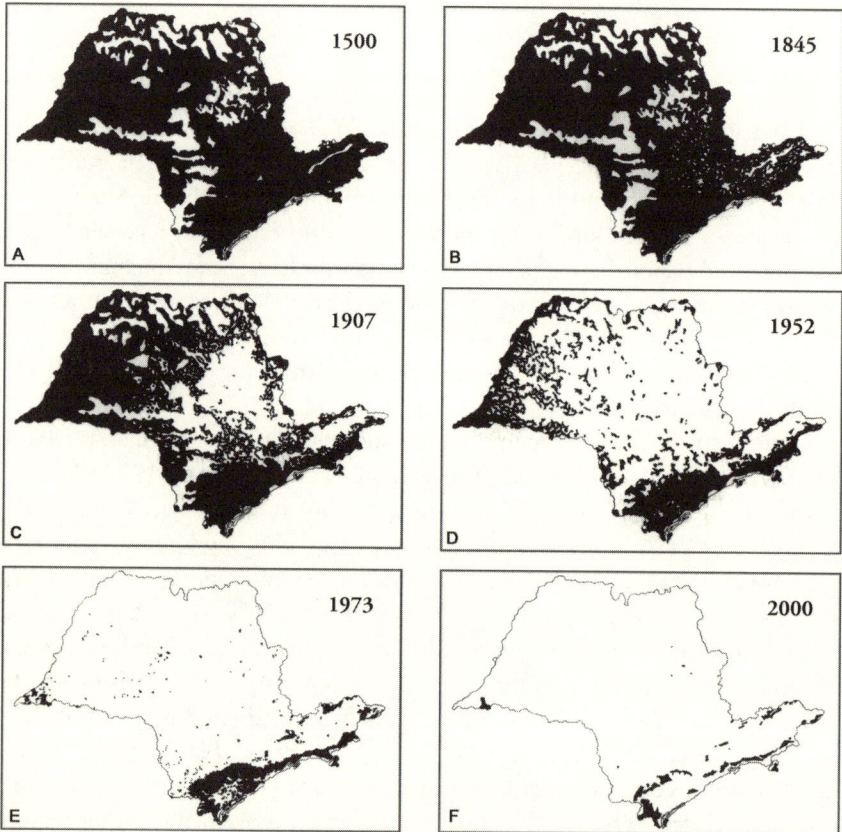

Figure 17.1. Depletion of the Atlantic Forest in the State of São Paulo, Brazil, from 1500 to 2000. Today only 8 percent of the original forest cover remains. In the Plateau region, 3 percent of the original forest remains, and most of the forest remnants are in the Pontal do Paranapanema Region (adapted from Shafer 1990).

Table 17.1. Size and number of protected forest fragments in the Plateau region of the State of São Paulo.

Size of Forest Patch (ha)	Number of Forest Patches
< 100	5
100–500	10
500–1,000	5
1,000–5,000	7
5,000–10,000	0
> 10,000	2
Total	29

Source: Viana and Tabanez (1996).

forest cover and is considered one of the poorest and most underdeveloped areas of the state. The majority of these forests are privately owned; protected reserves account for less than 1 percent of the total area of São Paulo State and officially protect 26 percent of the remaining forest (SOS Mata Atlântica and INPE 1993). Legislation demands that landholdings retain 20 percent of their land under the original forest cover; however, laws protecting forest fragments often are ineffective and beyond the enforcement capability of the state.

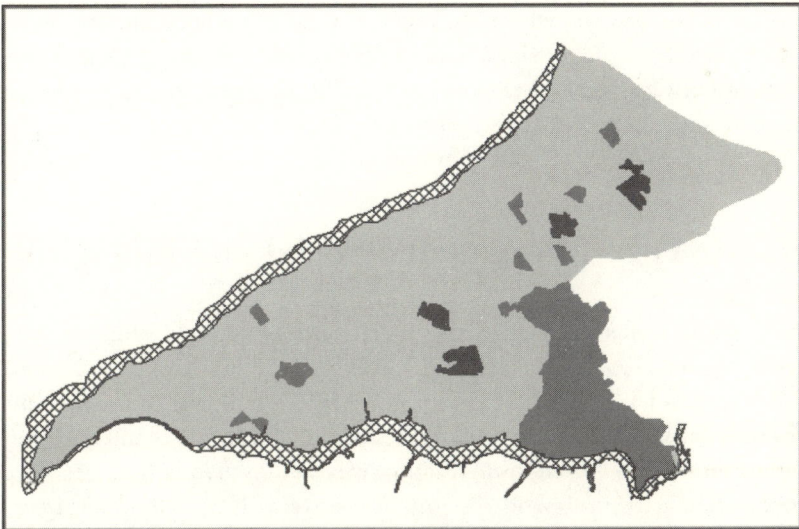

Figure 17.2. Pontal do Paranapanema Region in the State of São Paulo, Brazil. In right lower corner is the Morro do Diabo State Park (37,000 ha), surrounded by forest fragments. The patches in the map are the only forest fragments larger than 500 ha that remain in the region. The darker patches are forest fragments where the Instituto de Pesquisas Ecológicas concentrates most of its conservation activities, including the "green hugs" and stepping-stones.

Despite regulations to preserve Atlantic Forest fragments, activities of local people are depleting resources and accelerating environmental degradation, mainly because of current land reform and occupation patterns. The Pontal do Paranapanema is included in the Federal Decree 750 from 1993, which legally defines the Atlantic Forest and regulates its uses. Decree 750 prohibits deforestation in all primary Atlantic Forest in Brazil, although regulations are minimally enforced. Land concentration, speculation, and landlessness are the main causes of degradation in areas where traces of Atlantic Forest remain. This land tenure system results in the exploitation of forest remnants and threatens remaining habitat (Cullen et al. 2000; Cullen, Bodmer, and Valledares-Padua 2001; Cullen, Bodmer, Valledares-Padua, and Morato 2001).

Since 1997, the São Paulo state government, working with landowners, has developed a negotiation process in which the landowners donate 30–70 percent of cleared land to members of the Landless People's Movement (Movimento Sem Terra) of the Pontal do Paranapanema region in exchange for official titles to the remaining property (Cullen, Bodmer, Valledares-Padua, and Morato 2001). However, the land redistribution process lacked a comprehensive program to provide these newly landed families with the skills and technological assistance needed to make productive use of their small farms. Much of the donated land is marginal and borders on sensitive forest fragments. Consequently, forest edge perturbations are recognized as a main problem for the remaining forests in the Pontal. Illegal hunting and cattle grazing in the forests, the use of pesticides, and edge effects such as wind desiccation and invasion of the forest borders by fire, vines, and weeds degrade the forests close to rural communities. Over time, these impacts modify the forest structure, adversely affecting ecological processes and causing significant losses of animal and plant species (see also Chapter 2, this volume).

Atlantic Forest Fragments: Social and Biological Values

Recent publications emphasize the social and biological values of forest patches (Shafer 1995; Schelhas and Greenberg 1996; Turner and Corlett 1996; Viana and Tabanez 1996; Viana et al. 1997). Cultural and social values of forest patches often are particularly great among indigenous and traditional communities. Local communities that have existed for a long time near forest patches often have livelihood systems that are closely tied to the forest. In these cases forest fragments can be important economically, socially, and spiritually and are often valued, managed, and protected by local people (Jacobson 1995; Lyon and Horwich 1996). As Browder (1996) states, "Forest patches can function as social spaces shaped by human uses and values, seldom isolated and unused fragments of habitat, owing their permanence and existence to the value placed on them by local people" (288).

However, in other cases the benefits and use options of forest patches go unrecognized, and they are not considered socially and economically productive resources but mere physical spaces. This latter situation appears to prevail in the case of the Mata do Planalto, especially in the Pontal do Paranapanema region. Many forest patches exist on private land, where they are mostly regarded as unproductive areas. The social values placed on forest patches are small and are currently recognized mainly by private conservation institutions engaged in environmental education and conservation training (Padua 1991, 1997). About 70 percent of farmers in the Pontal are unfamiliar with lowland tropical forests and therefore lack traditional forest knowledge (Ferrari Leite 1998), and most have no tradition as subsistence hunters and gatherers. Around 20 percent are people who have spent most of their lives in urban centers, facing periods of marginal jobs and unemployment.

In contrast to their low social value, Atlantic Forest patches maintain high biological values for a variety of reasons:

- They protect regional biodiversity and provide source populations of forest fauna and flora for recolonization of deforested and degraded areas (Ditt 2000).
- They provide ecosystem services; for example, some forest fragments are gallery forests providing protection to riverbanks and watersheds and stabilizing potentially erodible soils.
- They may increase landscape connectivity by functioning as stepping stones for the dispersal of local organisms and provide wintering grounds for local and long-distance migratory birds (Powell and Bjork 1995; Greenberg 1996).
- They represent the only and last remaining building blocks of these endangered ecosystems that can be used for forest restoration programs.

Conservation of Rural Landscapes in the Atlantic Forest: Two Methodological Approaches

In response to the problems caused by fragmentation, innovative conservation strategies for highly fragmented rural landscapes, such as the Atlantic Forest, are an urgent conservation priority (see Chapter 1, this volume). Such strategies must define appropriate land uses that are both socially and ecologically sustainable. A useful approach to the development of such land uses is adaptability analysis, where farmers learn through participation in trials how best to implement new farming options and how to adapt them to specific local conditions and needs. On-farm researchers learn from these adaptation efforts how to design future technologies aimed at similar farmers and how to better interpret and extend the results of on-farm research (Hildebrand and Russel 1996).

In the following section we discuss two approaches that were specifically designed to meet the conservation needs and current land use problems in the Pontal region: agroforestry buffer zones and agroforestry stepping stones.

Agroforestry buffer zones are linear agroforestry plantings that are strategically located and designed to reduce the dependence of the farmers on forest resources and at the same time reduce edge effects by surrounding primary forest with forested systems instead of completely open pasture or cropland (see Chapter 2, this volume). Agroforestry stepping stones are plantings that are intended to reduce genetic isolation of forest fragments by promoting animal and plant dispersal between fragments (see Chapters 3 and 12, this volume).

Both approaches are used in a project focusing on one of the most important forest fragments in the Pontal region. Located adjacent to the Ribeirão Bonito settlement, this 350-ha forest fragment, designated as the legally stipulated 20 percent forest reserve of the settlement, functions as a corridor linking the 37,000-ha Morro do Diabo State Park to a 2,000-ha fragment, the Tucano Forest (Figure 17.2). This corridor is critical for maintaining the Tucano fragment as viable habitat and for movement of animals between the park and fragment. It is known to be used by three groups of peccaries (*Tayassu tajacu*), three ocelots (*Leopardus pardalis*), and two tapirs (*Tapirus terrestris*). Farms of newly settled members of the Landless People's Movement surround this forest corridor. Each family, averaging five people, owns approximately 18 ha of land. Half the land is used for growing subsistence and cash crops (e.g., maize, cotton, coffee, cassava, rice, and beans), and the other half is used to maintain dairy animals. These farms do not retain any forest vegetation, which is all concentrated in the 20 percent legal reserve. Because of soil constraints and lack of appropriate management practices and technical support, agricultural production is extremely low, and the majority of families struggle to meet basic needs.

Agroforestry Buffer Zones

Very little attention has been given to the potential role agroforestry systems might play in protecting forest fragments by serving as a buffer zone. Diversified agroforestry belts around forest fragments have only recently been considered as potential buffers for biodiversity reserves or as land bridges for fragmented habitats in the tropics (Wilson and Diver 1991; Gajaseni et al. 1996). Reforestation using agroforestry can promote habitat and species conservation and ensure community commitment to reforestation. Surrounding forest edges with agroforestry buffers instead of open pasture or cropland can help to reduce edge effects by creating an environment adjacent to forest fragments that is similar to forest habitat (Cullen, Bodmer, Valledares-Padua, and Morato 2001; Chapter 2, this volume).

The Green Hug Project (Projeto Abraço Verde, PAV) was initiated in 1997 by the Brazilian nongovernment organization Institute for Ecological Research (Instituto de Pesquisas Ecológicas, IPÊ). With technical assistance from the project, communities living around forest fragments work to establish an agroforestry buffer zone on the farms bordering forest fragments, specifically at the interface between the open land and the forest edge. At the same time, the Green Hug agroforestry systems are intended to raise family living standards and establish viable alternatives for income generation. The systems are designed to sustain a supply of firewood, timber, fruits, grains, and animal fodder to local farmers, thus relieving the pressure to advance further into forest fragments.

A buffer zone consists of a linear agroforestry planting (40–80 m wide and 1–2 km long) at the interface between a forest fragment and the open matrix (Figure 17.3a). In these buffer zones, trees are planted at a spacing of 3–4 m

Figure 17.3. (a) Agroforestry buffer (40–80 m wide and 1–2 km long) implemented at the interface between a forest fragment and the open matrix to reduce dependence of the farmers on the forest fragment for timber and nontimber products and reduce edge effects. (b) Native vegetation undergrowth developing under a planted agroforestry buffer strip of *Eucalyptus* intercropped with fruit trees and other multipurpose trees and shrubs.

within rows and 4–5 m between rows, providing 700 trees per hectare on average. In most systems, 50 percent of the trees planted are introduced *Acacia mangium* and *Eucalyptus* spp., and the remaining 50 percent are native timber and fruit trees (Table 17.2). Farmers also practice intercropping (maize, rice,

Table 17.2. Major crops and trees used in plantation crop combinations in the Pontal region of São Paulo, Brazil.[a]

Species used	Major Uses and Functions
CROPS	
Maize (*Zea mays*)	F, A
Pigeon pea (*Cajanus cajan*)	A, F, GM, N, SC
Cassava (*Manihot esculenta*)	F, A, M
Sweet potato (*Ipomoea batatas*)	F, SC
Pineapple (*Ananas comosus*)	A, M, SC
Papaya (*Carica papaya*)	F, A, M, FA
Cotton (*Gossypium* spp.)	FI, O
Groundnut (*Arachis hypogaea*)	F, M
TREES	
Exotic Tree Species	
Acacia (*Acacia mangium*)	FD, FW, N, PW, SB, SC, T
Eucalyptus (*Eucalyptus* spp.)	FW, CT, M, O, PW, SB, ST, T, FA
Native Tree Species	
Angico (*Anadenathera* sp.)	CT, FW, M, SB, SC, ST
Cedro (*Cedrela fissilis*)	CT, FW, M, SB, SC, ST
Gurucaia (*Peltophorum dubium*)	BF, CT, FA, FW, SC, ST
Inga (*Inga laurina*)	BF, CT, F, FW, N, ST, T, SB, FA
Inga liso (*Inga uruguensis*)	BF, CT, F, FW, N, ST, T, SB, FA
Jacarandá (*Jacaranda cuspidifolia*)	Or, FA, T, ST
Ipê (*Tabebuia* sp.)	BF, CT, FA, FW, M, OR, ST
Louro pardo (*Cordia trichotoma*)	BF, CT, N, SB, SC, ST
Monjoleiro (*Acacia polyphylla*)	BF, CT, SB, SC, ST
Mutambo (*Guazuma ulmifolia*)	A, BF, FA, FI, FW, GM, M, SC, ST
Sobrasil (*Columbrina glandulosa*)	CT, FW, SB, T
Tamboril (*Enterolobium contortisiliquum*)	CT, FA, OR, ST, SC
Fruit Trees	
Avocado (*Persea americana*)	F, SB, FA
Cashew (*Anacardium occidentale*)	F, SB, O, FW, ST, FA
Guava (*Psidium guajava*)	F, FA
Macadamia (*Macadamia integrifolia*)	F, O, SB
Mango (*Mangifera indica*)	F, A, SB, FA
Nectarine (*Prunus persica*)	F, SB, FA, ST
Orange (*Citrus sinensis*)	F, SB, FA
Tamarind (*Tamarindus indica*)	F, A, SC, FA, FW, T, SB

Abbreviations: A, animal feed; BF, bee forage; CT, construction timber; F, food (human consumption); FA, fauna use; FI, fiber; FW, fuelwood; G, gum; GM, green manure; M, medicine; N, nitrogen-fixing ability; O, oil; OR, ornamental; PC, pest control; PW, pulpwood; SB, shelterbelts; SC, soil conservation; ST, shade tree (over plantation crops); T, timber.

and beans) between the tree rows in the first years after planting. Observations suggest that, if established on degraded sites along the forest edge, these buffer zone plantations can act as catalysts for recolonization by the native flora through their influence on microclimate, soil fertility, suppression of dominant grasses (e.g., *Panicum, Imperata*), and habitat for seed-dispersing wildlife (Figure 17.3b). These effects are presently being studied in experimental plots within the Green Hug plantations.

In summary, the overall premise of the project approach is that by producing a regular supply of forest products such as firewood, timber, fruits, and fodder from the agroforestry buffer zones, while protecting forest edges, people will be able to live adjacent to forest fragments without incurring further loss of forest area or biodiversity.

Agroforestry Stepping Stones

Restoring ecological connectivity through private smallholdings between protected areas may be critical to ecoregional conservation efforts. One possible way to help mitigate the effects of fragmentation is to create continuous corridors between fragments (see Chapter 3, this volume). Stepping stones consisting of small agroforestry parcels that increase connectivity between forest fragments but are not necessarily connected to the forests can also contribute to the genetic flux of many species by allowing animal and plant dispersal to occur (see Chapter 12, this volume). Stepping stones allow the mixing of populations and the sharing of genes, thereby reducing problems of inbreeding depression and demographic and genetic stochasticity in fragmented populations (Gerlach and Musolf 2000; see Chapter 2, this volume).

In the last 3 years, 65 stepping stones of approximately 1 ha each have been created by the project to connect several of the forest fragments. stepping stones consist of homegardens and other agroforestry plantings such as small groups of trees, with an emphasis on flowering and fruiting trees in an essentially linear arrangement between the much larger fragments (Figure 17.4). The corridors enrich the local matrix dominated by completely deforested pasture land, enhancing local biodiversity and facilitating the movement of organisms between forest fragments. Individuals dispersing from the Morro do Diabo Park could also help to replenish the populations located in the smaller fragments. The use of the stepping stones by birds and bats is being monitored.

For certain groups, edge habitat provided by corridors and stepping stones could also serve as permanent habitat; for example, most butterflies are attracted to edge habitat because of the dominance of flowering plants with abundant nectar (Haddad 2000). Butterfly taxonomy and diversity have been well studied and described in the Morro do Diabo State Park. Of 426 butterfly species found in the park, 160 are common, and 134 of those species are

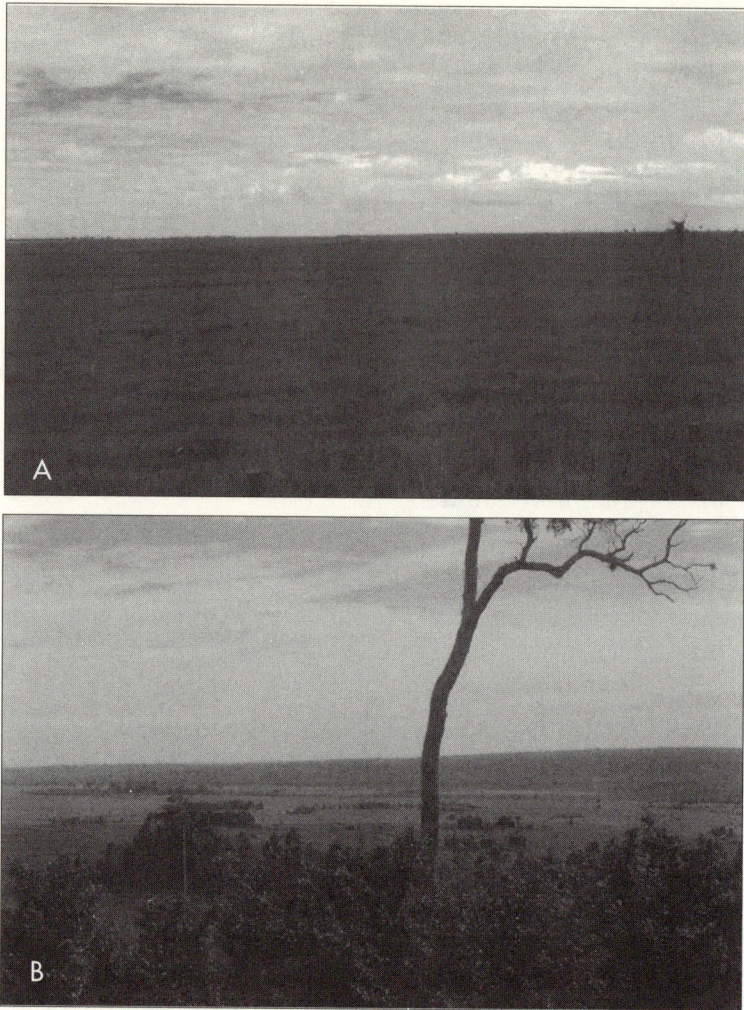

Figure 17.4. (a) A typical scenario in the Pontal do Paranapanema region, where large and totally deforested pasture lands dominate the landscape. (b) After the introduction of agroforestry stepping stones, consisting of small patches of trees with an emphasis on flowering and fruiting species, the matrix becomes less hostile to forest organisms, and connectivity between forest fragments increases.

specialists in disturbed areas of primary forest (Mielke and Casagrande 1997). These insects are highly vagile and conspicuous and disperse freely across small (100- to 300-m) gaps. By providing edge habitat, stepping stones may augment insect abundance and diversity, both locally and regionally. The charisma of butterflies makes them an important icon to advocate conservation.

Finally, we are only beginning to fathom the long-term effects on migratory bats, birds, and butterflies of having fewer nectar plants to forage and fewer safe roost sites available as stopovers in largely deforested areas (Buchmann and Nabhan 1997). News of declining birds and bees does not make many farmers happy, let alone those who grow a variety of crops that benefit from—and in some cases need—cross-pollination for high yields.

Community Involvement and Implementation

Because of the level of community support that has been built over the 4 years in which the project has operated, both project approaches have accomplished a great deal in this region. To date, 37 families, self-organized in small groups, are involved in every stage of the project, from training and extension to project planning, implementation, monitoring, and evaluation. Initial training and agroforestry extension are provided by short courses in which community members learn and experience the various benefits of agroforestry systems. Slides and video presentations show how agroforestry systems can improve microclimate (e.g., wind protection by shelterbelts), enhance nutrient cycling, and increase soil fertility and soil conservation. Benefits such as reduced pest and disease pressure and weed control are also discussed.

Farmers may reject agroforestry innovations because learning new technologies may be difficult, and positive results do not occur without some trial and error (Hildebrand and Russel 1996). Therefore, selection and design of agroforestry practices around forest fragments must be discussed carefully with each farmer. The goal is to plan *with* the farmers and not *for* the farmers. Through participatory diagnosis and design (Raintree 1990), collaborating farmers and researchers learn how best to implement new cropping options and how to modify and adapt them to specific local conditions to maximize benefits. It is important to note that only about 10–15 percent of the total land area belonging to a farmer is considered for agroforestry systems. However, the remaining farmed area, usually under subsistence and cash crops or dairy production, is likely to receive direct and indirect benefits from the agroforestry zone, such as soil conservation and wind protection.

Although no standard agroforestry prescriptions are imposed on the farmers, some general agroforestry practices are suggested. The farmers are then free to develop and adapt these practices to their own preferences and needs. General guidelines sometimes are necessary to direct each farmer during the design and implementation process, especially because the majority of small landholders have little experience with agroforestry practices. The following two agroforestry practices are likely to meet the conservation needs of the region as well as the farmers' preferences; that is, they have promise for enhancing agriculture and livestock production while protecting and conserving forest fragments:

- *Plantation crop combinations:* In heavily populated areas, farmers usually integrate annual crops and animal production with perennial crops, primarily to meet food needs. A list of crops grown in the region and those with potential for use in plantation crop combinations is presented in Table 17.2. The table also lists native and exotic trees suitable for agroforestry combinations that have potential for improving soil conditions and buffering forest edges. Although a large number of other trees and crops are available and have potential in agroforestry systems, the list presents only those most likely to perform well under local soil and climate conditions.

- *Silvopastoral systems:* In these systems, livestock graze on herbaceous plants grown under the shade of trees that are planted to provide shade to livestock, promote grass growth, and provide fodder or other products while sheltering forest edges (Payne 1985). For example, livestock may graze under *Eucalyptus* and *Acacia* plantations that specifically serve for timber and firewood production and land restoration. Silvopastoral systems have been widely used in temperate regions, and some combinations have shown great potential for use in the tropics (Payne 1985; Oliveira et al. 1986; Lima 1996).

Some alien species, including several *Acacia* and *Eucalyptus* species, may become invaders of natural and seminatural ecosystems (see Chapter 15, this volume); however, *Acacia* and *Eucalyptus* spp. have been widely planted for decades in the Pontal region but have not invaded forests or open landscapes. Because of the extremely high demand for timber and firewood in the region, these species have been kept in check by human harvest. Our observations also suggest that these species are shade intolerant and unlikely to germinate and grow in the dense, shaded understory of the forest fragments.

Community Promoters and Agroforestry Nurseries

The project carries out training courses and establishes community-based agroforestry nurseries in biologically important areas to supply the agroforestry plots. So far, the experience of agroforestry training courses has reached approximately 400 families settled on more than 12,000 ha of land. Educational programs and extension visits are conducted throughout the region.

Project staff found that an excellent way to enhance participation in agroforestry programs was to hire a member of each community as a liaison between the community and project staff. Called promoters, these people are chosen for their leadership skills and ability to organize and motivate others. All promoters are long-standing community members with strong service records. Promoters attend monthly meetings with project staff and are responsible for informing the community about project goals and activities. They also serve as information sources in the community, answering questions and dispelling misconceptions as needed.

With the assistance of promoters and the enthusiasm of community members, the project has successfully established 17 community agroforestry nurseries as sources of planting stock for use in buffer zones and stepping stones. It provides technical assistance and training in the construction and management of nurseries. In return, community members agree to plant at least 60 percent of their allotted tree seedlings in the project area; the remaining 40 percent can be used in other parts of the farm or sold in the local market.

Agroforestry has received much attention in the Pontal region as a promising form of sustainable land use that is adaptable to the needs of small-scale producers. The project hopes to develop and promote a local culture of agroforestry that relies on on-farm demonstration to convince farmers to participate in the program. Effective programs can begin by encouraging villagers to establish simple demonstration plots or experiments and to evaluate and share their results with others. Although they are in the early stages of development, these landscape conservation approaches already serve as examples of the ecological, social, and economic benefits of agroforestry.

Lessons Learned

According to the experiences of this project, the following elements are important in applying similar approaches in other regions facing similar challenges:

- Establish trust before initiating the process to ensure positive communication between all participants.
- Understand the needs of participants; be contextual.
- Keep it simple and flexible.

The issue of trust is an important element to many policy problems. Without trust, participants often are suspicious of others' actions and may be apprehensive about completely participating in the decision process. Prior environmental education programs in the community and collaboration with large landowners in ecological studies increased the visibility of the program in the communities and helped establish a high level of trust. Involving potential participants in less controversial activities can build trust and respect among all parties, which is especially important when participants' well-being and wealth are affected. The agroforestry buffers and stepping stones caused little controversy because they affected only a limited part of the farmers' land area, and the project helped in their implementation by providing tree seedlings and preparing the soil.

The next step requires understanding participants' needs. Community members will continue to participate only if the program meets their goals and addresses issues they believe are important. Once agroforestry was selected as the backbone of the program, further discussion was needed to understand which techniques would work best for the farmers (Cullen, Bodmer, Valladeras-Padua, and Morato 2001). By listening to the needs of the farmers,

IPÊ was able to introduce agroforestry techniques that were straightforward and provided quick returns.

Finally, it is important to keep the model simple and flexible so that farmers' suggestions can be incorporated easily. By starting small, a program can address problems as they develop. For example, the Green Hug Project started with 15 families and has grown to more than 30. We incorporated all families of the Landless People's Movement and large landowners into the decision process but selected a single forest fragment for the initial module. The project continues to grow, and more nurseries are being built to accommodate more farms and forest fragments. New techniques are disseminated quickly through the community advocate program.

Conclusions

Remaining forest fragments in the Plateau region of the State of São Paulo are sad reminders of the once widespread Atlantic Forest. Protecting these fragments is the only way to ensure the survival of many forest species and the long-term conservation of the ecosystem. In an approach that combines recent results from conservation biology and sustainable land use methods, the project described in this chapter attempts to introduce innovative management schemes into the Atlantic Forest region that could help to create a more secure future for both farmer communities and remaining natural ecosystems. Although the project is still young and no definite conclusions can be drawn, its overall approach seems to hold promise for other highly fragmented tropical landscapes also.

Acknowledgments

This project is funded by grants from the Beneficia Foundation; The Liz Claiborne Art Ortenberg Foundation; the Wildlife Trust (WT-USA); The Wildlife Preservation Trust Canada (WPTC); the Conservation, Food and Health Foundation; the Durrel Wildlife Preservation International; the Rain Forest Alliance; and the United States Agency for International Development (USAID). Institutional support was also provided by the Forestry Institute of São Paulo and Cooperativa dos Assentados da Reforma Agrária do Pontal do Paranapanema (COCAMP), São Paulo. The authors are deeply indebted to Kent Redford, Richard Bodmer, Giselda Durigan, and Ralph Vetters for valuable reviews of earlier drafts of this manuscript.

References

Alonso, M. T. A. 1977. *Vegetação. região sudeste*. Rio de Janeiro: Geografia do Brasil, Instituto Brasileiro de Geografia e Estatística.
Browder, J. O. 1996. Reading colonist landscapes: social interpretations of tropical forest

patches in an Amazonian agricultural frontier. Pages 285–299 in J. Schelhas and R. Greenberg (eds.), *Forest patches in tropical landscapes*. Washington, DC: Island Press.

Buchmann, S., and G. P. Nabhan. 1997. *The forgotten pollinators*. Washington, DC: Island Press.

Cullen, L., Jr., R. E. Bodmer, and C. Valladares-Padua. 2000. Effects of hunting in habitat fragments of the Atlantic forests, Brazil. *Biological Conservation* 95:49–56.

Cullen, L., Jr., R. E. Bodmer, and C. Valladares-Padua. 2001. Ecological consequences of hunting in Atlantic forest patches, São Paulo, Brazil. *Oryx* 35:137–144.

Cullen, L., Jr., M. Schmink, C. Valladares-Padua, and I. Morato. 2001. Agroforestry benefit zones: a tool for the conservation and management of Atlantic Forest fragments, São Paulo, Brazil. *Natural Areas Journal* 21:346–356.

Dean, W. 1995. *With broadax and firebrand: the destruction of the Brazilian Atlantic Forest*. Chicago: University of Chicago Press.

Ditt, E. H. 2000. *Diagnóstico da conservação e das ameaças a fragmentos florestais no Pontal do Paranapanema*. Master's dissertation, Programa de Pós-Graduação em Ciências Ambientais, Universidade de São Paulo, Brazil.

Eiten, G. 1974. An outline of the vegetation of South America. *Symposium Congress of the International Primatology Society* 5:529–545.

Ferrari Leite, J. 1998. *A ocupação do Pontal do Paranapanema*. São Paulo, Brasil: Editora Hucitec, Fundação UNESP.

Fonseca, G. A. B. 1985. The vanishing Brazilian Atlantic Forest. *Biological Conservation* 34:17–34.

Gajaseni, J., R. Matta-Machado, and C. F. Jordan. 1996. Diversified agroforestry systems: buffers for biodiversity reserves, and landbridges for fragmented habitats in the tropics. Pages 506–513 in R. C. Szaro and D. W. Johnston (eds.), *Biodiversity in managed landscapes: theory and practice*. Oxford, UK: Oxford University Press.

Gerlach, G., and K. Musolf. 2000. Fragmentation as of landscapes as a cause for genetic subdivision in bank voles. *Conservation Biology* 14:1066–1074.

Greenberg, R. 1996. Managed forest patches and the diversity of birds in southern México. Pages 154–174 in J. Schelhas and R. Greenberg (eds.), *Forest patches in tropical landscapes*. Washington, DC: Island Press.

Haddad, N. M. 2000. Corridor length and patch colonization by a butterfly, *Junonia coenia*. *Conservation Biology* 14:738–745.

Hildebrand, P. E., and J. T. Russel. 1996. *Adaptability analysis: a method for the design, analysis and interpretation of on-farm research-extenuation*. Ames: Iowa State University Press.

Hueck, K. 1972. *As florestas da América do Sul*. São Paulo, Brazil: Editora Polígono and Editora Universidade de Brasília.

Jacobson, S. K. 1995. *Conserving wildlife: international education and communication approaches*. New York: Columbia University Press.

Lima, W. P. 1996. *Impacto ambiental do eucalipto*. São Paulo, Brazil: Editora Universidade de São Paulo.

Lorenzi, H. 1992. *Árvores Brasileiras: manual de identificação e cultivo de plantas arbóreas nativas do Brasil*. São Paulo, Brazil: Editora Plantarum.

Lyon, J., and R. H. Horwich. 1996. Modification of tropical forest patches for wildlife protection and community conservation in Belize. Pages 205–230 in J. Schelhas and R. Greenberg (eds.), *Forest patches in tropical landscapes*. Washington, DC: Island Press.

Mielke, O. H. H., and M. M. Casagrande. 1997. Papilionoidea e Hesperioidea (Lepidoptera) do Parque Estadual do Morro do Diabo, Teodoro Sampaio, São Paulo, Brasil. *Revista Brasiliera Zoologica* 14:967–1001.

Oliveira, L. P., W. Perdoncini, and A. Bonnemann. 1986. Sistemas agrosilviculturais. *Manual Técnico Florestal* 1:217–325.

Padua, S. M. 1991. *Conservation awareness through an environmental education school program at Morro do Diabo State Park, São Paulo State, Brazil.* Master's thesis, University of Florida, Gainesville.

Padua, S. M. 1997. Uma pesquisa em educação ambiental: a conservação do mico-leão-preto *Leontopithecus chrysopygus.* Pages 34–51 in C. Valladares-Padua, R. E. Bodmer, and L. Cullen Jr. (eds.), *Manejo e conservação de vida silvestre no Brasil.* Belém, Brazil: MCT-CNPq-Sociedade Civil Mamirauá.

Passos, F. C. 1992. *Hábito alimentar do mico-leão-preto* Leontopithecus chrysopygus *(Mikan, 1823) (Callitrichidae, Primates) na estação ecológica dos Caetetus, Município de Gália, SP.* Master's thesis, Universidade Estadual de Campinas, Campinas, São Paulo, Brazil.

Payne, W. J. A. 1985. A review of the possibilities for integrating cattle and tree crop production systems in the tropics. *Forest Ecology and Management* 12:1–36.

Powell, G. V. N., and R. Bjork. 1995. Implications of intratropical migration on reserve design: a case study using *Pharomachrus mocinno. Conservation Biology* 9:354–362.

Quintela, C. E. 1990. An SOS for Brazil's beleaguered Atlantic Forest. *Nature Conservation Magazine* 40:14–19.

Raintree, J. B. 1990. Theory and practice of agroforestry diagnosis and design. Pages 58–97 in K. G. Mac Dicken and N. T. Vergara (eds.), *Agroforestry: classification and management.* New York: Wiley.

Schelhas, J., and R. Greenberg. 1996. Introduction. Pages xv–xxxvi in J. Schelhas and R. Greenberg (eds.), *Forest patches in tropical landscapes.* Washington, DC: Island Press.

Shafer, C. L. 1995. Values and shortcomings of small reserves. *BioScience* 45:80–88.

Smith, J. H., J. T. Williams, D. L. Plucknett, and J. P. Talbot. 1992. *Tropical forests and their crops.* New York: Cornell University Press.

SOS Mata Atlântica and INPE. 1993. *Evolução dos remanescentes florestais e ecossistemas associados do domínio da Mata Atlântica.* São Paulo, Brazil: SOS Mata Atlântica and Instituto de Pesquisas Espaciais.

Turner, I. M., and R. T. Corlett. 1996. The conservation value of small, isolated fragments of lowland tropical rain forest. *Trends in Ecology and Evolution* 11:330–333.

Viana, V. M., and A. A. J. Tabanez. 1996. Biology and conservation of forest fragments in the Brazilian Atlantic moist forest. Pages 151–167 in J. Schelhas and R. Greenberg (eds.), *Forest patches in tropical landscapes.* Washington, DC: Island Press.

Viana, V. M., A. A. J. Tabanez, and J. L. F. Batista. 1997. Dynamics and restoration of forest fragments in the Brazilian Atlantic moist forest. Pages 351–365 in W. F. Laurance and R. O. Bierregaard Jr. (eds.), *Tropical forest remnants: ecology, management, and conservation of fragmented communities.* Chicago: University of Chicago Press.

Wilson, R. J., and S. G. Diver. 1991. The role of birds in agroforestry systems. Pages 18–21 in E. Garrett (ed.), *Proceedings of the Second Conference on Agroforestry in North America.* Springfield, MO: School of Natural Resources.

Chapter 18

Agroforestry and Biodiversity: Improving Conservation Outcomes in Tropical Northeastern Australia

Nigel I. J. Tucker, Grant Wardell-Johnson,
Carla P. Catterall, and John Kanowski

The spatial and temporal distribution of biodiversity in tropical ecosystems is in stark contrast to the agricultural and forestry monocultures that often replace them. For example, Australia's tropical forests show local heterogeneity in composition and structure, associated with variation in physical conditions such as moisture, elevation, topography, and soil fertility, and support many taxa, including many primitive angiosperms, considered relicts of the ancient Gondwanan rainforests (Tracey 1982; Adam 1992).

Although Australia is a wealthy and educated nation, its tropical forests have fared only marginally better than those of many poorer countries, having been extensively fragmented over the 200-year period of European settlement. Closed forests (i.e., rainforests) and their associated sclerophyll forests have been subjected to especially high rates of clearing on the coastal plain and upland plateaus. Intact closed forest cover is now largely limited to mangroves and rainforests growing on steep mountain ranges. As a result of clearing, around 42 percent of the plant communities recognized in the wet tropics bioregion are classified as "endangered" (less than 10 percent of original extent) or "of concern" (10–30 percent of original extent; Goosem et al. 1999). Conversion of native forests for agriculture still continues in the region but is now controlled by state legislation (Vegetation Management Act of 1999), which regulates clearing of "endangered" and "of concern" ecosystems through a permit system.

In northeastern Australia, broad-scale clearing has produced a landscape mosaic typical of much of the world's tropics; in the wet tropics bioregion there may be up to 10,000 fragments of various sizes (Crome and Bentrupperbaumer 1993), embedded in a landscape matrix of agriculture and expanding urban settlements. The region's protected area network features the 900,000-ha Wet Tropics World Heritage Area (WTWHA) and a number of parks and crown reserves that are outside the WTWHA boundary. Many of these reserves, and some parts of the WTWHA, were gazetted for a single feature (e.g., lakes, feature trees, volcanic craters, and waterfalls), and the forests surrounding these features often were clearcut, leaving many of the parks, reserves, and fragments that are situated on private land isolated in a sea of pasture. Isolated fragments have lost many of their more specialized fauna (Pahl et al. 1988; Laurance 1990, 1991) and are at risk of losing a significant proportion of their species over the longer term as a consequence of reductions in population size, increased barriers to dispersal, and the influence of hostile forces from the surrounding landscape (Lamb et al. 1997; see also Chapter 2, this volume). Consequently, we argue that strategies to improve or maintain tree cover on private land are a crucial plank in a regional biodiversity conservation strategy.

Adoption of farm forestry as a land use by private landholders is a promising means of increasing tree cover in fragmented tropical landscapes, including those of northeastern Queensland. The resulting areas of forested land may contribute to biodiversity conservation goals in a variety of ways. First, they may provide habitat for indigenous flora and fauna, increasing population numbers and reducing the chance of local or regional extirpation. Second, they may protect or enhance the ecological capacity of the formal habitat reserve network (Hobbs 1993) by providing buffer zones to ameliorate edge effects and acting as corridors or stepping stones to increase the probability of dispersal between scattered remnants (see Chapters 2 and 3, this volume). In addition, they may be simply increasing the regional or landscape-scale percentage of land area that is forested, which may be important in its own right (Andrén 1994). Third, by protecting catchments and streambanks they may enhance the habitat quality of riparian and in-stream environments (Bunn et al. 1999) and thus contribute to the conservation of aquatic biota.

These functions may be modified by the specific manner in which a farm forestry planting is established and managed. Various aspects of plantation design, including the species matrix, tree spacing, and degree of suppression of understory shrubs, are likely to exert a strong influence on the habitat value of a forest plot (Catterall 2000). Plantations that are intensively managed for rapid timber growth (through thinning, tree pruning, and understory suppression) are less likely to provide suitable habitat for forest-dependent species than those where a densely packed and diverse mix of indigenous trees and shrubs exists (Bentley et al. 2000). Furthermore, tree felling for timber extraction sets an ultimate limit to the long-term habitat value of an individual forest plot, an effect largely dependent on the style of tree harvesting (e.g., selective logging or clearcutting) and the length of the harvesting cycle. However, over wider areas even clearcut plantations may help sustain regional flora and fauna populations because there could be mosaic stability at the landscape scale, with some part of the land area always under older plantations at a given time.

In this chapter we consider the potential contribution of farm forestry plantings to sustaining indigenous biota in the specific context of northeastern Queensland's wet tropics. We discuss recent developments in the nature and extent of agroforestry in the bioregion, identify opportunities and constraints for improving the habitat and biodiversity value of reforested plots, present data from a case study that examines some of these issues, and describe recommendations for improving the contribution of agroforestry to local and regional biodiversity conservation. We do this from a broad perspective that seeks the best possible outcome for persistence of rare and common indigenous species and ecological processes, at both local and regional scales, in landscapes that are settled for productive use by humans. This includes the acknowledgment that the formal conservation reserve system is not capable of conserving the region's biodiversity in the long term and that there may be trade-offs between productive use and habitat retention and recreation that allow off-reserve forests to prevent or mitigate further losses.

Forestry and Agroforestry Development in Australia's Wet Tropics

In northeastern Queensland, timber plantations were established by state forest agencies on crown holdings by the 1930s. State-owned plantations were made up mostly of the exotic Caribbean pine (*Pinus caribaea* var. *hondurensis*) and the native conifer hoop pine (*Araucaria cunninghamii*), although some other local cabinet timbers were also trialed, including red cedar (*Toona ciliata*), Queensland maple (*Flindersia brayleyana*), and Queensland kauri (*Agathis robusta*). However, until recently there was very little encouragement for the development of plantations on private land or farm forestry because of the near total reliance on rainforest logging, inflated estimates of the yield from managed forests, and extensive forest clearing for agriculture that produced low-cost timber. Future timber supplies in the region generally were seen as coming from the residual managed natural forests.

By the 1970s, however, signs of land degradation on cleared land were obvious and the role of trees on farms was publicly promoted, although there was little adoption by farmers (Gilmour and Riley 1970). Small stands of hoop pine were established on dairy farms on the southeastern Atherton Tableland, although these were largely unmanaged and planted in windbreak configurations. This was followed by 1- to 2-ha plantings of Caribbean pine in the 1960s through the 1980s, but these plots were also largely unmanaged, and no market currently exists for these farm-grown logs, damaging local landholder perceptions of farm forestry (Herbohn et al. 2000). Kent and Tanzer (1983a, 1983b) identified more than 30,000 ha of freehold land on the cleared uplands of the Atherton Tablelands as unsuitable for intensive agriculture and well suited to timber production, showing that potential existed for development of farm forestry.

In the 1980s public awareness of environmental issues grew rapidly in Australia, largely among urban populations in response to issues such as the damming of the Franklin River in Tasmania and logging and road building in the wet tropics of northeastern Queensland. With this increased environmental awareness came a desire by many city residents to reestablish themselves in rural areas for lifestyle reasons, bringing about a shift in attitudes toward natural resource management in some rural areas. Many of the new settlers purchased small holdings of 2–4 ha that had previously been dairy or beef properties or purchased entire farms in depressed rural property markets. A clear manifestation of the attitudinal shift under way was an interest by such new rural landholders in tree planting for reasons other than timber production.

Increased awareness of conservation issues, including many high-profile antilogging campaigns, led to a large upswing of interest in tree planting for a range of reasons including land restoration, catchment protection, wildlife habitat, windbreaks and shelterbelts, improving land values, and aesthetic reasons (Tracey 1986). Evidence of this interest is shown in the growth of membership in community nature conservation groups with a focus on ecological restoration projects. For example, membership of the community tree-planting group Trees for the Evelyn and Atherton Tablelands Inc. (TREAT) rose from 30 landholders in 1984 to more than 600 by 1990 before stabilizing at 500 by the mid-1990s. The growth of community interest in plantings for biodiversity was in contrast to the decline of the local logging industry, increasingly curtailed by a dwindling allocation of logs from the managed forest estate.

Because of this decline, the lack of cleared government land for plantations, and a lack of investment interest by industrial forestry groups, federal and state governments throughout Australia began an active promotion of farm forestry initiatives including, for the first time, heavily subsidized schemes to maximize landholder participation. In 1988 the WTWHA was declared, and there was a subsequent ban on logging in the area's tropical forests. This was accompanied by a government-funded structural adjustment package aimed at providing alternative forest-related employment for workers who had lost their source of income when logging ceased.

Arising from the structural adjustment package that followed the logging ban was the Community Rainforest Reforestation Program (CRRP), which commenced in 1992. The main objectives of this scheme were to establish mixed-species farm forestry plantations on private land for timber production and the amelioration of land degradation and to provide training and employment opportunities (QDPI Forest Service 1994; Lamb et al. 1997). This program was one of the first attempts by governments to meet such a broad range of objectives, albeit with a primary focus on farm forestry. This approach was adopted partly to attract landholders with interests in tree planting for reasons other than timber production, planting around 2,000 ha, representing 0.1 percent of the wet tropics bioregion. The majority of the plantations are small (1–2 ha) and have relied on a narrow species pool, based on *Eucalyptus, Flindersia,* and *Araucaria* (Lamb et al. 1997; Catterall 2000). Tree spacing typically was around 3–4 m (ca. 1,000 stems per hectare), denser than typical commercial plantings but sparser than biodiversity-oriented ecological restoration plantings, where trees typically are spaced 1.5–2 m apart (see Table 18.1). The wider spacing of the CRRP plantations reduces the likelihood of competition but greatly increases the time taken for canopy closure, resulting in a greater maintenance effort to reduce weed competition. Although many landholders fully embraced the CRRP program and accepted their role in plantation maintenance, others did not, and many plots experienced mortality and subsequent weed invasion. CRRP records indicate that by 1998 (10–12 years after establishment), at least 15 percent of plantations had failed because of poor maintenance and cattle damage (Vize and Creighton 2001). The CRRP program has been discontinued.

Despite the many reports of successful plant establishment, this program appears to have met with only moderate success in achieving good outcomes for biodiversity, an outcome common to many farm forestry projects in the tropics (Haggar et al. 1997). This is unfortunate because the many local landholders planting trees appear to be doing so with biodiversity as a key focus (Herbohn et al. 2000). Aspects of the CRRP that limit their value to biodiversity are discussed later in this chapter.

Opportunities and Constraints for Improving the Habitat and Biodiversity Values of Farm Forestry in North Queensland

As yet, there are no scientific assessments of the extent to which multipurpose tropical cabinet timber plantations whose species mix, plant spacing, and maintenance regime were designed with timber production as a major goal are also effective in meeting biodiversity outcomes. The lack of an effective research program and subsequent extension of this research have been previously identified as impediments to the broad-scale adoption of farm forestry in tropical Queensland for any goal (Vize and Creighton 2001). Nevertheless, we argue that the value of farm forestry for biodiversity depends on a range of factors, including those related to the initial establishment success of the plots, which determine their value as habitat or dispersal corridors for indigenous plants and animals. Here we consider the limitations of the biodiversity values of CRRP farm forestry plantings.

Species Selection

Although about 200 species were trialed in the CRRP, most trees planted were of three *Eucalyptus* species, two species in the Araucariaceae, and smaller numbers of three *Flindersia* species. Of the 16 main species planted by the CRRP, 6 are exotic, and with the exception of one fleshy-fruited species (*Elaeocarpus angustifolius*), the remainder are wind dispersed. The use of species with dry, dehiscent fruits from a narrow range of taxa would limit the resource utility of these plots to frugivorous (and hence seed-dispersing) vertebrate wildlife and therefore may reduce their ability to recruit seedling immigrants of other native rainforest species. Although any tall tree can provide a perch and act as a focus for recruitment (Aide and Cavelier 1994), fruiting trees are likely to be used more often by frugivorous species. Furthermore, although the thin crowns of *Eucalyptus* spp. favor grass retention and allow grazing to continue (as desired by some landholders), in ungrazed plots the thin crowns transmit light levels that favor ground-level dominance by woody weeds.

In addition to the narrow range of species used in the majority of plantings, species selection was not optimal from a conservation perspective. For example, when native species were used, trees were not always from local provenances, or they were planted outside their normal range or ecological situation, and some exotics were used, including species with known weed potential. Examples include the use of lowland riparian species on upland basalts, establishment of blackwood (*Acacia melanoxylon*) sourced from temperate Tasmanian provenances rather than local genotypes, and planting of the exotic East Indian mahogany (*Chukrasia velutina*), now spreading in plantations in the Gadgarra State Forest and showing similar weed potential in other areas.

Plot Location

To a large extent, land management agencies and groups involved in tree planting on private lands, such as the CRRP, are bound by landholder wishes when choosing areas to establish plots. This often means that projects are not established in areas where native tree cover is most urgently needed, including areas where clearing has reduced plant community distribution to "endangered" or "of concern" status, or areas where interfragment distance is large, so in many areas a homogenous agricultural landscape remains. In north Queensland's wet tropics, much of the coastal plain and upland plateaus are occupied by agriculture, yet these landholders are generally less aware of biodiversity issues, and the natural ecosystems are especially diverse but greatly reduced in area.

Plot Design and Management

In many CRRP plantings there was insufficient consideration of plot design features and the siting of species in different zones of the plantation, for example, based around species performance in other tree-planting projects. Rigorous design would include consideration of the growth rates, spacings and configurations of species in rows, crown architecture, and attributes such as windbreak utility, frost tolerance, and attractiveness to local wildlife. For example, hoop pine is a light-demanding species that has excellent windbreak qualities. It has the ability to protect species in the plantation interior and grow well on the exposed margin, but it has been largely underused in this role in the CRRP plots. Blue quandong (*Elaeocarpus angustifolius*) is a high-value wildlife species that grows much more quickly in a mixed-species, high-density planting than as a monoculture in widely spaced grazing regimes, where it has sometimes been established. In contrast, some high-value species such as silky oak (various species of the family Proteaceae), walnut (*Beilschmiedia* spp., *Endiandra* spp.), and tulip oak (*Argyrodendron* spp.) are likely to perform best in areas with minimal edge effects, toward the interior of the plantation.

Furthermore, most CRRP plots were established in linear configurations and were very small, maximizing edge effects and minimizing the development of a zone of interior forest habitat. Many tropical plant and animal species are intolerant of edge effects, so linear plots are likely to provide suboptimal habitats for these species. The extensive edge of linear plantings is also favorable to many weeds. Although some woody weeds may hasten local rainforest succession by outcompeting the exotic pasture grasses, the majority of weeds are undesirable, and their persistence is likely to be favored by edge-affected habitats. It does not follow that all farm forestry plots are likely to exclude all weeds; rather, the establishment of tree cover is likely to alter the structure and competitive effect of the weed community away from woody shrubs toward soft herbs and vines.

Although most farm forestry plots are still very young (less than 20 years), it is evident that the understory is heavily managed in grazed and nongrazed plantings. Habitat features such as fallen branches and other woody debris and regenerating shrubs, vines, and trees are generally absent in these developing systems. Standing dead trees, logs, old fenceposts, and other debris often are present on sites before planting and provide valuable niche features for a range of organisms, vertebrate and nonvertebrate, but are usually removed when the site is planted (Grove and Tucker 2000). Removal of these features when preparing sites for replanting reduces the area's potential to offer a more diverse range of food and cover resources to a wide range of organisms.

Case Study: Agroforestry on the Atherton Tablelands

This case study has been included to illustrate the biodiversity benefits of different styles of reforestation on the Atherton Tablelands in north Queensland. Reforestation in this case includes three intervention techniques—ecological restoration, agroforestry, and commercial plantations—and natural regrowth as a passive, noninterventionist technique. Information relating to numbers of species used, stem density, management, and costs of the intervention techniques discussed in this case study is listed in Table 18.1.

The Atherton Tablelands are an upland plateau of varying topography with moderate- to high-fertility soils of volcanic origin. The area was extensively cleared for dairy pasture from the early 1900s. Clearing was nonrandom, and reserve selection often was based on scenic appeal only, resulting in a patchwork of fragments generally isolated within extensive pasture. With the decline of the dairy industry from the 1960s onward, some cleared land reverted to regrowth, particularly on steeper slopes and creek lines. However, despite this regrowth and the extensive tree planting that has occurred over the past 20 years, the area remains highly fragmented.

The Cooperative Research Centre for Tropical Rainforest Ecology and Management (Rainforest CRC) is a joint federal and state government initiative that brings together researchers from universities and land management agencies, initiating and coordinating research into aspects of forest conservation and management. Researchers under the auspices of the Rainforest CRC established the case study described in this section.

The primary question addressed in this case study was, "To what extent have sites supporting the four different styles of reforestation recovered the integrity of their plant assemblages?" To examine this question, 50 quadrats

Table 18.1. Reforestation techniques used in this case study.

Technique	Average Number of Species per Plot	Stem Density per Hectare	Management	Cost per Hectare
Farm forestry (CRRP)[a]	10 (native and exotic)	400–1,700	12–18 mo of weed exclusion around stems (herbicide)	$5,000–$10,000
Ecological restoration	50 (local native spp.)	3,000	18–24 mo of total weed exclusion (herbicide)	$15,000–$20,000
Commercial plantations[b]	1 (native)	600–1,000 (thinned to 400 stems per hectare at harvest; 50 years)	12 mo of exclusion around stems (herbicide), pruning, and thinning at 28 yr	$3,000

[a]After Erskine (2002).
[b]After Keenan et al. (1997).

were established on sites representing a range of land cover types on basalt-derived soils on the Atherton Tablelands. All sites were at least 2 ha in area (most were 4 ha) and at least 5 years old. Sites were selected in seven broad categories ranging from grazed pasture to mature rainforest. They included high-density restoration plantings, usually of a wide range of local rainforest species (ecological plantings [E], 10 sites), mixed-species farm forestry plantings (CRRP plantings [C], 5 sites), and commercial monoculture timber plantations. The commercial plantations included 15 sites, 5 of which were classed as young plantations (YP), being equivalent in age to the C and E sites (5–15 years). All young plantations were of hoop pine with differing management histories. Ten old plantation sites (OP) were selected, including monocultures of three locally indigenous species: Queensland kauri, red cedar, and hoop pine. These sites were aged 38–70 years. In the study, we also included 10 forest reference sites (F), 5 pasture sites (P), and 5 areas supporting natural regeneration, principally along riparian zones that had developed after reduction or cessation of cattle grazing (regrowth sites [R]). Here we consider floristic patterns associated with 48 of these sites (data were not available for one R and one C site).[1]

Figure 18.1 demonstrates the effect of time and planting style (i.e., the diversity of planted species, spacing, and management) on the plant assemblages of

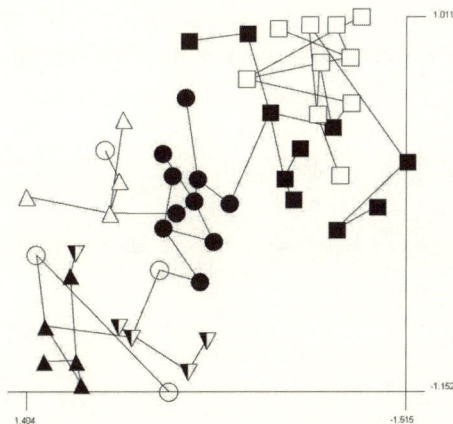

Figure 18.1. Two-dimensional ordination (SSH-MDS, Stress = 0.158) and minimum spanning tree for 48 sites with a range of land cover types on basalt-derived soils on the Atherton Tablelands, northeast Queensland. Analysis is based on floristic composition (presence or absence) of all vascular plant species occurring in more than two quadrats (265 species). Predefined groupings of the seven land cover types are shown symbolically: rainforest reference sites, open square; old plantations, closed square; regrowth sites, open triangle; pasture, closed triangle; young plantations, half closed triangle; ecological restorations, closed circle; CRRP plantings, open circle.

reforested sites. After 15 years, the plant assemblages of ecological restoration plantings had moved substantially toward the "forest" state, whereas farm forestry and other plantations remained more similar to pasture sites. Old plantation and forest sites were most similar to each other and most different from other land cover types, as portrayed in cluster analysis, ordination, and network analysis. Colonization by local rainforest species has taken place in almost all the old plantation monoculture sites examined in this study, despite wide variation in management history, undoubtedly because of a range of factors including the proximity of native forest to all the old plantations and the likely persistence of a soil seed bank and rootstock of rainforest plants through the clearing and establishment phase.

A pattern similar to that observed in ordination can be discerned based on cluster analysis (Figure 18.2). Broadly speaking, forest reference sites and old plantation sites are very similar floristically. Pasture sites and young plantation sites also form a broad group, very distinct from the old plantation and reference sites. A third group between these two includes the regrowth, ecological restoration, and CRRP sites. At a finer level, the ecological plantings more closely resemble the reference sites than do the CRRP plantings.

Even at this finer seven-group scale, most groups differed significantly ($p <$ .05) from one another based on their plant assemblages. The exception was the grassy OP site, the group that consisted of a single site (OP1), which was not significantly different from any other group. Similarly, all seven predefined groups differed ($p < .05$) from one another. Thus, even though the old com-

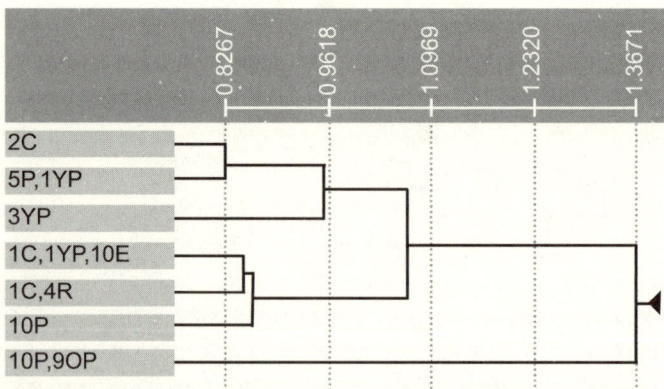

Figure 18.2. Groupings of the 48 sites shown in Figure 18.1 through cluster analysis (UPGMA, Czechanowski Metric, $B = -0.1$) showing membership of Site Groups 1–7 by land cover type. Land cover types are 5 pastures (P), 4 CRRP plantings (C), 4 regrowth sites (R), 10 ecological restorations (E), 10 old plantations (OP), 5 young plantations (YP), and 10 rainforest reference (F) sites.

mercial timber plantations overlapped with the rainforest reference sites in the ordination plot (Figure 18.1) and grouped together in the cluster analysis (Figure 18.2), the ANOSIM (Analysis of Similarity) test showed that they differed significantly in their plant species assemblages. In the plantations, few species other than the planted species occur in the upper third of the canopy, whereas the canopy of mature rainforest includes a large array of species. This may also lead to differences in the assemblages of other organisms affected by canopy structure and diversity. Studies are under way to examine the use of reforested sites by a range of wildlife and to test how well floristic data correlate with data on other taxa.

Other components of biodiversity and several ecosystem processes are also under study at these sites, including avifauna, reptile, and invertebrate assemblages, decomposition rates, and seed predation. Our preliminary analysis of these parameters has revealed that different components of the ecosystem do not necessarily respond in the same manner or at the same rates to reforestation and also show some differences in response to various planting styles (Proctor et al. 2003). However, the farm forestry sites generally were less similar to the forest reference sites than were the ecological restoration sites at 5–10 years of age. Environmental features of farm forestry plots that may be associated with this difference include a more open canopy and less complex structure, lower plant diversity, and the relative rarity of certain habitat features such as vines, epiphytes, woody debris, and fleshy fruits, all characteristics of rainforest (see Kanowski at al. 2003 for a comparison based on structure of different reforestation types). Paradoxically, landholder motivation for establishing these plots was driven at least partly by concerns for environmental protection rather than commercial returns. This suggests that a number of the limitations identified here must be addressed if a farm forestry ethic is to be nurtured and developed in the region.

Recommendations for Improving the Contribution of Agroforestry to Local and Regional Biodiversity Conservation in North Queensland

Since the late 1990s, commercial farm forestry in tropical Australia appears not to have expanded significantly, whereas ecological restoration initiatives such as the TREAT program continue to flourish. This is despite the availability of a number of government-funded programs to promote the full array of tree-planting approaches, including farm forestry (Herbohn et al. 2000). Lack of local interest in farm forestry has been attributed to a range of factors, including landholder resistance to establishing plantations with no prospect of economic returns for 20–30 years, uncertainty over future harvesting rights, and landholder or institutional perceptions that farm forestry is a poor use of productive agricultural lands for which other crops are far better suited (Emtage et al. 2001). In short, this can be described as lack of a farm forestry ethic or culture (Herbohn et al. 2000).

However, the CRRP program has clearly shown that many landholders are interested in timber production and establishing trees for environmental purposes. This suggests that revising farm forestry policy to achieve better conservation outcomes for local and regional biodiversity and articulating this revision to landholders may improve the adoption and success of farm forestry in tropical Australia. We suggest that a blend of farm forestry and ecological restoration techniques is appropriate for this region and assert that restoration forestry has a greater potential for adoption by landholders and will improve landscape biodiversity values at a range of spatial scales. In this context restoration forestry can be described as the establishment, in unproductive areas, of mixed indigenous species with attributes that promote biodiversity preservation and provide a future source of additional property income.

A number of actions can be taken to improve biodiversity conservation outcomes of local farm forestry plots. These include planting of trees that are used by wildlife, planting of local endemics, better matching of species to sites, conservation plantings among timber plantations, and better location, design, and management of plots. There are likely to be some disadvantages to production from some of these actions, and the extent to which landholders are prepared to accept trade-off options will determine the degree of integration into standard farm forestry practice. The degree of uptake in trade-off situations varies depending on landholder attitude and the ability of extension staff to inform landholders of the options.

Use of Trees Valuable to Wildlife in Plantations

Increasing the range and choice of species planted would significantly improve the value of these plots to wildlife, particularly the frugivorous species that depend on fruit resources that are patchily distributed in both space and time. Crome (1975) lists 10 key plant families important in sustaining frugivorous birds on a year-round basis, and incorporating some or all of these families is likely to promote the structural complexity and resource heterogeneity of farm forestry plots. Table 18.2 provides a list of local species on the Atherton Tableland that have both timber and wildlife values, the inclusion of which would provide more year-round resources to invertebrate and vertebrate wildlife, particularly frugivorous seed dispersers. The inclusion of one or more *Ficus* species in the plantation will also add significantly to its wildlife value (Goosem and Tucker 1995). Although

Table 18.2. Commercial timber species with wildlife conservation values that could be included in restoration forestry plantings.

Family	Species	Common Name	Notes
Anacardiaceae	*Pleiogynium timorense*	Burdekin plum	Large-fruited species eaten by many frugivores, including cassowaries; grows well on poorly drained lowlands; novel food crop.
Combretaceae	*Terminalia sericocarpa*	Damson plum	Keenly sought by many frugivorous birds; leaves eaten by coppery brushtail possums; host tree for common oakblue, narcissus jewel, copper jewel, and emperor moth butterflies.
Elaeocarpaceae (many other Elaeocarpaceae are also high-value wildlife species)	*Elaeocarpus angustifolius*	Blue quandong	High value to many larger frugivores including flying foxes and cassowaries; leaves eaten by Lumholtz tree kangaroo; known food plant for the rare Herbert River ringtail possum.
	Elaeocarpus bancroftii	Kuranda quandong	Large-fruited species; tasty nut, novel food crop.
Fabaceae	*Castanospermum australe*	Black bean	Profuse flowering attracts parrots and other nectar feeders; leaves eaten by Lumholtz tree kangaroo; known food plant for the rare Herbert River ringtail possum.
Lauraceae	*Beilschmiedia bancroftii*	Yellow walnut	Large-fruited taxon; food plant for the lemuroid ringtail possum.
	Beilschmiedia obtusifolia	Blush walnut	Attracts frugivores, especially Torresian imperial pigeon; food plant for coppery brushtail and lemuroid ringtail possums.
	Cryptocarya hypospodia	Northern laurel	Eaten by many frugivores; food plant for Lumholtz tree kangaroo; host tree for Macleays swallowtail, blue triangle, common oakblue, and banded red-eye butterflies.
	Cryptocarya oblata	Zig zag laurel	Larger-fruited taxon favored by cassowaries.
	Endiandra hypotephra	Rose walnut	Eaten by many frugivores.
	Endiandra insignis	Hairy walnut	Large-fruited taxon; leaves eaten by Lumholtz tree kangaroo.
	Litsea leefeana	Brown bollywood	Eaten by many frugivores including larger pigeons and cassowaries; food plant for Lumholtz tree kangaroos, coppery brushtail, green and lemuroid ringtail possums; purple brown-eye and blue triangle butterfly host plant.
Meliaceae	*Dysoxylum muelleri*	Miva mahogany	Eaten by many frugivores.
	Dysoxylum parasiticum	Yellow mahogany	Eaten by many frugivores.

(continues)

Table 18.2. *Continued*

Family	Species	Common Name	Notes
Myrtaceae	*Acmena resa*	Red Eungella satinash	Frugivorous birds; leaves eaten by Herbert River and lemuroid ring tail possums.
	Syzygium gustavoides	Water gum	Large-fruited taxon accessed by rodents.
	Syzygium johnsonii	Rose satinash	Frugivorous birds.
	Syzygium kuranda	Kuranda satinash	Large-fruited taxon accessed by rodents.
	Syzygium sayeri	Pink satinash	Favored by flying foxes; flowers very popular nectar source for vertebrates and invertebrates.
Proteaceae	*Alloxylon flammeum*	Pink silky oak	Vulnerable species; high timber value; flowers may also be marketable.
	Athertonia diversifolia	Atherton oak	Tasty nut; marketable foliage.
Rutaceae	*Flindersia brayleyana*	Queensland maple	Leaves eaten by coppery brushtail and lemuroid ringtail possums.
	Flindersia pimenteliana	Maple silkwood	Leaves and flowers eaten by Lumholtz tree kangaroo.
	Flindersia schottiana	Tropical ash	Widely distributed, rapid growth rates; food plant for coppery brushtail possums.
Sapindaceae	*Castanospora alphandii*	Brown tamarind	Flying fox and cassowary food plant; leaves eaten by coppery brushtail, lemuroid, and Herbert River ringtail possums.
Sapotaceae	*Palaquium galactoxylum*	Cairns pencil cedar	Large-fruited taxon eaten by many frugivores and rodents.
Sterculiaceae	*Argyrodendron* spp.	Tulip oak	Late-successional species; popular food plant for green ringtail and coppery brushtail possums and Lumholtz tree kangaroo.
Sapotaceae	*Planchonella obovoidea*	Yellow boxwood	Attracts frugivorous birds; leaves eaten by Lumholtz tree kangaroo and coppery brushtail possums.
Verbenaceae	*Gmelina fasciculiflora*	White beech	Larger-fruited taxon eaten by many frugivores.

landowners may request that certain species be included in the plantation, government extension staff and nongovernment organizations usually are free to suggest and use other species and can therefore exert a significant influence on the planting species pool at many sites. The principal trade-off is likely to be the slower growth rates in some species, but selected harvesting, thinning, or brushing of the understory may ameliorate this problem, although this will of course reduce some plant diversity.

Use of Local Endemics and Large-Fruited Taxa

Where possible, consideration could be given to local endemics, species with rare or patchy distributions, and large-fruited taxa whose dispersal mechanisms have been largely extirpated (Tucker and Murphy 1997; Tucker 2000). Several of the valuable timber species within Lauraceae, including yellow walnut (*Beilschmiedia bancroftii*), rose walnut (*Endiandra hypotephra*), and zig zag laurel (*Cryptocarya oblata*), are all examples of large-fruited endemics with limited distributions and dispersers. Also worthy of inclusion are many of the late-successional wind-dispersed species with exceptionally high timber values. Many of these species appear unable to invade disturbed or restored systems (Tucker and Murphy 1997), and whereas some species such as pink silky oak (*Alloxylon flammeum*) are rare, others, including brown tulip oak (*Argyrodendron peralatum*), are common and conspicuous elements of local canopy flora. The indications are that if these species are not anthropogenically reintroduced to disturbed areas, they are likely to be confined to the larger fragments in the landscape. A trade-off is again likely to be in the differential growth rates encountered, although this may be overcome by different approaches to plantation design. Larger-fruited Lauraceae often demonstrate very erratic and recalcitrant germination, and this would also necessitate research, but other species listed germinate readily and reliably.

Better Matching of Species to Sites

In concert with a more diverse species pool must come a greater attention to the ecological niche needs of planted stems. Soil, rainfall, drainage, altitude, and microsite preference are key determinants of species' natural distribution, and adequate information is available to ensure that these parameters are matched when planting stock is selected. Such an approach may effectively preclude *Eucalyptus* from the majority of Atherton Tableland sites. However, alternatives include the rare *Stockwellia quadrifida*. This species occurs in two small populations in a very wet zone of the wet tropics and is considered to represent a primitive species that may have given rise to some of the more widespread yet recent elements of the sclerophyllous Australian flora (i.e., *Eucalyptus, Syncarpia, Lophostemon*). Visually, the wood of this species is similar to that of many *Eucalyptus* species in its long, dense, and hard grain. Limited planting of the species in a variety of sites has revealed moderate to fast growth on a variety of Tableland soils, particularly in wetter zones and on basalt soils, where *Eucalyptus* typically performs poorly (N. I. J. Tucker, pers. obs., 1995). Trials using this rare species would assist in securing its future and may provide an alternative to *Eucalyptus* in very wet areas. It is difficult to envisage any negative outcomes from adopting an ecologically based approach to species selection and the siting of microsite planting. Research is needed to examine the utility of *Stockwellia quadrifida*.

Incorporating Conservation Plantings among Monoculture Timber Plantations

In larger farm forestry blocks, managers should consider incorporating ecological restoration areas through the plantation to improve the ability of native plants and animals to move through suboptimal plantation habitat. Where possible, this planting style should strategically use key landscape features such as riparian zones and ridgelines to maximize site heterogeneity and provide a refuge for fauna and flora that are displaced during logging and a nucleus from which populations could expand and recolonize after logged areas are replanted. This reduces the area under production, a significant trade-off. However, the total loss of cover on harvesting is likely to severely affect local wildlife populations. Staggered harvest regimes, based on individual species or subcompartments within plantations, are a further means of improving the temporal continuity of wildlife habitat, obviously a major research question for any tropical farm forestry plantation.

Plot Location

Plot location could also be refined to improve outcomes for biodiversity conservation (Hobbs 1993). Establishing plots close to existing fragments provides shelter for newly established seedlings and provides an extension of and buffer for the existing native vegetation. A more immediate contribution can be achieved by installing farm forestry plots as close as possible to native forest, guaranteeing rapid colonization by native species. In recent years much attention has focused on the catalyzing effect of forest plantations in enhancing successional processes (Parrotta et al. 1997 and references therein), resulting in the development of species-rich understories beneath the plantation canopy. The natural regeneration of these secondary forests, reflected in the rate and direction of the succession, may be limited by a range of factors, including site disturbance history, distance to a primary seed source and their dispersers, the plantation species matrix, and management history. However, there is no doubt that the establishment of plantations adjacent to native forest can have rapid and long-term benefits for the conservation of local biodiversity, as evidenced by the rapid species accretion in old plantation sites discussed in the case study presented earlier.

Extension staff are needed who can work as closely as possible with landholders in agricultural areas where tree cover is particularly low. In areas of north Queensland previously supporting complex notophyll and mesophyll vine forests on basalt and alluvium, the effects of agriculture have been especially severe, and landholders in these areas should be a priority target for farm forestry extension officers. Stepping stone plots of native species in these areas would improve the permeability of this homogenized and largely hostile landscape, allowing native species easier movement between fragments and enhancing the ecological services they provide (see also Chapter 17, this volume).

Riparian zones are areas of high biological productivity, and well-vegetated catchments make a large contribution to soil stability and water quality in addition to providing a hierarchical system of interconnected habitats. Although riparian zones are clearly a priority for ecological restoration, this does not preclude using areas extending outward from the top bank for farm forestry plantations. This buffer zone can also reduce edge effects along restored riparian zones. This technique has been adopted at Donaghy's Corridor, a 1.2-km by 100-m restored habitat linkage reconnecting a 498-ha fragment to adjacent intact forest on the Atherton Tableland (Tucker 2000). On both sides of the linkage, three rows of hoop pine have been established to reduce edge effects, provide an additional source of farm timber and income, supply perching and nesting resources to pasture-based granivorous birds, and provide shade and shelter for grazing livestock. As these trees mature they should add value to the farm and the quality of habitat in the adjacent linkage restoration. This form of management is the optimum approach for these zones, providing the full array of agroecological services.

Plot Design

Plot topography and shape can be improved by establishing plots along contour lines and, wherever possible, establishing them as consolidated patches, such as circles or squares, to minimize edge effects (Murcia 1995). The interior of square and circular plots is likely to be the most appropriate position for many of the more sensitive rainforest species. These configurations may be more likely to resist weed invasion and thereby reduce maintenance inputs. Larger plots are also more likely to promote a zone of core interior habitat and support greater diversity of life forms and species.

Weed invasion and management inputs could be minimized by reducing plant spacing in and between rows. Increasing per hectare stem density from 1,500 to 2,000 would lessen management inputs such as weed control and pruning by hastening canopy closure and encouraging natural shedding of lower branches. In the wet tropics, the use of *Acacia* species such as *A. aulacocarpa*, *A. crassicarpa*, and *A. mangium* should be considered in this style of planting because of their rapid growth rates, ability to provide cover, and timber value. This form of higher-density planting may be more costly during establishment, but reduction in weed growth and natural shedding are compensatory factors. Where fallen logs and branches do not seriously impede the use or management of the area, they could also remain in place as habitat features. Restoration forestry at these spacings is unsuited to areas where grazing is the main land use.

Caveats

All of these strategies are predicated on changes to other management practices. No native forest should be cleared to facilitate farm forestry establishment, and all habitat features such as dead logs and rock piles should remain undisturbed in sites to be planted. Planting stock should be collected from as close as possible to the planting site to maintain genetic integrity, taken from a number of individuals to ensure a wide genetic base. The use of native species should be promoted wherever possible, and any exotic species included must be carefully and independently evaluated to minimize the risk of weed invasion into adjacent crops, pastures, or forests (see also Chapter 15, this volume).

Nontimber Production Benefits of Reforestation

With clear indications that local farm foresters appear likely to embrace multiple-use goals of reforestation, incorporating both biodiversity and production, the articulation and communication of other benefits may also increase participation rates (Emtage et al. 2001).

There is increasing evidence to suggest that increasing tree cover has important productivity benefits in a range of local agricultural crops. Trees planted in a windbreak configuration on the Atherton Tablelands increased yield in potatoes (up to 4.8 percent increase) and peanuts (up to 11.9 percent) (RIRDC 2001). Silver (1987) demonstrated that a dairy cow with access to shade can produce 1.5–2.2 L extra milk daily, compared with cows in unshaded pastures (see also Chapter 19, this volume). Trees used in both these trials were established for purposes other than timber production or local biodiversity conservation. However, by manipulating the species pool the same benefits could be achieved, in addition to timber production and biodiversity conservation.

Research has also shown that trees can significantly improve sugarcane yields by shading out the grasses and weeds that form the bulk of the diet of the canefield rat (*Rattus sordidus*), a major pest of sugarcane. Rodents cause up to $10 million (Australian dollars) damage to sugarcane crops in Queensland annually (Wilson and Whisson 1993), despite cane being only 20 percent of the rodent's diet. Traditional methods of control included herbicide application, burning of refuge areas (unproductive areas of grass and weeds adjacent to the crop that form the rodents' principal habitat), and baiting programs using a range of rodenticides. These actions are costly and repetitive, and rodent predators such as owls (*Tyto* spp.) and pythons (*Morelia* spp.) are susceptible to secondary baiting effects. Ecological restoration of rodent refuge areas leads to rapid and sustained decreases in rodent damage (Story et al. in preparation), and this could also be achieved using timber plantations to manipulate rodent habitat. As grasses and weeds are shaded out, grassland rodents are replaced by forest dwellers, species that do not consume sugarcane (Tucker 2001). In ecologically restored areas, rodent populations have declined by up to 80 percent within 12 months, although this may take longer in areas where stem density is lower and canopy closure slower.

Growers using this technique have reported savings in herbicide costs of A$20,000 per year (F. S. Gatti, pers. comm., 2001), in addition to cost savings in rodenticide and higher cane production. This approach is also being tried in a commercial macadamia nut (*Macadamia integrifolia*) orchard where rodent damage is costly, again related to availability of harborage areas (White et al. 1997). Preliminary results indicate potential reductions in rodent-damaged nuts of about 60 percent through habitat manipulation, leading to significant and sustained increases in farm productivity (White et al. 1998; Ward et al. in press).

Increasing the income generation potential of farm forestry is likely to increase adoption by local landholders. Many native species also have other potential commercial uses as food plants or food additives, firewood, foliage, or flowers, and these features can add to the value of farm forestry plantations. To date, there has been little evaluation of these secondary benefits. A more intensive research and development program will be needed not only to evaluate secondary benefits but also to address a number of the trade-off issues identified in this chapter.

Conclusions

Despite the significant loss of habitat in north Queensland's wet tropics, the situation is not as serious as in other parts of Australia. The more benign environment and natural recuperative abilities of tropical forest offer some hope for the wet tropics. Landscape-scale initiatives to improve tree cover can certainly assist the natural regeneration process, and a modified approach to farm forestry could hasten and direct this regeneration. Given the commitment by local landholders to improve the environmental and ecological qualities of the local landscape, such initiatives offer the possibility of improving both ecological and production outcomes.

Endnote

1. We sampled vascular plant assemblages at each site by scoring the presence or absence of plant species in an area of around 390 m², made up of five circular plots, each of 5 m radius, regularly spaced along a 100-m transect. Plants were assessed separately in three height strata (ground, midstory, and canopy). Data for the three strata were combined and the 612-species matrix reduced to a 265-species matrix by removing species that occurred at just one or two sites (singletons, 258; doubletons, 89).

References

Adam, P. 1992. *Australian rainforests.* Melbourne, Australia: Oxford University Press.

Aide, T. M., and J. Cavelier. 1994. Barriers to lowland tropical forest restoration in the Sierra Nevada de Santa Marta, Colombia. *Restoration Ecology* 2:219–229.

Andrén, H. 1994. Effects of habitat fragmentation on birds and mammals in landscapes with different proportions of suitable habitat: a review. *Oikos* 71:355–366.

Bentley, J. M., C. P. Catterall, and G. C. Smith. 2000. Effects of fragmentation of araucarian vine forest on small mammal communities. *Conservation Biology* 14:1075–1087.

Bunn, S. E., P. M. Davies, and T. D. Mosisch. 1999. Ecosystem measures of river health and their response to riparian and catchment degradation. *Freshwater Biology* 41:333–345.

Catterall, C. P. 2000. Wildlife biodiversity challenges for tropical rainforest plantations. Pages 191–195 in *Proceedings of the Biennial Conference of the Australian Forest Growers,* September 2000, Cairns, Australia.

Crome, F. H. J. 1975. The ecology of fruit pigeons in tropical northern Queensland. *Australian Wildlife Research* 2:155–185.

Crome, F. H. J., and J. Bentrupperbaumer. 1993. Special people, a special animal and a special vision: the first steps to restoring a fragmented tropical landscape. Pages 267–279 in D. A. Saunders, R. J. Hobbs, and P. R. Ehrlich (eds.), *Nature conservation 3. The reconstruction of fragmented ecosystems.* Chipping Norton, New South Wales, Australia: Surrey Beatty & Sons.

Emtage, N. F., S. R. Harrison, and J. L. Herbohn. 2001. Landholder attitudes to and participation in farm forestry activities in sub-tropical and tropical eastern Australia. Pages 195–210 in S. R. Harrison, J. L. Herbohn, and K. F. Herbohn (eds.), *Sustainable small-scale forestry.* St. Lucia, Australia: University of Queensland Press.

Erskine, P. 2002. Land clearing and forest rehabilitation in the wet tropics of north Queensland, Australia. *Ecological Management and Restoration* 3:135–137.

Gilmour, D. A., and J. J. Riley. 1970. Productivity survey of the Atherton Tableland and suggested land use changes. *Annual Journal of the Australian Institute of Agricultural Science* 38:254–272.

Goosem, S. P., G. Morgan, and J. E. Kemp. 1999. Regional ecosystems of the wet tropics. Pages 1–73 in P. S. Sattler and R. D. Williams (eds.), *The conservation status of Queensland's bioregional ecosystems.* Brisbane, Australia: Environmental Protection Agency.

Goosem, S. P., and N. I. J. Tucker. 1995. *Repairing the rainforest. Theory and practice of rainforest re-establishment in north Queensland's wet tropics.* Cairns, Australia: Wet Tropics Management Authority.

Grove, S. J., and N. I. J. Tucker. 2000. Importance of mature timber habitat in forest management and restoration: what can insects tell us? *Ecological Management and Restoration* 1:62–64.

Haggar, J. P., K. Wightman, and R. F. Fisher. 1997. The potential of plantations to foster woody regeneration within a deforested landscape in lowland Costa Rica. *Forest Ecology and Management* 99:55–64.

Herbohn, K. F., S. R. Harrison, and J. L. Herbohn. 2000. Lessons from small-scale forestry initiatives in Australia: the effective integration of environmental and commercial values. *Forest Ecology and Management* 128:227–240.

Hobbs, R. J. 1993. Can revegetation assist in the conservation of biodiversity in agricultural areas? *Pacific Conservation Biology* 1:29–38.

Kanowski, J., C. P. Catterall, G. W. Wardell-Johnson, H. Proctor, and T. Reis. 2003. Development of forest structure on cleared rainforest land in north-eastern Australia under different styles of reforestation. *Forest Ecology and Management* 183:265–280.

Keenan, R., D. Lamb, O. Woldring, A. K. Irvine, and R. Jensen. 1997. Restoration of plant diversity beneath tropical tree plantations in northern Australia. *Forest Ecology and Management* 99:117–131.

Kent, D. J., and J. M. Tanzer. 1983a. *Evaluation of agricultural land, Atherton Shire, north Queensland.* Brisbane, Australia: Queensland Department of Primary Industries.

Kent, D. J., and J. M. Tanzer. 1983b. *Evaluation of agricultural land, Eacham Shire, north Queensland.* Brisbane, Australia: Queensland Department of Primary Industries.

Lamb, D., J. A. Parrotta, R. Keenan, and N. I. J. Tucker. 1997. Rejoining habitat remnants: restoring degraded rainforest lands. Pages 366–385 in W. F. Laurance and R. O. Bierregaard (eds.), *Tropical forest remnants: ecology, conservation, and management of fragmented communities.* Chicago: University of Chicago Press.

Laurance, W. F. 1990. Comparative responses of five arboreal marsupials to tropical forest fragmentation. *Journal of Mammalogy* 71:641–653.

Laurance, W. F. 1991. Ecological correlates of extinction proneness in Australian tropical rainforest mammals. *Conservation Biology* 5:79–89.

Murcia, C. 1995. Edge effects in fragmented forests: implications for conservation. *Trends in Ecology and Evolution* 10:58–62.

Pahl, L. I., J. W. Winter, and G. Heinsohn. 1988. Variation in responses of arboreal marsupials to fragmentation of tropical rainforest in north eastern Australia. *Biological Conservation* 46:71–82.

Parrotta, J. A., J. W. Turnbull, and N. Jones. 1997. Catalysing native forest regeneration on degraded tropical lands. *Forest Ecology and Management* 99:1–7.

Proctor, H. C., J. Kanowski, G. Wardell-Johnson, T. Reis, and C. P. Catterall. 2003. Does diversity beget diversity? A comparison between plant and leaf-litter invertebrate richness from pasture to rainforest. Pages 51–65 in A. D. Austin, D. A. Mackay, and S. Cooper (eds.), *Proceedings of the 5th Invertebrate Biodiversity and Conservation Conference,* Records of the South Australian Museum, Supplementary Series, Adelaide, Australia.

QDPI (Queensland Department of Primary Industries) Forest Service. 1994. *Community Rainforest Reforestation Program. Annual report 1993/94.* Brisbane, Australia.

RIRDC (Rural Industries Research and Development Corporation). 2001. *Windbreaks: increasing crop growth on the Atherton Tablelands.* Short Report No. 67. Canberra, Australia: Rural Industries Research and Development Corporation.

Silver, B. A. 1987. Shade is important for milk production. *Queensland Agricultural Journal* 113:31–33.

Storey, P., N. I. J. Tucker, and A. D. Brodie. In preparation. Habitat manipulation to control rodent pests in sugar cane crops.

Tracey, J. G. 1982. *The vegetation of the humid tropical region of north Queensland.* Melbourne, Australia: CSIRO.

Tracey, J. G. 1986. *Trees on the Atherton Tableland: remnants, regrowth and opportunities for planting.* CRES Working Paper 1986/35. Canberra, Australia: Centre for Resource and Environmental Studies, Australian National University.

Tucker, N. I. J. 2000. Linkage restoration: interpreting fragmentation theory for the design of a rain forest linkage in the humid wet tropics of north-eastern Queensland. *Ecological Management and Restoration* 1:39–45.

Tucker, N. I. J. 2001. Wildlife colonisation on restored tropical lands: what can it do, how can we hasten it and what can we expect? Pages 279–295 in S. Elliott, J. Kerby, D. Blakesley, K. Hardwicke, K. Woods, and V. Anusarnsunthorn (eds.), *Forest restoration for wildlife conservation*. Chiang Mai, Thailand: International Tropical Timber Organisation and the Forest Restoration Research Unit, Chiang Mai University.

Tucker, N. I. J., and T. M. Murphy. 1997. The effect of ecological rehabilitation on vegetation recruitment: some observations from the wet tropics of north Queensland. *Forest Ecology and Management* 99:133–152.

Vize, S. M., and C. Creighton. 2001. Institutional impediments to farm forestry. Pages 241–255 in S. R. Harrison, J. L. Herbohn, and K. F. Herbohn (eds.), *Sustainable small scale forestry*. Cheltenham, UK: Edward Elgar.

Ward, D. J., N. I. J. Tucker, and J. Wilson. In press. Cost-effectiveness of revegetating degraded riparian habitats adjacent to macadamia plantations in reducing damage. *Crop Protection*.

White, J., K. Horskins, and J. Wilson. 1998. The control of rodent damage in Australian macadamia orchards by manipulation of adjacent non-crop habitats. *Crop Protection* 17:353–357.

White, J., J. Wilson, and K. Horskins. 1997. The role of adjacent habitats in rodent damage levels in Australian macadamia orchard systems. *Crop Protection* 16:727–732.

Wilson, J., and D. Whisson. 1993. *The management of rodents in north Queensland canefields*. Brisbane, Australia: Sugar Research and Development Corporation Research Publication BS 16S.

Chapter 19

Silvopastoral Systems: Ecological and Socioeconomic Benefits and Migratory Bird Conservation

Robert A. Rice and Russell Greenberg

The last decade has brought an upsurge of interest in exploring more environmentally sound ways to practice agriculture in deforested regions of the tropics. Increasingly, the notion that some of the original biodiversity can be recovered or maintained in agricultural systems has been discussed in the ecology and conservation biology literature (Pimentel et al. 1992; Vandermeer and Perfecto 1997). Much of the earlier interest in ecologically sustainable tropical agricultural practices focused on systems using traditional shifting cultivation strategies (Thrupp et al. 1997; see also Chapter 8, this volume) or diverse cropping systems that incorporate trees (Nair 1990). It is thought that some 2.9 billion ha worldwide are devoted to shifting cultivation practices, involving 1 billion people in some fashion (Thrupp et al. 1997). Although pasturelands are the only other land use that rivals shifting cultivation in terms of area, some major tropical cash crops have taken center stage recently where biodiversity maintenance is concerned (see Chapters 9 and 10, this volume). A shift in emphasis from subsistence agroforestry practices to those that produce at least one major, globally traded commodity has held out the promise that sustainable practices could more readily be scaled up to affect large expanses of land and hence be a meaningful and more-than-symbolic effort at conservation (Perfecto et al. 1996; Rice and Greenberg 2000).

However, even the most important of these crops, coffee and cocoa, are grown on limited amounts of tropical lands. The worldwide land coverage for both these crops sums to only about 15 million ha. The looming, generally unmentioned issue in the arena of tropical land use—at least as far as connecting agricultural land use with biodiversity maintenance is concerned—is the widespread presence and continued development of pastureland for livestock. Developing countries today report more than 2.2 billion ha of pasture. The amount of land used for cattle pasture in Central America alone is 93.6 million ha, which accounts for 72 percent of all the agricultural land (FAO 2000). However, the attention paid to developing sustainable and biodiversity-friendly approaches to livestock rearing in the tropics typically has not been commensurate with the importance of this land use. Perhaps this is because clearing forestland for cattle production often is considered the most destructive land use for tropical soils, particularly those in the humid lowlands. Any recommendations for mitigating the environmental impact of cattle ranching may be perceived as endorsing a practice that is widely considered ecologically inappropriate.

However, research and conservation interest in sustainable pastureland development are slowly expanding (Naranjo 2000; see also Chapters 11 and 18, this volume). Slowly but surely, a literature is developing that explores the ecological sustainability of different pasture management practices. In this chapter we will briefly outline the general approaches that have been emerging and then focus our discussion on lessons to be learned from a silvopastoral system based on the ecological properties of native savanna trees that has developed in several areas of Mesoamerica.

Conservation of Bird Diversity on Neotropical Pastures

A large portion of the research on pasture management and biodiversity has focused on birds, particularly migratory birds (Lynch 1992; Saab and Petit 1994; Warkentin et al. 1995; Greenberg et al. 1996; Siegel and Centeno 1997). Birds make a particularly attractive focus for such discussion because so much is known about their distribution and ecological needs. Furthermore, migratory bird conservation is one area in which reliance on preserves of natural habitat alone will not be a sufficient approach to maintaining global population numbers.

Active pasture with introduced grasses supports very low levels of avian diversity. Avian diversity increases dramatically on pastures where shrub and tree cover is developed. The following three practices are particularly important for the development of the shrub and tree components of pastures.

Fallowing

Leaving areas fallow allows the rapid development of a shrub assemblage that supports an avifauna typical of a variety of early successional communities. In a study in Belize, Saab and Petit (1994) found a two-and-a-half times greater species richness of the avifauna in fallow pastures than in regularly mowed and burned pasture. However, the number of species involved in the more diverse pastures was very low compared with that of most other habitats in the region

Table 19.1. The number of total species and forest species of birds detected on 25 1-km transect surveys of different pasture and nonpasture habitats in the Selva Lacandona, Chiapas, Mexico.

Habitat	Total Species	Forest Species
Active pasture	47	0
Fallow pasture	67	2
Pasture riparian corridors	100	10
Pasture with second growth	120	26
High-graded forest	121	50
Primary forest	123	62
Clearing in primary forest	143	51

Source: R. Greenberg, unpublished data.

(39 species). Systematic surveys of different pasture and natural habitats in the Selva Lacandona (near Chajul, Chiapas, Mexico) also show higher numbers of bird species in fallow pasture (Table 19.1). However, the gain in richness is small compared with that of other (even human disturbed) habitats, and few forest birds were observed using the fallows. Given the contribution of extensive fallowing to the areas dedicated to pasture systems, it may be better to focus on management systems that result in greater gains in diversity and increase the efficiency of cattle production.

Maintenance of Riparian Corridors

Trees or shrubs often are left along small stream courses through pastures (see also Chapter 18, this volume). This is a traditional practice found throughout the tropics and probably is a small effort to maintain water quality and prevent streamside erosion or is the result of the inaccessibility of the slopes of stream arroyos (i.e., dry or intermittent creeks) and canyons. The riparian band is generally quite narrow (less than 20 m wide), but the three-dimensional structure and floristic composition can vary with origins and subsequent management of the zone. In the Selva Lacandona, Warkentin et al. (1995) found that arroyo vegetation varied markedly between different communally owned pasture areas (ejidos) in the same region. At one extreme, arroyo vegetation can be made up of remnants of the original forest vegetation, where at least some of the larger trees are left standing. At the other extreme, it can be burned frequently, resulting in a combination of shrubs and small fire-resistant trees. Finally, a mature canopy can be maintained but the understory cleared for grass and herb growth all the way to the stream boundary.

The conservation value of these remnant or secondary woods along streamsides has received almost no attention. Survey data from these riparian zones and pastures that incorporate patches of young second growth show substantially greater bird richness (Table 19.1). Small populations of forest generalists have been found in these habitats, even where they have been isolated from forest for long periods (Table 19.1; Warkentin et al. 1995; Siegel and Centeno 1996). A high diversity of migratory birds often is associated with these formations as well. J. Salgado (unpublished data) found that variation in structural and floristic diversity had a large impact on bird diversity. Specifically, he found that an increase in the layering of riparian vegetation resulted in a 30–40 percent increase in species numbers. Given that these corridors often are the only substantial wooded vegetation remaining in many areas, further research on the details of vegetation management and its effect on biodiversity in these zones should be a high research priority.

Silvopastoral Practices

Finally, incorporating trees in pasture systems adds structural diversity and resources that enhance the biodiversity of pastures. Many tropical pasture systems incorporate a low density of trees simply to provide shade for livestock. However, the management of trees for fodder, building material, and other products is also widespread. Studies of the contribution of silvopastoral systems to biodiversity are few but promising. For example, Naranjo (1992) found a substantial increase of species associated with trees incorporated into ranching systems in the Cauca Valley of Colombia (see also Chapter 11, this volume).

It is important to realize that silvopastoral systems are not limited to small experimental enterprises, the resulting benefits of which can be disseminated through agricultural extension somewhere down the road. Rather, silvopastoral systems have developed within the ranch communities themselves and in some areas represent common practices. Any strategy to promulgate the use of such practices should first fully understand their costs and benefits in areas where they are already being applied.

The presence of trees or woodlots in pasturelands, especially in arid or semiarid zones, is a common practice in a number of regions throughout the neotropics. In southern Mexico, Nicaragua, Cuba, Colombia, and Bolivia, we find beef and dairy production taking place on pasturelands in conjunction with leguminous trees, which themselves offer a number of agronomic, environmental, and socioeconomic benefits (Durr 1992; Murgueitio 1999; Purata et al. 1999; Naranjo 2000; Botero 2001). The use of pods as supplemental feed (especially during extended dry season periods), the use of the wood from the trees for items such as fenceposts, firewood, tools, and building material, and the potential habitat provided by the woodlots themselves are all included in the benefits list of such trees. Other tropical regions also use such associations in agricultural production (Hashim 1994; Viswanath et al. 2000) and even for human consumption of specific plant parts, such as the flowers or flower buds of *Acacia acatlensis* (Hersch-Martinez et al. 1999).

Two tree genera are particularly well known for their incorporation into livestock systems: *Prosopis* (mesquite) and *Acacia*. In this chapter we will focus on silvopastoral systems that, while incorporating the use of a number of native tree species, have developed primarily around the uses of carbón (or "huizache" in Mexico), *Acacia pennatula*.

The Carbón System

Carbón (Spanish for "charcoal") is a tree-sized member of the *Acacia* genus that occurs in semiarid, subtropical regions from Mexico to Colombia. Although it usually shows an elevational restriction to 800–1,200 m above sea level, it regenerates well in association with cattle and other livestock and often achieves high abundance, forming large single-species stands. Researchers have documented its role in cattle production in the highlands of the state of Veracruz, Mexico (Chazaro 1977; Purata et al. 1999), the Ocosingo Valley of eastern Chiapas (Greenberg et al. 1997), and the highlands of north-central Nicaragua (the focus of this chapter). A member of the family Mimosaceae, carbón is a small to medium-sized tree, reaching 6–12 m (Figure 19.1). Compound leaves are pinnate, alternate, and up to 20 cm long, with 20–40 pairs of pinnae. Leaflets are 2 mm long and about 1 mm wide, with some 18–40 per pinna. An extrafloral nectary is present at the base of each petiole. Defoliation often occurs before flowering, in which small yellow to yellowish-orange spherical flowers cover the tree in raceme groupings up to 10 cm long. Fruits are indehiscent pods 9–12 cm long, deep coffee-colored when mature, that remain on the tree for some time before falling (Salas 1993).

Carbón is thought to have traits that evolved to take advantage of an abundant and diverse megafauna that persisted in Mesoamerica through most of the Pleistocene (Janzen and Martin 1982). These traits are shared with a number of species that, even now, grow abundantly in areas with natural assemblages of large mammalian herbivores in India and Africa (Coe and Beentje 1991). The fruit is a hard indehiscent pod, the seeds of which are deposited on the ground after mastication and digestion of a large animal. It must also be transported away from the shade of the parent tree to germinate, primarily because it is shade intolerant but also to evade bruchid seed predators. Furthermore, the seedlings benefit from the differential grazing of potential competitor plants in the pioneer growth of abandoned pastures. The spines, which are particularly large and dense on young plants and along the trunk of older specimens, discourage both herbivory and incidental damage by large animals. Because of the combination of a large mammal-dispersed fruit and spines, carbón spreads rapidly in the presence of cattle and the absence of fire. The propensity to spread rapidly and form impenetrable thickets makes it similar

Figure 19.1. Carbón (*Acacia pennatula*) in pasture in north-central Nicaragua during (a) the rainy season (note oaks on the hillslope in the background) and (b) the dry season.

to *Prosopis juglans* in the southern United States, several species of *Acacia* in Africa, and *Acacia cavens* in Argentina. All of these species have a reputation for being weeds that can consume rangeland, causing problems for safe livestock husbandry. Although the local reputation of carbón vacillates between being a pest and an almost sacred tree (similar to that of *Prosopis juliflora* in several regions; see Chapter 15, this volume), its life history and ecology have not received detailed study. However, empirical evidence supports the view that *A. pennatula* is easily and quickly dispersed by cattle and that it does well in areas opened to pastures.

Therefore, carbón management focuses less on its propagation on rangeland and more on controlling where it grows to maximize the benefits and minimize its interference with other activities. Often carbón is restricted to a high density of small trees in woodlots and a low density of large trees in savanna-like pasture (Figure 19.1). These formations are maintained by fencing off cattle from the woodlots (and hence the seed source) or allowing cattle access but hiring laborers to cut or use herbicides on seedlings that sprout in open pastures.

Carbón-Based Silvopastoral Systems in North-Central Nicaragua

Nicaragua's north-central region—here considered to be the area made up of some portions of the departments of Estelí, Jinotega, and Matagalpa—offers a landscape in which silvopastoral practices are commonly observed. This chapter focuses on the area stretching east and south from the town of Estelí, a region characterized by a disturbed landscape, often used as cattle pastures (Figure 19.2). Within an elevational range of about 100–900 m above sea level, *A. pennatula* dots the pasture landscape and plays an integral part of many ranchers' management practices. Often, nearby slopes and hilltops display mixed hardwood forests dominated by local oaks (*Quercus sapitofolia* and *Q. peduncularis*), offering a notable vegetational contrast to the single-species woodlots of carbón found in the pastures.

Although *A. pennatula* is the principal tree found in pasturelands in the Estelí region, other tree species also are present. Moreover, ranchers take advantage of species to varying degrees, making use of the fruits and foliage for farm animals. Ranchers use genízaro (*Samanea saman*), guacimo (*Guazuma ulmifolia*), vainilla (*Senna atomaria*), nacascolo (*Caesalpinia coriaria*), and, to a

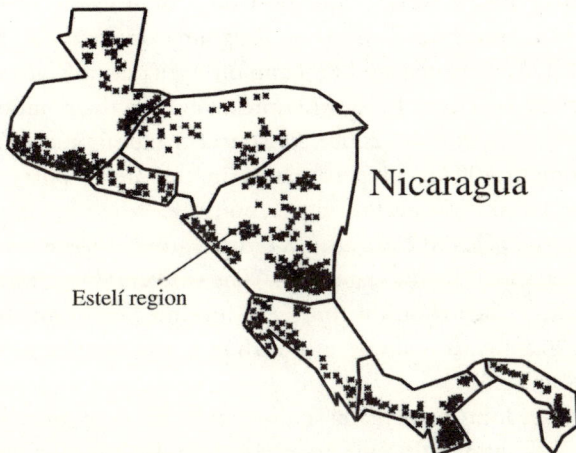

Figure 19.2. Distribution of pasture and silvopastoral systems in Central America (based on Winograd and Farrow 1999).

limited degree, guanacaste (*Enterolobium cyclocarpum*) in some way as supplemental food sources for cattle and other farm animals such as chickens or rabbits. However, it is carbón that constitutes the major association with pasturelands and has coevolved in socioecological terms to form an integral part of the pasture landscape in north-central Nicaragua.

Carbón is found in woodlots and scattered patterns in many of the pastures. The other species are found along pasture edges and disturbed areas, often as single individuals or groups of several individuals of the same species. However, no other species occurs at such high density as that observed for *A. pennatula* (Figure 19.1). The range of carbón in Nicaragua historically has been confined to the central portion of the country and seems to be expanding through the narrow bottleneck between Lake Nicaragua and Lake Managua toward the Pacific side (Salas 1993). Rainfall for the Estelí region totals 900–1,000 mm per year, distributed temporally such that January to May constitute an extremely marked dry season, called the "critical period" by local ranchers because of the sparseness and low nutritional value of pasture grass and the general lack of water regionwide.

The Socioeconomic Benefits of Carbón

During this critically dry time of the year, pods of *A. pennatula* are collected and fed to cattle as a feed supplement. Aside from pods, carbón provides foliage for fresh fodder and wood for a number of farm uses. This section presents some of the ways in which the silvopastoral system based on carbón serves both socioeconomic and environmental needs.

Attitudes differ with respect to the usefulness of the tree and its byproducts, but such discrepancies may speak to whether someone is from the region and how closely someone works with the land. For instance, the nonlocal director of a local program known as Programa Integral de Desarrollo Agropecuario (PIDA, the Integrated Program on Agricultural/Cattle Development) characterized the tree as being widespread, even to the point of being "a pest" (S. Sandoval, pers. comm., 2000). By contrast, a local rancher who owns and oversees some 1,100 ha of pastureland in the area reports that local campesinos have a healthy respect for the carbón tree. When this large rancher sends workers to the fields to clear out overgrown areas in pastures, their attitude toward the species is highly respectful: "The workers always leave the acacia seedlings intact. They chop out near and around them, but they always leave the seedlings alone. It is almost as if carbón is a sacred tree to them" (H. Torres, pers. comm., 2001).

The various uses local residents have discovered for *A. pennatula* speak to its long history with humans in the region. As might be expected, the wood is an important source of firewood and, as its common name implies, charcoal. The foliage is harvested during the dry season as a source of fodder for cattle, with ranchers large and small cutting leaves and delivering them to cattle for feed. And all carbón trees in pastures show the classic browse line, determined by the maximum height the cattle can reach in their free-ranging activities, attesting to the palatable foliage of the trees.

Construction material for buildings (beams or columns) is another commonly cited use of carbón. Posts for fencing, especially from pieces containing heartwood, are highly prized for their durability. More than one rancher referred to the use of *A. pennatula* wood for axles in ox-carts, a use that illustrates the wood's toughness. Similarly, some ranchers train or prune specific branches in a tree to obtain the correct shape for an ox-drawn plow. Less common uses of the tree were discovered during informal interviews with a number of ranchers in the area. Poor rural families use the thorny branches as makeshift fences around their houses. And the pod itself was called a "country toothbrush" by one rancher, who explained that the rural poor used the fibrous pod to scrub their teeth (S. Torres and H. Torres, pers. comm., 2001).

The Use of A. pennatula and Other Species as Cattle Feed

Producers both large and small tend to incorporate native tree species into their animal husbandry practices. A short survey of 19 of the larger ranchers in the area, conducted through the Asociación de Ganaderos in Estelí, together with a similar survey collected from 138 small ranchers represented by the Unión Nicaragüense de Agricultores y Ganaderos (UNAG) provide information about the important role of these trees in farm production (Table 19.2). The general profile of small and large producers reveals that although average area (24 and 142 ha, respectively) and average cattle wealth (12 and 132 head, respectively) differ by an order of magnitude, milk production is quite similar (3 to 4 L per cow per day in the dry season and about 6 L per cow per day in the rainy season for both groups).

The use of carbón pods by ranchers takes a number of forms. As in southern Mexico, many growers simply allow cattle to consume the pods in free-range foraging (Purata et al. 1999). Some growers collect the pods and break

Table 19.2. Percentage of large and small ranchers using native tree species' fruits as animal feed.

Species	Large Ranchers	Small Ranchers	
	Cattle	Cattle	Chickens
Carbón (*Acacia pennatula*)	74	49	49
Genízero (*Samanea saman*)	15	15	4
Nacascolo (*Caesalpinia coriaria*)	10	5	0
Vainilla (*Senna atomaria*)	5	15	6
Guácimo (*Guazuma ulmifolia*)	36	30	76
Guanacaste (*Enterolobium cyclocarpum*)	21	17	7

them up in a minimal way that allows cattle to consume (and digest) them more easily. Still others take collected pods and mill them into a flour that is added to fodder and other materials during the dry season in an attempt to maintain milk production. Of all the small ranchers surveyed, 49 percent use carbón pods in some way, with 35 percent allowing cattle to find pods as they range about in the pasture. Both growers feeding partially broken pods to cattle and those taking the trouble to mill pods into flour represent 7 percent of all those interviewed.

By contrast, larger ranchers make greater use of the pods overall for cattle feed. Nearly three quarters of all those surveyed (74 percent) report using carbón pods in some way. Thirty-two percent allow the cattle to find the pods on their own, and about 5 percent take the trouble to collect and break them up for the cattle. Another 37 percent grind the pods into flour to add to cattle feed during the dry season.

The Informal Economy of Pod Collection

Ranchers milling carbón pods into meal or flour for cattle feed must get the pods to a central site to make milling cost-effective. Some instruct their ranch workers to collect the pods, but most rely on an informal network of local residents who work every year in the provisioning of nearby ranchers with pods. No central collection sites exist near Estelí, but rural family members—usually some combination of women and their children—harvest the pods, pack them into large jute or plastic sacks, and sell them to interested ranchers.

The labor involved in collection and selling the pods is strenuous and often surreptitious, with gatherers entering pastures in which pod-laden trees have been left for free-ranging cattle. Collectors use long poles to knock the pods off the trees, often hurrying to gather the pods into buckets or sacks before cattle consume them. Once gathered, prior arrangements to sell them to interested ranchers involves a roadside weekly exchange of pods for money or, in rare cases, delivery of sacks to the rancher. From interviews conducted in this area, it was found that local residents can earn up to about US$200 over the 2-month pod production season. Pod theft does not seem to cause intense conflict between ranchers and local collectors but does speak to the value of these fruits as a resource and hence to the overall need to foster productive carbón management practices.

A more established business around the collection, delivery, and sale of carbón pods (and other species as well) exists in the town of Sébaco, a 40-minute drive south of Estelí. A large rancher there buys pods from local residents each year, using these native species for his own cattle and selling the meal or flour from ground pods to other ranchers. The species include all those mentioned earlier: *A. pennatula*, *Samanea saman*, *Caesalpinia coriaria*, *Senna atomaria*, and *Guazuma ulmifolia*. Regardless of the degree of commercial establishment around the use of the pods, from a policy perspective it is obviously an intriguing arena for development of local, regional, and even international market potential.

Nutritional Value of the Native Tree Species' Fruits

For any rancher using the fruits from native trees as feed supplement, a major concern is the nutritional value of the forage and its potential harm from possible toxins. For instance, a number of ranchers report that although cattle consume and seem generally unaffected by the pods of guanacaste (*Enterolobium cyclocarpum*) trees, something in the pods has an abortive property for pregnant cows. Therefore, pregnant cows are kept from eating pods from this species. A number of sources report the nutritional value of *A. pennatula* pods (Sotelo 1981; Durr 1992; Purata et al. 1999). Table 19.3 presents nutritional data on *A. pennatula* and some of the other species featured in this chapter. As seen in these data, the percentage of crude protein and the percentage of soluble (digestible) protein are quite high for most of the species listed, comparing well with common silages (e.g., sorghum, maize, soybeans) used in dairy feed. Moreover, the percentage of total digestible nutrients also ranks quite high for these native tree fruits. And finally, the iron content of these feed supplements is listed to show that some species are an excellent source of that mineral. Although iron deficiency generally is not a problem for bovine cattle, iron is a critical ingredient in blood transport and (for ruminants especially) vital organ development and maintenance. And where porcine cattle are involved, anemia of newborns is a common problem (L. Osegueda, pers. comm., 2001).

Noteworthy are the values for genizero (*Samanea saman*), which show the

Table 19.3. Nutritional properties of fruits from native tree species used as cattle feed supplement in north-central Nicaragua in comparison to common types of silage.[a]

	Crude Protein (% of dry weight)	Soluble Protein (% of crude protein)	Total Digestible Nutrients (%)	Iron (mg kg⁻¹)
Maize silage	8.0	49	68	190
Sorghum silage	9.8	44	57	282
Soybean silage	19.0	51	58	492
Acacia pennatula[b]	11.1	46	64	83
Caesalpinia coriara[b]	5.0	40	80	62
Senna atomaria[b]	11.4	47	62	122
Samanea saman[b]	17.2	51	74	494
Guazuma ulmifolia[b]	6.1	39	66	679

[a]Analysis of pods or seeds was carried out by Dairy One's DHI Forage Testing Lab, Ithaca, NY. Figures for maize, sorghum, and soybean silage were provided by M. Reuter, Dairy One.

[b]All tree species except *G. ulmifolia* (Stericuliaceae) are legumes.

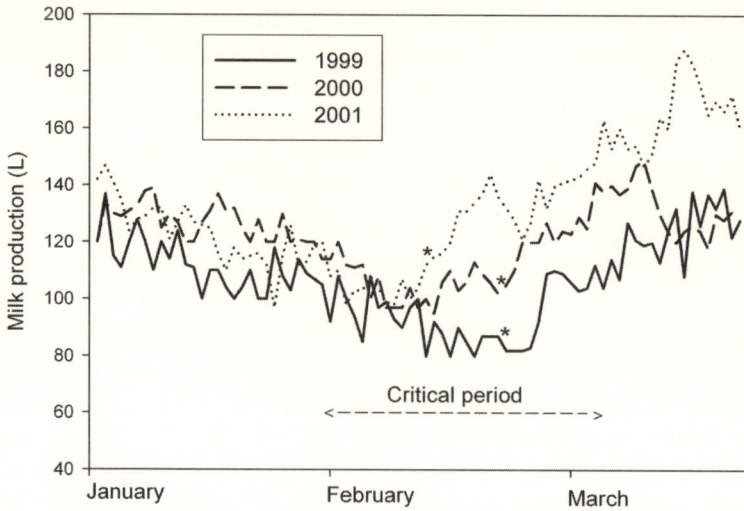

Figure 19.3. Effect of *Acacia pennatula* pod meal as a feed supplement on milk production on Finca Ajenjal, Estelí, Nicaragua (1999–2001). The asterisks indicate the initiation of supplement feeding.

highest levels of protein and second-highest levels of nutrient digestibility and iron content. Obviously, silvopastoral systems that incorporate two or more of these species would offer a mix of nutritional resources for ranchers deciding to use native species as well as mixed stands or galleries of trees for biodiversity maintenance in pastures. Tests on the toxicity levels of *A. pennatula* seeds show that the mycotoxin contents pose no reason for concern.

Figure 19.3 shows 3 years of data on milk production on a ranch, depicting the overall effect of a feed supplement containing one-fourth meal from *A. pennatula*. During the critical period of the dry season, when grass and other naturally available forage materials are difficult to find and generally of low nutritional value, milk production for the 40 milk cows climbed up to rainy season levels shortly after the carbón meal and other ingredients were added to the diet. The benefit of such nutrient-rich native tree species for milk production in seasonally arid regions positions trees such as carbón quite well as a resource ranchers can exploit and researchers can further study.

Native Trees as Avian Habitat

The avifauna associated with *A. pennatula* was systematically surveyed in both natural and anthropogenic habitats along an altitudinal gradient in eastern Chiapas from the Selva Lacandona to the area around San Cristobal de las Casas (Greenberg et al. 1997). In the surveys, carbón stands in the Ocosingo Valley supported the highest density and diversity of migratory birds (from the temperate zone) of any of the 18 habitats sampled (Table 19.4). This included the highest densities of 12 species. The overall density of birds was also high, although species richness was not remarkable. The low overall species richness was the result of a general lack of all but the most generalized resident species. The composition of the avifauna was strongly dominated by small foliage-gleaning species, such as magnolia and black-throated green warblers (*Dendroica magnolia* and *Dendroica virens*), least flycatcher (*Empidonax minimus*), American redstart (*Setophaga ruticilla*), and blue-gray gnatcatcher (*Polioptila cerulea*). In mixed stands of oaks and other trees, a disproportionate number of foraging migrants (not residents) were found in carbón trees. The clear domination by foliage insectivores and the preference for the tree in mixed associations led us to hypothesize that *A. pennatula* provided an abundant source of arthropods to support birds overwintering in pasture or woodland habitats dominated by this tree species.

Bird use of *A. pennatula* was further studied in north-central Nicaragua. At this site we conducted a series of year-round surveys of bird use of carbón and other trees in woodlots and pastures along with intensive sampling of arthropod abundance and analyses of foliage chemistry. We established 20 km of survey transects in the oak-carbón ecotone at the edge and in estab-

Table 19.4. The abundance and species richness of migratory birds in *A. pennatula*–dominated and other habitats in the Selva Lacandona–Ocosingo region of Chiapas, Mexico.[a]

Habitat	Individuals Observed	Species	Individuals per Survey Point
A. pennatula woodlot	73	19	9.0
A. pennatula savanna	21	11	6.0
Low-elevation pasture	70	3	0.7
Low-elevation second growth	100	6	1.3
Low-elevation riparian strip	187	13	4.7
Low-elevation forest	102	5	1.2
Mid-elevation pasture	70	5	0.7
Mid-elevation second growth	100	4	2.1
Mid-elevation slash-and-burn field	70	8	1.2
Shade coffee	212	14	5.1
Mid-elevation riparian strip	52	13	5.2
Mid-elevation pine-oak-liquidambar forest	100	3	1.3
Mid-elevation pine-oak forest	82	7	3.2
Mid-elevation pine forest	70	3	1.1
High-elevation pine forest	70	4	1.8
High-elevation shifting cultivation plots	70	3	1.1
High-elevation second growth	100	2	1.3
High-elevation pine-oak forest	50	4	1.3

[a]Based on fixed radius point counts per habitat (Greenberg et al. 1997). Species count includes only those with abundance > 0.1 individual per survey point.

lished cattle pastures. These were surveyed at four times during the year, at which time we clipped foliage from 20 specimens of the dominant oak (*Quercus sapitofolia*) and carbón. All insectivorous birds showed a strong preference for feeding in carbón (the average foliage cover of carbón was 17 percent, and its average use was 70 percent for migrants and 50 percent for resident species).

Overall insect abundance was substantially higher for *A. pennatula;* this includes all of the major orders except roaches and Orthoptera. The difference in abundance was particularly strong for small arthropods. Carbón supports very high abundances of very small lepidoptera, beetles, and spiders. Larger arthropods were nearly as common in oaks, and the very large roaches and Orthoptera were found primarily in oaks (Table 19.5). Consistent with the higher overall abundance (and estimated biomass) of arthropods was the plant chemistry. Carbón foliage has more than twice the protein and significantly higher levels of digestible carbohydrates than oaks. Furthermore, the concentration of condensed tannins, thought to be particularly important for discouraging herbivory, was significantly lower in carbón than in the oak foliage. We detected high concentrations of cyanogenic compounds in the carbón foliage, which is thought to deter herbivory, particularly by vertebrates, which lack the ability to sequester or neutralize these compounds.

Based on our observation of birds and sampling of arthropods and vegetation, we offer the following hypothesis for why carbón supports such high densities of insectivorous birds: *A. pennatula* foliage shows properties that are often found in pioneer tree species in tropical environments (Greenberg and Bichier in press). The foliage is soft and palatable, with a low ratio of cell wall to digestible materials. It also has low concentrations

Table 19.5. The number of arthropods of different size and taxonomic classes in *A. pennatula* and oak (individuals per 100 g dry foliage).

Arthropod Type	A. pennatula	Oak
Total	37.0	12.0
> 5 mm	1.5	0.5
Orthoptera (+ roaches)	1.0	1.2
Spiders	5.6	2.0
Homoptera	3.9	1.2
Hemiptera	0.7	0.4
Lepidoptera	5.9	0.9
Coleoptera	12.5	4.2
Hymenoptera	2.2	1.2
Diptera	1.8	0.7

Source: Based on Greenberg and Bichier (in press).1

of more expensive chemical defensive compounds. Moreover, the high growth rate and the fact that carbón has nitrogen-fixing symbiotic bacteria contribute to the high protein content of its foliage. Aside from these characteristics, the tree is highly adapted to coexistence with large mammalian herbivores. Incidental damage and herbivory from these animals is minimized by the presence of large spines. The increased expense of a mechanical defense of thorns may further limit energy available for more general chemical defenses that could otherwise deter insect herbivory. Taken together, the adaptation for rapid growth for success in early successional competition and the investment in mechanical defenses specialized on large mammalian herbivores may result in foliage that is highly palatable to arthropods and hence lead to high arthropod densities, which support the foraging of insectivorous birds.

However, the arthropod fauna is characterized by small species. This probably reflects the small leaflet size of the plant, which affords little protection from predators and the harsh environment of a tropical savanna. The preponderance of small arthropods supports an insectivorous avifauna that is, itself, made up of small species. In the northern neotropics, migrants from the north dominate the small foliage-gleaning guild, hence the value of the carbón savannas for migratory birds.

The Multiple Roles of *A. pennatula* as a Keystone Species

A. pennatula probably deserves the designation "keystone species" in its role in modern pasture savanna ecosystems and its likely ancient role in natural savannas in Central America. Keystone plant species are those that play a critical role in determining composition and abundance in animal assemblages (Howe and Westley 1998; Peres 2000). Keystone species generally are those that produce fruit or nectar that supports a wide variety of animals during times of resource shortage. Therefore, their absence would severely depress the diversity and abundance of frugivorous or omnivorous species. It is likely that the high abundance of pods produced during the severe dry seasons in savanna regions supported many large grazing and browsing animals, particularly ungulates, before the episodes of Pleistocene extinctions (Greenberg et al. 1997). In the modern silvopastoral systems, this resource is a critical one for allowing livestock access to a reliable source of protein, digestible carbohydrates, and critical trace elements.

In discussing the potential importance of *A. pennatula*–based silvopastoral systems in the New World, it is interesting to consider the critical role that ecologically similar *Acacia* species have in both natural and managed ecosystems of the Old World in India and from the Middle East through the Sahel and Eastern and Southern Africa:

Since acacias nearly always flower just before or during rains and pods reach maturity during the dry season, the robust and generally thick-walled pods of the indehiscent species provide an important dry season food source. This availability of high quality food during a period of general shortage has led the pastoral peoples throughout the range of these trees to devise hooked sticks to shake down the ripe pods manually. The sound of falling pods will bring their animals from several hundred metres away. Indeed, the sound of falling pods has traditionally been used by tribal peoples in central India to kill wild herbivores. They are similarly attracted to the sound of the pods being shaken down, since the sound mimics that made when Langur monkeys feed in these trees. (Coe and Beentje 1991)

However, carbón also appears to provide an unusual example of a "food tree" in which the resource provided to consumers is a high abundance of arthropods rather than plant material itself. We believe that the presence of *A. pennatula* greatly enhances resources and allows increased survival of insectivorous birds through the severe dry season. We demonstrated that high arthropod abundance was maintained at the peak of the dry season, even when most or all of the foliage had dropped. Finally, *A. pennatula* has an important and as yet unstudied role in secondary succession on fallow pastures. Where its seedlings and saplings are not removed, they appear to protect seedlings of other shrub species from cattle grazing. A small, diverse community of shrubs and tree seedlings grows up in the thorny branches of the *A. pennatula* seedlings. The long-term effect of such protection should be the subject of further investigation.

Despite the fact that *A. pennatula* trees are the center of foraging activity for most insectivorous birds in the mixed habitats we studied, preliminary observations suggest that some species remain specialized on other tree species (primarily oaks) and that most birds moved out of carbón stands and into oaks during the hot, sunny, and exceedingly windy afternoons. Therefore, *A. pennatula* and broadleaf trees appear to provide complementary resources even for the species that forage primarily in carbón. Silvopastoral strategies that recognize the importance of particular tree species, while managing for a diversity of tree species, will undoubtedly support the highest diversity and abundance of birds (and other organisms as well).

Conclusions

A. pennatula plays two important roles in the conservation of bird populations in Mesoamerican pasturelands. First, we believe that it increases the carrying capacity of the pastures for supporting insectivorous birds, particularly long-distance migrants from the temperate zone. We need more research to assess the survivorship and condition of birds that winter in carbón-based habitats, but some preliminary data suggest that individuals of territorial species persist at a high rate throughout the winter (Greenberg et al. 1997).

The second (and more indirect) role that carbón plays is to increase the ability of cattle ranches to provide their own resources. Aside from the obvious foliage-as-forage benefits and the respite from heat and direct sun provided by its shade, carbón supplies wood to ranchers. The wood plays a critical role in providing fencing and other materials that might otherwise be harvested from oak woodlands and mesophyllous forests less able to withstand this use in a sustained manner. In addition, farms where the cattle are fed or have access to carbón pods are less dependent on maize and commercial feed mixtures that themselves have substantial environmental impacts.

Despite the fact that in some areas *A. pennatula* is a valuable resource for livestock management that provides resources for insectivorous birds, the system can be expanded. The expansion should be focused on the management of the silvopastoral system that maximizes both agronomic and ecological benefits. Fieldwork in Chiapas, Mexico, demonstrated that the greatest abundance and diversity of birds was found where carbón was managed in woodlots rather than as isolated trees in pastures (Greenberg et al. 1997). The continuous canopy of the woodlots helps create a microclimate that supports a distinct understory fauna. Management of *A. pennatula* in fenced woodlots would minimize the random transport of seeds into pasture and reduce the intensive labor (or herbicide use) needed to prevent the unwanted spread of seedlings. Collection and processing of seeds (as is done throughout the neotropics for *Prosopis;* Silbert 1988) would also help to reduce the spread of carbón. On the other hand, Purata et al. (1999) found that seed pod production was reduced by tree density, so a research program focused on optimizing the biodiversity and production value of the tree should be undertaken. The research in Nicaragua (Greenberg and Bichier in press) demonstrates that the benefits for bird diversity are greatest where carbón occurs in mixed stands with oaks and other native trees. Further research into the development of mixed systems with other pod-producing trees should also be a high priority for further research.

Incorporating riparian corridor protection and management into the *A. pennatula* silvopastoral system may be the most promising approach to maximizing biodiversity of pastoral systems in ecologically appropriate regions. Having developed a case for the expansion and refinement of the system in the regions where *A. pennatula* is native, we would strongly argue against introducing this species into regions where it is not native. The invasive qualities of related species (notably *Acacia farnesiana* and several *Prosopis* species) in areas where they have been transplanted throughout the tropics (see Chapter 15, this volume) suggests that a similar spread of *A. pennatula* would be a dangerous and uncontrolled ecological experiment.

For policy and decision makers in areas where tree species such as carbón might be integrated into the pasture management scheme, the agronomic benefits are obvious. To the degree that cottage industries or even regional and international economies could develop around such silvopastoral practices, the ecological attraction would be a bonus. Planners understanding the connection between the economic and biodiversity benefits could take advantage of these links, making bold statements by supporting research and model farms. Issues related to planting density and spatial arrangement of trees in pastures are little understood, but the vast area devoted to pasture throughout the tropics warrants coordinated efforts on the part of local governments, international lenders, and nongovernment bodies.

Acknowledgments

We thank the small grants program of the American Bird Conservancy and the Abbott Fund of the Smithsonian Institution for their support of our fieldwork.

References

Botero, J. A. 2001. *Menú técnico para la intensificación ganadera en Colombia, Nicaragua y Costa Rica.* Report by consultants to the GEF PO72929 project, Regional integrated silvopastoral approaches to ecosystem management, October 2001.

Chazaro, M. 1977. El huizache, *Acacia pennatula* (Schlecht. & Cham.) Benth. Una invasora del Centro de Veracruz. *Biotica* 2:1–18.

Coe, M., and H. Beentje. 1991. *A field guide to the acacias of Kenya.* Oxford, UK: Oxford University Press.

Durr, P. 1992. *Manual de arboles forrajeros de Nicaragua.* Estelí, Nicaragua: Ministerio de Agricultura y Ganadería, Región I.

FAO (Food and Agriculture Organization of the United Nations). 2000. United Nations Food and Agriculture Organization Web site for Land Use Statistics: http://apps.fao.org/page/collections?subset=agriculture.

Greenberg, R., and P. Bichier. In press. Determinants of tree species preference of birds in oak-acacia woodlands in Central America. *Journal of Tropical Ecology.*

Greenberg, R., P. Bichier, and J. Sterling. 1997. Acacia, cattle, and migratory birds in southeastern Mexico. *Biological Conservation* 80:235–247.

Hashim, I. M. 1994. Assessment of haraz (*Acacia albida*) pods as a potential supplement for cattle. *Tropical Grasslands* 28:127–128.

Hersch-Martinez, P., M. M. González, and A. Fierro-Alvarez. 1999. *Acacia acatlensis:* an alimentary resource in southwest Puebla and north of Guerrero, Mexico. *Economic Botany* 53:448–450.

Howe, H. F., and L. C. Westley. 1998. *Ecological relationships of plants and animals.* Oxford, UK: Oxford University Press.

Janzen, D. H., and P. Martin. 1982. Neotropical anachronisms: the fruit that gomphotheres ate. *Science* 215:19–27.

Lynch, J. 1992. Distribution of overwintering birds in the Yucatan Peninsula. II. Use of native and human-modified vegetation. Pages 178–195 in J. Hagen and D. Johnston (eds.), *Ecology and conservation of neotropical migrant landbirds.* Washington, DC: Smithsonian Institution Press.

Murgueitio, R. E. 1999. *Sistemas agroforestales para la producción ganadera en Colombia.* Online: http://www.cipav.org.co/redagrofor/memorias99/murgueit.htm.

Nair, P. K. R. 1990. *The prospects for agroforestry in the tropics.* World Bank Technical Paper no. 131. Washington, DC: World Bank.

Naranjo, L. G. 1992. Estructura de la avifauna en un área ganadera en el Valle del Cauca, Colombia. *Caldasia* 17:55–66.

Naranjo, L. G. 2000. Recovering paradise: making pasturelands productive for people and biodiversity. In *Proceedings of the First International Workshop on Bird Conservation in Livestock Production Systems,* Arlie Conference Center, Virginia, April 13, 2000.

Peres, C. A. 2000. Identifying keystone plant resources in tropical forests. *Journal of Tropical Ecology* 16:287–317.

Perfecto, I., R. Rice, R. Greenberg, and M. van der Voort. 1996. Shade coffee: a disappearing refuge for biodiversity. *BioScience* 46:598–609.

Pimentel, D., U. Stachow, D. A. Takacs, H. W. Brubaker, A. R. Dumas, J. J. Meaney, J. A. S. O'Neil, D. E. Onsi, and D. B. Corzilius. 1992. Conserving biological diversity in agricultural/forestry systems. *BioScience* 42:354–362.

Purata, S. E., R. Greenberg, V. Barrientos, and J. López-Portillo. 1999. Economic potential of the huizache, *Acacia pennatula* (Mimosoideae) in central Veracruz, Mexico. *Economic Botany* 53:15–29.

Rice, R., and R. Greenberg. 2000. Cacao cultivation and the conservation of biological diversity. *Ambio* 29:167–173.

Saab, V., and D. Petit. 1994. Impact of pasture development on winter bird communities in Belize. *Condor* 94:66–71.

Salas, J. B. 1993. *Arboles de Nicaragua.* Managua: Instituto Nicaragüense de Recursos Naturales y del Ambiente.

Siegel, R. B., and M. V. Centeno. 1996. Neotropical migrants in marginal habitats on a Guatemalan cattle ranch. *Wilson Bulletin* 108:166–170.

Silbert, M. S. 1988. *Mesquite pod utilization for livestock feed: an economic development alternative in central Mexico.* Master's thesis, University of Arizona, Tucson.

Sotelo, A. 1981. Leguminosas silvestres, reserva de proteinas para la alimantacion del futuro. *Información Científica y Tecnológica* 3:28–32.

Thrupp, L. A., S. B. Hecht, and J. O. Browder. 1997. *The diversity and dynamics of shifting cultivation: myths, realities, and policy implications.* Washington, DC: World Resources Institute.

Vandermeer, J., and I. Perfecto. 1997. The agroecosystem: a need for the conservation biologist's lens. *Conservation Biology* 11:1–3.

Viswanath, S., P. K. R. Nair, P. K. Kaushik, and U. Prakasam. 2000. *Acacia nilotica* trees in rice fields: a traditional agroforestry system in central India. *Agroforestry Systems* 50: 157–177.

Warkentin, I., R. Greenberg, and J. Salgado Ortiz. 1995. Songbird use of gallery wood-lands in recently cleared and older settled landscapes of the Selva Lacandona, Chiapas, Mexico. *Conservation Biology* 9:1095–1106.

Winograd, M., and A. Farrow. 1999. *Agroecosystem assessment for Latin America.* Prepared for World Resources Institute, Washington, DC, as part of the PAGE project.

Chapter 20

Agroforestry and Climate Change–Integrated Conservation Strategies

Lee Hannah

Mounting evidence indicates that global climate is changing, that species ranges are shifting in response to warming, and that current conservation strategies must be revised to be effective in the face of future changes (Hughes 2000; IPCC 2001b; Hannah, Midgely, Lovejoy, et al. 2002). Past climate change has resulted in alterations to species' ranges that have necessitated migration over tens or hundreds of kilometers (Huntley and Webb III 1989; Clark et al. 1998). In fully natural landscapes, these range shifts would present a serious challenge to species' survival. In today's world, in which natural habitat is heavily fragmented, rapid range shifts may be impossible for many species, dramatically reducing the prospect of successful natural adjustment to climate change (Hannah, Midgely, Lovejoy, et al. 2002). Habitat loss alone may be on the verge of causing a major mass extinction episode (Myers et al. 2000; see also Chapter 1, this volume). Climate change may well provide the impetus that makes such a mass extinction inevitable. Because past conservation strategies have assumed a stable climate, new strategies are urgently needed that can maintain biodiversity in heavily fragmented landscapes and a dynamic climate.

Biodiversity will be most strongly affected by climate change in tropical montane settings as species move upslope and to higher latitudes (Peters 1991; Bush 2002). Lowland tropical species will face less chance of range displacement because they are adapted to warmer and wetter conditions likely to be typical of future climates, whereas temperate areas will experience many range displacements, but in a biota not nearly as rich as the tropics. Cloud forest species in the tropics may be at particular risk because both global climate change and regional deforestation lead to changes in cloud cover and cloud base height (Still et al. 1999). Species face extinction when mountaintop habitat such as cloud forest shifts upward, when there is no habitat in areas that become climatically suitable in the future, or when no habitat exists between areas of suitable present and future climate. It is therefore in restoring and maintaining tree cover and seminatural habitat in high-biodiversity areas of the tropics that agroforestry has its greatest potential to benefit conservation of biodiversity as species' ranges shift.

Some of the most species-rich areas of the planet (the biodiversity hotspots; see Chapter 1, this volume), such as the tropical Andes, are heavily montane areas and therefore at elevated risk from climate change. These hotspots have already experienced high levels of habitat loss and are likely to be under increasing pressure for agricultural development as frost lines rise with climate change. Already, agroforestry is an important land use in upland areas where crops such as coffee and tea are prevalent. Climate change will alter the dynamic between species and cropping patterns in these areas, and agroforestry will play a role in determining the net impact of these alterations on biodiversity.

Conservation strategies that respond to climate change have been called climate change–integrated conservation strategies, or a CCS (Hannah, Midgely, Millar, et al. 2002). A major component of a CCS is the active management of the matrix of land uses outside and between parks to accommodate range shifts and other changes caused by climate variation. Agroforestry is a major land use option that can be important in conservation efforts in the matrix. Not all agroforestry systems work well in a CCS, however; some are valuable, whereas others have little advantage over other cropping systems. In this chapter, we examine the conservation challenges posed by climate change, then describe the attributes of agroforestry systems that can contribute to conservation of biodiversity in a changing climate.

Conservation Challenges

Climate has changed over the past century, leading to biotic changes (Walther et al. 2002). Mean global temperature has increased approximately 0.6°C since 1900, and models of global climate (general circulation models [GCMs]) predict further warming of 2.5°C to 6.7°C by 2100 (IPCC 2001b). Biological responses to the warming already under way include species range shifts and changes in abundance and phenology (Parmesan and Yohe 2003; Root et al. 2003). Major impacts on biodiversity from these changes are expected in tropical montane areas and at high latitudes. Agroforestry will be most important in helping manage impacts in tropical areas. Montane agroforestry crops such as coffee will play an especially important role (see Chapter 9, this volume).

Our understanding of the conservation implications of climate change stems in large part from similar changes in the past. Understanding the role of agroforestry entails understanding the conservation challenges that result from biotic response to climate change, as well as emerging strategic responses.

Range shifts have affected many species during past climate changes (Huntley and Webb 1989). Past shifts have been recorded in both plants and animals (Graham and Grimm 1990; Ponel 1995; Webb 1995). Paleoecologists have used pollen deposited in lakes, fossil and semifossil small mammal nests, and numerous other tools to detect past vegetation changes. Fossil remains have been used to show changes in distribution in mammals, birds, marine shellfish, and other species after global, local, or regional changes in climate (Riddle 1996).

Changes in abundance and patterns of abundance of individual species may occur in the absence of or in tandem with range shifts. For example, extensive pollen core data show shifting relative and absolute abundance of plant species in the transition from cooler climates to the warmer climate of the present (Huntley, Cramer, et al. 1995). This is most evident in temperate recolonization after the last glacial maximum but is also well supported in tropical regions, especially tropical mountains (Webb 1995; Flenley 1998).

The speed of biotic response to past climate change has often been remarkable. Recolonization after glacial retreat commonly exceeds rates calculated from mean dispersal and growth time to reproductive maturity (King and Herstrom 1995). Plant and animal species have tracked major temperature changes (about 10°C) occurring on time scales of decades or centuries without major extinction episodes (Roy et al. 1996). Figure 20.1 illustrates rapid climate changes recorded in a Greenland ice core temperature proxy. Outlier pockets of vegetation have been implicated in vegetation responses to rapid changes such as these. For example, forests on the South Island of New Zealand dominated the landscape within a few centuries after the decline of glacial conditions, despite the presence of a water barrier separating the known forest refugia on the North Island (McGlone 1995). It has been suggested that microrefuges in the complex montane relief of the South Island maintained small pockets of forest that subsequently expanded when suitable climatic conditions returned (McGlone 1995). Rapid dispersal may also be dependent on rare long-distance dispersal events, so maintaining healthy populations of possible long-distance dispersal agents such as bats, birds, and large mammals will increase the probability of maintaining the natural capacity for range migration (Clark et al. 1998). Whether from microrefugia or by long-distance dispersal, the rapid range shifts of the past will be greatly constrained by current levels of habitat loss. Many species able to make rapid range adjustments in a fully natural landscape will face extinction in future rapid changes.

Changing frequency and intensity of extreme events have been associated with past climate changes (Easterling et al. 2000). High-intensity storms, El Niño events, and droughts are examples. These events may play important roles in determining biodiversity patterns (Connell 1978). Rapid temperature transitions in northern European climate, for instance, are associated with increased dust storm intensity in Asia, indicating both a global signature to

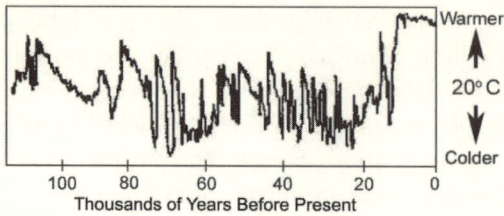

Figure 20.1. Rapid climate changes over 150,000 years as recorded in a Greenland ice core temperature proxy.

climate changes and changes in extreme events associated with variations in temperature (Broecker 1999). Drought plays a major direct role in shaping species distributions and abundance in tropical forests (Condit 1998). It is associated with greater fire risk in temperate, mediterranean, and tropical systems (Clark 1990; Simmons and Cowling 1996; Bond 1997; Cochrane et al. 1999). The frequency of El Niño events may change with climate and create major regional changes in biodiversity through drought, fire, and other factors (Trenberth and Hoar 1997).

Climate Change–Integrated Conservation Strategies

The sum of range shifts, changes in abundance, and other factors will determine the net biotic impact of climate change. To manage the impact of these changes on biodiversity, Climate Change–Integrated Conservation Strategies (CCS) have been proposed. A CCS consists of five elements:

- Regional climate and biotic modeling
- Expansion of protected areas
- Management of the matrix outside protected areas
- Regional coordination of management actions
- Financial and technical resource transfer

Use of regional climate change models and models of species response are an essential first step in generating scenarios of future species distributions. Regional climate models such as MM5 (developed by the National Center for Atmospheric Research and Pennsylvania State University) and RAMS (Regional Atmospheric Model System, developed by Pielke) are fine-scaled models imbedded in a GCM. They take GCM parameters at the boundary of the region and use them to drive much higher-resolution projections of regional climate. They are particularly important in representing changes in albedo and convective dynamics caused by land use change and regional topography in ways that are impossible in GCMs (e.g., western North America from the Sierra Nevada to the Rocky Mountains is treated as a giant plateau in most GCMs). Regional climate models currently exist for many parts of the world. They are normally used for weather forecasting or agricultural studies and, more recently, for biodiversity studies. Regional climate models exist or are under development for Brazil, South Africa, China, and other tropical countries.

Projections produced by a regional climate model may be used to drive regional biotic models. One of the most advanced instances of such use is under way in the Cape Mediterranean Region of South Africa. There, an MM5-based regional climate model, originally developed for climatology studies, is being used to drive models of biome, species, and plant functional type changes in the region (Midgley et al. 2001). Figure 20.2 illustrates projected biome changes in the two biodiversity hotspots of this region. Other regional analyses are under way in other parts of the world, although they often use projections from GCMs rather than regional climate models (Huntley, Berry, et al. 1995a; Peterson et al. 2002).

The regional modeling phase of a CCS culminates in a sensitivity analysis that incorporates as much site- and species-specific information as possible. Integrative sensitivity analysis examines the effects of climate change on

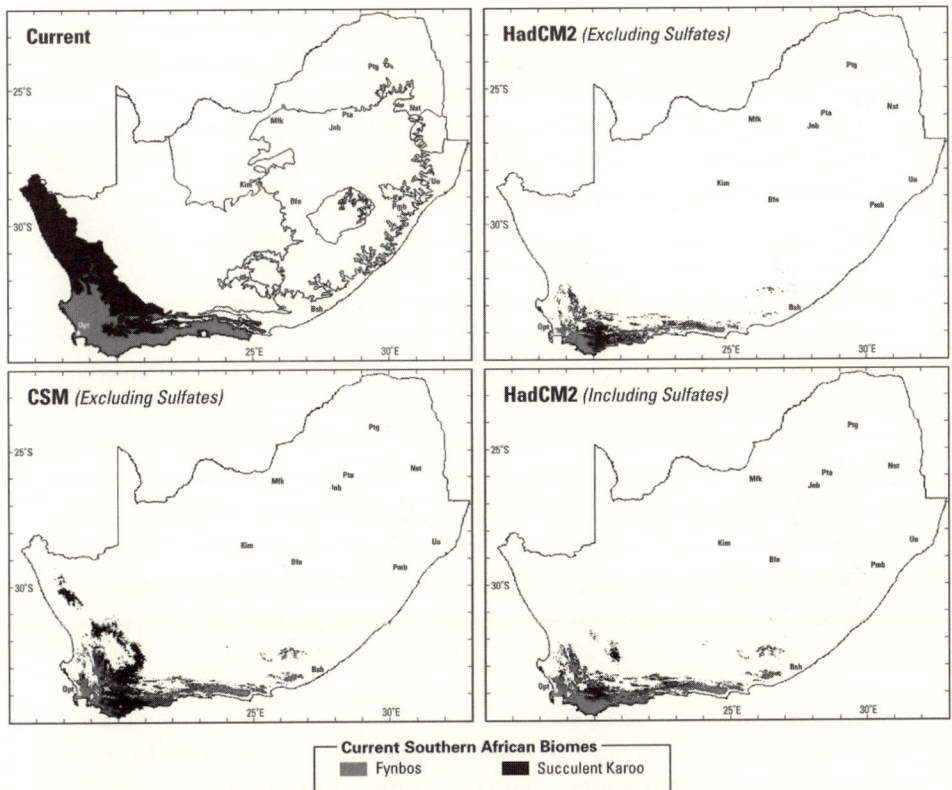

Figure 20.2. Projected biome shifts in the Succulent Karoo and Cape Floristic Region biodiversity hotspots projected by regional modeling.

individual species—especially rare, threatened, and climate-sensitive species—and ecological processes. Processes include disturbance regimes that may affect vegetation dominance (e.g., fire, drought), migrations, and other factors. Habitats that are sensitive to climate change (e.g., wetlands) and effects on the distribution of montane species receive special consideration. The result of modeling and sensitivity analysis is a series of scenarios of possible climate change effects and management options and may include spatially explicit mapping of possible changes in species ranges. Based on these scenarios, a conservation strategy is developed that includes expansion of protected areas, management of the land use matrix between protected areas, and regional coordination of conservation efforts.

Matrix Management

Biodiversity-friendly land uses, such as appropriately designed agroforestry systems, may provide habitat for many species, increasing the chance of movement through the matrix when climate change alters habitat in parks. Conversely, the wrong mix of land uses in the matrix can mean little chance of species moving outside of park boundaries and can increase penetration of a microclimate typical for open agricultural areas and plant invasion, causing a retreating forest edge in the reserves themselves (Gascon et al. 2000).

Efforts to connect protected areas with natural or seminatural land uses may be called corridors, landscape conservation, or ecoregional conservation (see Chapters 1 and 3, this volume). Connections provide continuity between habitat fragments and can help maintain viable populations in multiple-use landscapes. They also have value in allowing species range shifts with climate change, but unless they are specifically designed for this using regional modeling results, they will not realize their full potential. CCS efforts therefore complement existing connectivity planning by explicitly designing connectivity with climate change in mind.

The ability of species to exist and traverse the matrix becomes critical as climate change–driven range shifts occur. As changing conditions or extreme events alter vegetation in protected areas, the matrix may contain the only available habitat for some species or at some times. Predicting when the matrix would come into play carries many uncertainties, so one CCS element is to maximize biodiversity-friendly land uses in the matrix, including the option to revert human-oriented land uses to natural habitat.

Restoring habitat is an increasingly used option in biodiversity and may take on increasing importance with climate change. In particular, intentional maintenance of lands in seminatural uses such as agroforestry may be useful in keeping restoration options open in the future. For example, if modeling shows that a species' range is likely to shift along two potential routes but is unable to say with great certainty which route is more likely (because of differences in GCM projections or other sources of uncertainty), then a matrix management strategy might opt for keeping both routes in seminatural vegetation, monitoring the species in the field, and restoring habitat as the actual path of the range shift becomes clear. Such corridor management strategies have little or no track record, but they are conceptually feasible. The restoration principles involved are generally understood, among which limiting soil tillage is one of the most important. Applying these principles in a flexible land use management strategy will entail careful monitoring coupled with conservation agreements with landholders.

Conservation managers can prepare for future matrix habitat needs by preparing conservation agreements with landholders outside parks. Matrix conservation agreements may specify maintaining certain habitat qualities, such as a percentage of tree cover, or may include rights to future restoration for biodiversity management. This is an extension of the principle of purchasing packets of rights to land (in this case the right to future use) rather than outright full acquisition of land (see also Chapter 7, this volume).

An example of planning the matrix for both conservation and production with possible agroforestry applications is a system being implemented in the wine-growing region of South Africa near Cape Town. Wine is a major land use in the region and often occupies hillsides that neighbor upland parks. Here, joint planning between conservationists and landholders is helping generate a biodiversity-friendly land use matrix. The Botanical Society, a local conservation organization, has developed a geographic information system (GIS) to help vineyards plan future plantings while maximizing conservation. Landowners and conservationists will use analysis of future consumer demand and possible climate changes to jointly plan location and scale of new plantings. GIS will be used to identify suitable areas for planting while conserving natural habitat.

Similar tools could be applied in agroforestry. Like grape vines, trees take years to mature, so planning horizons are on the order of decades, making such operations good partners for planning with climate change, which also unfolds on decadal time scales. The productivity, product quality, and susceptibility to drought and diseases of both vineyards and agroforestry enterprises are intimately linked to climatic conditions, so that joint planning with state-of-the-art regional climate models is an advantage to the land user as well.

Agroforestry in the Matrix

Agroforestry, particularly in the biodiversity-rich tropics, has characteristics important for matrix management and conservation during climate change. Agroforestry initiatives may be well suited to provide increased habitat value and connectivity in the matrix because they involve increased tree cover, often of native species, relative to conventional agriculture or pasture. By promoting or maintaining tree cover, agroforestry provides habitat, may leave land better suited to future conversion to natural habitat, creates a microclimate suitable for many plant and animal species, and may provide food sources outside reserves for many seed dispersers and pollinators (Lord et al. 1997). These elements are in turn important in providing source populations, transitional habitat, seed dispersal to new areas, and destination habitat for species moving because of climate change.

Agroforestry may play a role in broader regional climate maintenance as well. Forest denuded to grassland may warm rapidly, causing hot updrafts that lead to convective rainfall. At a medium scale, this effect results in increased rainfall intensity. Over larger areas, it is believed to lead to regional drying (Pitman et al. 2000). Either effect can compound climate change–related increases in storm frequency or drying. By providing a landscape with substantial tree cover, agroforestry in conjunction with major protected areas under natural forest cover may help moderate rainfall disruptions and regional drying. This in turn reduces regional climatic changes that may affect species' ranges and abundance through changes in food resource availability, reproductive timing, or survival (especially during drought or fire).

Habitat provided by agroforestry is important for some species. For example, shade coffee has been shown to provide habitat to bird species that unshaded coffee (or "sun coffee") does not (see Chapter 9, this volume). If the ranges of these species shift because of climate change, shade coffee plantations provide a semipermeable matrix land use that may facilitate natural movements. Shade coffee often is practiced by smallholders with low capital inputs, meaning that the marginal cost of maintaining these conservation values is low. In contrast, intensive coffee culture offers little alternative habitat, may pollute the soil with pesticides, and has high cost of conversion to conservation set-asides, if such conversion is even biologically possible.

Agroforestry can help provide food sources outside reserves that maintain populations of large seed-dispersing birds and some mammals that facilitate range shifts. It can also facilitate long-distance foraging when food is less plentiful in nature reserves. The importance of providing food sources to seed dispersers is illustrated by the brown-cheeked hornbills (*C. cylindricus*) of the Dja Reserve in Central Africa (Whitney et al. 1998). These birds are resident in the largest park in Cameroon and one of the largest in Africa yet depend on food sources outside the park in certain seasons. Radio tracking in small planes has shown that they move hundreds of kilometers and across international borders in search of off-season food, despite the fact that the Dja Reserve is half the size of Belgium (Figure 20.3). Movements such as these can be critically important to forest plant species whose ranges are shifting because of climate change. Hornbills are important seed dispersers, and in times of rapidly changing climate, getting seeds to areas with newly suitable climates is critical to the ability of forest

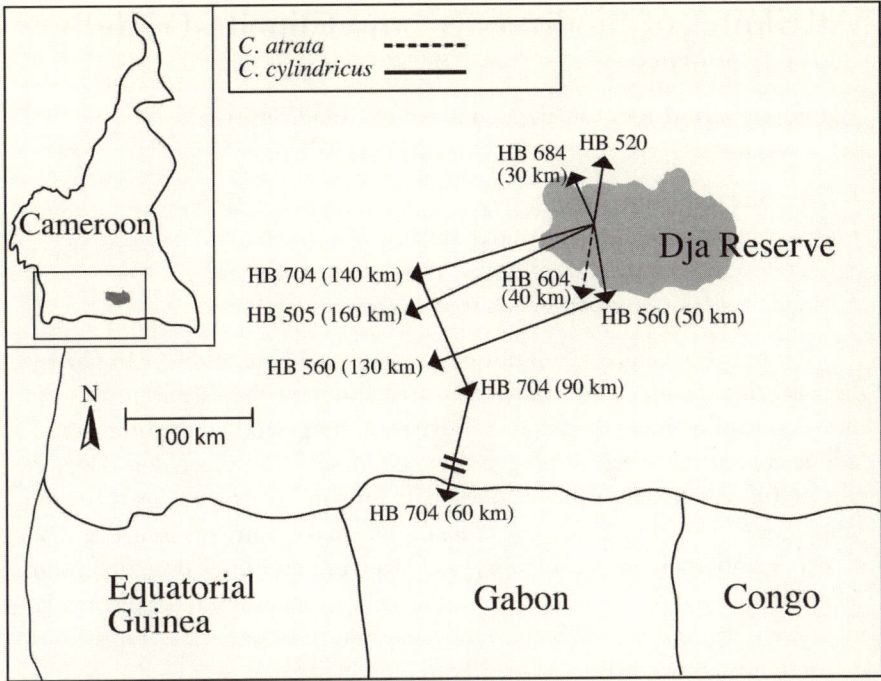

Figure 20.3. Foraging by *Certogymna cylindricus* in the Dja Region, Central Africa, as determined by radiotelemetry. (Figure courtesy of Thomas B. Smith.)

tree species to shift their ranges. Without food sources outside reserves, these movements would be reduced, greatly decreasing the chance of long-distance dispersal for some forest plants. In landscapes in which forest is being increasingly converted to agriculture, agroforestry can therefore play a pivotal role in maintaining trees that provide off-season food resources for large dispersers such as hornbills.

A final example illustrates the potential of agroforestry to foster forest regeneration. Cocoa is an agroforestry crop often favored by smallholders. Cocoa must be replanted periodically to maintain productivity, providing an economic window in which conversion to conservation set-asides could take place. Apart from the habitat value of native plants, the cocoa and associated trees maintain many soil components and understory microclimate conditions necessary for forest regeneration (see Chapter 10, this volume). Restoring forest on denuded land can be extremely difficult and slow because of the changes in soil and the light and moisture regime that occur with total clearing. Cocoa, though in no way comparable to natural forest, does provide conditions for forest regeneration far superior to those of cleared land. If intentionally managed as part of a CCS, these attributes may help maintain future options for biodiversity management.

Attributes of Biodiversity- and Climate-Friendly Agroforestry

Agroforestry land uses can be a constructive, biodiversity-friendly part of a CCS when they

- Maximize soil conservation,
- Maintain high levels of native tree cover,
- Include appropriate levels of capital investment, and thereby
- Maintain options for conversion to conservation land uses

Options that minimize soil disturbance and avoid pollution are important because they maintain soil biota that are critical to the establishment and movement of natural vegetation. Soil bacteria, fungi, and mycorrhizal associations are critical to establishing and maintaining native vegetation (Perry et al. 1990). Forests and other native plant formations depend on these soil microorganisms to break down minerals and make nutrients available. Soil fauna is important for maintaining soil structure and litter decomposition. High-intensity agriculture that breaks the soil and reduces soil organic matter negatively affects these soil biota, and it may take years or decades for them to become reestablished (Perry et al. 1990).

As climate change causes changes in vegetation, soil biota may be the limiting factor in plants becoming established in areas that have become newly climatically suitable (Perry et al. 1990). Agroforestry land uses that minimize soil tillage, avoid soil pollution with pesticides, and maintain soil organic matter through soil cover and litter inputs thus foster soil microflora and fauna, which may play an important role in facilitating plant range shifts in the matrix in response to climate change.

Tree cover maintained by agroforestry is directly correlated with habitat values and the ability of the land use to maintain regional climate. Agroforestry trees may provide habitat for birds, epiphytes, and other species and may provide a corridor of transit for species such as primates even when not suitable as primary habitat (see Chapter 3, this volume). Trees may provide indirect benefits to many more species in the surrounding region by helping to maintain regional climate. Loss of tree cover over large areas may result in reduced moisture recycling and regional drying. Specific levels of tree cover necessary to maintain regional climate may be derived from climatic modeling, as illustrated by Woodward et al. (1998). Tree cover beneficial for the conservation of regional climate can be a mix of native and introduced trees if no invasive species are used (see Chapter 15, this volume). Depending on their size and planting density, tree crops could provide microclimate conditions similar to those of forests; however, their value as habitat for forest wildlife usually is greater if native tree species are used. Extensive monocultures, even of tree crops, cannot provide these biodiversity benefits.

Future management adjustments necessitated by climate change are difficult to predict precisely, making land uses that keep options open an asset in conservation efforts. Highly capital-intensive land uses are unlikely to be available for future conversion to conservation set-asides. Conversely, lack of capital investment resulting in soil degradation and erosion may also make future conversion costly. On the other hand, land uses with moderate levels of capital investment may give landowners more flexibility in considering conservation agreements in which land is reverted to natural forest in return for payments (see Chapter 7, this volume) while maintaining soil properties in a state that allows forest reestablishment.

In addition to outright conversion of agroforestry land to conservation through reforestation or allowing natural regeneration, areas with flexible land use, such as agroforestry, may also be valuable in conservation planning involving trade-offs between conservation and production objectives. For example, an agroforestry concession that could help meet wood demand through use of shaded tree crops or silvopastoral systems with timber trees might be valuable in allowing some production forest to be retired to conservation, as part of a CCS.

Impact of Climate Change on Agroforestry

Climate change also affects human land uses, and these changes may in turn affect biodiversity. Current agricultural areas may become less productive as climate change alters regional growing conditions (IPCC 2001a). This may force landholders to expand their agricultural land, resulting in encroachment on natural habitats, an effect that should be considered when planning agroforestry in support of conservation.

Land use planning should consider the impacts of climate change on farming systems in a region in parallel with likely biological changes caused by climate shifts. By anticipating declining agricultural production caused by a drier climate, for example, planners can help provide appropriate levels of capital input and crop germplasm adapted to future climatic conditions, thereby maintaining income and food security and fostering social stability in areas that might otherwise resort to desperation clearing of natural forest to compensate for income losses. Agroforestry crops and planting designs should be selected based on future as well as current growing conditions; consequently, long-term plans are needed to ensure transitions in land use practices that maintain productivity and stability of land use in the face of climate change.

Conclusions

Not all agroforestry systems have biodiversity benefits, either in the near term or as future climate changes. Systems consisting of exotic tree species are less valuable for biodiversity conservation than agroforests possessing a closed canopy and forest structure (see Chapter 10, this volume). Replacing natural forest with any type of agroforestry system entails huge losses of biodiversity and should be considered only where conversion to less desirable land uses would be the inevitable alternative. In these cases, agroforestry systems that avoid soil disturbance and pollution, maintain tree cover, incorporate native species, have appropriate levels of capital inputs, and thereby keep options open for the regeneration of natural forest in the future have the greatest potential for matrix management as part of a CCS.

Application of agroforestry in CCSs is a new concept. Opportunities to test agroforestry in this context should expand in the future as these use strategies are more widely implemented in the matrix of tropical mosaic landscapes. The potential roles for agroforestry include providing transitional habitat for species whose ranges are shifting, supplemental habitat for species in decline, food sources outside protected areas, microclimatic conditions appropriate for many species, and flexible options for future regeneration of natural forest.

References

Bond, W. J. 1997. Ecological and evolutionary importance of disturbance and catastrophes in plant conservation. Pages 87–107 in G. M. Mace, A. Balmford, and J. R. Ginsberg (eds.), *Conservation in a changing world*. Cambridge, UK: Cambridge University Press.

Broecker, W. S. 1999. What if the conveyor were to shut down? Reflections on a possible outcome of the great global experiment. *GSA Today* 9:1–5.

Bush, M. 2002. Distributional change and conservation on the Andean flank: a palaeoecological perspective. *Global Ecology and Biogeography* 11:475–484.

Clark, J. S. 1990. Fire and climate change during the last 750 yr in northwestern Minnesota USA. *Ecological Monographs* 60:135–160.

Clark, J. S., C. Fastie, G. Hurtt, S. T. Jackson, C. Johnson, G. A. King, M. Lewis, J. Lynch, S. Pacala, C. Prentice, E. W. Schupp, T. Webb III, and P. Wyckoff. 1998. Reid's paradox of rapid plant migration: dispersal theory and interpretation of paleoecological records. *BioScience* 48:13–24.

Cochrane, M. A., A. Alencar, M. D. Schulze, C. M. Souza, D. C. Nepstad, P. Lefebvre, and E. A. Davidson. 1999. Positive feedbacks in the fire dynamic of closed canopy tropical forests. *Science* 284:1832–1835.

Condit, R. 1998. Ecological implications of changes in drought patterns: shifts in forest composition in Panama. *Climate Change* 39:413–427.

Connell, J. 1978. Diversity in tropical rain forests and coral reefs. *Science* 199:1302–1310.

Easterling, D. R., G. A. Meehl, C. Parmesan, S. A. Changnon, T. R. Karl, and L. O. Mearns. 2000. Atmospheric science: climate extremes—observations, modeling, and impacts. *Science* 289:2068–2074.

Flenley, J. R. 1998. Tropical forests under the climates of the last 30,000 years. *Climatic Change* 39:177–197.

Gascon, C., G. B. Williamson, and G. A. B. da Fonseca. 2000. Receding forest edges and vanishing reserves. *Science* 288:1356–1358.

Graham, R. W., and E. C. Grimm. 1990. Effects of global climate change on the patterns of terrestrial biological communities. *Trends in Ecology and Evolution* 5:289–292.

Hannah, L., G. F. Midgley, T. E. Lovejoy, W. J. Bond, M. Bush, J. C. Lovett, D. Scott, and F. I. Woodward. 2002. Conservation of biodiversity in a changing climate. *Conservation Biology* 16:264–268.

Hannah, L., G. F. Midgley, and D. Millar. 2002. Climate change–integrated conservation strategies. *Global Ecology and Biogeography* 11:485–495.

Hughes, L. 2000. Biological consequences of global warming: is the signal already apparent? *Trends in Ecology and Evolution* 15:56–61.

Huntley, B., P. M. Berry, W. Cramer, and A. P. McDonald. 1995. Modelling present and potential future ranges of some European higher plants using climate response surfaces. *Journal of Biogeography* 22:967–1001.

Huntley, B., W. Cramer, A. V. Morgan, H. C. Prentice, and J. R. M. Allen. 1995. *Past and future rapid environmental changes: the spatial and evolutionary responses of terrestrial biota.* Berlin: Springer-Verlag.

Huntley, B., and T. Webb III. 1989. Migration: species' response to climatic variations caused by changes in the earth's orbit. *Journal of Biogeography* 16:5–19.

IPCC (Intergovernmental Panel on Climate Change). 2001a. *Climate change 2001: impacts, vulnerability and adaptation.* Third Assessment Report, Intergovernmental Panel on Climate Change, Working Group 2, Port Chester, New York. Cambridge, UK: Cambridge University Press.

IPCC (Intergovernmental Panel on Climate Change). 2001b. *Climate change 2001: the scientific basis.* Third Assessment Report, Intergovernmental Panel on Climate Change, Working Group 1, Port Chester, New York. Cambridge, UK: Cambridge University Press.

King, G. A., and A. A. Herstrom. 1995. Holocene tree migration rates objectively determined from fossil pollen data. Pages 91–102 in B. Huntley, W. Cramer, A. V. Morgan, H. C. Prentice, and J. R. M. Allen (eds.), *Past and future rapid environmental changes: the spatial and evolutionary responses of terrestrial biota.* Berlin: Springer-Verlag.

Lord, J., J. Egan, T. Clifford, E. Jurado, M. Leishman, D. Williams, and M. Westoby. 1997. Larger seeds in tropical floras: consistent patterns independent of growth form and dispersal mode. *Journal of Biogeography* 24:205–211.

McGlone, M. S. 1995. The responses of New Zealand forest diversity to Quaternary climates. Pages 73–80 in B. Huntley, W. Cramer, A. V. Morgan, H. C. Prentice, and J. R. M. Allen (eds.), *Past and future rapid environmental changes: the spatial and evolutionary responses of terrestrial biota.* Berlin: Springer-Verlag.

Midgley, G. F., L. Hannah, D. J. MacDonald, and J. Alsopp. 2001. Have Pleistocene climatic cycles influenced species richness patterns in the greater Cape Mediterranean region? *Journal of Mediterranean Ecology* 2:137–144.

Myers, N., R. A. Mittermeier, C. G. Mittermeier, G. A. B. da Fonseca, and J. Kent. 2000. Biodiversity hotspots for conservation priorities. *Nature* 403:853–858.

Parmesan, C., and G. Yohe. 2003. A globally coherent fingerprint of climate change impacts across natural systems. *Nature* 421:37–42.

Perry, D. A., J. G. Borchers, S. L. Borchers, and M. P. Amaranthus. 1990. Species migrations and ecosystem stability during climate change: the belowground connection. *Conservation Biology* 4:266–274.

Peters, R. L. 1991. Consequences of global warming for biological diversity. Pages 99–118 in R. L. Wyman (ed.), *Global climate change and life on Earth.* New York: Routledge, Chapman & Hall.

Peterson, A. T., M. A. Ortega-Huerta, J. Bartley, V. Sanchez-Cordero, J. Soberon, R. H. Buddemeier, and D. R. Stockwell. 2002. Future projections for Mexican faunas under global climate change scenarios. *Nature* 416:626–629.

Pitman, A., R. Pielke, R. Avissar, M. Claussen, J. Gash, and H. Dolman. 2000. The role of land surface in weather and climate: does the land surface matter? *IGBP Newsletter* 39:4–24.

Ponel, P. 1995. The response of Coleoptera to late-Quaternary climate changes: evidence from north-east France. Pages 143–152 in B. Huntley, W. Cramer, A. V. Morgan, H. C. Prentice, and J. R. M. Allen (eds.), *Past and future rapid environmental changes: the spatial and evolutionary responses of terrestrial biota.* Berlin: Springer-Verlag.

Riddle, B. R. 1996. The molecular phylogeographic bridge between deep and shallow history in continental biotas. *Trends in Ecology and Evolution* 11:207–211.

Root, T., J. T. Price, K. R. Hall, S. H. Schneider, C. Rosenzweig, and J. A. Pounds. 2003. Fingerprints of global warming on wild animals and plants. *Nature* 421:57–60.

Roy, K., J. W. Valentine, D. Jablonski, and S. M. Kidwell. 1996. Scales of climatic variability and time averaging in Pleistocene biotas: implications for ecology and evolution. *Trends in Ecology and Evolution* 11:458–463.

Simmons, M. T., and R. M. Cowling. 1996. Why is the Cape Peninsula so rich in plant species: an analysis of the independent diversity components. *Biodiversity and Conservation* 5:551–573.

Still, C. J., P. N. Foster, and S. H. Schneider. 1999. Simulating the effects of climate change on tropical montane cloud forests. *Nature* 398:608–610.

Trenberth, K. E., and T. J. Hoar. 1997. El Niño and climate change. *Geophysical Research Letters* 24:3057–3060.

Walther, G., E. Post, P. Convey, A. Menzel, C. Parmesan, T. J. C. Beebee, J. Fromentin, O. Hoegh-Guldberg, and F. Bairlein. 2002. Ecological responses to recent climate change. *Nature* 416:389–395.

Webb, T. I. 1995. Spatial response of plant taxa to climate change: a palaeoecological perspective. Pages 55–72 in B. Huntley, W. Cramer, A. V. Morgan, H. C. Prentice, and J. R. M. Allen (eds.), *Past and future rapid environmental changes: the spatial and evolutionary responses of terrestrial biota.* Berlin: Springer-Verlag.

Whitney, K. D., M. K. Fogiel, A. M. Lamperti, K. M. Holbrook, D. J. Stauffer, B. D. Hardesty, V. T. Parker, and T. B. Smith. 1998. Seed dispersal by *Ceratogymna* hornbills in the Dja Reserve, Cameroon. *Journal of Tropical Ecology* 14:351–371.

Woodward, F. I., M. R. Lomas, and R. A. Betts. 1998. Vegetation-climate feedbacks in a greenhouse world. *Philosophical Transactions of the Royal Society of London B Biological Sciences* 353:29–39.

Conclusion: Agroforestry and Biodiversity Conservation in Tropical Landscapes

Götz Schroth, Gustavo A. B. da Fonseca,
Celia A. Harvey, Claude Gascon, Heraldo L. Vasconcelos,
Anne-Marie N. Izac, Arild Angelsen, Bryan Finegan,
David Kaimowitz, Ulrike Krauss, Susan G. W. Laurance,
William F. Laurance, Robert Nasi, Lisa Naughton-Treves,
Eduard Niesten, David M. Richardson, Eduardo Somarriba,
Nigel I. J. Tucker, Grégoire Vincent, and David S. Wilkie

Following the World Agroforestry Center, in this book we define agroforestry as a dynamic, ecologically based natural resource management practice that, through the integration of trees and other tall woody plants on farms and in the agricultural landscape, diversifies production for increased social, economic, and environmental benefits. The 20 chapters in this book analyzed ways in which agroforestry could best contribute to one particular group of environmental benefits, namely the conservation of tropical biodiversity. In the Introduction we proposed three hypotheses on how agroforestry could help conserve tropical biodiversity: by reducing the pressure to deforest remaining forestland and degrade forest through the unsustainable extraction of its resources, by providing suitable habitat for forest-dependent plant and animal species, and by creating a biodiversity-friendly matrix to facilitate movements between existing patches of natural habitat and buffer them against more hostile land uses. In this Conclusion, we briefly review these three hypotheses in light of the contributions in this volume, identify opportunities where existing knowledge can be applied, and pinpoint knowledge gaps where further research is needed. We conclude with a list of the most immediate research needs.

Agroforestry as a Means of Protecting Natural Forest

Agroforestry can help reduce pressure to deforest additional land for agriculture if adopted as an alternative to more extensive and less sustainable land use practices, or it can help the local population cope with limited availability of forestland and resources, such as near effectively protected parks (the agroforestry-deforestation hypothesis).

This hypothesis should not be misunderstood to suggest that the promotion of agroforestry in forest frontier areas as such would have a general forest-conserving effect. Both historical and economic arguments reviewed in this book show that this is not a simple causal relationship. A key argument in the early discussion of the effects of agroforestry on deforestation was that by being more sustainable than alternative land use practices, agroforestry may reduce deforestation as a means to create new fertile crop lands to substitute for degraded agricultural lands (Chapter 5). However, although it is true that agroforestry often is more sustainable than alternative agricultural land uses, this does not necessarily lead to less deforestation, for two reasons. First, even potentially sustainable land use practices can be used in an unsustainable manner if the prevailing socioeconomic conditions favor the occupation of new land rather than investments in sustaining the productivity of existing fields, pastures, and plantations. For example, converting new forest areas for agriculture may be a way for farmers (e.g., immigrants) to establish use rights, defend traditional use rights against other land users or the government, or simply expand their farmed land. Furthermore, the profitability of tree crops such as cocoa and rubber usually is highest if they are grown in newly deforested areas, which provides a further incentive for farmers to establish new plantations in primary forest rather than replant already cultivated land as long as forestland is readily available (Chapter 6). Second, if agroforestry land use, such as growing commercial tree crops, is more profitable than alternative activities in the same or nearby regions, such as slash-and-burn agriculture with subsistence crops, there will be a tendency for agroforestry to expand, including into forested areas, if socioeconomic factors (access to land, labor, and capital) or biophysical factors (soil conditions and pest and disease pressures) do not pose obstacles to such expansion (Chapter 5).

In combination, these two factors explain why some perennial crops, which form the basis of the most biodiversity-rich agroforests, have also contributed significantly to the expansion of the agricultural frontier into primary forest. Chapter 6 demonstrates this for cocoa; other examples include rubber in lowland Sumatra (Chapter 5) and coffee in Colombia (Chapter 7) and elsewhere. Global markets for these commodities, their (former) profitability, and an abundant labor force of immigrants, for example in Sumatra and the West African rainforest zone, are among the factors that have permitted this expansion of tree crops, sometimes grown in agroforests, into primary forest areas (Chapters 5 and 6).

On the other hand, there are combinations of farmer characteristics, types of agroforestry practices, and market and tenure conditions under which agroforestry is likely to pull labor and capital resources away from the forest frontiers and thereby help to reduce deforestation. In particular, techniques that necessitate long-term investments in the land (e.g., through the planting of trees with a long productive life), that are labor or capital intensive, and that reduce production risks and thus the need to clear excessive land as a form of insurance are likely to have this effect. Cases in which agroforestry adoption has helped to stabilize the forest frontier include the introduction of coffee in slash-and-burn systems in Nicaragua (Chapter 5). Also, in some regions traditional shaded cocoa systems reduced forest conversion compared with modern monoculture cocoa by extending the useful life of the cocoa tree and, critically, facilitating replanting on the same site (Chapter 6).

However, as Angelsen and Kaimowitz (Chapter 5) point out, it will often be difficult to promote the techniques that have the highest forest-saving potential in a forest frontier situation. Because farmers tend to prefer technologies that increase rather than limit their options (including the option to move into new forest areas if this appears advantageous), they will often be reluctant to adopt labor- or capital-intensive agroforestry technologies when these factors are scarce but land is abundant, as is typically the case at open forest frontiers, especially where these frontiers are continuously being opened up by road construction and logging development (Chapter 2). Therefore, rather ironically, agroforestry practices or technologies that have the highest forest-saving potential are more likely to be adopted once the forest is gone (Chapter 5) or has become inaccessible to farmers.

In isolation, promoting agroforestry therefore will not usually be an effective means of reducing deforestation. Rather, the evidence and experiences reviewed in this book suggest that agroforestry can make its greatest contribution to forest conservation when it is combined with other, more direct forest-conserving measures, such as the declaration and enforcement of protected areas and other environmental legislation, or approaches that provide farmers with net benefits from forest conservation, such as conservation concessions (Chapter 7), access to special markets, ecotourism revenues, payments for watershed functions, or perhaps carbon credits (Chapter 4).

Agroforestry can complement such direct forest conservation measures in several ways. Agroforestry practices that allow the sustainable intensification of land use in deforested areas and increase the profitability per unit area, such as through the introduction of tree crops in pasture or slash-and-burn areas or valuable secondary crops and timber trees in tree crop systems, can help populations cope with reduced land availability in a closed-frontier situation, such as the buffer zone of a park. Under such conditions of limited land availability, the potential for sustainability of agroforestry practices is more likely to be realized and labor-intensive practices more likely to be adopted. By providing timber and nontimber products that would otherwise be taken from the forest, agroforestry in buffer zones can also help reduce pressure on forest resources and make legal restrictions on their extraction from the forest more acceptable to the local population (Chapter 10). Where populations live within the limits of less strictly protected areas, tree-dominated land uses such as complex agroforests may be the land use option that is most acceptable to both farmers and park managers and thereby help avoid conflict (Chapter 13).

Agroforestry is a suitable land use option not only in areas surrounding legally protected forests but also where land users voluntarily opt for conservation set-asides in exchange for direct payments, investments in health or educational infrastructure, or other benefits, as in the conservation concession approach (Chapter 7). In the long term, the protected status of conservation set-asides will depend on the sustainable use of the remaining agricultural land, or at least the cost of setting aside land will increase if unsustainable land use in the surroundings increases the pressure on the forest protected by a concession. Therefore, sustainable agroforestry practices that also reduce farmers' dependency on forest products probably can contribute to the long-term viability of conservation set-asides. As Hannah (Chapter 20) suggests, the greater flexibility of agroforestry compared with other land uses for eventual future conversion into natural habitat, which results from native tree cover, soil conservation, and extensive management of agroforestry areas, is an added advantage in the proximity of conservation set-asides and legally protected areas (by contrast, agroforestry practices using exotic tree species can be a serious threat to such areas, as discussed later).

Including agroforestry land uses in conservation concession agreements could sometimes make a critical difference in farmers' decision to adopt agroforestry in exchange for simpler agricultural practices. Transition to agroforestry often entails long-term investments (e.g., in tree planting), or agroforestry may be more sustainable but produce lower yields than nonagroforestry alternatives, at least in the short term, as is often the case with shaded compared with unshaded tree crops (Chapters 6 and 9). Consequently, by adopting agroforestry, farmers often incur immediate additional costs for which they may be rewarded only in the longer term. Some biodiversity-friendly agroforestry practices may even necessitate permanent subsidies to be competitive with conventional agricultural land uses. Through conservation concessions, land users can be compensated for such investments in biodiversity and sustainability in return for conserving additional forest or retiring agricultural land, thereby increasing the opportunities for integrated land use planning (Chapters 7 and 20). In a similar way, farmers could be rewarded for conserving forest and managing agricultural land sustainably through revenues from ecotourism, access to specialty markets (ecolabeling), payments for maintaining watershed functions, or carbon trading (Chapter 4).

As mentioned earlier, agroforestry may help maintain the integrity of forests by providing timber and nontimber forest products that would otherwise be collected from forests, often in an unsustainable manner. Empirical evidence is emerging that farmers who possess agroforests are less dependent on forest resources in general (Chapter 10). Whether this also applies to hunting is less clear, however. Tropical farmers commonly hunt in agroforests, where wildlife may be attracted by fruits and seeds and where hunting may serve the purpose of crop protection and subsistence (Chapter 13). In some cases, farmers even conserve tree species in cultivated areas specifically to attract wildlife (Chapter 8). In principle, hunting in agroforestry buffers could reduce hunting pressure in primary habitat; however, the extent to which this is the case and how source populations in primary habitat are affected by possibly high hunting pressure in cultivated areas, which could act as population sinks, have not been studied adequately.

In summary, although it would be naive to expect that promoting agroforestry as such would lead to reduced forest clearing, agroforestry can make significant contributions to the political acceptability and long-term stability of forest frontiers if they are imposed or protected by other, more direct mechanisms and may reduce pressures on forest resources. Forest conservation in parks or conservation concessions and the promotion of agroforestry land uses in the surroundings are likely to have synergistic effects. The development of such synergies depends to a large extent on good governance, especially sound environmental legislation and its effective enforcement, and institutions that allow and engage in integrated approaches to conservation and rural development planning. Thus, under particular scenarios and in combination with other measures, the agroforestry-deforestation hypothesis appears to be valid, although more empirical work is needed to clarify the range of social and economic conditions under which its validity is maintained.

Agroforestry as Habitat for Native Plant and Animal Species

Agroforestry systems can provide habitat and resources for partially forest-dependent native plant and animal species that would not be able to survive in a purely agricultural landscape (the agroforestry-habitat hypothesis).

As several contributions to this volume have shown, tropical agroforestry systems such as shifting cultivation, shaded tree crops, and complex agroforests contain or contribute to supporting many species and varieties of plants and animals that are not present in agricultural monocrops and pastures. Some of these species and varieties belong to the planned components of a system, especially crops, planted shade trees or tree crops, and domestic animals; others are remnants from prior plant and animal communities or are present as populations or metapopulations distributed over, or using, both natural and agroforestry-based patch types in the landscape (unplanned components). Both groups may contain species or varieties that are threatened and in need of conservation and others that have low conservation priority but may contribute to the productivity and stability of the land use system.

Some of the planned components of an agroforestry system may be widespread and often exotic crop and tree species, whereas others may be of native origin, including threatened crop species and varieties (Chapter 8). Although not usually considered in biodiversity conservation projects, the active conservation of such local crop species and varieties may be important for many reasons, including site adaptation and use in future breeding programs, but also as insurance of cropping systems against yield failure, especially under changing climatic conditions, which could force impoverished farmers to clear more forest or migrate to new forest frontiers (Chapter 20). Among the wild species present in a system, some common, disturbance-adapted species, including weeds, may contribute to ecosystem processes such as nutrient retention, pollination, and biological pest control but have low conservation value, whereas others may be rare forest species. Of the latter, some may be actively managed to fulfill essential functions, for example as shade trees, whereas others may be exploited occasionally for timber or nontimber products, and yet others may be merely tolerated or not noticed at all. Some common weedy species may also provide important food resources and habitat for other species more in need of conservation (Chapter 19).

As land use systems are intensified, wild species tend to be increasingly suppressed (e.g., by weeding and application of pesticides with cascading effects on higher trophic levels) or be replaced by a smaller number of planned species (e.g., forest remnant trees by planted, often exotic shade trees and tree crops, spontaneous weed communities by introduced cover crops, or diversified native by monospecific planted fallows; Chapters 8 and 9). Consequently, where high diversity of wild plant and animal species occurs in agroforestry systems or landscapes, this is usually the result of extensive management or even temporary abandonment of cultivated areas, that is, a result of the tolerance or unintentional maintenance (as in fallows) of biodiversity in production systems rather than specific management to promote its persistence.

This is most obvious in shifting cultivation, the most widespread and among the oldest forms of agroforestry. As Finegan and Nasi (Chapter 8) point out, the length of the fallow period (i.e., the time during which land is not managed or is managed only very extensively to enable it to recover from previous cropping) is a key factor determining the accumulation of wild species both at the plot and at the landscape scales. Consequently, the widespread phenomenon of shortening fallow periods in the tropics as a result of increasing land use intensity and the concomitant increase of disturbance through vegetation cutting and burning threatens not only the sustainability of these systems but also their potential to conserve native forest species. This is particularly true for forest plants, which tend to be excluded from short-rotation fallows through burning and weeding during the cropping phase and perhaps through competition from pioneers during the fallow phase. Forest animals, on the other hand, are often able to move between different habitat patches in shifting cultivation landscapes and tend to be less affected, although they will suffer if shortening of fallow periods leads to smaller areas of older fallows, or primary vegetation, in the landscape. Some groups may even be more abundant or diverse in structurally heterogeneous fallow landscapes than in more homogeneous primary forest, as has been shown for some raptors, certain primates, and small to medium-sized mammals, whereas other groups such as terrestrial and understory insectivorous forest birds generally are less abundant (Chapter 8).

Extensive management is also a key factor influencing the habitat value of more permanent agroforestry practices, including shaded tree crop systems and complex agroforests. Shaded, and especially extensively managed, rustic coffee plantations have been shown to provide habitat for numerous species of mammals and birds that are absent from intensively managed sun coffee or other treeless areas and also to provide critical overwintering sites for migrant birds (Chapter 9). Complex agroforests, such as jungle rubber in Indonesia or shaded cocoa systems in Africa and Latin America, are the most forest-like of all agricultural systems and have been shown to provide at least temporary habitat to many forest species, including some threatened cat, primate, and bird species. Extensive management of the whole system or of certain sections or vegetation strata, or even abandonment of the system during certain phases (e.g., at times of low commodity prices), often allow the persistence or recolonization by native plant and animal species (Chapter 10). However, even such extensively managed, forest-like systems are no substitutes for native habitat because certain species groups tend to be underrepresented or absent. This is particularly true for forest interior species, large herbivores, and top predators (high hunting pressure clearly contributes to the declines especially of larger animals; Chapter 14). Furthermore, many forest animals that are observed in agroforestry systems or fallows may depend on the existence of nearby native habitat and use these managed systems only sporadically or as stepping stones between patches of natural vegetation (Chapter 10). The degree to which some threatened species depend on the presence of natural forest in agroforestry landscapes is a key question for future research because it will influence the rates of depletion of local populations as natural forest is replaced by even the most forest-like agricultural systems.

Although the same biodiversity benefits arising from rustic coffee systems or complex cocoa or rubber agroforests cannot be expected from isolated trees, windbreaks, or hedgerows, such agroforestry elements in managed landscapes have been shown to permit the presence of several species of birds, bats, and small mammals that would not be present in treeless fields or pastures, including occasional forest interior species (Chapter 11). Woodlots of native trees in pasture areas have been shown not only to provide supplementary fodder resources for livestock and increase their carrying capacity during periods of low forage availability but also to provide crucial habitat for insects and migratory birds (Chapter 19). In wider wooded corridors connected to native forest, substantial colonization by forest plant, bird, and mammal communities has occurred within a few years of establishment, suggesting rapid development of suitable habitat conditions (Chapter 18).

Whether agroforestry systems provide habitat for forest species depends to a large degree on their management, especially whether they are managed on a short rotational or semipermanent basis, the structural complexity and diversity of their shade canopy and understory, and the degree of weeding, pollarding, and pesticide use. In addition, the size and location of the system within the landscape, particularly its proximity and degree of connectivity to remaining forest cover, strongly influence the abundance and diversity of plant and animal species present (Chapters 10 and 11).

Another key factor that acts directly on the animal diversity and abundance of agroforestry landscapes is hunting pressure. Wildlife is hunted in most tropical regions for food, income, medicine, and trophies, to control crop pests and predators of domestic livestock, and to reduce threats to human safety (Chapter 14). Although as a result of moderate levels of disturbance agroforestry landscapes tend to be more productive for certain wildlife species than undisturbed forest, present levels of wildlife consumption in many tropical regions are unsustainable and risk driving many species to extinction, particularly large and slowly reproducing species. In more densely populated tropical regions, the potential of agroforestry landscapes to serve as habitat for forest wildlife therefore also depends on changes in local consumption preferences and attitudes toward wildlife (Chapter 14). Furthermore, wildlife species that pose threats to crops or the safety of humans or domestic livestock are unlikely to be tolerated in inhabited areas (Chapter 13), stressing again the need for large and undisturbed areas of natural habitat even in regions where agroforestry is a dominant land use.

In summary, certain agroforestry practices have a significant potential to provide habitat for many species of forest-dependent flora and fauna and can probably play a crucial role in reducing species extinctions in regions where the area of remnant native habitat has been greatly reduced. However, because even in the most diversified and extensively managed agroforests certain groups of forest species are missing or underrepresented, agroforestry systems cannot be seen as a substitute but only as a complement to areas of natural habitat. Furthermore, increasing intensification of land use practices in tropical regions, such as shortening of fallow periods or reduction and simplification of shade canopies in tree crop plantations, reduces the habitat value of these systems for native species (Chapters 8 and Chapter 9). Therefore, it will be increasingly important to find ways to increase the profitability of traditional agroforestry systems while maintaining as much as possible of their biodiversity benefits.

Agroforestry as a Benign Matrix Land Use for Fragmented Landscapes

In landscapes that are mosaics of agricultural areas and natural vegetation, the conservation value of the natural vegetation remnants (which may or may not be protected) is greater if they are embedded in a landscape dominated by agroforestry elements than if the surrounding matrix consists of crop fields and pastures largely devoid of tree cover (the agroforestry-matrix hypothesis).

It has been suggested that the type of land cover and land use in the matrix of managed areas around and between remnant forests and parks has an important influence on ecosystem processes and population dynamics within these patches; therefore, matrix management has become an emergent topic in the design and implementation of biodiversity conservation strategies (Chapter 1). Key functions of the matrix that were identified in this book include providing a smooth transition between open agricultural areas and forest boundaries that reduces edge effects and the incursion of fire into forest areas (Chapter 2), providing connectivity between patches of primary habitat (Chapter 3), and providing alternative or supplementary habitat and resources for forest species (Chapter 8). Agroforestry systems may also provide a supply of timber and nontimber forest products that reduces the dependency of the local population from forest resources, as discussed earlier. Evidence supporting the hypothesis that the conservation value of habitat fragments is greater if they are embedded in a matrix of agroforestry than in less diversified and structurally simpler agricultural land uses is still mainly indirect, although direct evidence is slowly accumulating.

In several tropical regions, complex agroforests form the transition between human settlements and intensively used agricultural areas on one hand and natural forest on the other (Chapter 10). In this situation, depending on the height, structure, and extension of such agroforests, their buffering effect on microclimate and wind can be expected to result in lower edge-related mortality of forest trees than in unprotected forest, as has been shown for buffers of tall secondary forest; however, empirical evidence that such an effect also occurs with agroforests is lacking. Similarly, it is very likely but has not yet been demonstrated that forests buffered in this way are less affected by fire incursions because farmers will attempt to avoid damage to their tree crops. Despite the lack of proof of such effects, pioneer projects have started to implement agroforestry buffers around forest fragments (Chapter 17), which will in due course also serve as an empirical test of their basic hypotheses.

Some direct and substantial indirect evidence exists for the role of different types of agroforestry in increasing the connectivity of landscape mosaics for forest species. Shaded coffee ecosystems and pastures invaded by native legume tree species in Central America regularly host migrant birds arriving from North America, thereby providing a terminus to migrations as well as stopovers and, in a sense, ensuring connectivity on a continental scale (Chapters 9 and 19). Species of large cats that are known to use large territories, such as tiger and puma, have been seen in damar agroforests in Sumatra and cocoa agroforests in Costa Rica, respectively, although the degree to which such forested habitat in the agricultural landscape is necessary to ensure connectivity remains to be established. In Bahia, Brazil, lion tamarins (an endangered primate species) have been recorded in cocoa agroforests but are believed to be dependent on primary forest and to use the agroforests only as temporary habitat (Chapter 10). Remnant trees and windbreaks in fields and pastures often are used as stepping stones and dispersion paths for birds and flying mammals and may occasionally allow the passage of howler monkeys or other larger fauna (Chapters 11 and 17). In wider agroforestry corridors, such as 100-m-wide wooded strips connecting forest fragments, forest species have become established within only a few years, providing direct evidence of the value of such corridors as dispersion paths (Chapters 3 and 18).

Recent research has also shown a surprising degree of connectivity on a landscape scale for trees in forest fragments that were previously assumed to be "living dead" for lack of nearby mating partners. Agroforestry trees in the matrix may cross-pollinate with trees of the same species in forest fragments or may facilitate movements of pollinators and seed dispersers across the landscape. Consequently, such fragments are genetically less isolated than previously expected. However, the possibility that a few highly productive remnant trees ("superadults") in the open landscape dominate the pollen pool in habitat fragments and reduce effective population sizes warrants further consideration (Chapter 12).

However, there are further risks associated with the use of agroforestry in the matrix around natural habitat. One of these is the use of invasive alien tree species. Many species of timber, fruit, and service trees that are commonly used in agroforestry outside their native home range have occasionally been found to invade natural habitat, often after disturbance (Chapter 15). Because forests bordering on agricultural areas are particularly prone to disturbance, for example through logging and fire, the chances of invasion by agroforestry trees grown in the matrix may be greater. Mutualists such as pollinators, seed dispersers, mycorrhizal fungi, and nitrogen-fixing bacteria (also often aliens) in and around agroforestry systems may also facilitate invasions. In some genera, introduced species may also hybridize with native species, thereby threatening native gene pools (Chapter 12), or exotic species may hybridize with each other, thereby increasing their invasiveness (Chapter 15). Nonnative trees (and crops) may also host exotic pests and diseases to which the natural vegetation is not adapted, resulting in severe epidemics that may fundamentally alter the composition of natural communities or may be susceptible to native pests and diseases to which they have no tolerance or resistance (Chapter 16). Although these risks can be avoided by using native tree species, this may not be without cost and additional effort for the land user, who may forgo income opportunities from faster-growing, more valuable, or simply more available and better-known exotic species, which are often promoted by "diversification projects" using "promising" exotic species (Chapters 15 and 16).

For farmers, growing agroforestry crops in buffer zones and mosaic landscapes may involve both opportunities and risks, and the perception of these will influence their attitudes toward the conservation of primary habitat and wildlife. As the agricultural matrix becomes more hospitable and permeable to wildlife, there will also be greater risks of wildlife damaging crops and threatening domestic livestock or even humans (Chapter 13). For some farmers living close to the forest boundary and possessing the ecological knowledge to hunt larger fauna, benefits from the presence of wildlife may exceed crop losses, but for others it may not. To avoid conflict, land managers could help farmers plant unpalatable crops such as tea or coffee along park boundaries, thereby repelling wildlife from crop fields and channeling it into less sensitive corridors, or develop schemes through which farmers are compensated for crop losses to wildlife (e.g., from tourism revenues). Open communication between stakeholders about their respective objectives is clearly a condition for such integrated, landscape-scale planning and management (Chapter 13).

Although crop damage by wildlife generally is higher in proximity to primary habitat, little is known about how the exposure of crops to pests and diseases is affected by mosaic landscapes compared with more open and homogeneous areas. Although examples of both positive and negative effects exist, there is evidence of greater potential for biological pest and disease control in mosaic landscapes, which adds to the value of natural vegetation for the in situ conservation of resistance genes. However, where agricultural areas are taken out of production as part of conservation set-asides, they may turn into sources of disease inoculum, and intervention may be needed to avoid damage (Chapter 16).

In summary, the available evidence suggests that managing the agricultural matrix for soft transitions between cultivated and protected areas and increased connectivity through a diversified, structurally heterogeneous, and interconnected network of agroforestry elements, preferentially dominated by trees from the regional species pool, could make a substantial contribution to the long-term viability of plant and animal populations in tropical mosaic landscapes. Whether and at what cost this type of matrix management can be implemented in real tropical landscapes depend critically on the perceptions of farmers who live in such biodiverse landscapes and who will experience both the synergies and the trade-offs between their private goals and the objective of biodiversity conservation. The benefits of greater access to forest products and ecosystem services, such as water supply, pollination, and biological pest and disease control, and the costs of not planting exotic and potentially invasive crop and tree species, of tolerating wildlife on one's farmland, and possibly of coping with pests and diseases that cross-infect between native vegetation and agricultural crops and trees are all parts of an equation that determines whether farmers will perceive the biodiversity of tropical mosaic landscapes as an asset or a liability for which they need to be compensated by those who value this biodiversity and the ecosystem services connected to it.

Outlook and Research Needs

This book provides evidence in support of all three mechanisms through which agroforestry can help conserve tropical biodiversity:

- By helping to reduce deforestation and pressure on forest resources through synergies between agroforestry land use and direct measures of forest protection
- By increasing habitat for native species in cultivated areas, which is most important where natural habitat has been severely reduced
- By providing a more benign and permeable matrix for patches of primary habitat in land use mosaics

This suggests that agroforestry has an important role to play in biodiversity conservation strategies for the tropics, complementing and supporting the essential role of natural habitat within and outside parks and other protected areas. In conservation strategies that integrate agroforestry with other conservation measures, the relative importance of these three mechanisms depends on the specific situation of an area, particularly the degree of human colonization of a landscape and the availability of native habitat. In largely forested wilderness areas, biodiversity will be most effectively conserved by maintaining a maximum of forest cover; consequently, the most important role of agroforestry will be to help reduce the pressure on forestland and forest resources, and the direct habitat role of agroforestry areas will be of secondary importance. In areas where natural habitat has already become fragmented through human colonization, this role will be complemented by the creation of a matrix that maintains connectivity and softens transitions between forest and agricultural areas. The direct habitat role of agroforestry areas will be most important in regions where natural habitat has been greatly reduced so that agroforestry areas provide some of the last remaining habitat for those forest-dependent species that tolerate a certain level of disturbance.

The effective integration of agroforestry into conservation strategies is a major policy and institutional challenge that necessitates integrated approaches to natural resource management and rural development across traditional disciplinary, ministerial, and departmental divisions. It will also take substantial research and development efforts on biological, agronomic, socioeconomic, institutional, and political levels to close critical knowledge gaps, as well as substantial efforts on the part of scientists to communicate more effectively with decision makers and policymakers.

We conclude with a list of some of the most immediate research needs (in no particular order of priorities), hoping to inspire increased research efforts that may help develop agroforestry into an effective tool to complement ongoing conservation efforts in tropical landscapes.

- Develop, in a participatory manner, indicators and effective monitoring systems to assess the ecological services fulfilled by agroforestry systems, especially in mosaic landscapes comprising both agroforestry and natural forest areas, as both a guide for decision making in land management and a basis for valuing these services and creating political support for biodiversity-friendly land uses.
- Determine the degree to which threatened forest species present in agroforestry systems depend on natural forest within the landscape because this will determine which species will be able to maintain viable populations in agroforestry landscapes in the absence of natural habitat.
- Determine the ecological, social, economic, and political conditions under which shifting cultivation is a stable land use form in biodiversity-rich forested areas; in research on improved fallows include analyses of biodiversity values of longer, extensively managed tree-based fallows and their often substantial use values to local people.
- Assess the ability of different agroforestry types, when planted along forest boundaries, to reduce forest degradation through edge effects, including wind and microclimatic disturbance, invasions by weedy plant and animal species, and fire incursions, as a basis for the strategic use of such systems in buffer zones and mosaic landscapes.
- Determine the effectiveness of different types of agroforestry systems as biological corridors or stepping stones for wildlife (particularly fragmentation-sensitive or edge-avoiding species), the influence of species composition, structure, size, landscape position, and management of these systems, and their associated costs and benefits for farmers and society at large.
- Assess the conditions for hunting sustainability (ecological and economic) in mosaic landscapes, including source-sink dynamics of wildlife populations between agroforestry areas and natural habitat from where overhunted

populations may be replenished; in particular, study the effect of hunting in agroforestry buffer zones on wildlife populations in protected areas, both as it reduces hunting pressure in core areas and as it creates a drain on source populations of hunted wildlife.

- Develop methods of increasing the productivity of traditional diversified agroforestry systems while maintaining biodiversity benefits, and quantify trade-offs between productivity and biodiversity as a basis for designing incentives for farmers to maintain or adopt practices that optimize these trade-offs.

- Design agroforestry systems that optimize trade-offs between biological control of pests and diseases (e.g., through the right species combinations, degree of shade, and structural complexity) and profitability. At the landscape scale, study the effect of diversified mosaic landscapes on the dynamics and interactions of pests, diseases, and their natural enemies and the associated costs and benefits for farmers.

- Identify more indigenous tree species suitable for shading tree crops such as coffee and cocoa that optimize agronomic (labor needs, establishment, growth), economic (e.g., fruits, timber), and ecological benefits (habitat for wild species, nutrient cycling); currently an insufficient number of species (mostly exotics) are used in shaded plantations worldwide.

- Formulate robust strategies to avoid tree invasions and introduction of exotic pests and diseases arising from agroforestry where the use of alien species cannot be avoided, including improved screening and quarantine procedures for alien species and management options for reducing invasiveness.

- Determine the properties of agroforestry systems that increase the acceptance of restricted access to forest resources in protected areas by the local population (e.g., range of products provided by agroforestry, labor, and capital inputs).

- Determine which agroforestry practices, under which circumstances, necessitate initial or even permanent compensations or subsidies to be economically viable for tropical farmers; research the kinds of incentives (e.g., carbon credits, ecotourism, payments for watershed services) that could lead land users to adopt sustainable and biodiversity-friendly land uses and the institutional mechanisms that would ensure that those who fulfill the environmental services of these land uses also benefit from the rewards.

- Develop marketing channels for products from biodiversity-friendly land use systems that ensure that tropical farmers benefit from increased prices of such products in developed countries, including certification systems that are transparent to consumers and that the farmers can afford.

- Use participatory methods to devise effective ways of undertaking land use planning by integrating disciplines that are traditionally spread over several government departments and ministries, such as agriculture, forestry, infrastructure development, tourism, and biodiversity conservation, taking into account current trends toward decentralization of natural resource management in many tropical countries and empowerment of local authorities and civil society.

About the Contributors

Arild Angelsen is Associate Professor in Development and Resource Economics at the Agricultural University of Norway and Associate Scientist of the Center for International Forestry Research (CIFOR), Indonesia. His main research interests are causes of tropical deforestation, impacts of new technologies and macroeconomics policies, and poverty and sustainability.

François Anthony is a specialist of coffee genetics at Institut de Recherche pour le Développement (IRD), France. His main research interests are in physiological and genetic mechanisms of the resistance of coffee to root-knot nematodes.

Tiago Pavan Beltrame is Project Coordinator and Forestry Extensionist at the Institute for Ecological Research (IPÊ), a nonprofit organization in Brazil. He develops extension programs for involving rural communities in landscape restoration.

Kaycie A. Billmark is a Research Associate based at the Department of Earth and Planetary Sciences at the University of Tennessee at Knoxville, USA. Her primary research interests include regional and global atmospheric pollution, nutrient cycling, and global change, with most of her current research focused in southern Africa and Antarctica.

Pierre Binggeli is a researcher and international consultant in woody plant ecology with special emphasis on the sustainable use and management of natural resources. He has investigated invasive plants and agroforestry systems in Africa, the Pacific and western Europe.

David H. Boshier is a Senior Research Associate at the Oxford Forestry Institute, Department of Plant Sciences, University of Oxford, UK. His research focuses on the genetics of tropical tree populations, human impacts on such populations, and applications to their sustainable use and conservation.

Carla P. Catterall teaches ecology at the Faculty of Environmental Sciences, Griffith University, and leads the Rehabilitation and Restoration Program of the Rainforest Cooperative Research Centre, Australia. Her research is centered on the responses of wildlife to alterations in the quality and quantity of their habitat, especially changes in forest cover.

Laury Cullen Jr. is Research Coordinator at the Institute for Ecological Research (IPÊ), a non-profit organization in Brazil. His main research interests are in agroforestry, community development, and restoration of degraded areas in the Brazilian Atlantic Forest.

Alejandro Estrada is a Senior Research Scientist at the field station Los Tuxtlas, Universidad Nacional Autónoma de México, located in southern Mexico. His main research interest is in documenting the demographic and behavioral responses of mammals to landscape changes in the wet tropics of Northern Mesoamerica.

Bryan Finegan has been involved in research, teaching, and training in ecological and conservation biological aspects of tropical forest management since 1985. He coordinates the Chair of Ecology in the Management of Tropical Forests at the Tropical Agricultural Research and Higher Education Center (CATIE), Costa Rica, but his work currently takes him to seven neotropical countries.

Gustavo A. B. da Fonseca is Executive Vice President for Programs and Science with Conservation International, USA, and Full Professor of Zoology at the Federal University of Minas Gerais, Brazil. He holds a Ph.D. in Wildlife Ecology from the University of Florida and has published extensively on community ecology, wildlife management, and interdisciplinary approaches to biodiversity conservation.

Claude Gascon is Senior Vice President for Regional Programs at Conservation International (CI), USA, where he oversees large-scale conservation and strategic planning projects around the world. Prior to joining CI he was Project Director and Scientific Coordinator at the Biological Dynamics of Forest Fragments Project at the National Institute for Research in the Amazon (INPA), Manaus, Brazil, where he directed research and conservation projects focusing on vertebrate species. He is a Visiting Professor with the Department of Ecology at INPA, Research Associate with the Biodiversity Program at the National Museum of Natural History of the Smithsonian Institution, USA, and recipient of the Order of the Golden Ark of the Netherlands.

Jorge González is a wildlife biologist based at the Tropical Agricultural Research and Higher Education Center (CATIE), Costa Rica. His main interest is in the conservation of biodiversity (especially birds and small mammals) in tropical agricultural landscapes.

Russell Greenberg is the Director of the Smithsonian Migratory Bird Center at the National Zoological Park, Washington, DC, USA. He conducts research on migratory birds and bird ecology in temperate and tropical regions, including sites in Panama, Mexico, Guatemala, and the Caribbean. Research interests include habitat selection, the ecology and evolution of migration, interspecific interaction, and use of human-modified tropical habitats.

Lee Hannah is Senior Fellow in Climate Change Biology with the Center for Applied Biodiversity Science at Conservation International, USA, where he heads efforts to develop conservation responses to climate change with a focus on Southern Africa. His further research interests include methods of corridor design, the global extent of wilderness, and the role of communities in protected area management.

Celia A. Harvey is an Associate Professor in the Department of Agriculture and Agroforestry at the Tropical Agricultural Research and Higher Education Center (CATIE), Costa Rica, and coordinates its Masters Program on Tropical Agroforestry. She has a Ph.D. in Ecology, with a concentration in Conservation and Sustainable Development, from Cornell University. Her research focuses on understanding patterns of biodiversity in fragmented landscapes and the roles of agroforestry systems in conserving both local and regional biodiversity.

Anne-Marie N. Izac is the Scientific Director of the Centre de Coopération Internationale en Recherche Agronomique pour le Développement (CIRAD), France. Previously she was Director of Research at the International Centre for Research in Agroforestry (ICRAF), Kenya, for more than 5 years, and worked in the CGIAR system for 14 years in Africa, Latin America, and Southeast Asia. She also worked as a professor of environmental and natural resource economics in universities in Australia and the United States.

David Kaimowitz is Director General of the Center for International Forestry Research (CIFOR) in Bogor, Indonesia. His main research interests are forest policy, the causes of deforestation, decentralization of natural resource management, and violent conflicts in forested regions.

John Kanowski is a Research Fellow in the Faculty of Environmental Sciences, Griffith University, Australia. His research interests include plant-animal interactions in rainforest and restoration ecology.

Ulrike Krauss is Project Coordinator for CABI Bioscience, UK, and Assistant Professor at the Tropical Agricultural Research and Higher Education Center (CATIE), Costa Rica. She is a tropical plant pathologist and is particularly interested in biological control of plant diseases and technology transfer.

Susan G. W. Laurance is a Mellon Postdoctoral Fellow at the Smithsonian Tropical Research Insitute in Panama. Her main research interests include the effects of habitat fragmentation on plants and wildlife and how corridors can facilitate dispersal in a modified landscape.

William F. Laurance is a Staff Research Scientist at the Smithsonian Tropical Research Institute in Panama and serves on the Board of Directors of the Biological Dynamics of Forest Fragments Project in Manaus, Brazil. His main interests concern the impacts of intensive land uses, such as habitat fragmentation, logging, and fires, on Amazonian, Australasian, and African tropical ecosystems.

Robert J. Lee directs the Indonesia Program for the Wildlife Conservation Society (WCS). His main research interests are in tropical wildlife ecology and sustainable hunting and wildlife trade.

Jefferson Ferreira Lima is Project Coodinator and Agroforestry Extentionist at the Institute for Ecological Research (IPÊ), a nonprofit organization in Brazil. He develops extension programs for involving rural communities in landscape restoration.

Susana Mourato is Senior Lecturer in Environmental Economics at the Department of Environmental Science and Technology, Imperial College London, UK. Her main research interest is nonmarket valuation of environmental resources.

Robert Nasi is Principal Scientist in the Environmental Services and Sustainable Use of Forests Program of the Center for International Forestry Research (CIFOR), Indonesia. His main research interests are on biodiversity of managed forest ecosystems.

Lisa Naughton-Treves is an Associate Professor in the Department of Geography at the University of Wisconsin, Madison, USA. Her main research interests are in wildlife conservation and protected area management in the tropics.

Eduard Niesten is with the Conservation Economics Program in Conservation International's Center for Applied Biodiversity Science, USA. His work concentrates on the use of economic tools to design conservation incentive agreements, of which conservation concessions are one example.

David Pearce is Professor of Environmental Economics at University College London and Visiting Professor at Imperial College London, UK. He has written widely on cost-benefit analysis, environmental economics, and regulation and is currently completing a major volume on environment and economic development.

Shelley Ratay is currently attending the Stanford Graduate School of Business. Previously she worked on conservation land purchases in the United States with the Trust for Public Land and analyzed conservation tools for the Conservation Economics Program at Conservation International's Center for Applied Biodiversity Science, Washington, DC, USA.

Richard Rice is Chief Economist with the Center for Applied Biodiversity Science at Conservation International, USA. He has published widely on natural resource and public policy analysis, including the viability of sustainable forest management, and is currently working on the development and implementation of conservation concessions, an approach involving annual payments for the acquisition of development rights in priority habitats.

Robert A. Rice is a geographer at the Smithsonian Migratory Bird Center at the National Zoological Park, Washington, DC, USA. His research interests focus on the intersection between agriculture and the environment.

David M. Richardson is Associate Professor and Deputy Director at the Institute for Plant Conservation, University of Cape Town, South Africa. Most of his research deals with alien plant invasions, with special emphasis on the ecology and management of tree invasions in South Africa's fynbos biome.

François Ruf is an Agricultural Economist at the Centre de Coopération Internationale en Recherche Agronomique pour le Développement (CIRAD), France, and is currently a Visiting Scientist at the University of Ghana. His main research interests are in comparative economics and the history of cocoa boom-and-bust cycles.

Nick Salafsky is Codirector of Foundations of Success, USA, an organization that works with conservation practitioners around the world to define clear and practical measures of conservation success, determine sound guiding principles for using conservation strategies, and develop the knowledge and skills of individuals and organizations to do good adaptive management. He did his dissertation research on forest gardens in Indonesia and maintains a keen albeit amateur interest in agroforestry.

Mario Samper is Professor of History and Historical Geography as well as a Researcher in Agricultural History at the Universidad Nacional and Universidad de Costa Rica. His main interest is in the technological history of coffee cultivation in Central America.

Jim Sanderson is a Senior Research Scientist at the Center for Applied Biodiversity Science at Conservation International, USA. His research interests include small cats and camera trapping. He has active projects in more than a dozen countries around the world.

Götz Schroth is an agroforestry and tropical land use specialist with experience mostly in West Africa and Central and South America. He is presently an Associate Scientist with the Center of International Forestry Research (CIFOR) in Brazil.

Wes Sechrest is a Scientific Officer with IUCN—the World Conservation Union, where he is Lead Coordinator of the Global Mammal Assessment under the Species Survival Commission. His main research interests are in conservation, including biodiversity science, ecology, evolution, and environmental science. His current research includes analyzing global biodiversity patterns, evaluating the effectiveness of conservation management at regional and local levels, and examining the effects of global change on biodiversity.

Eduardo Somarriba is Professor of Agroforestry at the Tropical Agricultural Research and Higher Education Center (CATIE), Costa Rica. His main interest is in agroforestry with perennial crops and agroforestry farm planning in Latin America.

Charles Staver is an agricultural scientist with the Tropical Agricultural Research and Higher Education Center (CATIE) in Central America, working on the agroecology of shaded coffee, mulch-based annual cropping, and improved grazing systems. He is interested in the role of institutions in participatory approaches to strengthen farm household and rural community capacity for ecological reasoning and innovation.

Nigel I. J. Tucker is a Director of Biotropica Australia and a member of the Faculty of Biological Sciences at James Cook University, north Queensland, Australia. His research interests include tropical forest restoration and landscape ecology.

Heraldo L. Vasconcelos is Professor of Ecology and Conservation Biology at the Federal University of Uberlândia and former Director of the Biological Dynamics of Forest Fragments Project in Manaus, Brazil. His research is focused on the ecology of ant communities, especially ant-plant interactions and their responses to forest fragmentation.

Grégoire Vincent is a plant biologist at the Institut de Recherche pour le Développement (IRD), France, specialized in the ecology of Indonesian agroforests, with a focus on the role of traditional agroforests in landscape ecology and forest biodiversity conservation.

Grant Wardell-Johnson is based in the School of Natural and Rural Systems Management, University of Queensland, and is also a researcher within the Rehabilitation and Restoration Program of the Rainforest Cooperative Research Centre, Australia. His research interests include vegetation and wildlife ecology and management, with an emphasis on the composition and structure of plant assemblages.

David S. Wilkie is Chief Social Scientist in the Living Landscape Program of the Wildlife Conservation Society (WCS), USA. His primary research interests are in tropical forest conservation and the socioeconomic factors driving unsustainable hunting and wildlife trade.

Index